Lecture Notes in Computer Science 7876

Commenced Publication in 1973
Founding and Former Series Editors:
Gerhard Goos, Juris Hartmanis, and Jan van Leeuwen

T-H. Hubert Chan   Lap Chi Lau
Luca Trevisan (Eds.)

Theory and Applications of Models of Computation

10th International Conference, TAMC 2013
Hong Kong, China, May 20-22, 2013
Proceedings

 Springer

Volume Editors

T-H. Hubert Chan
The University of Hong Kong, China
E-mail: hubert@cs.hku.hk

Lap Chi Lau
The Chinese University of Hong Kong, China
E-mail: chi@cse.cuhk.edu.hk

Luca Trevisan
Stanford University, CA, USA
E-mail: trevisan@stanford.edu

ISSN 0302-9743   e-ISSN 1611-3349
ISBN 978-3-642-38235-2   e-ISBN 978-3-642-38236-9
DOI 10.1007/978-3-642-38236-9
Springer Heidelberg Dordrecht London New York

Library of Congress Control Number: 2013937226

CR Subject Classification (1998): F.2, F.3, F.4, G.2.2, H.1.1, E.1, G.4, I.1

LNCS Sublibrary: SL 1 – Theoretical Computer Science and General Issues

Typesetting: Camera-ready by author, data conversion by Scientific Publishing Services, Chennai, India

Printed on acid-free paper

Springer is part of Springer Science+Business Media (www.springer.com)

Preface

Theory and Applications of Models of Computation (TAMC) is an international conference series with an interdisciplinary character, bringing together researchers working in different areas of theoretical computer science and mathematics. TAMC 2013 was the tenth conference in the series, held during May 20–22 in Hong Kong, China. This year, there were 70 submissions, out of which 31 papers were selected by the Program Committee. There was also a poster session for researchers to illustrate and discuss their recent research work. We are very grateful to the Program Committee for their hard work, and to the authors who submitted their work for our considerations. The conference had invited talks by two leading researchers, Sanjeev Arora from Princeton University and Avi Wigderson from the Institute of Advanced Study.

We would like to thank the Department of Computer Science, The University of Hong Kong, for organizing the conference, and the "K.C. Wong Education Foundation" for the Conference Sponsorship Programme for providing financial support to Chinese scholars to attend this conference.

May 2013

Lap Chi Lau
Luca Trevisan

Invited Talks: Turing Lectures 2013

- **Randomness and Pseudorandomness**
  Avi Wigderson, Institute for Advanced Study
  Is the universe inherently deterministic or probabilistic? Perhaps more importantly — can we tell the difference between the two?
  Humanity has pondered the meaning and utility of randomness for millennia. There is a remarkable variety of ways in which we utilize perfect coin tosses to our advantage: in statistics, cryptography, game theory, algorithms, gambling... Indeed, randomness seems indispensable! Which of these applications survive if the universe had no randomness in it at all? Which of them survive if only poor-quality randomness is available, e.g., that arises from "unpredictable" phenomena like the weather or the stock market?
  A computational theory of randomness, developed in the past three decades, reveals (perhaps counterintuitively) that very little is lost in such deterministic or weakly random worlds – indeed, most application areas above survive! The main ideas and results of this theory are explained in this talk. A key notion is pseudorandomness, whose understanding impacts large areas in mathematics and computer science.

- **Towards Provable Bounds for Machine Learning: Three Vignettes**
  Sanjeev Arora, Princeton University
  Many tasks in machine learning (especially unsupervised learning) are provably intractable: NP-hard or worse. Nevertheless, researchers have developed heuristic algorithms to solve these tasks in practice. In most cases, there are no provable guarantees on the performance of these algorithms/heuristics —neither on their running time, nor on the quality of solutions they return. Can we change this state of affairs?
  This talk suggests that the answer is yes, and cover three recent works as illustration. (a) A new algorithm for learning topic models. This concerns a new algorithm for topic models (including the Linear Dirichlet Allocations of Blei et al. but also works for more general models) that provably works in theory under some reasonable assumptions and is also up to 50 times faster than existing software in practice. It relies upon a new procedure for non-negative matrix factorization. (b) What classifiers are worth learning? (c) Provable ICA with unknown Gaussian noise.
  (Based joint works with Rong Ge, Ravi Kannan, Ankur Moitra, Sushant Sachdeva.)

Organization

Conference Chair

Francis Chin — The University of Hong Kong, SAR China

Program Committee

Andrej Bogdanov	The Chinese University of Hong Kong, SAR China
T-H. Hubert Chan	The University of Hong Kong, SAR China
Ho-Lin Chen	National Taiwan University
Jianer Chen	Texas A&M University, USA
Ning Chen	Nanyang Technological University, Singapore
Wei Chen	Microsoft Research Asia
Xi Chen	Columbia University, USA
Marek Chrobak	UC Riverside, USA
Nicola Galesi	Università di Roma La Sapienza, Italy
Naveen Garg	Indian Institute of Technology, Delhi, India
Navin Goyal	Microsoft Research Asia
Nick Harvey	University of British Columbia, Canada
Rahul Jain	National University of Singapore
David Jao	University of Waterloo, Canada
Ken-Ichi Kawarabayashi	National Institute of Informatics, Japan
Jochen Koenemann	University of Waterloo, Canada
Amit Kumar	IIT Delhi, India
Lap Chi Lau (Co-chair)	The Chinese University of Hong Kong, SAR China
Jian Li	Tsinghua University, China
Huijia Rachel Lin	MIT and Boston University, USA
Pinyan Lu	Microsoft Research Asia
Mohammad Mahdian	Google Research
Seffi Naor	Technion University, Israel
Krzysztof Onak	IBM Research, USA
Periklis Papakonstantinou	Tsinghua University, China
Seth Pettie	University of Michigan, USA
Atri Rudra	University at Buffalo, State University of New York, USA
Alexander Russell	University of Connecticut, USA
Piotr Sankowski	University of Warsaw, Poland
Rahul Santhanam	University of Edinburgh, UK

Anastasios Sidiropoulos	University of Illinois at Urbana-Champaign, USA
Mohit Singh	Microsoft Research Redmond, USA
Man Cho Anthony So	The Chinese University of Hong Kong, SAR China
John Steinberger	Tsinghua University, China
Luca Trevisan (Co-chair)	Stanford University, USA
Laszlo Vegh	London School of Economics, UK
David Woodruff	IBM Research, USA
Ke Yi	Hong Kong University of Science and Technology, SAR China
Qin Zhang	IBM Research and Indiana University, USA

Steering Committee

Manindra Agrawal	Indian Institute of Technology, Kanpur, India
Jin-Yi Cai	University of Wisconsin-Madison, USA
Barry S. Cooper	University of Leeds, UK
John Hopcroft	Cornell University, USA
Angsheng Li	Chinese Academy of Sciences, China

Local Organizing Committee

T-H. Hubert Chan	The University of Hong Kong
H.F. Ting	The University of Hong Kong
Zhang Yong	The University of Hong Kong

External Reviewers

Georgios Barmpalias
Simone Bova
Christian Cachin
Lorenzo Carlucci
Yijia Chen
Andreas Feldmann
Zachary Friggstad
Lingxiao Huang
Jiongxin Jin
Raghav Kulkarni
Anna Labella
Duc-Phong Le
Troy Lee

Michael Naehrig
Li Ning
Kenta Ozeki
Attila Pereszlenyi
Giovanni Pighizzini
Janos Simon
Frank Stephan
Xiaoming Sun
Kanat Tangwongsan
Chengu Wang
Haitao Wang
Yajun Wang
Zhewei Wei

Xiaowei Wu

Guang Yang

Jonathan Yaniv

Penghui Yao

Chihao Zhang

Jialin Zhang

Peng Zhang

Zhichao Zhao

Table of Contents

Online Scheduling on a CPU-GPU Cluster 1
  Lin Chen, Deshi Ye, and Guochuan Zhang

Throughput Maximization for Speed-Scaling with Agreeable
Deadlines 10
  Eric Angel, Evripidis Bampis, Vincent Chau, and Dimitrios Letsios

Temperature Aware Online Algorithms for Minimizing Flow Time 20
  Martin Birks and Stanley Fung

Priority Queues and Sorting for Read-Only Data 32
  Tetsuo Asano, Amr Elmasry, and Jyrki Katajainen

$(1 + \epsilon)$-Distance Oracles for Vertex-Labeled Planar Graphs 42
  Mingfei Li, Chu Chung Christopher Ma, and Li Ning

Group Nearest Neighbor Queries in the L_1 Plane 52
  Hee-Kap Ahn, Sang Won Bae, and Wanbin Son

Modelling the Power Supply Network – Hardness and Approximation ... 62
  Alexandru Popa

Approximation Algorithms for a Combined Facility Location
Buy-at-Bulk Network Design Problem 72
  Andreas Bley, S. Mehdi Hashemi, and Mohsen Rezapour

k-means++ under Approximation Stability 84
  Manu Agarwal, Ragesh Jaiswal, and Arindam Pal

An Exact Algorithm for TSP in Degree-3 Graphs via Circuit Procedure
and Amortization on Connectivity Structure 96
  Mingyu Xiao and Hiroshi Nagamochi

Non-crossing Connectors in the Plane 108
  Jan Kratochvíl and Torsten Ueckerdt

Minimax Regret 1-Sink Location Problems in Dynamic Path
Networks 121
  Siu-Wing Cheng, Yuya Higashikawa, Naoki Katoh, Guanqun Ni,
  Bing Su, and Yinfeng Xu

A Notion of a Computational Step for Partial Combinatory Algebras ... 133
  Nathanael L. Ackerman and Cameron E. Freer

Selection by Recursively Enumerable Sets 144
  Wolfgang Merkle, Frank Stephan, Jason Teutsch, Wei Wang, and
  Yue Yang

On the Boundedness Property of Semilinear Sets..................... 156
  Oscar H. Ibarra and Shinnosuke Seki

Turing Machines Can Be Efficiently Simulated by the General Purpose
Analog Computer 169
  Olivier Bournez, Daniel S. Graça, and Amaury Pouly

Computing with and without Arbitrary Large Numbers 181
  Michael Brand

On the Sublinear Processor Gap for Parallel Architectures 193
  Alejandro López-Ortiz and Alejandro Salinger

On Efficient Constructions of Short Lists Containing Mostly Ramsey
Graphs 205
  Marius Zimand

On Martin-Löf Convergence of Solomonoff's Mixture 212
  Tor Lattimore and Marcus Hutter

Any Monotone Property of 3-Uniform Hypergraphs Is Weakly
Evasive 224
  Raghav Kulkarni, Youming Qiao, and Xiaoming Sun

The Algorithm for the Two-Sided Scaffold Filling Problem 236
  Nan Liu and Daming Zhu

Energy-Efficient Threshold Circuits Detecting Global Pattern in
1-Dimentional Arrays.. 248
  Akira Suzuki, Kei Uchizawa, and Xiao Zhou

Resolving Rooted Triplet Inconsistency by Dissolving Multigraphs...... 260
  Andrew Chester, Riccardo Dondi, and Anthony Wirth

Obnoxious Facility Game with a Bounded Service Range 272
  Yukun Cheng, Qiaoming Han, Wei Yu, and Guochuan Zhang

Efficient Self-pairing on Ordinary Elliptic Curves..................... 282
  Hongfeng Wu and Rongquan Feng

Grey-Box Public-Key Steganography 294
  Hirotoshi Takebe and Keisuke Tanaka

Linear Vertex-Kernels for Several Dense RANKING r-CONSTRAINT
SATISFACTION Problems 306
  Anthony Perez

On Parameterized and Kernelization Algorithms for the Hierarchical
Clustering Problem 319
    Yixin Cao and Jianer Chen

Vector Connectivity in Graphs.................................... 331
    Endre Boros, Pinar Heggernes, Pim van 't Hof, and Martin Milanič

Trees in Graphs with Conflict Edges or Forbidden Transitions 343
    *Mamadou Moustapha Kanté, Christian Laforest, and
    Benjamin Momège*

Author Index 355

Online Scheduling on a CPU-GPU Cluster⋆

Lin Chen, Deshi Ye, and Guochuan Zhang

College of Computer Science, Zhejiang University, Hangzhou 310027, China

Abstract. We consider the online scheduling problem in a CPU-GPU cluster. In this problem there are two sets of processors, the CPU processors and GPU processors. Each job has two distinct processing times, one for the CPU processor and the other for the GPU processor. Once a job is released, a decision should be made immediately about which processor it should be assigned to. The goal is to minimize the makespan, i.e., the largest completion time among all the processors. Such a problem could be seen as an intermediate model between the scheduling problem on identical machines and unrelated machines. We provide a 3.85-competitive online algorithm for this problem and show that no online algorithm exists with competitive ratio strictly less than 2. We also consider two special cases of this problem, the balanced case where the number of CPU processors equals to that of GPU processors, and the one-sided case where there is only one CPU or GPU processor. We provide a $(1 + \sqrt{3})$-competitive algorithm for the balanced case, and a 3-competitive algorithm for the one-sided case.

Keywords: Online scheduling, Competitive ratio, CPU-GPU cluster, Unrelated machine scheduling.

1 Introduction

The fast development of technology makes it possible for a graphics processing unit (GPU) to handle various of tasks in a more efficient way than the central processing unit (CPU). For example, tasks like video processing, image analysis and signal processing are usually processed on GPU. Nevertheless, CPU is still more suitable for a wide range of tasks, and it is well possible that some task can only be processed by CPU.

The model of identical machine scheduling fails to capture the difference between CPU processors and GPU processors, while the model of unrelated machine scheduling seems to make the problem complicated in an unnecessary way, indeed, there is even no online algorithm of constant competitive ratio for the unrelated machine scheduling problem [2]. Thus it is worth investigating the scheduling problem in a CPU-GPU cluster, i.e., scheduling on two kinds of different machines.

There are a lot of research towards the above mentioned model as well its different variants. Verner et al. [17] studied data stream scheduling problems on the

---

⋆ Research was supported by in part by NSFC(11071215,11271325).

system that contains multiple CPU processors and a single GPU processor. Luk et al. [16] studied mapping computations on processing elements on a CPU/GPU machine. Yang et al. [19] address the load balancing problem among multiple GPU and CPU processors of TianHe-1 by running the Linpack. However, most of these studies are experimental.

In this paper, we consider online algorithms on the heterogenous CPU/GPU system and analyze their performances in terms of competitive ratio. Formally speaking, we consider the scheduling problem where processors (machines) are divided into two sets, the set G_1 consists of all the GPU processors and G_2 consists of all CPU processors. The processing time of job j is p_{j1} on a GPU processor and p_{j2} on a CPU processor. In our problem, jobs are released one by one and once a job is released, the scheduler need to decide immediately which machine it should be assigned to. The goal is to minimize the makespan (the largest completion time of machines). Specifically, if $|G_1| = |G_2|$, it is called as the balanced case. If $|G_1| = 1$ or $|G_2| = 1$, it is called as the one-sided case.

Related Work: Our problem is related to the unrelated machine scheduling problem, where the processing time of job j on machine i is p_{ji}. The offline version of this problem admits an elegant 2-approximation algorithm, which is proposed by Lenstra et al. [14]. Specifically, an FPTAS (Fully Polynomial Time Approximation Scheme) exists for this problem if the number of machines is a constant [11]. The online version of unrelated machine scheduling problem admits an $O(\log m)$-competitive algorithm where m is the number of machines. The algorithm is proposed by Aspens et al. [1], and it is shown to be the best possible up to a constant [2].

Another closely related problem is the related machine scheduling problem, in which each processor i is associated with a speed s_i, and the processing time of job j is p_j/s_i where p_j is its workload. The offline version of related machine scheduling problem admits a PTAS [10]. For the online version, the first algorithm of constant competitive ratio is given by Aspens et al. [1]. The current best online algorithm has a competitive ratio of $3 + 2\sqrt{2} \approx 5.828$ due to Berman et al. [3]. A special case of the related machine scheduling problem is similar to our one-sided case. In this case there are m machines, the speeds of $m - 1$ machines are all 1, while the remaining machine has a speed larger than 1. Cho and Sahni [6] prove that the list scheduling algorithm is at most $3 - 4/(m + 1)$ competitive, and this result is further improved to 2.45 by Cheng et al. [5]. Kokash [13] presents an online algorithm of competitive ratio 2, which is shown to be the optimal online algorithm [15].

Additionally, there is one related problem called scheduling with speedup resource. In this model, the processors are identical, while there is a renewable discrete resource which could be utilized to reduce the processing time of jobs. Indeed, the processing time of job j is p_j if it is processed directly, and becomes p'_j, however, if it is processed on a machine with the resource. Xu et al. [18] give an online algorithm with competitive ratio 1.781 for this problem when the number of machines is 2 and show that the lower bound is 1.686. A variant of this problem where machines are dedicated is considered by Kellerer et al. [12] and a 3/2-approximation algorithm is presented for the offline version.

Our Result: We are the first to propose online algorithms for scheduling in a CPU-GPU cluster and analyze their competitive ratios. The main contribution is an online algorithm of competitive ratio 3.85 where the numbers of CPU and GPU processors are both arbitrary inputs. This algorithm is quite sophisticated and we also present a much more simplified algorithm with competitive ratio 4. We also consider the two special cases of this problem, namely the balanced model and the one-sided model. We provide a $(1 + \sqrt{3})$-competitive algorithm for the balanced model, and a 3-competitive algorithm for the one-sided model.

The paper is organized as follows. We first give the preliminary in Section 2 and then show the lower bounds of this problem in Section 3. We start with the balanced case and provide a $(1 + \sqrt{3})$-competitive algorithm as well as a simple 3-competitive algorithm in Section 4. Then we consider the one-sided case in Section 5. In Section 6 we study the general case and provide the 3.85-competitive algorithm.

2   Preliminary

We interpret the problem as a scheduling problem on two kinds of machines, and let G_1 be the set of machines of one kind and G_2 be the other. Let $m_1 = |G_1|$ and $m_2 = |G_2|$. Without loss of generality, we may assume that $m_1 \geq m_2$. The processing time of job j on every machine of G_i is p_{ji} where $i = 1, 2$. The load of a machine is the total processing time of jobs allocated to this machine. The objective is to minimize the makespan, i.e. the maximum load over all the machines.

Competitive analysis [4] is used in this paper to evaluate online algorithms. An online algorithm is said to be ρ-competitive if $C_{max}(A(I)) \leq \rho OPT(I)$ for any job list I, where $C_{max}(A(I))$ and $OPT(I)$ are the makespans given by an online algorithm A and an optimal offline algorithm, respectively. The supremum value of ρ is defined to be the *competitive ratio* of an algorithm, i.e., $R_A = \sup_I \{C_{max}(A(I))/OPT(I)\}$. The competitive ratio measures the difference of the makespans between an online algorithm and the optimal offline algorithm in the worst case. Under the worst case study, we may restrict that the completion time of the last job in the job list achieves the makespan of the online algorithm.

To design an online algorithm, it is natural to consider a greedy method. The greedy algorithm for the classical identical machine scheduling problem is called *list scheduling* (LS) [8]. Once a job is released, the algorithm always assigns it to the machine with the least load. The competitive ratio of list scheduling is 2 for the classical problem. However, the list scheduling for our model has two versions, one is to assign a job to the machine with the least load, and the other is to assign it to a machine such that the completion time of this job is minimized. It is easy to see that the former version fails to achieve a constant competitive ratio. Nevertheless, the latter one is also not favorable, as we will show in Section 3.

Despite this fact, our algorithms in this paper could be viewed as modified list scheduling algorithms. Indeed, each of the algorithms is a combination of list scheduling and a set of selecting rules R, where every selecting rule in R decides that under a certain condition, whether the current job should be assigned to G_1 or G_2. Once we decide that a job should be assigned to the group G_i, we schedule this job onto machines of G_i according to the list scheduling, i.e., we schedule this job onto the machine with the least load in G_i. We break ties arbitrarily.

We use the notation $j \rightarrow G_i$ to indicate that job j is assigned to G_i. We write $j \rightarrow LS(G_i)$ to indicate that job j is assigned to G_i and scheduled on machines of G_i according to list scheduling.

3   Lower Bound for Online Algorithms

In this section, we show that the competitive ratio of any online algorithm for our problem is at least 2 even in the most special case that $|G_1| = |G_2| = 1$. We further show that if we directly apply the list scheduling to our problem, then the competitive ratio is not a constant.

Theorem 3.1. *The competitive ratio of any online algorithm is at least 2.*

Proof. We consider the special case that $m_1 = 1$ and $m_2 = 1$. Let A be an arbitrary online algorithm.

Suppose the processing time of the first job released is 1 on either machine and this job is assigned by the algorithm A to machine i where $i = 1, 2$. Then, the next job is released with processing time of 1 on machine i, and processing time of 2 on the other machine. Thus, no matter on which machine the new job is scheduled, the makespan is at least 2. However, the makespan of the optimal solution is 1 and the theorem follows immediately.   □

Theorem 3.2. *The list scheduling algorithm, if applied directly to our problem, is of competitive ratio $\Omega(m)$ even in the special case that $m_1 = 1$ and $m_2 = m$.*

Proof. Recall that the list scheduling algorithm always assigns a job to a machine such that the completion time of this job is minimized, as we have mentioned.

Let ε be a sufficient small positive number. Suppose the following $m + 1$ jobs are released one by one. The processing time of the first job is $p_{11} = 1$ on G_1 and $p_{12} = 1 + \varepsilon$ on G_2. While the processing times of the next m jobs are the same, which are $p_{j1} = \frac{1}{m^2}$ and $p_{j2} = 1$ for $j = 2, \ldots, m+1$. According to the list scheduling algorithm, the first job is assigned to the one machine in G_1, and the other m jobs are assigned to machines of G_2, one for each. Notice that after the assignment of these $m + 1$ jobs, the load of each machine is 1. The next $m + 1$ jobs are the same as the first $m + 1$ jobs, and it can be easily seen that they are scheduled in the same way by the algorithm and the load of each machine is 2.

We continue to release $m - 2$ additional copies of the first $m+1$ jobs and it can be easily seen that the load of each machine is m by list scheduling. However, in the optimal solution, we can assign all the jobs such that $p_{j1} = 1/m^2$ and

$p_{j2} = 1$ to the machine in G_1, recall that there are in all m^2 such kind of jobs and the load of this machine is 1. On the other hand, there are m jobs such that $p_{j1} = 1$ and $p_{j2} = 1 + \varepsilon$, and each of them is assigned to a different machine of G_2. Thus, the makespan of the optimal solution is $1 + \varepsilon$ and the competitive ratio of list scheduling is $\Omega(m)$. □

4   The Balanced Case

We consider the special case that $|G_1| = |G_2| = m$ in this section. Notice that the lower bound of 2 holds even for such a special case.

Consider the following simple algorithm Al_1.

Algorithm Al_1:

Upon the arrival of job j, it is assigned as below.

- if $p_{j,1} \leq p_{j,2}$, then $j \rightarrow LS(G_1)$.
- if $p_{j,1} > p_{j,2}$, then $j \rightarrow LS(G_2)$.

We have the following theorem.

Theorem 4.1. *The competitive ratio of algorithm Al_1 is 3.*

Proof. Let A (B) be the set of jobs scheduled in G_1 (G_2) according to the algorithm, and $A_1 \subset A$ ($B_1 \subset B$) be the set of jobs in G_1 (G_2) in both the algorithm Al_1 and the optimal solution. Let λ_A (λ_B) be total processing time of jobs in A_1 (B_1). Let v_A (v_B) be the total processing time of jobs in $A \setminus A_1$ ($B \setminus B_1$).

Consider the optimal solution and let OPT be its makespan. Notice that the optimal solution would put jobs of $A \setminus A_1$ in G_2 and $B \setminus B_1$ in G_1. Furthermore, for any job of $A \setminus A_1$, its processing time in G_1 is no greater than its processing time in G_2, thus we have

$$OPT \geq (\lambda_B + v_A)/m.$$

Similarly we have $OPT \geq (\lambda_A + v_B)/m$.

Let $C_{max}(Al_1)$ be the makespan of the solution produced by Al_1, then we have

$$C_{max} \leq \max\{(\lambda_A + v_A)/m, (\lambda_B + v_B)/m\} + p_{max}$$

where p_{max} is the largest processing time of jobs scheduled due to Al_1. Obviously $OPT \geq p_{max}$. Thus it follows directly that $C_{max}(Al_1) \leq 3OPT$.

It can be easily seen the upper bound 3 is tight. In the worst case, each job has a similar (or even the same) processing time in both G_1 and G_2, however, they are all scheduled in G_1 or G_2 due to Al_1. □

To give an improved algorithm, we should take into consideration that the load balancing between machines of G_1 and G_2. Based on this observation, we have the following improved algorithm Al_2 for $0 < \alpha < 1$.

Algorithm Al_2:

Once job j is released, it is scheduled according to the following rules.

- Rule 1: $p_{j,1} \leq \alpha p_{j,2}$, then $j \to LS(G_1)$
  or $p_{j,1} \geq \frac{1}{\alpha} p_{j,2}$, then $j \to LS(G_2)$.
- Rule 2: $\alpha p_{j,2} < p_{j,1} < 1/\alpha p_{j,2}$,
    - if $L(G_1) \leq L(G_2)$, $j \to LS(G_1)$
    - if $L(G_1) > L(G_2)$, $j \to LS(G_2)$.

Here $L(G_i)$ is the total processing time of jobs scheduled in G_i among job 1 to job $j-1$, i.e., the load of G_i just before job j is released.

We have the following theorem, and the detailed proof is omitted due to space limited.

Theorem 4.2. *The competitive ratio of algorithm Al_2 is at most $1+\sqrt{3} \approx 2.732$ by setting $\alpha = \sqrt{3} - 1$.*

5   The One-Sided Case

In this case we assume $m_1 = m$ and $m_2 = 1$. The algorithm we derive in the previous section fails to handle this case. Indeed, no matter how one manipulate the parameter α in Al_2, its competitive ratio tends to infinity if m goes to infinity. It seems that one need a better idea to keep load balancing between machines of the two groups, while still take into consideration the extreme jobs whose processing time on G_1 differs greatly with that on G_2.

We provide Al_3 as follows.

Algorithm Al_3:

Once job j is released, it is assigned according to the following rules.

- Rule 1: If $p_{j,1} \geq L(G_2) + p_{j,2}$, then $j \to LS(G_2)$. Otherwise apply Rule 2.
- Rule 2: $p_{j,1}/m \leq p_{j,2}$, then $j \to LS(G_1)$.
-         $p_{j,1}/m > p_{j,2}$, then $j \to LS(G_2)$.

Here $L(G_2)$ is the load of the machine in G_2 when j is released. We have the following theorem, and the detailed proof will be given in the full version of this paper.

Theorem 5.1. *The competitive ratio of algorithm Al_3 is at most 3.*

6   The General Case

We consider the general case in this section. Using the ideas from the previous section, we give the following algorithm Al_4.

Algorithm Al_4:

Once job j is released, it is assigned according to the following rules.

- Rule 1: If $p_{j,1} \geq C_{min}(G_2) + p_{j,2}$, then $j \to LS(G_2)$. Otherwise apply Rule 2.
- Rule 2: $p_{j,1}/m_1 \leq p_{j,2}/m_2$, then $j \to LS(G_1)$.
-         $p_{j,1}/m_1 > p_{j,2}/m_2$, then $j \to LS(G_2)$.

Here $C_{min}(G_2)$ is the load of the least loaded machine in G_2 when j is released. We have the following theorem.

Theorem 6.1. *The competitive ratio of algorithm Al_4 is at most 4.*

Proof. We let A (B) be the set of jobs scheduled in G_1 (G_2), and classify them into three subsets, namely Λ_A, U_A, V_A (Λ_B, U_B, V_B). All the definitions are the same as that in the proof of Theorem 5.1. Again, we know $U_A = \emptyset$.

If $U_B \neq \emptyset$, we consider the job in U_B whose completion time is the last among all the jobs in U_B and let it be job j_0. Once this job comes, $j_0 - 1$ jobs are scheduled and we know that $p_{j_0,1} \geq C_{min}(G_2) + p_{j_0,2} \geq u_B/m_2$ since j_0 is assigned to G_2. Since j_0 is scheduled in G_1 in the optimal solution, we know that $u_B/m_2 \leq OPT$.

Meanwhile, the total processing time of jobs in V_B is at least $m_1 v_B/m_2$ in the optimal solution, while the total processing time of jobs in V_A is at least $m_2 v_A/m_1$ in the optimal solution. Thus we have

$$OPT \geq \frac{\lambda_A + m_1 v_B/m_2}{m_1} = \lambda_A/m_1 + v_B/m_2,$$

$$OPT \geq \frac{\lambda_B + m_2 v_A/m_1}{m_2} = \lambda_B/m_2 + v_A/m_1.$$

Again we have

$$C_{max}(Al_4) \leq \max\{\frac{\lambda_A + v_A}{m_1} + p_{max}(1), \frac{\lambda_B + u_B + v_B}{m_2} + p_{max}(2)\}$$

where $p_{max}(i)$ is the processing time of the largest job in G_i in the solution produced by the algorithm.

Consider $p_{max}(1)$ and let this job be job k. If job k is also scheduled in G_1 in the optimal solution then $p_{max}(1) \leq OPT$. Otherwise k is scheduled in G_2 in the optimal solution and $OPT \geq p_{k,2}$. Notice job k would have been assigned to G_2 if $p_{k,1} \geq C_{min}(G_2) + p_{k,2}$ when it is released. This implies that

$$p_{k,1} \leq \frac{\lambda_B + u_B + v_B}{m_2} + p_{k,2} \leq \frac{\lambda_B + u_B + v_B}{m_2} + OPT$$

which implies that

$$\frac{\lambda_A + v_A}{m_1} + p_{max}(1) \leq \frac{\lambda_A + v_A}{m_1} + \frac{\lambda_B + u_B + v_B}{m_2} + OPT \leq 4OPT.$$

On the other hand, consider $p_{max}(2)$ and let it be job k'. If this job is also in G_2 in the optimal solution then $p_{max}(2) \leq OPT$. Otherwise k' is in G_1 in the optimal solution and $OPT \geq p_{k',1}$. If job k' is assigned to G_2 according to Rule 1, then obviously $OPT \geq p_{k',1} \geq p_{k',2}$. Otherwise k' is assigned with Rule 2. Then $p_{k',1}/m_1 > p_{k',2}/m_2$. Since $m_1 \geq m_2$, we still have $OPT \geq p_{k',1} \geq p_{k',2}$, thus $p_{max}(2) \leq OPT$ always holds.

Thus again we have $\frac{\lambda_B + u_B + v_B}{m_2} + p_{max}(2) \leq 4OPT$.

So, $C_{max}(Al_4) \leq 4OPT$. $\square$

Again introducing parameters into Al_4 does not seem to improve the the algorithm in terms of competitive ratio. Nevertheless, recall that by using the idea of load balancing, we improve Al_1 with competitive ratio of 3 to Al_2 whose competitive ratio is $1 + \sqrt{3}$, and thus we try to use the same idea here.

We have the following modified algorithm Al_5.

Algorithm Al_5:

Once job j is released,

- Rule 1: If $p_{j,1} \geq \beta(C_{min}(G_2) + p_{j,2})$, then $j \to LS(G_2)$.
- Rule 2: $p_{j,1}/m_1 \leq \theta p_{j,2}/m_2$, then $j \to LS(G_1)$.
-       $p_{j,1}/m_1 \geq \lambda p_{j,2}/m_2$, then $j \to LS(G_2)$.
- Rule 3: $\theta p_{j,2}/m_2 < p_{j,1}/m_1 < \lambda p_{j,2}/m_2$,
    - $L_3(G_1)/m_1 \leq \phi L_3(G_2)$, $j \to LS(G_1)$.
    - $L_3(G_1)/m_1 > \phi L_3(G_2)$, $j \to LS(G_2)$.

Here $L_3(G_i)$ is the total processing time of jobs scheduled in G_i due to Rule 3 when job j is released. Notice that it is different from Al_2. In Al_2, we try to make load balancing for the whole set of jobs, while in Al_5, we only try to make load balancing for jobs scheduled according to Rule 3.

We have the following theorem with detailed proof in the full version of this paper.

Theorem 6.2. *The competitive ratio of algorithm Al_5 is at most 3.85 by setting $\lambda \approx 1.69$, $\beta \approx 0.80$, $\theta \approx 1.04$ and $\phi \approx 0.64$.*

Acknowledgement. We thank Huajingling Wu for useful communications.

References

1. Aspnes, J., Azar, Y., Fiat, A., Plotkin, S., Waarts, O.: On-line routing of virtual circuits with applications to load balancing and machine scheduling. Journal of the ACM 44(3), 486–504 (1997)
2. Azar, Y., Naor, J.S., Rom, R.: The competitiveness of on-line assignments. In: Proc. 3rd Symp. on Discrete Algorithms (SODA), pp. 203–210 (1992)
3. Berman, P., Charikar, M., Karpinski, M.: On-line load balancing for related machines. Journal of Algorithms 35, 108–121 (2000)
4. Borodin, A., El-Yaniv, R.: Online Computation and Competitive Analysis. Cambridge University Press (1998)
5. Cheng, T., Ng, C., Kotov, V.: A new algorithm for online uniform-machine scheduling to minimize the makespan. Information Processing Letters 99(3), 102–105 (2006)
6. Cho, Y., Sahni, S.: Bounds for list schedules on uniform processors. SIAM Journal on Computing 9(1), 91–103 (1980)
7. Ebenlendr, T., Sgall, J.: A lower bound on deterministic online algorithms for scheduling on related machines without preemption. In: Solis-Oba, R., Persiano, G. (eds.) WAOA 2011. LNCS, vol. 7164, pp. 102–108. Springer, Heidelberg (2012)
8. Graham, R.: Bounds for certain multiprocessing anomalies. Bell System Technical J. 45, 1563–1581 (1966)

9. Grigoriev, A., Sviridenko, M., Uetz, M.: Machine scheduling with resource dependent processing times. Mathematical Programming 110(1), 209–228 (2007)

10. Hochbaum, D., Shmoys, D.: A polynomial approximation scheme for scheduling on uniform processors: Using the dual approximation approach. SIAM Journal on Computing 17(3), 539–551 (1988)

11. Jansen, K., Porkolab, L.: Improved approximation schemes for scheduling unrelated parallel machines. In: Proceedings of the 31st Annual ACM Symposium on Theory of Computing (STOC), pp. 408–417 (1999)

12. Kellerer, H., Strusevich, V.: Scheduling parallel dedicated machines with the speeding-up resource. Naval Research Logistics 55(5), 377–389 (2008)

13. Kokash, N.: An efficient heuristic for online scheduling in a system with one fast machine. M.S. thesis, Belorussian State University (2004)

14. Lenstra, J., Shmoys, D., Tardos, E.: Approximation algorithms for scheduling unrelated parallel machines. Mathematical Programing 46(1), 259–271 (1990)

15. Li, R., Shi, L.: An on-line algorithm for some uniform processor scheduling. SIAM Journal on Computing 27(2), 414–422 (1998)

16. Luk, C., Hong, S., Kim, H.: Qilin: exploiting parallelism on heterogeneous multiprocessors with adaptive mapping. In: Proceedings of the 42nd Annual IEEE/ACM International Symposium on Microarchitecture (MICRO), pp. 45–55 (2009)

17. Verner, U., Schuster, A., Silberstein, M.: Processing data streams with hard real-time constraints on heterogeneous systems. In: Proceedings of the International Conference on Supercomputing (ICS), pp. 120–129 (2011)

18. Xu, H., Chen, L., Ye, D., Zhang, G.: Scheduling on two identical machines with a speed-up resource. Information Processing Letters 111, 831–835 (2011)

19. Yang, C., Wang, F., Du, Y., Chen, J., Liu, J., Yi, H., Lu, K.: Adaptive optimization for petascale heterogeneous cpu/gpu computing. In: Proceedings of the IEEE International Conference on Cluster Computing (CLUSTER), pp. 19–28 (2010)

Throughput Maximization for Speed-Scaling with Agreeable Deadlines*

Eric Angel[1], Evripidis Bampis[2], Vincent Chau[1], and Dimitrios Letsios[1]

[1] IBISC ; Université d'Évry, Évry, France
{Eric.Angel,Vincent.Chau,Dimitris.Letsios}@ibisc.univ-evry.fr
[2] LIP6 ; Université Pierre et Marie Curie; Paris, France
Evripidis.Bampis@lip6.fr

Abstract. We are given a set of n jobs and a single processor that can vary its speed dynamically. Each job J_j is characterized by its processing requirement (work) p_j, its release date r_j and its deadline d_j. We are also given a budget of energy E and we study the scheduling problem of maximizing the throughput (i.e. the number of jobs which are completed on time). We show that the problem can be solved by dynamic programming when all the jobs are released at the same time in $O(n^4 \log n \log P)$, where P is the sum of the processing requirements of the jobs. For the more general case of agreeable deadlines, where the jobs can be ordered such that for every $i < j$, both $r_i \leq r_j$ and $d_i \leq d_j$, we propose a dynamic programming algorithm solving the problem optimally in $O(n^6 \log n \log P)$. In addition, we consider the weighted case where every job j is also associated with a weight w_j and we are interested in maximizing the weighted throughput. For this case, we prove that the problem becomes $\mathcal{NP}$-hard in the ordinary sense and we propose a pseudo-polynomial time algorithm.

1 Introduction

The problem of scheduling n jobs with release dates and deadlines on a single processor that can vary its speed dynamically with the objective of minimizing the energy consumption has been first studied in the seminal paper by Yao, Demers and Shenker [3]. In this paper, we consider the problem of maximizing the throughput for a given budget of energy. Formally, we are given a set of n jobs $J = \{J_1, J_2, \ldots, J_n\}$, where each job J_j is characterized by its processing requirement (work) p_j, its release date r_j and its deadline d_j. (For simplicity, we suppose that the earliest released job is released at $t = 0$.) We assume that the jobs have to be executed by a single speed-scalable processor, i.e. a processor which can vary its speed over time (at a given time, the processor's speed can be any non-negative value). The processor can execute at most one job at each time. We measure the processor's speed in units of executed work per unit of

---

* This work has been supported by the ANR project TODO (09-EMER-010), by PHC CAI YUANPEI (27927VE) and by the ALGONOW project of the THALES program.

T-H.H. Chan, L.C. Lau, and L. Trevisan (Eds.): TAMC 2013, LNCS 7876, pp. 10–19, 2013.

time. If $s(t)$ denotes the speed of the processor at time t, then the total amount of work executed by the processor during an interval of time $[t, t')$ is equal to $\int_t^{t'} s(u)du$. Moreover, we assume that the processor's power consumption is a convex function of its speed. Specifically, at any time t, the power consumption of the processor is $P(t) = s(t)^\alpha$, where $\alpha > 1$ is a constant. Since the power is defined as the rate of change of the energy consumption, the total energy consumption of the processor during an interval $[t, t')$ is $\int_t^{t'} s(u)^\alpha du$. Note that if the processor runs at a constant speed s during an interval of time $[t, t')$, then it executes $(t' - t) \cdot s$ units of work and it consumes $(t' - t) \cdot s^\alpha$ units of energy. Each job J_j can start being executed after or at its release date r_j. Moreover, we allow the preemption of jobs, i.e. the execution of a job may be suspended and continued later from the point of suspension. Given a budget of energy E, our objective is to find a schedule of maximum throughput whose energy does not exceed the budget E, where the throughput of a schedule is defined as the number of jobs which are completed on time, i.e. before their deadline. Observe that a job is completed on time if it is entirely executed during the interval $[r_j, d_j)$. By extending the well-known 3-field notation by Graham et al. [2], this problem can be denoted as $S1|pmtn, r_j| \sum U_j(E)$. We also consider the weighted version of the problem where every job j is also associated with a weight w_j and the objective is no more the maximization of the cardinality of the jobs that are completed on time, but the maximization of the sum of their weights. We denote this problem as $S1|pmtn, r_j| \sum w_j U_j(E)$. In what follows, we consider the problem in the case where either all jobs have a release date equal to 0 and for an important family of instances, the agreeable instances for which the jobs can be ordered such that for every $i < j$, both $r_i \leq r_j$ and $d_i \leq d_j$.

1.1 Related Works and Our Contribution

Up to the best of our knowledge no work exists for the off-line case of our problem. On the contrary, some works exist for some online variants of throughput maximization: the first work that considered throughput maximization and speed scaling in the online setting has been presented by Chan et al. [9]. They considered the single processor case with release dates and deadlines and they assumed that there is an upper bound on the processor's speed. They are interested in maximizing the throughput and minimizing the energy among all the schedules of maximum throughput. They presented an algorithm which is $O(1)$-competitive with respect to both objectives. In [8] Bansal et al. improved the results of [9], while in [13], Lam et al. studied the 2-processor environment. In [11], Chan et al. defined the energy efficiency of a schedule to be the total amount of work completed in time divided by the total energy usage. Given an efficiency threshold, they considered the problem of finding a schedule of maximum throughput. They showed that no deterministic algorithm can have competitive ratio less than Δ, the ratio of the maximum to the minimum jobs' processing requirement. However, by decreasing the energy efficiency of the online algorithm the competitive ratio of the problem becomes constant. Finally, in [10], Chan et al. studied the problem of minimizing the energy plus a rejection penalty. The rejection penalty is a cost incurred for each job which is not completed on time and each job is associated with a value which is its importance. The authors proposed an $O(1)$-competitive algorithm for the case where the speed is unbounded and they showed that no $O(1)$-competitive algorithm exists for the case where the speed is bounded.

The paper is organized as follows: we first present an optimal algorithm for the case where all the jobs are released at time 0, and then we present another algorithm for the more general case with agreeable deadlines. The reason of presenting both these cases is that in the first case we have a complexity of $O(n^4 \log n \log P)$ which is better than the one in the second case where the complexity becomes $O(n^6 \log n \log P)$. Finally, we consider the weighted case where we are interested in maximizing the weighted throughput. For this case, we prove that the problem is $\mathcal{NP}$-hard in the ordinary sense and we propose a pseudo-polynomial time algorithm.

2   Preliminaries

Given that the processor's speed can be varied, a reasonable distinction of the scheduling problems that can be considered is the following:

- **FS** (Fixed Speed): The processor has a fixed speed which implies directly a processing time for each job. In this case, the scheduler has to decide which job must be executed at each time. This is the classical scheduling setting.
- **CS** (Constant Speed): The processor's speed is not known in advance but it can only run at a single speed during the whole time horizon. In this context, the scheduler has to define a single value of speed at which the processor will run and the job executed at each time.
- **SS** (Scalable Speed): The processor's speed can be varied over the time and, at each time, the scheduler has to determine not only which job to run, but the processor's speed as well.

3   Properties of the Optimal Schedule

Among the schedules of maximum throughput, we try to find the one of minimum energy consumption. Therefore, if we knew by an oracle the set of jobs J^*, $J^* \subseteq J$, which are completed on time in an optimal solution, we would simply have to apply an optimal algorithm for $S1|pmtn, r_j, d_j|E$ for the jobs in J^* in order to determine a minimum energy schedule of maximum throughput for our problem. Based on this observation, we can use in our analysis some properties of an optimal schedule for $S1|pmtn, r_j, d_j|E$.

Let $t_1, t_2, \ldots, t_k$ be the time points which correspond to release dates and deadlines of the jobs so that for each release date and deadline there is a t_i value that corresponds to it. We number the t_i values in increasing order, i.e. $t_1 < t_2 < \ldots < t_k$. The following theorem comes from [3].

Theorem 1. *A feasible schedule for $S1|pmtn, r_j, d_j|E$ is optimal if and only if all the following hold:*

1. *Each job J_j is executed at a constant speed s_j.*
2. *The processor is not idle at any time t such that $t \in (r_j, d_j]$, for all $J_j \in J$.*
3. *The processor runs at a constant speed during any interval $(t_i, t_{i+1}]$, for $1 \le i \le k - 1$.*
4. *A job J_j is executed during any interval $(t_i, t_{i+1}]$ ($1 \le i \le k - 1$), if it has been assigned the maximum speed among the speeds of the jobs $J_{j'}$ with $(t_i, t_{i+1}] \subseteq (r_{j'}, d_{j'}]$.*

Theorem 1 is also satisfied by the optimal schedule of $S1|pmtn, r_j| \sum U_j(E)$ for the jobs in J^*.

4   Agreeable Deadlines

For the special case of the problem $S1|pmtn, r_j| \sum U_j(E)$ where the deadlines of the jobs are agreeable we propose an optimal algorithm which is based on dynamic programming. As mentioned before, among the schedules of maximum throughput, our algorithm constructs a schedule of minimum energy consumption. Next, we describe our dynamic program and we elaborate on the complexity of our algorithm.

Initially, we consider the problem $1|pmtn, r_j| \sum U_j$ which is a classical scheduling problem where we are given a set of jobs $J = \{J_1, J_2, \ldots, J_n\}$ that have to be executed by a single processor. Each job J_j is associated with a processing time p_j, a release date r_j and a deadline d_j. The objective is to find a schedule of maximum throughput. We refer to the problem as FS. This problem is polynomially-time solvable and the fastest known algorithm for general instances is in $O(n^4)$ [1]. When all the release dates are equal, this problem can be solved in $O(n \log n)$ with Moore's algorithm [5]. Finally, if the jobs have agreeable deadlines, the time complexity is also in $O(n \log n)$ using Lawler's algorithm [7].

Next, we consider another problem which we denote as CS. In this problem we are given a set of jobs $J = \{J_1, J_2, \ldots, J_n\}$, where each job J_j has a processing requirement p_j, a release date r_j and a deadline d_j, that have to be executed by a single speed scalable processor. Moreover, we are given a value of throughput k. The objective is to find the minimum energy schedule which completes at least k jobs on time so that all jobs that are completed on time are assigned equal speed and the jobs not completed on time have zero speed. For notational convenience, we denote the problem $S1|pmtn, r_j| \sum U_j(E)$ as SS.

The inspiration for our dynamic programming for the special case of the SS where the deadlines are agreeable was the fact that the problem CS can be solved in polynomial time by repeatedly solving instances of the problem FS. In fact, if we are given a candidate speed s for the CS problem, we can find a schedule of maximum throughput w.r.t. to s simply by setting the processing time of each job J_j equal to $\frac{p_j}{s}$ and applying an optimal algorithm for the FS problem. So, in order to get an optimal algorithm of the CS problem, it suffices to establish a lower and upper bound on the speed of the optimal schedule. A naive choice is $s_{min} = 0$ and $s_{max} = \infty$. Then, it suffices to binary search in $[s_{min}, s_{max}]$ and find the minimum speed s^* in which k jobs are completed on time.

Property 1. There exists an optimal solution in which all jobs are scheduled according to EDF (Earliest Deadline First) order and without preemption.

This property comes from the fact that the algorithm of [3] is optimal and that we have an agreeable instance.

In the following, we assume that the jobs $J_1, J_2, \ldots, J_n$ are sorted according to the EDF order, i.e. $d_1 \le d_2 \le \ldots \le d_n$.

4.1   Special Case When $r = 0$

For a subset of jobs $S \subseteq J$, a schedule which involves only the jobs in S will be called a S-schedule.

Definition 1. Let $J(k) = \{J_j | j \le k\}$ be the set of the first k jobs according to the EDF order. For $1 \le u \le |J(k)|$, we define $E(k, u)$ as the minimum energy consumption of an S-schedule such that $|S| = u$ and $S \subseteq J(k)$. If such a schedule does not exist, i.e. when $u > |J(k)|$, then $E(k, u) = +\infty$.

Definition 2. We define $B(t', t, \ell)$ as the minimum energy consumption of an S-schedule such that $|S| = \ell$, $S \subseteq \{J_j | t' < d_j \le t\}$ and such that all these jobs are scheduled only within the interval $[t', t]$, and with a constant common speed. If such a schedule does not exist, then $B(t', t, \ell) = +\infty$.

Proposition 1. $B(d_j, d_k, \ell)$ can be computed in $O(n \log n \log P)$ time, for any j, k, ℓ, with $P = \sum_j p_j$.

Proof. In order to compute $B(d_j, d_k, \ell)$, we consider the set of jobs $\{J_{j'} | d_j < d_{j'} \le d_k\}$. For each job in this set, we modify its release date to d_j. Since we want the minimum energy consumption and there is only one speed, we search the minimum speed such that there are exactly ℓ jobs scheduled. This minimum speed can be found by performing a binary search in the interval $[0, s_{max}]$, with $s_{max} = P/(d_k - d_j)$. For every speed s, the processing time of a job J_j is $t_j = p_j/s$, and we compute the maximum number m of jobs which can be scheduled using Moore's algorithm [5] in $O(n \log n)$. If $m < \ell$ (resp. $m > \ell$) the speed s must be increased (resp. decreased). □

Proposition 2. One has

$$E(k, u) = \min\{E(k - 1, u), B(0, d_k, u), \min_{\substack{1 \le j < k \\ 1 \le \ell < u}} \{E(j, \ell) + B(d_j, d_k, u - \ell)\}\}.$$

Proof. Let S be an optimal schedule associated with $E(k, u)$. We can assume that this schedule satisfies the properties of Theorem 1 and Property 1.

If $J_k \notin S$, then $E(k, u) = E(k - 1, u)$. If $J_k \in S$, then there are two cases to consider. The first case is when all the jobs in S are scheduled at the same speed.

This case is equivalent to the CS problem, and one has $E(k, u) = B(0, d_k, u)$. The second case is when the schedule S has at least two different speeds. Let C_j be the completion time of job J_j in the schedule S. Let $t = \min_j \{C_j|$ all the jobs scheduled after J_j (at least one job) are executed with the same speed $\} = C_{j^*}$. Necessarily, job J_{j^*} is executed with a different speed. This means that at time C_{j^*} the processor is changing its speed, and using Property 3. of Theorem 1 we can deduce that $C_{j^*} = d_{j^*}$. Now we consider the subschedule S_1 obtained from S by considering only the tasks executed during the interval $[0, d_{j^*})$. Let us assume that there are ℓ^* tasks in this subschedule. Then, necessarily the energy consumption of S_1 is equal to $E(j^*, \ell^*)$, otherwise by replacing S_1 with a better subschedule with energy consumption $E(j^*, \ell^*)$ we could obtain a better schedule than S. Now we consider the subschedule S_2 obtained from S by considering only the tasks executed from time d_{j^*} until the end of the schedule. In a similar way, the energy consumption of S_2 is equal to $B(d_{j^*}, d_k, u - \ell^*)$.

Notice that since the jobs involved in $E(j, \ell)$ have a deadline smaller or equal to d_j, whereas the jobs involved in $B(d_j, d_k, u - l)$ have a deadline greater than d_j, those two sets of jobs are always distinct, and therefore the schedule associated with $E(j, \ell) + B(d_j, d_k, u - \ell)$ is always feasible. $\square$

Theorem 2. The problem $S1|pmtn, r_j = 0| \sum U_j(E)$ can be solved in $O(n^4 \log n \log P)$ time.

Proof. We use a dynamic program based on Proposition 2, with $E(0, u) = +\infty$, $\forall u > 0$. The maximum throughput is equal to $\max\{u|E(n, u) \leq E\}$.

The number of values $B(d_j, d_k, \ell)$ is $O(n^3)$. They can be precomputed with a total processing time $O(n^4 \log n \log P)$, using Proposition 1. The number of values $E(k, u)$ is $O(n^2)$, and the complexity to calculate each $E(k, u)$ value is $O(n^2)$ (we have to look for $O(n^2)$ values for j, ℓ and we assume that the previous $E(., .)$ values have already been computed). Thus the overall complexity is $O(n^4 \log n \log P)$. $\square$

4.2 Agreeable Deadlines

Definition 3. We define $E_k(t, u)$ as the minimum energy consumption of an S-schedule, such that $|S| = u$, $S \subseteq J(k, t) = \{J_j|j \leq k, r_j < t\}$ and such that all these jobs are executed within the interval $[r_{min}, t]$. If such a schedule does not exist, then $E_k(t, u) = +\infty$.

Definition 4. We define $A(t', t, \ell, j, k)$ as the minimum energy consumption of a S-schedule such that $|S| = \ell$, $S \subseteq \{J_j, \ldots, J_k\}$, and such that all these jobs are scheduled within the interval $[t', t]$, and with a constant common speed.

Proposition 3. $A(t', t, \ell, j, k)$ can be computed in $O(n \log n \log P)$ time, for any t', t, ℓ, j, k.

Proof. In order to compute $A(t', t, \ell, j, k)$, we change the release date of job J_j to t' if $r_j < t'$, and the deadline of job J_j to t if $d_j > t$. The set $\{J_j, \ldots J_k\}$ still has agreeable deadlines. Then we proceed as in the proof of Proposition 1 using a binary search over the interval $[0, s_{max}]$, with $s_{max} = P/(t - t')$. Note that in this case, we use Lawler's algorithm in [7]. □

Proposition 4. *One has*

$$E_k(t, u) = \min_{\substack{r_{min} \leq t' \leq t, 0 \leq j < k \\ 0 \leq \ell \leq u}} \left\{ E_j(t', \ell) + A(t', t, u - \ell, j + 1, k) \right\}.$$

Proof. Let S be an optimal schedule associated with $E_k(t, u)$. We can assume that this schedule satisfies the properties of Theorem 1 and Property 1.

If $J_k \notin S$, then $E_k(t, u) = E_{k-1}(t, u)$. In that case, $t' = t$, $j = k - 1$ and $\ell = u$ in the above expression. If $J_k \in S$, then there are two cases to consider. The first case is when the optimal schedule S has one speed. In that case $t' = r_{min}$, $\ell = 0$, $j = 0$ in the above expression. This case is equivalent to the CS problem. The second case is when the optimal schedule S has at least two speeds. In that case we proceed as in the Proposition 2, we split the schedule S into two subschedules S_1 and S_2 (see the figure below).

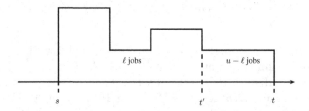

There exists t' with $r_{min} < t' < t$, such that all the jobs scheduled after t' are scheduled with a common speed, and this is the subschedule S_2. The subschedule S_1 (resp. S_2) has an energy consumption equal to $E_j(t', \ell)$ (resp. $A(t', t, u - \ell, j + 1, k)$). Notice that we have to guess the value of j and ℓ in the first subschedule, and the sets of jobs in the second subschedule depend on the first one. □

Theorem 3. *The problem $S1|pmtn, agreeable| \sum U_j(E)$ can be solved in $O(n^6 \log n \log P)$ time.*

Proof. We use a dynamic program based on Proposition 4. Notice that the important dates are included in the set $\Theta = \{r_j | 1 \leq j \leq n\} \cup \{d_j | 1 \leq j \leq n\}$. This comes from the Property 1 and Theorem 1, i.e. the changes of speed of the processor occur only at some release date or some deadline. Therefore we can always assume that $t', t \in \Theta$. Notice also that $|\Theta| = O(n)$.

We define $E_0(t, 0) = 0 \ \forall t \in \Theta$, and $E_0(t, u) = +\infty \ \forall u > 0, t \in \Theta$. The maximum throughput is equal to $\max\{u | E_n(d_{max}, u) \leq E\}$.

The number of values $A(t', t, \ell, j, k)$ is $O(n^5)$. They can be precomputed with a total processing time $O(n^6 \log n \log P)$, using Proposition 3. The number of values $E_k(t, u)$ is $O(n^3)$. To compute each value, we have to look for the $O(n^3)$ cases (for each value of t', j, ℓ). In each case, we pick up two values which are already computed. Thus the $E_k(t, u)$ values are computed in $O(n^6)$ time. The overall complexity is $O(n^6 \log n \log P)$.   □

4.3 Weighted Version

Next we consider the weighted version of our problem, i.e.

$S1|pmtn, r_j| \sum_j w_j U_j(E)$. In this version a job J_j is defined by its release date r_j, its deadline d_j, its amount of work p_j and its weight w_j. We want to maximize the total weight of the jobs scheduled. We first show that the problem is $\mathcal{NP}$-hard even in the case where all the jobs are released at the same time and have equal deadlines. Then, we present a pseudo-polynomial algorithm for the case where the deadlines are agreeable.

Theorem 4. *The problem* $S1|| \sum_j w_j U_j(E)$ *is* $\mathcal{NP}$ *hard.*

Proof. In order to establish the $\mathcal{NP}$-hardness of $S1|| \sum_j w_j U_j(E)$, we present a reduction from the KNAPSACK problem which is known to be $\mathcal{NP}$-hard. In an instance of the KNAPSACK problem we are given a set I of n items. Each item $i \in I$ has a value v_i and a capacity c_i. Moreover, we are given a capacity C, which is the capacity of the knapsack, and a value V. In the decision version of the problem we ask whether there exists a subset $I' \subseteq I$ of the items of total value not less than V, i.e. $\sum_{i \in I'} v_i \geq V$, whose capacity does not exceed the capacity of a knapsack, i.e. $\sum_{i \in I'} c_i \leq C$.

Given an instance of the KNAPSACK problem, we construct an instance of $S1|| \sum_j w_j U_j(E)$ as follows. For each item i, $1 \leq i \leq n$, we introduce a job J_i with $r_i = 0$, $d_i = 1$, $w_i = v_i$ and $p_i = c_i$. Moreover, we set the budget of energy equal to $E = C^\alpha$.

We claim that the instance of the KNAPSACK problem is feasible iff there is a feasible schedule for $S1|| \sum_j w_j U_j(E)$ of total weighted throughput not less than V.

Assume that the instance of the KNAPSACK is feasible. Therefore, there exists a subset of items I' such that $\sum_{i \in I'} v_i \geq V$ and $\sum_{i \in I'} c_i \leq C$. Then we can schedule the jobs in I' with constant speed $\sum_{i \in I'} c_i$ during $[0, 1]$. Their total energy consumption of this schedule is no more that C^α since the instance of the Knapsack is feasible. Moreover, their total weight is no less than V.

For the opposite direction of our claim, assume there is a feasible schedule for $S1|| \sum_j w_j U_j(E)$ of total weighted throughput not less than V. Let J' be the jobs which are completed on time in this schedule. Clearly, due to the convexity of the speed-to-power function, the schedule that executes the jobs in J' with constant speed during the whole interval $[0, 1]$ is also feasible. Since the latter schedule is feasible, we have that $\sum_{j \in J'} p_j \leq C$. Moreover, $\sum_{j \in J'} w_j \geq V$. Therefore, the items which correspond to the jobs in J' form a feasible solution for the KNAPSACK.   □

In this part, we propose a pseudo-polynomial time algorithm based on a dynamic programming algorithm for the KNAPSACK problem.

Definition 5. *We redefine $E_k(t, w)$ to be the minimum energy consumption of a S-schedule, with $S \subseteq J(k, t) = \{J_j | j \le k, r_j < t\}$, such that all the jobs in S are scheduled within the interval $[r_{min}, t]$ and such that the sum of their weight is at least w. If such a schedule does not exist, then $E_k(t, w) = +\infty$.*

We redefine $A(t', t, w, j, k)$ to be the minimum energy consumption of a S-schedule such that $S \subseteq \{J_j, \ldots, J_k\}$, $w(S) \ge w$ and such that these jobs are scheduled within the interval $[t', t]$, and with a constant common speed.

Proposition 5. *$A(t', t, w, j, k)$ can be computed in $O(nW \log P)$ time, where W is the sum of weights of the jobs.*

Proof. The proof is similar to Proposition 3. In this case, we use Lawler's algorithm in [6]. □

Lemma 1. *One has*

$$E_k(t, w) = \min_{\substack{r_{min} \le t' \le t, 0 \le j < k \\ 0 \le \ell \le w}} \left\{ E_j(t', \ell) + A(t', t, w - \ell, j + 1, k) \right\}.$$

Proof. The proof is similar to the Proposition 4. □

Theorem 5. *The problem $S1|pmtn, agreeable| \sum_j w_j U_j(E)$ can be solved in $O(n^5 W^2 \log P)$ time.*

Proof. We use a dynamic program based on Proposition 5, with $E_0(t, 0) = 0$ $\forall t \in \Theta$ and $E_0(t, w) = +\infty \, \forall w > 0, t \in \Theta$. The maximum weighted throughput is obtained with $\max\{w | E_n(d_{max}, w) \le E\}$. The number of values $A(t', t, \ell, j, k)$ is $O(n^4 W)$. They can be precomputed and finally it takes $O(n^5 W^2 \log P)$ time. The number of values $E_k(t, u)$ is $O(n^2 W)$. To compute each value, we have to look for the $O(n^2 W)$ cases (for each value of t', j, ℓ). In each case, we pick up two values which are already computed. Thus the $E_k(t, u)$ values are computed in $O(n^4 W^2)$ time. Thus the overall complexity is $O(n^5 W^2 \log P)$. □

5   Future Work

While the throughput maximization problem is polynomially-time solvable for agreeable deadlines its complexity remains open for general instances. This is a challenging open question for future research.

References

1. Baptiste, P.: An $O(n^4)$ algorithm for preemptive scheduling of a single machine to minimize the number of late jobs. Operations Research Letters 24(4), 175–180 (1999)

2. Graham, R.L., Lawler, E.L., Lenstra, J.K., Rinnooy Kan, A.H.G.: Optimization and Approximation in Deterministic Sequencing and Scheduling: A Survey. Annals of Operations Research 5, 287–326 (1979)
3. Yao, F.F., Demers, A.J., Shenker, S.: A Scheduling Model for Reduced CPU Energy. In: Symposium on Foundations of Computer Science (FOCS), pp. 374–382 (1995)
4. Kise, H., Ibaraki, T., Mine, H.: A Solvable Case of the One-Machine Scheduling Problem with Ready and Due Times. Operations Research Letters 26(4), 121–126 (1978)
5. Moore, J.M.: An n job, one machine sequencing algorithm for minimizing the number of late jobs. Management Science 15, 102–109 (1968)
6. Lawler, E.L.: A dynamic programming algorithm for preemptive scheduling of a single machine to minimize the number of late jobs. Annals of Operations Research 26, 125–133 (1990)
7. Lawler, E.L.: Knapsack-like scheduling problems, the Moore-Hodgson algorithm and the 'tower of sets' property. Mathematical and Computer Modeling, 91–106 (1994)
8. Bansal, N., Chan, H.-L., Lam, T.-W., Lee, L.-K.: Scheduling for Speed Bounded Processors. In: Aceto, L., Damgård, I., Goldberg, L.A., Halldórsson, M.M., Ingólfsdóttir, A., Walukiewicz, I. (eds.) ICALP 2008, Part I. LNCS, vol. 5125, pp. 409–420. Springer, Heidelberg (2008)
9. Chan, H.-L., Chan, W.-T., Lam, T.W., Lee, L.-K., Mak, K.-S., Wong, P.W.H.: Energy Efficient Online Deadline Scheduling. In: SODA 2007, pp. 795–804 (2007); ACM Transactions on Algorithms
10. Chan, H.-L., Lam, T.-W., Li, R.: Tradeoff between Energy and Throughput for Online Deadline Scheduling. In: Jansen, K., Solis-Oba, R. (eds.) WAOA 2010. LNCS, vol. 6534, pp. 59–70. Springer, Heidelberg (2011)
11. Chan, J.W.-T., Lam, T.-W., Mak, K.-S., Wong, P.W.H.: Online Deadline Scheduling with Bounded Energy Efficiency. In: Cai, J.-Y., Cooper, S.B., Zhu, H. (eds.) TAMC 2007. LNCS, vol. 4484, pp. 416–427. Springer, Heidelberg (2007)
12. Li, M.: Approximation Algorithms for Variable Voltage Processors: Min Energy, Max Throughput and Online Heuristics. In: Dong, Y., Du, D.-Z., Ibarra, O. (eds.) ISAAC 2009. LNCS, vol. 5878, pp. 372–382. Springer, Heidelberg (2009)
13. Lam, T.-W., Lee, L.-K., To, I.K.K., Wong, P.W.H.: Energy Efficient Deadline Scheduling in Two Processor Systems. In: Tokuyama, T. (ed.) ISAAC 2007. LNCS, vol. 4835, pp. 476–487. Springer, Heidelberg (2007)

Temperature Aware Online Algorithms
for Minimizing Flow Time

Martin Birks and Stanley Fung

Department of Computer Science, University of Leicester, Leicester LE1 7RH,
United Kingdom
{mb259,pyfung}@mcs.le.ac.uk

Abstract. We consider the problem of minimizing the total flow time of
a set of unit sized jobs in a discrete time model, subject to a temperature
threshold. Each job has its release time and its heat contribution. At each
time step the temperature of the processor is determined by its temperature at the previous time step, the job scheduled at this time step and a
cooling factor. We show a number of lower bound results, including the
case when the heat contributions of jobs are only marginally larger than
a trivial threshold. Then we consider a form of resource augmentation
by giving the online algorithm a higher temperature threshold, and show
that the Hottest First algorithm can be made 1-competitive, while other
common algorithms like Coolest First cannot. We also give some results
in the offline case.

1 Introduction

Motivation. Green computing is not just trendy, but is a necessity. For example,
data centers around the world consume an enormous amount of energy. Very
often, this energy consumption and the associated issue of heat dissipation is
the biggest factor affecting system design from data centers to handheld devices.
Many ways to tackle the issue have been explored. Among them, the design of
energy-efficient algorithms is an active area of research; we refer to [1] for an
introduction.

In this paper we are interested in controlling the temperature of a microprocessor. Temperature is an important issue in processor architecture design: high
temperature affects system reliability and lifespan, but a powerful processor in-
evitably comes with a high energy consumption and hence high temperature.
It was proposed in [6] that, instead of slowing down the processor to control
the temperature, one can use proper scheduling algorithms to help as well. Since
then a number of papers [2,3,6] have worked on this model. We explain the model
below.

The Model. Time is split into discrete time steps. For an integer t, we refer to
the time interval between the time instants t and $t + 1$ as the *time step* t. A
total of n jobs arrive. Each job J has a release time r_J, a heat contribution
h_J, and a unit length processing time. All release times are integers. Thus each

T-H.H. Chan, L.C. Lau, and L. Trevisan (Eds.): TAMC 2013, LNCS 7876, pp. 20–31, 2013.

job fits into one time step. The temperature of a processor changes depending on the heat contribution of the jobs it executes, and a *cooling factor* $R > 1$: specifically, if at time t the temperature is τ_t and a job J is executed at this time step then the temperature at the next time step is given by $\tau_{t+1} = \frac{\tau_t + h_J}{R}$. This is a discrete approximation of the actual cooling which is a continuous time process governed by Faraday's Law, and the unit length jobs represent slices of processes given to a processor. The initial temperature can be set at 0 without loss of generality. The temperature can never exceed the *temperature threshold* T. This can be set to 1 without loss of generality. A job J is therefore *admissible* at time t if $\tau_t + h_J \leq R$. This means that any job with $h_J > R$ can never be admissible; without loss of generality we thus assume all jobs have $h_J \leq R$.

One way of quantifying the performance of temperature-aware scheduling algorithms is to optimize some Quality of Service (QoS) measure subject to the temperature threshold. Arguably the most widely used QoS measure for processor scheduling is the *flow time* (or *response time*). The flow time of a job J, denoted $|J|$, is defined as the difference between its release time and its completion time. We can consider the total (or average) flow time of all jobs, or the maximum flow time. In this paper we focus on the total flow time.

The scheduling algorithm is *online*, i.e. it is not aware of jobs not yet released. This is of course a natural way to model jobs arriving at a microprocessor. We use the standard *competitive analysis* to analyze the effectiveness of online algorithms: an online algorithm $\mathcal{A}$ is c-competitive if the objective value returned by $\mathcal{A}$ (for a minimization problem) is at most c times that of an offline optimal algorithm OPT, on all input instances.

There are several common and simple algorithms that can be used in this temperature model:

- Coolest First (CF): at every time step, schedule the coolest job among all admissible jobs, breaking ties arbitrarily.
- Hottest First (HF): at every time step, schedule the hottest job among all admissible jobs, breaking ties arbitrarily.
- FIFO: at every time step, schedule the earliest released job among all admissible jobs, breaking ties arbitrarily.

They all belong to a natural group of algorithms called *non-idling algorithms*, i.e. they do not idle when they have an admissible job pending. This is a weaker notion than that of *reasonable algorithms* as defined in [6], as reasonable algorithms are non-idling algorithms, but with stricter restrictions on which job must be scheduled.

Related Work. Without temperature constraints, the flow time problem is well-studied. It is well known that the SRPT (shortest remaining processing time first) algorithm is 1-competitive with preemption. Since in our case all jobs are of unit length, there is no issue of preemption and therefore (if without temperature) any non-idling algorithm is 1-competitive.

With temperatures, we are not aware of any prior work on flow time, although there were research on other objective functions. One of them is to maximize

throughput when each job has a deadline. This was shown to be NP-hard in the offline case [6], even if all jobs have identical release times and identical deadlines. It was observed that the same proof showed that minimizing the maximum or total flow time is NP-hard as well. They further showed that in the online case, all 'reasonable' algorithms are 2-competitive for $R = 2$ and this is optimal. This was generalised to all range of R [3], the weighted jobs case [4] and the longer jobs case [5].

Bampis et al. [2] considered the objective of minimizing the makespan on $m > 1$ processors when all jobs are released at time 0. They presented a generic 2ρ-approximate algorithm using a ρ-approximate algorithm for classical makespan scheduling as a subroutine, and a lower bound of $\frac{4}{3}$ on the approximability. For a single processor the algorithm gives an approximation ratio of 2. They also considered other objectives when there is no temperature threshold: they minimize the maximum temperature or the average temperature of the schedule instead (subject to a bound on the finishing time of jobs).

Our Contributions. We consider three different cases in this paper:

(1) **Bounded job heat:** Since the problem is trivial without temperature constraints, it is tempting to believe that the problem is still tractable when the jobs are not very hot; we therefore consider limiting the maximum permissible heat of a job, h_{max}. When h_{max} is allowed to be exactly R then it can be trivially shown that no algorithm can give a bounded competitive ratio. On the other hand, if $h_{max} \leq R - 1$ then it can be easily shown that any algorithm is 1-competitive (details are in Section 2). Therefore we consider the case where $h_{max} = R - \epsilon$ for some $0 < \epsilon < 1$. Unfortunately it turns out that positive results remain rather unlikely. The problem remains NP-hard, and we show that the competitive ratio approaches infinity as ϵ approaches 0. We also show that non-idling algorithms have an unbounded competitive ratio for all $\epsilon < 1$.

(2) **Increased temperature threshold:** In view of the above, we instead give online algorithms a bit more power by allowing them to have a higher temperature threshold of $1 + \epsilon$ while the offline algorithm still has a threshold of 1. This can correspond to the case where, for example, new technologies make the system more resistant to higher temperatures. Note that when $\epsilon \geq \frac{1}{R-1}$ a 1-competitive upper bound is trivial (see Section 3). In this model we can give a positive result: HF is 1-competitive if $\epsilon \geq \frac{R^2+R+1}{(R-1)(R+1)^2}$. We also show a number of lower bounds as in the bounded job heat case; in particular we show that CF cannot be even constant competitive given any non-trivial higher threshold. This is in stark contrast with the throughput case [6] where CF is optimal but HF can be shown to be not.

For easier illustration, consider the case $R = 2$: our results show that HF is 1-competitive whenever $\epsilon \geq \frac{7}{9}$ (any non-idling algorithm is trivially 1-competitive if $\epsilon \geq 1$), while there are no 1-competitive algorithms whenever $\epsilon < \frac{1}{4}$.

(3) The offline complexity: We show that no polynomial-time algorithms can have an approximation ratio better than $O(\sqrt{n})$. We also give an approximation algorithm with approximation ratio 2.618 for the case where all jobs have release time 0.

All of the results here are for the case of minimizing total flow time. In the full paper we show some negative results for maximum flow time.

Notations. We denote the offline optimal algorithm by OPT and the online algorithm being considered by $\mathcal{A}$. They will also denote the schedules of the algorithms whenever this is possible without confusion. For a set of jobs A or a schedule $\mathcal{A}$, $|A|$ and $|\mathcal{A}|$ denote the total flow time of their jobs. We use τ_t and τ'_t to denote the temperatures of $\mathcal{A}$ and OPT, respectively, at time t. We describe jobs that are pending for an algorithm as being stored in a queue. We denote the queues of $\mathcal{A}$ and OPT at a time step t as Q_t and Q'_t respectively. This refers to the time instant when all jobs that are released at t have arrived, but before any jobs have been scheduled for that time step. The number of jobs in Q_t is denoted as $|Q_t|$. We drop the subscript t if we are referring to the queue in general and not a particular time step.

Due to space constraints, some proofs are omitted from this extended abstract.

2   Bounded Maximum Job Heat

First we consider the online case and where the heat contribution of any job is at most $R - \epsilon$ for some $0 < \epsilon < 1$. This is the only ϵ range that gives non-trivial results. No algorithm can give a bounded competitive ratio when h_{max} is allowed to be exactly R. This is because after scheduling any job with a non-zero heat contribution, any algorithm will have a positive temperature which means that the algorithm will never be able to schedule a job with heat R, and so that job will end up with an infinite flow time. If, on the other hand, h_{max} is restricted to be at most $R - 1$ then any job can be scheduled at any time. This is because the maximum temperature of an algorithm is 1 and if the maximum heat of a job is $R - 1$ then after running any job the temperature of any algorithm will be no higher than $\frac{1 + (R-1)}{R} = 1$, which means that the temperature threshold can never be violated and that all jobs are always admissible. It is therefore equivalent to the case without temperature constraints where any non-idling algorithm is optimal.

We first note that the problem remains NP-hard even if the heat contributions are just above $R - 1$.

Theorem 1. *If $h_{max} = R - 1 + \delta$ for any $\delta > 0$, the offline problem remains NP-hard.*

2.1 Lower Bounds

Before proceeding we need an observation: if there are two pending jobs and both are admissible, it is preferable to schedule the hotter one first. This includes the case when one of them is a zero-heat job (i.e. an idle time step). The reason is because it leads to a lower resulting temperature than the other way round (see e.g. [8]), so it cannot harm the subsequent schedule; moreover if one of them is an idle step then scheduling the real job earlier can only reduce the flow time.

Theorem 2. *For any integer $k \geq 2$, if $h_{max} = R - \epsilon$ where $\epsilon \leq \frac{R-1}{R^k}$ then any deterministic algorithm is at least k-competitive.*

Proof. Fix a deterministic algorithm $\mathcal{A}$. At time 0 release a job J_1 with $h_{J_1} = R - 1$ and a job J_2 with $h_{J_2} = \frac{R-1}{R^{k-1}}$. $\mathcal{A}$ will eventually start the jobs, and let the earlier one be started at time t. We analyze the two cases.

Case 1: $\mathcal{A}$ starts J_1 first. Then $\tau_{t+1} = \frac{R-1}{R}$. At time $t+1$ we release another job J_3 with $h_{J_3} = R - \frac{R-1}{R^k}$. In order to start J_3 the temperature of $\mathcal{A}$ must be no higher than $\frac{R-1}{R^k}$ and therefore J_3 becomes just admissible at time $t + k$. (This assumes J_2 is not scheduled yet; if J_2 is scheduled before J_3 then the additional heat contribution will mean J_3 can only start even later.) Let u be the time $\mathcal{A}$ starts J_3.

Case 1a: $u \geq t + k + 1$. In this case we gift $\mathcal{A}$ by assuming J_2 has already been scheduled with flow time 0 and no heat contribution. Starting at time $u + 1$, we release a copy of a job J_4 with $h_{J_4} = R - \frac{1}{R^{k-1}}$ every k time steps, for a large enough number of copies. As $\frac{1}{R^{k-1}} > \frac{R-1}{R^k}$, this heat contribution is below h_{max}. No other jobs are released. We can assume each J_4 job is scheduled as soon as it becomes admissible because they are all identical and there will never be other pending jobs.

Observe that for a J_4-job to be admissible the temperature must be no higher than $\frac{1}{R^{k-1}}$, and that they are always admissible after $k - 1$ idle steps. We now show that $\mathcal{A}$ indeed needs $k - 1$ idle steps before being able to schedule each of these J_4-jobs. First, $\tau_{u+1} > 1 - \frac{R-1}{R^{k+1}}$ due to J_3 (and J_1). After $k - 2$ idle steps, $\tau_{u+k-1} > (1 - \frac{R-1}{R^{k+1}})/R^{k-2}$ and it can be easily verified that this is greater than $\frac{1}{R^{k-1}}$. Therefore after $k - 2$ idle steps $\mathcal{A}$ is still too hot to schedule the first J_4 job.

Now consider the rest of the J_4 jobs. The first J_4 job is scheduled at time $u + k$. If we allow only $k - 2$ idle steps, then $\tau_{u+2k-1} = (\tau_{u+k} + h_{J_4})/R^{k-1} = \tau_{u+k-1}/R^k + (R - \frac{1}{R^{k-1}})/R^{k-1}$. This is at least τ_{u+k-1} if and only if $\tau_{u+k-1} \leq \frac{1}{R^{k-2}}$, which is true noting that $\tau_{u+1} \leq 1$. Since we know J_4 is not admissible at $u + k - 1$, and $\tau_{u+2k-1} \geq \tau_{u+k-1}$, J_4 is also not admissible at $u + 2k - 1$. The same argument applies to all subsequent J_4 jobs and so each J_4 job requires $k - 1$ idle steps before they can be scheduled.

OPT schedules jobs J_2, J_3 and J_1 at t, $t+1$ and $t+2$ respectively. This gives $\tau'_{t+3} = 1$ and thus $\tau'_{u+1} \leq \tau'_{t+k+2} = \frac{1}{R^{k-1}}$. OPT can therefore schedule each J_4 job as soon as it is released, reaching a temperature of exactly 1 afterwards, and then repeat the same for the next J_4 job.

If in total x copies of J_4 are released then $|OPT| = x$ counting only the J_4 jobs while $|\mathcal{A}| \geq kx$. As x can be made arbitrarily large, we can ignore the flow time of the first three jobs and get a competitive ratio of k.

Case 1b: $u = t+k$. In this case J_2 cannot be scheduled before J_3, and $\tau_{u+1} = 1$. We release a copy of a job J_4 with $h_{J_4} = R - \frac{1}{R^{k-1}}$ every k steps starting at time $u + 2$, for a large enough number of copies. No other jobs are released.

Suppose first that J_2 is scheduled before any J_4 job. Clearly it is best to schedule J_2 as early as possible, i.e. at time $u + 1$. Then at time $u + k$ the temperature is strictly higher than $\frac{1}{R^{k-1}}$, making J_4 jobs inadmissible. Therefore the first J_4 job can be scheduled earliest at $u + k + 1$. Similar to case 1a we can argue that all subsequent J_4 jobs also require $k - 1$ idle steps before they can be scheduled.

Now suppose J_2 is not started before the first J_4 job, then the first J_4 job can be (and will be, as it is always better to schedule early) scheduled at $u + k$. Moreover $\tau_{u+k+1} = 1$. Therefore all subsequent J_4 jobs follow the same pattern: if J_2 remains unscheduled by $\mathcal{A}$ in $[u, u+ik)$, for some $i \geq 1$, then $\tau_{u+ik} = \frac{1}{R^{k-1}}$, and so the next J_4 job will be scheduled at $u + ik$, giving $\tau_{u+ik+1} = 1$. If J_2 is scheduled in some interval $[u + ik + 1, u + (i+1)k + 1)$, then the next J_4 job is not admissible until time $u + (i+1)k + 1$. From this point onwards there must again be $k-1$ idle steps between two J_4 jobs. We will show below that J_2 cannot be indefinitely delayed.

OPT schedules in the same way as in Case 1a. If in total x copies of J_4 are released after J_2 is scheduled, then $|OPT| = x$ counting only these J_4 jobs while $|\mathcal{A}| \geq kx$, so again this gives a competitive ratio of k. If J_2 gets postponed after x J_4 jobs, then the flow time of J_2 in $\mathcal{A}$ alone is already at least kx, so no more J_4 jobs need to be released and the argument works as well.

Case 2: $\mathcal{A}$ starts J_2 first at time t, then starts J_1 at time $v \geq t+1$. At time $v+1$ we release a job J_3 with $h_{J_3} = R - \frac{R-1}{R^k}$. After $k-1$ idle steps, $\tau_{v+k} > \frac{h_{J_1}}{R^k} = \frac{R-1}{R^k}$. This means that J_3 cannot be started by $\mathcal{A}$ until $v + k + 1$ at the earliest. Let u be the time $\mathcal{A}$ starts J_3, $u \geq v + k + 1$.

Next we release jobs J_4 with $h_{J_4} = R - \frac{1}{R^{k-1}}$, starting at time $u + 1$ and repeating every k time steps for a large number of copies. At $u + 1$ the situation of $\mathcal{A}$ is the same as in Case 1a, i.e. due to the heat contribution of J_3 alone the first J_4 job requires $k - 1$ idle time steps before it can be scheduled. The same argument in Case 1a for the subsequent J_4 jobs also applies.

Meanwhile OPT will start J_2, J_3 and J_1 at time $t, t + 2$ and $t + 3$. As in Case 1, each J_4 job will be scheduled immediately by OPT. Again x can be made arbitrarily large to give a competitive ratio of k. $\square$

The above theorem only gives a non-trivial result when $\epsilon \leq \frac{R-1}{R^2}$. The next theorem gives a bound that is not as strong but holds for any $\epsilon < 1$.

Theorem 3. *Any deterministic algorithm is at least 2-competitive when $h_{max} = R - 1 + \delta$ for any $\delta > 0$.*

2.2   Non-idling Algorithms

We prove a lower bound that gives an unbounded competitive ratio but for non-idling algorithms. This lower bound holds for all non-trivial values of h_{max} (recall that a bound of $R - 1$ makes every job always admissible.)

Theorem 4. *When $h_{max} = R - 1 + \delta$ for any $\delta > 0$, any non-idling algorithm has a competitive ratio of $\Omega(n)$.*

We now show a trivial upper bound. Note that for any constant h_{max} (i.e. constant k) this bound is tight (by comparing with Theorem 4).

Theorem 5. *For any integer $k \geq 1$, if $h_{max} = R - \epsilon$ where $\epsilon \geq \frac{1}{R^k}$, any non-idling algorithm is $O(kn)$-competitive.*

Proof. First we note that any job with the maximum heat contribution of $R - \epsilon$ can be scheduled after k idle time steps. Clearly $|OPT| \geq n$ as the flow time of each job must be at least 1. $\mathcal{A}$ meanwhile will schedule at least one job every $k+1$ time steps, with the first job being completed immediately, so $|J_1| = 1, |J_i| \leq (i - 1)k + i$ for $i > 1$, and $|\mathcal{A}| = \sum_{i=1}^{n} |J_i| \leq 1 + \sum_{i=2}^{n}((k + 1)i - k) = O(kn^2)$. Therefore $|\mathcal{A}|/|OPT| = O(kn)$. □

3   Increased Temperature Threshold

In this section we consider a form of resource augmentation where the temperature threshold of the online algorithm is increased to $1 + \epsilon$ for some $0 < \epsilon < \frac{1}{R-1}$ but the temperature threshold of OPT remains at 1. (Note that the maximum heat contribution of a job is not limited any lower than R, unlike Section 2, as this is the hottest that OPT can schedule). We limit ϵ to $< \frac{1}{R-1}$ because if a larger value is allowed then, in a similar way to setting $h_{max} \leq R - 1$ in Section 2, any job is always admissible at any time step and therefore any non-idling algorithm is trivially 1-competitive.

3.1   Lower Bounds

We can prove the following lower bound on the threshold required to give competitive algorithms. The proof is similar to that of Theorem 2.

Theorem 6. *For any integer $k \geq 1$, if $\epsilon < \frac{R-1}{R^{k}+1}$ then no deterministic algorithm is better than k-competitive.*

3.2   Hottest First Is 1-Competitive

We now show that HF is 1-competitive when given a sufficiently high threshold, namely $\epsilon \geq \frac{R^2+R+1}{(R-1)(R+1)^2}$. (Note that this threshold is lower than the threshold $\frac{1}{R-1}$ that makes any algorithm 1-competitive.)

First we split all jobs into two classes: any job J with $\frac{R^2}{R+1} < h_J(\leq R)$ is called an H-job, and every other job is called a C-job. We now show three lemmas regarding the properties of H- and C-jobs.

Lemma 1. *OPT can never schedule two H-jobs consecutively.*

Proof. If two H-jobs J_1 and J_2 run consecutively, the temperature of OPT immediately after running the second job is at least $\frac{h_{J_1}}{R^2} + \frac{h_{J_2}}{R} > \frac{R^2/(R+1)}{R^2} + \frac{R^2/(R+1)}{R} = 1$, i.e. it exceeds the temperature threshold. The inequality is due to the minimum heat contribution of H-jobs. □

Lemma 2. *If $\epsilon \geq \frac{R^2+R+1}{(R-1)(R+1)^2}$, then immediately after scheduling a C-job, $\mathcal{A}$ can always schedule an H-job if some H-job is pending.*

Proof. A C-job J_1 and an H-job J_2 is always able to be scheduled consecutively by $\mathcal{A}$ if $(1 + \epsilon + h_{J_1})/R^2 + h_{J_2}/R \leq 1 + \epsilon$, and as $h_{J_1} \leq \frac{R^2}{R+1}$ and $h_{J_2} \leq R$ this is true if $\frac{1+\epsilon+R^2/(R+1)}{R^2} + \frac{R}{R} \leq 1 + \epsilon$, which is equivalent to $\epsilon \geq \frac{R^2+R+1}{(R-1)(R+1)^2}$. □

Lemma 3. *If $\epsilon \geq \frac{1}{R^2-1}$, a pending C-job is always admissible to $\mathcal{A}$.*

Proof. A C-job J is always admissible for $\mathcal{A}$ if $(1 + \epsilon + h_J)/R \leq 1 + \epsilon$, and as we have that $h_J \leq \frac{R^2}{R+1}$ this is true if $(1 + \epsilon + \frac{R^2}{R+1})/R \leq 1 + \epsilon$, which is equivalent to $\epsilon \geq \frac{1}{R^2-1}$. □

We refer to the number of C-jobs scheduled by $\mathcal{A}$ and OPT in $[0,t)$ (i.e. time steps $0, \ldots, t-1$) as c_t and c'_t respectively. The number of H-jobs scheduled by $\mathcal{A}$ and OPT in $[0,t)$ will similarly be referred to as h_t and h'_t.

Lemma 4. *If $\epsilon \geq \frac{1}{R^2-1}$, and if there exists some time t where $\mathcal{A}$ is idle, it must be that $c'_t \leq c_t$.*

Proof. Consider such a time t. $\mathcal{A}$ will always schedule an admissible job and by Lemma 3 all C-jobs are always admissible. Hence, as $\mathcal{A}$ idles at t, it must have scheduled all of the C-jobs released so far, so $c'_t \leq c_t$. □

Lemma 5. *If $\epsilon \geq \frac{R^2+R+1}{(R-1)(R+1)^2}$, at every time t it must be that $h'_t \leq h_t$.*

Proof. We prove this claim by induction on t. First we show two trivial base cases. Before time 0 neither algorithm will have scheduled any job and so $h'_0 = h_0 = 0$. If OPT has scheduled a hot job J at time 0, then this job must also be admissible for $\mathcal{A}$, and as $\mathcal{A}$ always schedules the hottest job possible either J or a hotter job will be scheduled by $\mathcal{A}$ and so $h'_1 \leq h_1$.

For a general $t \geq 1$, we use the induction hypotheses $h'_{t-1} \leq h_{t-1}$ and $h'_t \leq h_t$ to show that $h'_{t+1} \leq h_{t+1}$. Consider the following cases. If OPT schedules a C-job at t then $h'_{t+1} = h'_t \leq h_t \leq h_{t+1}$. If both OPT and $\mathcal{A}$ schedule an H-job at t then $h'_{t+1} = h'_t + 1 \leq h_t + 1 = h_{t+1}$. The only remaining case to consider is where OPT schedules an H-job at t but $\mathcal{A}$ does not. In this case we know by induction that $h'_{t-1} \leq h_{t-1}$. By Lemma 1 we know that OPT cannot schedule an H-job at $t-1$, so $h'_{t+1} = h'_{t-1} + 1$. Hence, if $h'_{t-1} \leq h_{t-1} - 1$ then it must be that $h'_{t+1} = h'_{t-1} + 1 \leq h_{t-1} \leq h_{t+1}$.

Otherwise $h'_{t-1} = h_{t-1}$. If $\mathcal{A}$ schedules an H-job at $t-1$ then $h_{t+1} = h_{t-1}+1$ and so $h'_{t+1} = h'_{t-1} + 1 = h_{t-1} + 1 = h_{t+1}$. Otherwise $\mathcal{A}$ did not schedule an H-job at $t-1$. However an H-job must be pending at t: as $h'_{t-1} = h_{t-1}$ and OPT schedules an H-job at t, therefore at least $h_{t-1}+1$ H-jobs have been released up to and including time t, and as $\mathcal{A}$ does not schedule an H-job at $t - 1$, at least one H-job is pending at t. $\mathcal{A}$ will always schedule a job when one is admissible though, and we know that $\mathcal{A}$ scheduled either a C-job or no job at $t - 1$ so by Lemma 2 this H-job or any hotter admissible job will be scheduled by $\mathcal{A}$ at t, contradicting the assumption that it does not. □

Theorem 7. *If* $\epsilon \geq \frac{R^2+R+1}{(R-1)(R+1)^2}$, *HF is 1-competitive.*

Proof. Note that $\frac{R^2+R+1}{(R-1)(R+1)^2} > \frac{1}{R^2-1}$ whenever $R > 1$. Thus for any $\epsilon \geq \frac{R^2+R+1}{(R-1)(R+1)^2}$, Lemmas 4 and 5 both hold.

For $|OPT|$ to be less than $|\mathcal{A}|$ there must exist some time s such that $|Q_s| > |Q'_s|$. Moreover there must also exist some $t < s$ such that $|Q_t| = |Q'_t|$, $\mathcal{A}$ idles at t and OPT does not idle at t. It follows from Lemmas 4 and 5 that if such a time t were to exist then $c'_t \leq c_t$ and $h'_t \leq h_t$. As $|Q_t| = |Q'_t|$, we have that $c_t + h_t = c'_t + h'_t$. Hence the two inequalities must in fact be equalities, i.e. $h'_t = h_t$ and $c'_t = c_t$. We now show that such a t cannot exist, specifically by showing that if OPT does not idle at t then $\mathcal{A}$ would not idle either. OPT cannot schedule an H-job at t, because this means $h'_{t+1} = h'_t + 1 > h_t = h_{t+1}$, contradicting Lemma 5. Suppose OPT schedules a C-job at t. As $c'_t = c_t$, if a C-job is pending for OPT at t, that one must also be pending for $\mathcal{A}$ at t. By Lemma 3 we know that this C-job must be admissible for $\mathcal{A}$ and $\mathcal{A}$ always schedules a job if one is admissible, contradicting that $\mathcal{A}$ is idle at t. □

3.3 Non-idling Algorithms

We can show that the more restricted group of non-idling algorithms have unbounded competitive ratio. The general approach of the proof is similar to Theorem 4.

Theorem 8. *If* $\epsilon < \frac{1}{R^2}$, *any non-idling algorithm is at least* $\Omega(n)$-*competitive.*

Since CF is a non-idling algorithm, Theorem 8 applies, but we can give a stronger bound that shows it has an unbounded competitive ratio for all non-trivial values of ϵ.

Theorem 9. *For any fixed* $\epsilon < \frac{1}{R-1}$, *CF is at least* $\Omega(n)$-*competitive.*

Note that Theorems 7 and 9 together imply that HF performs provably better than CF: HF is 1-competitive given higher threshold whereas CF can never be even constant competitive given any non-trivial temperature threshold. This is perhaps somewhat surprising, given that CF (being a reasonable algorithm) is optimal in maximizing throughput [3,6]. In contrast, HF is not a reasonable algorithm, and it can be shown that HF is not optimal for throughput.

Similar results can be proved for augmenting the online algorithm with a higher cooling factor (i.e. with a more powerful fan) instead of a higher threshold.

4 The Offline Case

4.1 Inapproximability

Theorem 10. *For any $\epsilon > 0$, there is no polynomial time approximation algorithm for minimizing total flow time with approximation ratio $O(n^{1/2-\epsilon})$, unless P=NP.*

The idea of the proof is similar to the $\Omega(\sqrt{n})$ proof in [7], in that we use very hot jobs (which require a lot of cooling time before it for it to be scheduled) to simulate a long job. However, quite intricate technical details are required (essentially to ensure that the temperatures are high enough so that the very hot jobs are indeed 'long' jobs).

Note that this also implies that no online polynomial-time algorithm can have competitive ratio better than $O(\sqrt{n})$, unless P = NP. However the proof uses very hot jobs (unlike in Section 2).

4.2 Identical Release Times

Despite Theorem 10, in the special case of identical release times we can give a 2.618-approximation. The algorithm is similar to the algorithm for minimizing makespan in [2]. (As in [2] we assume there is at most one job of heat R, otherwise the second job of heat R will have a flow time of infinity for any schedule). It works as follows: first it orders all the jobs in non-decreasing order of heat contribution i.e. $h_{J_1} \leq h_{J_2} \leq \ldots \leq h_{J_n}$. Next we split the jobs into two sets depending on their heat contributions. The set C contains all jobs with heat contributions at most $R - 1$, i.e. $C = \{J_1, J_2, ..., J_c\}$ where $c = |C|$. All other jobs are in set H, i.e. $H = \{J_{c+1}, J_{c+2}, ..., J_n\}$. For simplicity we refer to a job J_i in C (where $i \leq c$) as C_i and a job J_{c+i} in H (where $i \leq n - c$) as H_i. The algorithm first assigns the hottest job, H_{n-c} if $H \neq \emptyset$, to the first time step. For time steps 2 to $c+1$ the algorithm then assigns all the C jobs in descending order i. These jobs will always be admissible as their heat contributions are at most $R - 1$. All remaining jobs $J_i \in H - \{H_{n-c}\}$ are then scheduled in the coolest first order, where each job of heat contribution h_{J_i} is preceded by k_i idle time steps, where k_i is defined as follows: for each job J_i with a heat contribution $h_{J_i} > R - 1$, k_i is the largest k such that $h_{J_i} > \frac{R^k-1}{R^k-1}$.

We require two propositions from [2] (generalized to all values of R) that are restated here for completeness. The first proposition ensures the schedule described above is feasible.

Proposition 1. *Any schedule in which every job J_i is executed after at least k_i idle steps is feasible.*

Proposition 2. *For $R \geq 2$, in an optimal schedule, between the execution of two jobs J_j and J_i (where J_j is before J_i) of heat contributions $h_{J_j}, h_{J_i} > R-1$, there are at least $k_i - 1$ steps, which are either idle or execute jobs of heat contributions at most $R - 1$.*

Theorem 11. *The above algorithm achieves a 2.618-approximation for minimizing total flow time, for $R \geq 2$.*

Proof. If $c \geq n-1$ then $\mathcal{A}$ schedules a job every time step without idling and so must be optimal, therefore we only need to consider the case where $c < n-1$. It is clear that $|\mathcal{A}| = |C| + |H| = |C| + |H - \{H_{n-c}\}| + 1$. For each job C_i, $1 \leq i \leq c$, we have $|C_i| = i+1$. Therefore $|C| = \sum_{i=1}^{c} |C_i| = \frac{c^2+3c}{2}$.

Next we consider each job H_i for $1 \leq i < n-c$: $|H_1| = 1 + c + k_{c+1} + 1$ and $|H_i| = |H_{i-1}| + k_{c+i} + 1$. This gives $|H_i| = c + i + 1 + \sum_{j=1}^{i} k_{c+j}$. Thus

$$|H - \{H_{n-c}\}| = \sum_{i=1}^{n-c-1} |H_i| = \frac{n^2 + n - c^2 - 3c - 2}{2} + \sum_{i=1}^{n-c-1} k_{c+i}((n-c-1)-(i-1))$$

The total flow time of $\mathcal{A}$ is therefore

$$|\mathcal{A}| = 1 + \frac{c^2 + 3c}{2} + \frac{n^2 + n - c^2 - 3c - 2}{2} + \sum_{i=1}^{n-c-1} k_{c+i}(n-c-i) \qquad (1)$$

We now analyze the flow time of OPT by analysing the flow time of a virtual schedule OPT^* that must have a flow time of no more than that of OPT. OPT^* schedules the hottest job H_{n-c} at the first time step. OPT^* then assigns each job J_i in H a virtual processing time of k_i, and schedules them according to the order given by the Shortest Processing Time First rule. In each of these k_i processing steps, the first $k_i - 1$ is idle and the last is where J_i is executed. Finally, OPT^* assigns the jobs from C into the earliest possible idle steps in between each of the jobs from H (or after the last job from H if there are not enough idle steps). This virtual schedule of OPT^* may not be feasible, but by Proposition 2 and the optimality of the Shortest Processing Time First rule when temperature is not considered, it must be that $|OPT| \geq |OPT^*|$.

We denote the flow time, in the schedule of OPT^*, of a job J as $|J^*|$. Analogous definitions for $|C^*|, |H^*|$ and $|OPT^*|$ follow naturally. We now analyze this virtual schedule. It must be that $|OPT^*| = |C^*| + |H^*_{n-c}| + |(H - \{H_{n-c}\})^*|$, and that $|C^*| + |H^*_{n-c}| \geq |C| + |H_{n-c}| = 1 + (c^2 + 3c)/2$. As for H_i, $1 \leq i < n-c$, we have $|H^*_1| = 1 + k_{c+1}$ and $|H^*_i| = |H^*_{i-1}| + k_{c+i}$. This gives $|H^*_i| = 1 + \sum_{j=1}^{i} k_{c+j}$. The total flow time for the set $H - \{H_{n-c}\}$ is therefore

$$|(H - \{H_{n-c}\})^*| = \sum_{i=1}^{n-c-1} |H^*_i| = (n-c-1) + \sum_{i=1}^{n-c-1} k_{c+i}((n-c-1)-(i-1))$$

These can then be combined to bound the flow time for OPT:

$$|OPT| \geq |OPT'| = 1 + \frac{c^2 + 3c}{2} + (n-c-1) + \sum_{i=1}^{n-c-1} k_{c+i}(n-c-i) \qquad (2)$$

Combining Equations (1) and (2) gives us the approximation ratio:

$$\frac{|\mathcal{A}|}{|OPT|} \leq \frac{n^2 + n + 2\sum_{i=1}^{n-c-1} k_{c+i}(n-c-i)}{c^2 + c + 2n + 2\sum_{i=1}^{n-c-1} k_{c+i}(n-c-i)}$$

As $k_i \geq 1$ for every $J_i \in H$, it must be that $\sum_{i=1}^{n-c-1} k_{c+i}(n - c - i) \geq \sum_{i=1}^{n-c-1}(n - c - i) = (n - c - 1)(n - c)/2$. Putting back into the above formula and simplifying, this gives

$$\frac{|\mathcal{A}|}{|OPT|} \leq \frac{c^2 - (2n - 1)c + 2n^2}{2c^2 - (2n - 2)c + (n^2 + n)} < \frac{c^2 - 2nc + 2n^2}{2c^2 - 2nc + n^2}$$

Let $x = c/n$, then this ratio is equal to $\frac{x^2 - 2x + 2}{2x^2 - 2x + 1}$. For $0 \leq x \leq 1$, the maximum value of this ratio is equal to $(3 + \sqrt{5})/2$, attained at $x = (3 - \sqrt{5})/2$.   □

Note that this algorithm is almost like CF (except the first step), but the change is necessary to obtain a good approximation ratio. Consider the example with two jobs J_1 and J_2 with $h_{J_1} = R - 1$ and $h_{J_2} = R - \epsilon$. CF will schedule J_1 first and requires $\log_R \frac{R-1}{R\epsilon}$ idle steps before it can schedule J_2, whereas scheduling J_2 first followed immediately by J_1 gives a total flow time of 3.

Interestingly the algorithm given in [2] for minimizing makespan (which in the case of identical release times is equivalent to minimizing the maximum flow time) is almost equivalent to HF, and it gave a $(2 + \epsilon)$-approximation.

References

1. Albers, S.: Energy-efficient algorithms. Communications of the ACM 53(5), 86–96 (2010)
2. Bampis, E., Letsios, D., Lucarelli, G., Markakis, E., Milis, I.: On multiprocessor temperature-aware scheduling problems. In: Snoeyink, J., Lu, P., Su, K., Wang, L. (eds.) FAW-AAIM 2012. LNCS, vol. 7285, pp. 149–160. Springer, Heidelberg (2012)
3. Birks, M., Fung, S.P.Y.: Temperature aware online scheduling with a low cooling factor. In: Kratochvíl, J., Li, A., Fiala, J., Kolman, P. (eds.) TAMC 2010. LNCS, vol. 6108, pp. 105–116. Springer, Heidelberg (2010)
4. Birks, M., Cole, D., Fung, S.P.Y., Xue, H.: Online algorithms for maximizing weighted throughput of unit jobs with temperature constraints. In: Atallah, M., Li, X.-Y., Zhu, B. (eds.) FAW-AAIM 2011. LNCS, vol. 6681, pp. 319–329. Springer, Heidelberg (2011)
5. Birks, M., Fung, S.P.Y.: Temperature aware online algorithms for scheduling equal length jobs. In: Atallah, M., Li, X.-Y., Zhu, B. (eds.) FAW-AAIM 2011. LNCS, vol. 6681, pp. 330–342. Springer, Heidelberg (2011)
6. Chrobak, M., Dürr, C., Hurand, M., Robert, J.: Algorithms for temperature-aware task scheduling in microprocessor systems. Sustainable Computing: Informatics and Systems 1(3), 241–247 (2011)
7. Kellerer, H., Tautenhahn, T., Woeginger, G.J.: Approximability and nonapproximability results for minimizing total flow time on a single machine. SIAM Journal on Computing 28(4), 1155–1166 (1999)
8. Yang, J., Zhou, X., Chrobak, M., Zhang, Y., Jin, L.: Dynamic thermal management through task scheduling. In: Proc. IEEE Int. Symposium on Performance Analysis of Systems and Software, pp. 191–201 (2008)

Priority Queues and Sorting for Read-Only Data

Tetsuo Asano[1], Amr Elmasry[2], and Jyrki Katajainen[3]

[1] School of Information Science, Japan Advanced Institute
for Science and Technology
Asahidai 1-1, Nomi, Ishikawa 923-1292, Japan
[2] Department of Computer Engineering and Systems, Alexandria University
Alexandria 21544, Egypt
[3] Department of Computer Science, University of Copenhagen
Universitetsparken 1, 2100 Copenhagen East, Denmark

Abstract. We revisit the random-access-machine model in which the
input is given on a read-only random-access media, the output is to
be produced to a write-only sequential-access media, and in addition
there is a limited random-access workspace. The length of the input
is N elements, the length of the output is limited by the computation
itself, and the capacity of the workspace is $O(S + w)$ bits, where S is a
parameter specified by the user and w is the number of bits per machine
word. We present a state-of-the-art priority queue—called an adjustable
navigation pile—for this model. Under some reasonable assumptions, our
priority queue supports *minimum* and *insert* in $O(1)$ worst-case time and
extract in $O(N/S + \lg S)$ worst-case time, where $\lg N \leq S \leq N/\lg N$. We
also show how to use this data structure to simplify the existing optimal
$O(N^2/S + N \lg S)$-time sorting algorithm for this model.

1 Introduction

Problem Area. Consider a sequential-access machine (Turing machine) that
has three tapes: input tape, output tape, and work tape. In space-bounded computations the input tape is read-only, the output tape is write-only, and the aim
is to limit the amount of space used in the work tape. In this set-up, the theory of language recognition and function computation requiring $O(\lg N)$ bits[1]
of working space for an input of size N is well established; people talk about
log-space programs [25, Section 3.9.3] and classes of problems that can be solved
in log-space [25, Section 8.5.3]. Also, in this set-up, trade-offs between space and
time have been extensively studied [25, Chapter 10]. Although one would seldom
be forced to rely on a log-space program, it is still interesting to know what can
be accomplished when only a logarithmic number of extra bits are available.

In this paper we reconsider the space-time trade-offs in the random-access
machine. Analogous with the sequential-access machine, we have a read-only
array for input, a write-only array for output, and a limited workspace that
allows random access. Over the years, starting by a seminal paper of Munro and

---

[1] Throughout the paper we use $\lg x$ as a shorthand for $\log_2(\max\{2, x\})$.

T.-H.H. Chan, L.C. Lau, and L. Trevisan (Eds.): TAMC 2013, LNCS 7876, pp. 32–41, 2013.
© Springer-Verlag Berlin Heidelberg 2013

Paterson [20], the space-time trade-offs have been studied in this model for many problems including: sorting [4,12,24], selection [11,12], and various geometric tasks [2,3,7]. The practical motivation for some of the previous work has been the appearance of special devices, where the size of working space is limited (e.g. mobile devices) and where writing is expensive (e.g. flash memories).

An algorithm (or a data structure) is said to be *memory adjustable* if it uses $O(S)$ bits of working space for a given parameter S. Naturally, we expect to use at least one word, so $\Omega(w)$ is a lower bound for the space usage, w being the size of the machine word in bits. Sorting is one of the few problems for which the optimal space-time product has been settled: Beame showed [4] (see also [25, Theorem 10.13.8]) that $\Omega(N^2)$ is a lower bound, and Pagter and Rauhe showed [24] that an $O(N^2/S + N \lg S)$ running time is achievable for any S when $\lg N \leq S \leq N/\lg N$.

Model of Computation. We assume that the elements being manipulated are stored in a read-only media. Throughout this paper we use N to denote the number of elements burned on the read-only media. Observe that N does *not* need to be known beforehand. If an algorithm must do some outputting, this is done on a separate write-only media. When something is printed to this media, the information cannot be read or rewritten again.

In addition to the input media and the output media, a limited random-access workspace is available. The data on this workspace is manipulated wordwise as on the word RAM [14]. We assume that the word size is at least $\lceil \lg N \rceil$ bits and that the processor is able to execute the same arithmetic, logical, and bitwise operations as those supported by contemporary imperative programming languages—like C [17]. It is a routine matter [19, Section 7.1.3] to store a bit vector of size n such that it occupies $\lceil n/w \rceil$ words and any string of at most w bits can be accessed in $O(1)$ worst-case time. That is, the *time complexity* is proportional to the number of the primitive operations plus the number of element accesses and element comparisons performed in total.

We do *not* assume the availability of any powerful memory-allocation routines. The workspace is an infinite array (of words), and the space used by an algorithm is the prefix of this array. Even though this prefix can have some unused zones, the length of the whole prefix specifies the *space complexity* of the algorithm.

Our Results. In our setting the elements lie in a read-only array and the data structure only constitutes references to these elements. We assume that each of the elements appears in the data structure at most once, and it is the user's responsibility to make sure that this is the case. Also, all operations are position-based; the position of an element can be specified by a pointer or an index. Since the positions can be used to distinguish the elements, we implicitly assume that the elements are distinct. Consider a priority queue Q. Recall that a *priority queue* is a data structure that stores a collection of elements and supports the operations *minimum*, *insert*, and *extract* defined as follows:

$Q.minimum()$: Return the position of the minimum element in Q.

$Q.insert(p)$: Insert the element at position p of the read-only array into Q.

$Q.extract(p)$: Extract the element at position p of the read-only array from Q.

Table 1. The performance of adjustable navigation piles (described in this paper) and their competitors in the read-only random-access model; N is the size of the read-only input and S is an asymptotic target for the size of workspace in bits where $\lg N \le S \le N/\lg N$

Reference	Space	minimum	insert	extract
[5]	$\Theta(N \lg N)$	$O(1)$	$O(1)$	$O(\lg N)$
[16]	$\Theta(N)$	$O(1)$	$O(\lg N)$	$O(\lg N)$
[12]	$\Theta(S)$	$O(1)$	$O(N \lg N/S + \lg S)$	$O(N \lg N/S + \lg S)$
[24]	$\Theta(S)$	$O(N/S^2 + \lg S)$	$O(N/S + \lg S)$ amortized	$O(N/S + \lg^2 S)$
[this paper]	$\Theta(S)$	$O(1)$	$O(1)$	$O(N/S + \lg S)$

In the non-adjustable set-up, any priority queue—like a binary heap [26] or a queue of binary heaps [5] (that are both in-place data structures)—could be used to store positions of the elements instead of the elements themselves.

The main result of this paper is a simplification of the memory-adjustable priority queue by Pagter and Rauhe [24] that is a precursor of all later constructions. First, we devise a memory-adjustable priority queue that we call an *adjustable navigation pile*. Compared to navigation piles [16], that require $\Theta(N)$ bits, our adjustable variant can achieve the same asymptotic performance with only $\Theta(N/\lg N)$ bits. (Another priority queue that uses $\Theta(N)$ bits in addition to the input array was given in [9].) Second, we use this data structure for sorting. The algorithm is priority-queue sort like heapsort [26]: Insert the N elements one by one into a priority queue and extract the minimum from that priority queue N times. In Table 1 we compare the performance of the new data structure to some of its competitors. Note that the stated bounds are valid under some reasonable assumptions that are made explicit in Section 2.

We encourage the reader to compare our solution to that of Pagter and Rauhe [24]. In the basic setting, Pagter and Rauhe proved that the running time of their sorting algorithm is $O(N^2/S + N \lg^2 S)$ using $O(S)$ bits of workspace. Lagging behind the optimal bound for the space-time product by a logarithmic factor when $S = \omega(N/\lg^2 N)$, they suggested using their memory-adjustable data structure in Frederickson's adjustable binary heap [12] to handle subproblems of size $N \lg N/S$ using $O(\lg N)$ bits for each. In accordance, by combining the two data structures, the treatment achieves an optimal $O(N^2/S + N \lg S)$ running time for sorting, where $\lg N \le S \le N/\lg N$. In our treatment we avoid the complication of plugging two data structures together.

Related Models. The basic feature that distinguishes the model of computation we use from other related models is the capability of having random access to the input data. In the context of sequential-access machines, the input is on a tape that only allows single-pass algorithms. The so-called streaming model still enforces sequential access, but allows multi-pass algorithms (Munro and Paterson [20] considered this model). For some problems, the read-only random-access model is more powerful than the multi-pass sequential-access model (for example, for selection the lower bound known for the multi-pass streaming model [6] can be bypassed in the read-only random-access model [11]).

2   Memory-Adjustable Priority Queues

Assumptions. In this section two memory-adjustable priority queues are described. The first structure is a straightforward adaptation of a tournament tree (also called a selection tree [18, Section 5.4.1]) for read-only data. For a parameter S, it uses $O(S)$ words of workspace. The second structure is an improvement of a navigation pile [16] for which the workspace is $O(S + w)$ bits, $\lg N \leq S \leq N/\lg N$, where N is the size of the read-only input and w is the size of the machine word in bits. Both data structures can perform *minimum* and *insert* in $O(1)$ worst-case time and *extract* in $O(N/S + \lg S)$ worst-case time.

When describing the data structures, we tactically assume that

1. N is known beforehand.
2. The elements are extracted from the data structure in monotonic fashion. Accordingly, at any given point of time, we keep a single element as a boundary telling that the elements smaller than or equal to that element have been extracted from the data structure. We call such an element the *latest output*, and say that an element is *alive* if it is larger than the latest output.
3. The elements are inserted into the data structure sequentially—but still insertions and extractions can be intermixed—in streaming-like fashion starting from the first element stored in the read-only input.

These assumptions are valid when a priority queue is used for sorting. Actually, in sorting all insertions are executed before extractions; a restriction that is not mandated by the data structure. At the end of this section, we show how to get rid of these assumptions. The first assumption is not critical. But, when relaxing the second assumption, the required size of workspace has to increase by N bits. In addition, when relaxing the third assumption, the worst-case running time of *insert* will become the same as that required by *extract*.

Tournament Trees. For an integer S, we use $\bar{S}$ as a shorthand for $2^{\lceil \lg S \rceil}$. It suffices that $S \leq N/\lg N$; even if S was larger, the operations would not be asymptotically faster. The input array is divided into $\bar{S}$ *buckets* and a complete binary tree is built above these buckets. Each leaf of the tree *covers* a single bucket and each branch node covers the buckets covered by the leaves in the subtree rooted at that branch. We call the elements within the buckets covered by a node the *covered range* of this node. Note that the covered range of a node is a sequence of elements stored in consecutive locations of the input array. The actual data stored at each node is a pointer or an index specifying the position of the smallest alive element in the covered range of that node.

In its basic form, the data structure is an array of $2\bar{S} - 1$ positions (indices). To make the connection to our adjustable navigation piles clear, we store the positions in breadth-first order as in a binary heap [26]. We start the numbering

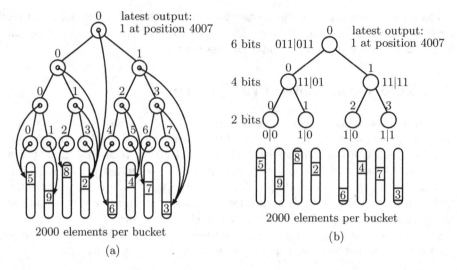

Fig. 1. (a) A tournament tree and (b) a navigation pile when $N = 16\,000$ and $\bar{S} = 8$. Only the smallest alive element in each bucket is shown

of the nodes at each level from 0. For the sake of simplicity, we maintain a *header* that stores the pointers to the beginning of each level (even though this information could be calculated). For a node number i, its left child has number $2i$ at the level below, the right child has number $2i + 1$ at the level below, and the parent has number $\lfloor i/2 \rfloor$ at the level above. When we know the current level and the number of a node, the information available at the header and these formulas are enough to get to a neighbouring node in constant time. In Fig. 1(a) we give an illustration of a tournament tree when $N = 16\,000$ and $\bar{S} = 8$.

To support *insert* efficiently, we partition the data structure into three components: the tournament tree, the submersion buffer, and the insertion buffer. The *submersion buffer* is the last full bucket that is being integrated with the tournament tree incrementally. The *insertion buffer* is the bucket that embraces all the new elements. Observe that one or both of these buffers can be empty. The idea is to insert the elements into a buffer and, first when the buffer gets full, integrate it with the tournament tree. This buffering technique has been used in other contexts as well (see, for example, [1,8]). The overall minimum can be in any of the three components. To support *minimum* in $O(1)$ worst-case time, we keep track of the position of the overall minimum.

In connection with *insert*, the next element from the read-only array becomes part of the insertion buffer. If the new element is smaller than the buffer minimum and/or the overall minimum, the positions of these minima are updated. If the insertion buffer becomes full, the submersion buffer must have been already integrated into the tournament tree. At this point, we treat the insertion buffer as the new submersion buffer and start a bottom-up submersion process updating the nodes of the tournament tree that cover this bucket. Every branch node inherits the position of the smaller of the two elements pointed to via its two children. As long as the submersion is not finished, each *insert* is responsible for continuing a constant amount of the submersion work. Since the work needed to update this path is $O(\lg S) = O(N/S)$, the process terminates before the insertion buffer is again full. Clearly, *insert* takes $O(1)$ worst-case time.

In *extract*, simple calculations are done to determine which bucket covers the element being extracted. The latest output is first set up to date. Since the smallest alive element of the present bucket must have been extracted, the bucket is scanned to find its new minimum. There are three cases depending on whether the present bucket is one of the buffers or covered by the tournament tree. If the present bucket is the insertion buffer, it is just enough to update the position of its minimum. If the present bucket is the submersion buffer, the submersion process is completed by recomputing the positions in the nodes of the tournament tree that cover the submersion buffer. Hereafter, the submersion buffer ceases to exist. If the present bucket is covered by the tournament tree, it is necessary to redo the comparisons at the branch nodes that cover the present bucket. In all three cases the position of the overall minimum is updated, if necessary. It is the scanning of a bucket that makes this operation expensive: The worst-case running time is $O(N/S + \lg S)$, which is $O(N/S)$ when $S \leq N/\lg N$.

Navigation Piles. In principle, a navigation pile [16] is a compact representation of a tournament tree. The main differences are (see Fig. 1(b)):

1. Only the nodes whose heights are larger than 0 have a counterpart. Hence, the number of nodes in the complete binary tree is $\bar{S} - 1$.
2. Any node only stores partial information about the position of the smallest alive element in the covered range of that node. Here our construction differs from that used in the navigation piles [16] and their precursors [24].

A branch node of height $h \in \{1, 2, \ldots, \lg \bar{S}\}$ covers 2^h buckets. As in the original navigation piles, we use h bits to specify in which bucket the smallest alive element is. A significant ingredient is the concept of a *quantile*. (A similar quantile-thinning technique was used in [24], but not in an optimal way, and later in [10].) For a branch node of height h, every covered bucket is divided into 2^h quantiles, and another h bits are used to specify in which quantile the smallest alive element is. That is, except that the last quantile can possibly be smaller, a quantile contains $\lceil N/(\bar{S} \cdot 2^h) \rceil$ elements. We need $2h$ bits per node; but if $2h \geq \lceil \lg N \rceil$, we do not use more than $\lceil \lg N \rceil$ bits (since this is enough to specify the exact position of the smallest alive element). To sum up, since there are $\bar{S}/2^h$ nodes of height h and since at each node we store $\min \{2h, \lceil \lg N \rceil\}$ bits, the total number of bits is bounded by

$$\sum_{h=1}^{\lg \bar{S}} \frac{\bar{S} \cdot \min \{2h, \lceil \lg N \rceil\}}{2^h} < 4\bar{S}.$$

The navigation bits are stored in a bit vector in breadth-first order. As before, we maintain a header giving the position of the first bit at each level. The space needed by the header is $O(\lg^2 \bar{S})$ bits. Inside each level the navigation information is stored compactly side by side, and the nodes are numbered at each level starting from 0. Since the length of the navigation bits is fixed for all nodes at the same level, using the height and the number of a node, it is easy to calculate the positions where the navigation bits of that node are stored.

Let us now illustrate how to access the desired quantile for a branch node in constant time. Let the number of the branch node be x within its own level, and assume that its height is h. The first element of the covered range is in position $x \cdot 2^h \cdot \lceil N/\bar{S} \rceil$. The first h bits of the navigation information gives the desired bucket; let this bucket index be b, so we have to go $b \cdot \lceil N/\bar{S} \rceil$ positions forward. The second h bits of the navigation information gives the desired quantile inside that bucket; let this quantile index be q, so we have to proceed another $q \cdot \lceil N/(\bar{S} \cdot 2^h) \rceil$ positions forward before we reach the beginning of the desired quantile. Obviously, these calculations can be carried out in constant time.

The priority-queue operations can be implemented in a similar way as for a tournament tree. To facilitate constant-time *minimum*, we can keep a separate pointer to the overall minimum (since the root of an adjustable navigation pile does not necessarily specify a single element). One subtle difference is that, when we update a path from a node at the bottom level to the root, we have to scan the elements in the quantiles specified for the sibling nodes of the nodes along the path. After updating the navigation bits of a node y, we locate its parent x and its sibling z. The navigation bits at z are used to locate the quantile that has the minimum element covered by z. This quantile is scanned and the minimum element is found and compared with the minimum element covered by y. From the bucket number and the position of the smaller of the two elements, the navigation bits at x are then calculated and accordingly updated. If the quantile for x has only one element, the position of this single element can be stored as such. The key is that for a node of height h, the size of the quantile is $\lceil N/(\bar{S} \cdot 2^h) \rceil$, so the total work done in the scans of the quantiles of the siblings along the path is proportional to $\lceil N/\bar{S} \rceil$ as it should be. It follows that the efficiency of the priority-queue operations is the same as for a tournament tree.

Getting Rid of the Assumptions. So far we have consciously ignored the fact that the size of the buckets depends on the value of N, and that we might not know this value beforehand. The standard way of handling this situation is to rely on global rebuilding [23, Chapter V]. We use an estimate N_0 and initially set $N_0 = 8$. We build two data structures, one for N_0 and another for $2N_0$. The first structure is used to perform the priority-queue operations, but insertions and extractions are mirrored in the second structure (if the extracted element exists there). When the structure for N_0 becomes too small, we dismiss the smaller structure in use, double N_0, and in accordance start building a new structure of size $2N_0$. We should speed up the construction of the new structure by inserting up to two alive elements into it at a time, instead of only one. This guarantees that the new structure will be ready for use before the first one is dismissed. Even though global rebuilding makes the construction more complicated, the time and space bounds remain asymptotically the same.

A possible scenario in applications is when extractions are no more monotonic. To handle this situation, we have to allocate one bit per array entry, indicating whether the corresponding element is alive or not. This increases the size of the workspace significantly if S is much smaller than N.

Since we have random-access capability to the read-only input, it is not absolutely necessary that elements are inserted into the data structure by visiting the input sequentially. We could insert the elements in arbitrary order. If this is the case, in connection with each *insert* we have to fix the information related to the present bucket as in *extract*. That is, we have to find the smallest alive element of the bucket and update the navigation information on the path from a node at the bottom level to the root. This means that the worst-case cost of *insert* becomes the same as that of *extract*, i.e. $O(N/S + \lg S)$.

3   Sorting

Priority-Queue Sort. To sort the given N elements, we create an empty adjustable navigation pile, insert the elements into this pile by scanning the read-only array from beginning to end, and then repeatedly extract the minimum of the remaining elements from the pile. See the pseudo-code in Fig. 2.

Analysis. From the bounds derived for the priority queue, the asymptotic performance can be directly deduced: The worst-case running time is $O(N^2/S + N \lg S)$ and the size of workspace is $O(S + w)$ bits. It is also easy to count the number of element comparisons performed during the execution of the algorithm. When inserting the N elements into the data structure, $O(N)$ element comparisons are performed. We can assume that after these insertions, the buffers are integrated into the main structure. In each *extract* we have to find the minimum of a single bucket which requires at most $N/\bar{S}$ element comparisons. In addition, we have to update a single path in the complete binary tree. At each level, the minimum below the current node is already known and we have to scan the quantiles of the sibling nodes. During the path update, we have to perform at most $N/\bar{S} + \lg \bar{S}$ element comparisons. We know that $S \leq \bar{S} \leq 2S$. Hence, the total number of element comparisons performed is bounded by $2N^2/S + N \lg S + O(N)$.

---

procedure: *priority-queue-sort*
input: A: read-only array of N elements; S: space target
output: stream of elements produced by the *print* statements
$P \leftarrow$ *navigation-pile*(A, S)
for $x \in \{0, 1, \ldots, N-1\}$
  $\mid$ $P.insert(x)$
repeat N times
  $\mid$ $y \leftarrow P.minimum()$
  $\mid$ $P.extract(y)$
  $\mid$ $print(A[y])$

Fig. 2. Priority-queue sort in pseudo-code; the position of an element is its index

4   Concluding Remarks

Summary. In the construction of adjustable navigation piles three techniques are important: 1) node numbering with implicit links between nodes, 2) bit packing and unpacking, and 3) quantile thinning. In addition to the connection to binary heaps [26], we pointed out the strong connection to tournament trees. We made the conditions explicit for when a succinct implementation of a priority queue requiring $o(N)$ bits is possible (when extractions are monotonic), so that algorithm designers would be careful when using the data structure.

Our sorting algorithm for the read-only random-access model is a heapsort algorithm [26] that uses an adjustable navigation pile instead of a binary heap. In spite of optimality, one could criticize the model itself since the memory-access patterns may not always be friendly to contemporary computers; and we are not allowed to move the elements. Navigation piles are slow [15] for two reasons: 1) The bit-manipulation machinery is heavy and index calculations devour clock cycles. 2) The cache behaviour is poor because the memory accesses lack locality. Unfortunately, the situation is not much better for adjustable navigation piles; the buckets are processed sequentially, but the quantiles lie in different buckets.

Other Data Structures for Read-Only Data. In our experience, very few data structures can be made memory adjustable as elegantly as priority queues. A stack is another candidate [3]. For example, a dictionary must maintain a permutation of a set of size N; this means that it is difficult to manage with much less than $N \lceil \lg N \rceil$ bits. However, when the goal is to cope with about N bits, a bit vector extended with rank and select facilities (for a survey, see e.g. [21]) is a relevant data structure. Two related constructions are the wavelet stack used in [11] and the wavelet tree introduced in [13] (for a survey, see [22]).

References

1. Alstrup, S., Husfeldt, T., Rauhe, T., Thorup, M.: Black box for constant-time insertion in priority queues. ACM Trans. Algorithms 1(1), 102–106 (2005)
2. Asano, T., Buchin, K., Buchin, M., Korman, M., Mulzer, W., Rote, G., Schulz, A.: Memory-constrained algorithms for simple polygons. E-print arXiv:1112.5904, arXiv.org, Ithaca (2011)
3. Barba, L., Korman, M., Langerman, S., Sadakane, K., Silveira, R.: Space-time trade-offs for stack-based algorithms. E-print arXiv:1208.3663, arXiv.org, Ithaca (2012)
4. Beame, P.: A general sequential time-space tradeoff for finding unique elements. SIAM J. Comput. 20(2), 270–277 (1991)
5. Carlsson, S., Munro, J.I., Poblete, P.V.: An implicit binomial queue with constant insertion time. In: Karlsson, R., Lingas, A. (eds.) SWAT 1988. LNCS, vol. 318, pp. 1–13. Springer, Heidelberg (1988)
6. Chan, T.M.: Comparison-based time-space lower bounds for selection. ACM Trans. Algorithms 6(2), 26:1–26:16 (2010)
7. De, M., Nandy, S.C., Roy, S.: Convex hull and linear programming in read-only setup with limited work-space. E-print arXiv:1212.5353, arXiv.org, Ithaca (2012)

8. Edelkamp, S., Elmasry, A., Katajainen, J.: Weak heaps engineered (submitted for publication)
9. Elmasry, A.: Three sorting algorithms using priority queues. In: Ibaraki, T., Katoh, N., Ono, H. (eds.) ISAAC 2003. LNCS, vol. 2906, pp. 209–220. Springer, Heidelberg (2003)
10. Elmasry, A., He, M., Munro, J.I., Nicholson, P.K.: Dynamic range majority data structures. In: Asano, T., Nakano, S.-i., Okamoto, Y., Watanabe, O. (eds.) ISAAC 2011. LNCS, vol. 7074, pp. 150–159. Springer, Heidelberg (2011)
11. Elmasry, A., Juhl, D.D., Katajainen, J., Satti, S.R.: Selection from read-only memory with limited workspace. In: Du, D.Z., Zhang, G. (eds.) COCOON 2013. LNCS. Springer, Heidelberg (2013)
12. Frederickson, G.N.: Upper bounds for time-space trade-offs in sorting and selection. J. Comput. System Sci. 34(1), 19–26 (1987)
13. Grossi, R., Gupta, A., Vitter, J.S.: High-order entropy-compressed text indexes. In: SODA 2003, pp. 841–850. SIAM, Philadelphia (2003)
14. Hagerup, T.: Sorting and searching on the word RAM. In: Morvan, M., Meinel, C., Krob, D. (eds.) STACS 1998. LNCS, vol. 1373, pp. 366–398. Springer, Heidelberg (1998)
15. Jensen, C., Katajainen, J.: An experimental evaluation of navigation piles. CPH STL Report 2006-3, Department of Computer Science, University of Copenhagen, Copenhagen (2006)
16. Katajainen, J., Vitale, F.: Navigation piles with applications to sorting, priority queues, and priority deques. Nordic J. Comput. 10(3), 238–262 (2003)
17. Kernighan, B.W., Ritchie, D.M.: The C Programming Language, 2nd edn. Prentice Hall, Englewood Cliffs (1988)
18. Knuth, D.E.: Sorting and Searching, The Art of Computer Programming, 2nd edn., vol. 3. Addison-Wesley, Reading (1998)
19. Knuth, D.E.: Combinatorial Algorithms: Part 1, The Art of Computer Programming, vol. 4A. Addison-Wesley, Boston (2011)
20. Munro, J.I., Paterson, M.S.: Selection and sorting with limited storage. Theoret. Comput. Sci. 12(3), 315–323 (1980)
21. Navarro, G., Providel, E.: Fast, small, simple rank/select on bitmaps. In: Klasing, R. (ed.) SEA 2012. LNCS, vol. 7276, pp. 295–306. Springer, Heidelberg (2012)
22. Navarro, G.: Wavelet trees for all. In: Kärkkäinen, J., Stoye, J. (eds.) CPM 2012. LNCS, vol. 7354, pp. 2–26. Springer, Heidelberg (2012)
23. Overmars, M.H.: The Design of Dynamic Data Structures. LNCS, vol. 156. Springer, Heidelberg (1983)
24. Pagter, J., Rauhe, T.: Optimal time-space trade-offs for sorting. In: FOCS 1998, pp. 264–268. IEEE Computer Society, Los Alamitos (1998)
25. Savage, J.E.: Models of Computation: Exploring the Power of Computing (2008), http://cs.brown.edu/~jes/book/home.html; the book is released in electronic form under a CC-BY-NC-ND-3.0-US license
26. Williams, J.W.J.: Algorithm 232: Heapsort. Commun. ACM 7(6), 347–348 (1964)

$(1 + \epsilon)$-Distance Oracles for Vertex-Labeled Planar Graphs

Mingfei Li[1,*], Chu Chung Christopher Ma[2], and Li Ning[2]

[1] Facebook, Menlo Park, CA
mfli@cs.hku.hk
[2] The Department of Computer Science, The University of Hong Kong
{cccma,lning}@cs.hku.hk

Abstract. We consider vertex-labeled graphs, where each vertex v is attached with a label from a set of labels. The vertex-to-label distance query desires the length of the shortest path from the given vertex to the set of vertices with the given label. We show how to construct an oracle for a vertex-labeled planar graph, such that $O(\frac{1}{\epsilon} n \log n)$ storing space is needed, and any vertex-to-label query can be answered in $O(\frac{1}{\epsilon} \log n \log \Delta)$ time with stretch $1 + \epsilon$. Here, Δ is the hop-diameter of the given graph. For the case that $\Delta = O(\log n)$, we construct a distance oracle that achieves $O(\frac{1}{\epsilon} \log n)$ query time, without changing space usage.

1 Introduction

The construction of distance oracles for vertex-labeled graphs was introduced in [5]. In this paper, we consider this problem for planar graphs. Given a graph $G = (V, E)$, the edges are undirected and each of them is assigned a non-negative weight. In addition, each vertex u is assigned a label $\lambda(u)$, by a labeling function $\lambda : V \to L$, where L is a set of labels. Let V_λ denote the set of vertices assigned label λ. The distance between two nodes $v, u \in V$, denoted by $\delta(v, u)$, is the length of the shortest path between v and u. The distance between a node $u \in V$ and a label $\lambda \in L$ is the distance between u and u's nearest neighbor with label λ, i.e. $\delta(u, \lambda) = \min_{v \in V_\lambda} \delta(u, v)$.

In applications, a vertex-to-label query asks for the distance between a node and a set of points with some common functionality, which is specified by a vertex label. These queries arise in many real-life scenarios. For example, one might want to find the restaurant in town that is nearest to their office, or to find the bus stop of a specific route that is closest to their home. Trivially, people can calculate in advance and store the answer for each pair of vertex and label. However, this might take $O(|V| \times |L|)$ space, which can be $O(n^2)$ in the worst case.

---

* This work was done when the author studied at the Department of Computer Science, the University of Hong Kong.

T-H.H. Chan, L.C. Lau, and L. Trevisan (Eds.): TAMC 2013, LNCS 7876, pp. 42–51, 2013.
© Springer-Verlag Berlin Heidelberg 2013

The aim of *vertex-to-label distance oracle* (VLDO) is to pre-calculate and store information using $o(|V| \times |L|)$ space, such that any vertex-to-label query can be answered efficiently. Approximate VLDO's answer queries with stretch greater than one. Specifically, an approximate VLDO with stretch t, denoted by t-VLDO, returns $d(u, \lambda)$ as the approximation to $\delta(u, \lambda)$, such that $\delta(u, \lambda) \leq d(u, \lambda) \leq t \cdot \delta(u, \lambda)$. In [5], Hermelin et al. generalized Thorup and Zwick's scheme for approximated vertex-to-vertex distance oracle (VVDO) [15] to construct a $(4k-5)$-VLDO with query time $O(k)$, storage of size $O(kn^{1+\frac{1}{k}})$ and preprocessing time $O(kmn^{\frac{1}{k}})$, where n and m denote numbers of vertices and edges of the given graph, respectively.

In a number of applications, the given graph can be drawn on a plane, and is thus called a *planar graph*. It has been shown that for many graph problems, there exist more efficient algorithms for planar graphs, and hence we are motivated to derive approximate VLDO's for planar graphs. Note that in the work by Hermelin et al., the space for storage is $O(kn^{1+\frac{1}{k}})$, which is $O(n^2)$ for $k = 1$. As this is as much as the space required by the trivial solution, their construction makes sense only when $k \geq 2$. Then, we can assume that the stretch of their distance oracle is at least 3. However, in this paper, we consider approximate VLDOs with stretch arbitrarily close to 1, with the sacrifice of increasing the query time to $polylog(n)$. Our approximate VLDO requires storage of size $O(n \log n)$, in which the big-O notation hides a parameter inversely proportional to the stretch.

1.1 Related Work

Vertex-to-Label Distance Oracles. The problem of constructing approximate VLDOs was formalized and studied by Hermelin et al. in [5]. Besides the construction that is adapted from Thorup and Zwick's scheme in [15], they also constructed $(2^k - 1)$-VLDOs with query time $O(k)$, storage of expected size $O(kn\ell^{\frac{1}{k}})$ (where $\ell = |L|$) and preprocessing time $O(kmn^{\frac{k}{2k-1}})$, as well as $(2 \cdot 3^k + 1)$-VLDOs that support label changes in $O(kn^{\frac{1}{k}} \log n)$ time, with query time $O(k)$ and storage of expected size $O(kn^{1+\frac{1}{k}})$. In [2], Chechik showed that Thorup and Zwick's scheme can also be modified to get $(4k - 5)$-VLDOs that support label changes in $O(n^{\frac{1}{k}} \log^{1-\frac{1}{k}} n \log \log n)$ time, with query time $O(k)$, storage of expected size $\tilde{O}(n^{1+\frac{1}{k}})$, and preprocessing time $O(kmn^{\frac{1}{k}})$.

VLDOs have also been studied for special classes of graphs. Tao et al. [12] have shown how to construct VLDOs for XML trees. For the case that each node is assigned exactly one label, their construction results in (exact) VLDOs with query time $O(\log n)$, storage of size $O(n)$ and preprocessing time $O(n \log n)$.

Vertex-to-Vertex Distance Oracles (VVDOs). In a seminal paper [15], Thorup and Zwick have introduced a scheme to construct a $(2k - 1)$-VVDO with query time $O(k)$, storage of expected size $O(kn^{1+\frac{1}{k}})$ and preprocessing time $O(kmn^{\frac{1}{k}})$. Wulff-Nilsen [16] improved the preprocessing time to $O(\sqrt{k}m + kn^{1+\frac{c}{\sqrt{k}}})$ for some universal constant c, which is better than $O(kmn^{\frac{1}{k}})$ except for very sparse graphs and small k. For (undirected) planar graphs, in [14] (whose conference version appeared in 2001), Thorup has shown how to construct a $(1+\epsilon)$-VVDO with query time $O(\frac{1}{\epsilon})$, and storage of size $O(\frac{1}{\epsilon}n\log n)$. The technique he designed is for directed graphs and hence the result for undirected graphs follows. In [7], Klein showed a simplified scheme that achieves the same properties, but only works for undirected graphs. Recently, Sommer et al. have derived more compact approximate VVDOs for planar graphs [11] (this paper also includes an detailed survey of VVDO results for planar graphs), and the linear-space approximate VVODs for planar, bounded-genus, and minor-Free Graphs [6].

Exact VVVDOs for planar graphs have been studied intensively. In [10], Mozes and Sommer have shown several constructions and surveyed other results on exact VVVDOs for planar graphs.

Shortest Paths. The construction of distance oracles often harnesses shortest path algorithms in the preprocessing stage. A *shortest path tree* of a graph G with respect to a vertex v is a spanning tree of G rooted at v, such that for any u in the given graph, the path from v to u in the tree is the shortest path from v to u in G. Given a vertex v, the *single source shortest path* problem requires one to calculate the shortest path tree with respect to v. In undirected graphs, the single source shortest path tree can be calculated in $O(m)$ time, where m is the number of the edges in the given graph. Such an algorithm is introduced by Thorup in [13].

1.2 Simple Solution for Doubling Metrics

If the metric induced by the given graph is doubling, the following procedure provides a simple solution to return $\delta(u, \lambda)$ with $(1 + \epsilon)$-stretch.

Preprocessing. Let $\epsilon' = \frac{\epsilon}{3}$. For $\epsilon < 1$, $(1 + \epsilon')^2 < 1 + \epsilon$. We first construct a distance oracle $\mathcal{O}$ introduced in [1] that supports $1 + \epsilon'$ approximate vertex-to-vertex queries. Suppose the distance between x and y returned by $\mathcal{O}$ is $d(x, y)$. Then, for each label $\lambda \in L$, we construct the approximate nearest neighbor (ANN) data structure introduced in [3] to support $(1 + \epsilon')$-ANN queries to the set of points with label λ, with d being the underlying metric. Note that in the construction and queries of the ANN data structure, we can use $\mathcal{O}$ to answer queries about distances between points.

Query. Given $u \in V$ and $\lambda \in L$, we find the $(1 + \epsilon')$-ANN of u among the nodes with label λ. Let v be this $(1 + \epsilon')$-ANN. Then, we query in $\mathcal{O}$ and then return the approximate distance $d(u, v)$ between u and v.

We show that $d(u, v) \leq (1 + \epsilon)\delta(u, \lambda)$. Let v' and v^* be the vertices with label λ such that $d(u, v') = d(u, \lambda)$ and $\delta(u, v^*) = \delta(u, \lambda)$, i.e. v' and v^* are the nearest neighbors of u with label λ under the metric d and δ. Then, we have $d(u, v) \leq (1 + \epsilon')d(u, v') \leq (1 + \epsilon')d(u, v^*) \leq (1 + \epsilon')^2\delta(u, v^*) \leq (1 + \epsilon)\delta(u, \lambda)$.

Space and Query Time. The data structure supporting ANN queries for λ can be constructed using $O(n_\lambda)$ space [3], where n_λ is the number of nodes with label λ. Since any node is allowed to be attached with only one label, the space used in total is $O(n)$. In addition, the ANN query can be answered in $O(\log n_\lambda) = O(\log n)$ time [3]. The vertex-to-vertex distance oracle can be constructed using $O(n)$ space and answer a query in $O(1)$ time [1].

1.3 Our Contribution

To the best of our knowledge, no construction of $(1+\epsilon)$-VLDOs for planar graphs has been shown. Theorem 1 and Theorem 2 are the main results of this paper. Recall that Δ is used to denote the hop-diameter of a graph.

Theorem 1. *Given a planar graph $G = (V, E)$ and a label set L, for any $0 < \epsilon < 1$, there exists a $(1 + \epsilon)$-VLDO with query time $O(\frac{1}{\epsilon} \log n \log \Delta)$, storage of size $O(\frac{1}{\epsilon} n \log n)$ and preprocessing time $O(\frac{1}{\epsilon} n \log^2 n)$.*

Theorem 2. *Given a planar graph $G = (V, E)$ with $\Delta = O(\log n)$ and a label set L, for any $0 < \epsilon < 1$, there exists a $(1+\epsilon)$-VLDO with query time $O(\frac{1}{\epsilon} \log n)$, storage of size $O(\frac{1}{\epsilon} n \log n)$ and preprocessing time $O(\frac{1}{\epsilon} n \log^2 n)$.*

In our construction, we first select portals on each separator of the *recursive graph decomposition* for each involved vertex, similar to the scheme in [7]. Then for each label $\lambda \in L$, we define its portals to be the union of portals for all label-λ nodes. Given a query of distance between a vertex $u \in V$ and a label $\lambda \in L$, the query algorithm works to find a portal z_u of u and a portal z_λ of λ, such that the shortest path that connects u to V_λ going through z_u and z_λ, approximates $\delta(u, \lambda)$ with stretch $1 + \epsilon$. In the case of VVDO, after fixing a separator, there are totally $O(\epsilon)$ portals for the source vertex and the destination vertex. Hence, by brute force comparisons, the time to find the pair of portals with shortest bypassing path, is at most a constant which depends only on ϵ. However, in the case of VLDO, since the destination is a set of nodes (nodes assigned the given label), there might be as many as $O(n)$ portals for the destination on a fixed separator. Hence, the brute force comparison does not work well. However, we show that in this case, the appropriate pair of portals on a fixed separator can be found by making use of range minimum queries, in $O(\frac{1}{\epsilon} \log \Delta)$ time. Furthermore, when $\Delta = O(\log n)$, we can improve the time to $O(\frac{1}{\epsilon})$. In our query algorithm, we consider all of the $O(\log n)$ separators involving u one by one, and hence the total query time is $O(\frac{1}{\epsilon} \log n \log \Delta)$ in general case, and $O(\frac{1}{\epsilon} \log n)$ for the case that $\Delta = O(\log n)$.

2 Preliminaries

Lipton Tarjan Separator [9]. Let T be a spanning tree of a planar embedded triangulated graph G with weights on nodes. Then there is an edge $e \notin T$, such that the strict interior and strict exterior of the simple cycle in $T \cup \{e\}$ each contains weight no more than $\frac{2}{3}$ of the total weight.

Recursive Graph Decomposition [7]. The recursive graph decomposition (RGD) of a given graph G is a rooted tree, such that each vertex p in RGD maintains

- a set $N(p)$ of nodes in G, in particular the root of RGD maintains (as a label) $N(p) = V(G)$, and
- p is a leaf of RGD *iff* $N(p)$ contains only one node of G. In this case, let $S(p) = N(p)$;
- if p is not a leaf of RGD, it maintains (as a label) an α-balanced separator $S(p)$ of G, balanced with respect to the weight assignment in which each node in $N(p)$ is assigned weight 1 and other nodes are assigned weight 0.

A non-leaf vertex v of the tree has two children p_1 and p_2, such that

- $N(p_1) = v \in N(p) \cap ext(\tilde{S}(p))$, and
- $N(p_2) = v \in N(p) \cap int(\tilde{S}(p))$,

where $\tilde{S}$ denotes the cycle corresponding to a separator S and $ext(\tilde{S})$ $(int(\tilde{S}))$ denotes the exterior (interior) part of $\tilde{S}$. For a leaf node p of RGD, $N(p)$ contains only one node of G. In practice, $N(p)$ may contain a small number of nodes, such that the distances in the subgraph induced by $N(p)$ for every pair of nodes in $N(p)$ are pre-calculated and stored in a table support $O(1)$-time look-up.

By the construction in [7,8], RGD could be calculated in time $O(n \log n)$.

Range Minimum Query. The range minimum query problem is to pre-process an array of length n in $O(n)$ time such that all subsequent queries asking for the position of a minimal element between two specified indices can be answered quickly. This can be done with constant query time and storage of linear size [4].

Notations. For label $\lambda \in L$, let V_λ denote the set of nodes assigned label λ. The given graph has hop-diameter Δ, if Δ is the minimum number, such that for any pair of vertices $v, u \in V$, the shortest path between them consists of at most Δ edges.

3   A $(1 + \epsilon)$-VLDO with $O(\frac{1}{\epsilon} \log n \log \Delta)$ Query Time

In this section we introduce an $(1 + \epsilon)$-VLDO that supports $O(\frac{1}{\epsilon} \log n \log \Delta)$ query time.

3.1   Preprocessing and Query Algorithm

Preprocessing. Given planar graph $G = (V, E)$ and label set L, the shortest path tree for any node $r \in V$ has height at most Δ. As the first step of preprocessing, we fix an arbitrary node $r \in V$ and compute the shortest-path tree T in G rooted at r. Then calculate the RGD based on T. In our oracle, we store

- an array of size n, in which for each node $v \in G$ there is exactly an entry storing the leaf node p of RGD, such that $v \in N(p)$;
- an array records the depth (i.e. hop distance from the root) for each node of RGD;
- a table T, in which for each pair of $v \in G$ and $p \in$RGD such that $v \in N(p)$, there is exactly one entry $T_{v,p}$. $T_{v,p}$ stores two sub-tables for the two paths P' and P'' forming the separator $S(p)$. In sub-table $T_{v,p}[P']$ (similarly for $T_{v,p}[P'']$), it stores a sequence of $O(\frac{1}{\epsilon})$ portals

$$z_{-q}, z_{-q+1}, \ldots, z_0, \ldots, z_{w-1}, z_w,$$

where z_i's are nodes on P', such that the distance property is satisfied, i.e. for any node u on P', there is a portal z in $T_{v,p}[P']$ such that $d(v, z) + d(z, w) \leq d(v, w)$. In addition, for each portal z_i, we record the distance from z_i to r and denoted it by $h(z_i)$.

The above structure is similar to that used in [7] and it needs $O(\frac{1}{\epsilon} n \log n)$ space. To support the vertex-to-label distance query, we store more information.

- **Group Portals for Labels.** We store a table $\widehat{T}$, in which for each pair of $\lambda \in L$ and $p \in$RGD such that $V_\lambda \cap N(p) \neq \emptyset$, there is exactly one entry $\widehat{T}_{\lambda,p}$. $\widehat{T}_{\lambda,p}$ stores two sub-tables for the two paths P' and P'' forming the separator $S(p)$. In sub-table $\widehat{T}_{\lambda,p}[P']$ (similarly for $\widehat{T}_{\lambda,p}[P'']$), it stores portals in $T_{v,p}[P']$ for all $v \in V_\lambda \cap N(p)$, in the increasing order of the distance between the portals and r. Since each node $v \in V$ is assigned exactly one label from L, this step dose not change the asymptotic usage of the storage. Since we want the portals to be sorted on a fixed separator and totally there are $O(\frac{1}{\epsilon} n \log n)$ portals, this step requires time $O(\frac{1}{\epsilon} n \log^2 n)$.
- **Hash.** For each node $v \in V$, we build a hash table to support $O(1)$-time query of the entry $T_{v,p}$ of table T. Since the number of nodes $p \in$RGD satisfying $v \in N(p)$ is at most $O(\log n)$, this step requires $O(n \log n)$ space and time in total. Similarly, for each label $\lambda \in L$, we build a hash table (in linear time) to support $O(1)$-time query of the entry $\widehat{T}_{\lambda,p}$ of table $\widehat{T}$, without changing the asymptotic usage of storage.
- **To support RMQ.** For any $\widehat{T}_{\lambda,p}[P']$ (similarly for $\widehat{T}_{\lambda,p}[P'']$), each stored portal z is associated with
  - a node v such that z is a portal of v (if there are several choices of v, we choose the one with minimum $\delta(v, z)$);
  - and a number $h(z)$, which is the distance from z to r.

  We construct a data structure to support $O(1)$-time RMQ query on portals $z \in \widehat{T}_{\lambda,p}[P']$ according to the value $\delta(v, z) + h(z)$, and a data structure to support $O(1)$-time RMQ query on portals $z \in \widehat{T}_{\lambda,p}[P']$ according to the value $\delta(v, z) - h(z)$. By [4], this can be achieved and the storage is asymptotically the same with the size of $\widehat{T}$. Since the construction of RMQ needs linear time, this step requires linear time in all.

Query. Given a node $u \in G$ and a label $\lambda \in L$, do as Algorithm 1.

Input: u, λ

Initialization: $d(u, \lambda) \leftarrow \infty$
for *Each* $p \in RGD$ *s.t.* $u \in N(p)$ *and* T_λ *has an entry for* p **do**
   for *Each path* P *of* $S(p)$ **do**
      for *Each path portal* z_u *of* u *on* P **do**
         $C^+ \leftarrow \{\lambda\text{'s portals on } P \text{ that are farther or equal than } z_u \text{ from } r\}$
         $\{z^+, v^+\} \leftarrow$ the portal of some λ labeled node v that achieves
         $\min\{\delta(v, z_v) + h(z_v)\}$ over C^+, and v
         $C^- \leftarrow \{\lambda\text{'s portals on } P \text{ that are closer or equal than } z_u \text{ from } r\}$
         $\{z^-, v^-\} \leftarrow$ the portal of some λ labeled node v that achieves
         $\min\{\delta(v, z_v) - h(z_v)\}$ over C^-, and v
         $d' \leftarrow \min\{\delta(u, z_u) + \delta(z_u, z^+) + \delta(v, z^+), \delta(u, z_u) + \delta(z_u, z^-) + \delta(v, z^-)\}$
         $d(u, \lambda) \leftarrow \min\{d', d(u, \lambda)\}$
      end
   end
end
Output: $d(u, \lambda)$

Algorithm 1. Query algorithm for approximate vertex-to-label distance

Lemma 1. *Given* u, λ, *let* v *be the node assigned label* λ *and satisfying* $\delta(u, v) = \delta(u, \lambda)$. *There exist a portal* z_u *of* u *and a portal* z_v *of* v *on the same path* P *of some separator, such that* $\delta(u, z_u) + \delta(z_u, z_v) + \delta(z_v, v) \leq (1 + \epsilon)\delta(u, \lambda)$.

Proof. Let p_u, p_v be the lowest pieces in RGD containing u, v, respectively, i.e. $u \in N(p_u)$ and $v \in N(p_v)$. Let p_{uv} be the lowest common ancestor (LCA) of p_u and p_v in RGD. Then $u \in N(p_{uv})$, $v \in N(p_{uv})$, and the shortest path from u to v crosses with $S(p_{uv})$. Denote the crossing point as c. There exists a u's portal z_u, such that $\delta(u, z_u) + \delta(z_u, c) \leq (1 + \epsilon)\delta(u, c)$, and a v's portal z_v, such that $\delta(v, z_v) + \delta(z_v, c) \leq (1 + \epsilon)\delta(v, c)$.
    Hence $\delta(u, z_u) + \delta(z_u, z_v) + \delta(z_v, v) \leq (1 + \epsilon)\delta(u, \lambda)$.

This lemma implies that the output of Algorithm 1 achieves the $(1 + \epsilon)$-approximation to $\delta(u, \lambda)$, since

- if z_v is farther than z_u from r, then $h(z_v) + \delta(z_v, v) \geq h(z^+) + \delta(z^+, v^+))$, and hence

$$\delta(u, z_u) + \delta(z_u, z^+) + \delta(v^+, z^+)$$
$$\leq \delta(u, z_u) + \delta(z_u, z_v) + \delta(v, z_v)$$
$$\leq (1 + \epsilon)\delta(u, \lambda);$$

- if z_v is closer than z_u from r, then $-h(z_v) + \delta(z_v, v) \geq -h(z^-) + \delta(z^-, v^-))$, and hence

$$\delta(u, z_u) + \delta(z_u, z^-) + \delta(v^-, z^-)$$
$$\leq \delta(u, z_u) + \delta(z_u, z_v) + \delta(v, z_v)$$
$$\leq (1 + \epsilon)\delta(u, \lambda).$$

To show the query time $O(\frac{1}{\epsilon} \log n \log \Delta)$, we only need to show that v^+ (v^-) and z^+ (z^-) can be found in $O(\log \Delta)$ time. Actually, this can be done by identifying the range of C^+ (C^-) of the portals of λ on the specified path, using binary search within $O(\log \Delta)$-time and locating v^+ (v^-) by RMQ, using $O(1)$ time.

Theorem 1. *Given a planar graph $G = (V, E)$ and a label set L, for any $0 < \epsilon < 1$, there exists a $(1+\epsilon)$-VLDO with query time $O(\frac{1}{\epsilon} \log n \log \Delta)$, storage of size $O(\frac{1}{\epsilon} n \log n)$ and preprocessing time $O(\frac{1}{\epsilon} n \log^2 n)$.*

Remark 1. Note that Theorem 1 applies for cases when $0 < \epsilon < 1$. The constraint $\epsilon < 1$ is used when we bound the number of portals of a node in V according to some separator path. However, for $\epsilon = 2$, to achieve stretch $1 + \epsilon = 3$, it is enough for any node v to have only one portal on a specified path of some separator. In particular, given node $v \in V$ and a path P' of the separator stored in node $p \in$RGD, we choose the node closest to v on path P' as the portal of v according to P'. Let z_v denote such node. Then for any node z on path P', we have $\delta(z, z_v) \leq \delta(v, z_v) + \delta(v, z) \leq 2\delta(v, z_v)$, and hence $\delta(v, z) \leq \delta(v, z_v) + \delta(z_v, z) \leq 3\delta(v, z_v)$. In this case there are at most $n \log n$ portals in total.

Given an undirected planar graph $G = (V, E)$ and a label set L, each vertex $v \in V$ is attached with one label in L. Consider the distance oracle that supports the 3-stretch, $O(\log n \log \Delta)$-query time, using space $O(n \log n)$. The *space, query time product* (suggested by Sommer [11]) is $O(n \log^2 n \log \Delta)$, which is better than the lower bound of $\Omega(n^{\frac{3}{2}}) \times O(1)$ storage for general graphs.

3.2 $O(1)$ Time to Identify C^+ (C^-) When $\Delta = O(\log n)$

In the case that $\Delta = O(\log n)$, the time to identify C^+ (C^-) is $O(\log \log n)$ by Theorem 1. We show that this can be improved to $O(1)$.

First, note that when we store the portals for a label λ, it is possible that a node serves as portal for different nodes. It is obvious that we can only store the one with the minimum portal-to-node distance. Thus after fixing a label λ, on a path of a separator, each node serves as at most one portal of λ. Using a word of $\Delta = O(\log n)$ bits, denoted by ω, it can be identified whether a node on the path is a portal, i.e. the i-th bit is 1 *iff* the i-th node on the path is a portal for λ. If the portals on a path for λ are stored in the increasing order of their positions on the path, its index can be retrieved by counting how many 1's there are before the i-th position of ω. Since any operation on a single word is assumed to cost $O(1)$ time, we achieve the $O(1)$ time method to identify C^+, with

- $O(n \log n)$ space to record the position on the path forming separator, for each portal; and
- $O(n \log n)$ space to store ω's for all labels.

Theorem 2. *Given a planar graph $G = (V, E)$ with $\Delta = O(\log n)$ and a label set L, for any $0 < \epsilon < 1$, there exists a $(1 + \epsilon)$-VLDO with query time $O(\frac{1}{\epsilon} \log n)$, storage of size $O(\frac{1}{\epsilon} n \log n)$ and preprocessing time $O(\frac{1}{\epsilon} n \log^2 n)$.*

3.3 Label Changes

In this section, we consider the cost to update the distance oracle, if a node v changes its label from λ_1 to λ_2.

Since the selection of v's portals is not affected by the label changes, we only need to consider the update of table $\widehat{T}$, hash tables and data structure to support RMQ.

Update of $\widehat{T}$ and RMQ. Note that for λ_1, all entries $T_{\lambda_1,p}$ satisfying $v \in N(p)$ should be updated, by removing v's portals. The case for λ_2 is similar. Since there are at most $O(\log n)$ such p's, we know that there are $O(\log n)$ entries of $\widehat{T}$ that should be updated. Note that for each sequence of sorting portals in one entry of $\widehat{T}$, there is a data structure construct to support RMQ. It has to be updated *iff* the sequence changes. Hence there are at most $O(\log n)$ data structure for RMQ that should be updated.

Update of Hash Table. Note that there are two hash tables constructed to support $O(1)$-time query of $\widehat{T}_{\lambda_1}$ and $\widehat{T}_{\lambda_2}$. They need to be updated if there exist $p \in$RGD, such that $V_{\lambda_1} \cap N(p)$ becomes empty after the label change or $V_{\lambda_2} \cap N(p)$ becomes non-empty after the label change.

Running Time of Update. Even though there are $O(\log n)$ entries of $\widehat{T}$ and RMQ data structures that need to be updated, the running time is $O(\max\{|V_{\lambda_1}|, |V_{\lambda_2}|\} \log n)$, since the hash tables and RMQ data structure can be constructed in linear time. Therefore, the time to update the distance oracle can be $O(n \log n)$, which is asymptotically the same with preprocessing time. We left here as a open problem whether we can improve the update time to $o(n \log n)$.

References

1. Bartal, Y., Gottlieb, L.-A., Kopelowitz, T., Lewenstein, M., Roditty, L.: Fast, precise and dynamic distance queries. In: SODA, pp. 840–853 (2011)
2. Chechik, S.: Improved distance oracles for vertex-labeled graphs. CoRR, abs/1109.3114 (2011)
3. Cole, R., Gottlieb, L.-A.: Searching dynamic point sets in spaces with bounded doubling dimension. In: STOC, pp. 574–583 (2006)
4. Fischer, J., Heun, V.: A new succinct representation of rmq-information and improvements in the enhanced suffix array. In: Chen, B., Paterson, M., Zhang, G. (eds.) ESCAPE 2007. LNCS, vol. 4614, pp. 459–470. Springer, Heidelberg (2007)
5. Hermelin, D., Levy, A., Weimann, O., Yuster, R.: Distance oracles for vertex-labeled graphs. In: Aceto, L., Henzinger, M., Sgall, J. (eds.) ICALP 2011, Part II. LNCS, vol. 6756, pp. 490–501. Springer, Heidelberg (2011)
6. Kawarabayashi, K.i., Klein, P.N., Sommer, C.: Linear-space approximate distance oracles for planar, bounded-genus, and minor-free graphs. CoRR, abs/1104.5214 (2011)
7. Klein, P.: Preprocessing an undirected planar network to enable fast approximate distance queries. In: Proceedings of the Thirteenth Annual ACM-SIAM Symposium on Discrete Algorithms, SODA 2002, pp. 820–827. Society for Industrial and Applied Mathematics, Philadelphia (2002)
8. Lipton, R.J., Tarjan, R.E.: Applications of a planar separator theorem. In: Proceedings of the 18th Annual Symposium on Foundations of Computer Science, pp. 162–170. IEEE Computer Society, Washington, DC (1977)
9. Lipton, R.J., Tarjan, R.E.: A separator theorem for planar graphs. SIAM Journal on Applied Mathematics 36(2), 177–189 (1979)

10. Mozes, S., Sommer, C.: Exact distance oracles for planar graphs. In: Proceedings of the Twenty-Third Annual ACM-SIAM Symposium on Discrete Algorithms, SODA 2012, pp. 209–222. SIAM (2012)
11. Sommer, C.: More compact oracles for approximate distances in planar graphs. CoRR, abs/1109.2641 (2011)
12. Tao, Y., Papadopoulos, S., Sheng, C., Stefanidis, K.: Nearest keyword search in xml documents. In: Proceedings of the 2011 International Conference on Management of Data, SIGMOD 2011, pp. 589–600. ACM, New York (2011)
13. Thorup, M.: Undirected single source shortest paths in linear time. In: Proceedings of the 38th Annual Symposium on Foundations of Computer Science, p. 12. IEEE Computer Society, Washington, DC (1997)
14. Thorup, M.: Compact oracles for reachability and approximate distances in planar digraphs. J. ACM 51, 993–1024 (2004)
15. Thorup, M., Zwick, U.: Approximate distance oracles. In: Proceedings of the Thirty-Third Annual ACM Symposium on Theory of Computing, STOC 2001, pp. 183–192. ACM, New York (2001)
16. Wulff-Nilsen, C.: Approximate distance oracles with improved preprocessing time. In: SODA 2012 (2012)

Group Nearest Neighbor Queries in the L_1 Plane[*]

Hee-Kap Ahn[1], Sang Won Bae[2], and Wanbin Son[1]

[1] Department of Computer Science and Engineering, POSTECH, Pohang, Republic of Korea
{heekap,mnbiny}@postech.ac.kr,
[2] Department of Computer Science, Kyonggi University, Suwon, Republic of Korea
swbae@kgu.ac.kr

Abstract. Let P be a set of n points in the plane. The k-nearest neighbor (k-NN) query problem is to preprocess P into a data structure that quickly reports k closest points in P for a query point q. This paper addresses a generalization of the k-NN query problem to a query set Q of points, namely, the *group* nearest neighbor problem, in the L_1 plane. More precisely, a query is assigned with a set Q of at most m points and a positive integer k with $k \leq n$, and the distance between a point p and a query set Q is determined as the sum of L_1 distances from p to all $q \in Q$. The maximum number m of query points Q is assumed to be known in advance and to be at most n; that is, $m \leq n$. In this paper, we propose two methods, one based on the range tree and the other based on the segment dragging query, obtaining the following complexity bounds: (1) a group k-NN query can be handled in $O(m^2 \log n + k(\log \log n + \log m))$ time after preprocessing P in $O(m^2 n \log^2 n)$ time and space, or (2) a query can be handled in $O(m^2 \log n + (k + m) \log^2 n)$ time after preprocessing P in $O(m^2 n \log n)$ time using $O(m^2 n)$ space. We also show that our approach can be applied to the group k-farthest neighbor query problem.

1 Introduction

The *nearest neighbor query* problem, also known as the proximity query or closest point query problem, is one of the fundamental problems in computer science. The problem is, for a set P of points in a metric space M, to preprocess P such that given a query point $q \in M$, one can find the closest point of q in P quickly. Many areas in computer science including computational geometry, databases, machine learning, and computer vision use the nearest neighbor query as one of the most primitive operations. The k-nearest neighbor query (shortly, k-NN) problem is a generalization of the nearest neighbor query problem the goal of which is to find the k closest points of the query in the data set P.

Various solutions to the nearest neighbor query problem have been proposed. A straightforward algorithm for this problem is the sequential search. Several tree-based data structures [5,12,21] have been proposed to increase efficiency. Approximate nearest neighbor algorithms have also been studied [2,11].

In this paper, we focus on a generalized version of the k-NN query problem, namely, the *group k-nearest neighbor query* problem, where a query is associated with a set Q

[*] This research was supported by NRF grant 2011-0030044 (SRC-GAIA) funded by the government of Korea.

T-H.H. Chan, L.C. Lau, and L. Trevisan (Eds.): TAMC 2013, LNCS 7876, pp. 52–61, 2013.
© Springer-Verlag Berlin Heidelberg 2013

of more than one query points and a positive integer k, and is to find k closest data points in P with respect to Q. The distance (or, the closeness) of a data point $p \in P$ with respect to the query set Q is determined to be the sum of distances between p and each $q \in Q$, that is, $\sum_{q \in Q} dist(p, q)$. We shall call this quantity the *sum-of-distance of p with respect to Q*. The goal is, for a given set P of n data points in the plane, and to preprocess P into a data structure that efficiently handles group k-NN queries.

To our best knowledge, the group nearest neighbor query problem has been first studied and coined by Papadias et al. [17]. They proposed a heuristic method in the Euclidean plane and showed several applications of the group k-NN query and the sum-of-distance function in GIS (Geographic information system), clustering and outlier detection [17]. Later, Papadias et al. [18] and Li et al. [14] considered a generalization, and Yiu et al. [24] studied the group nearest neighbor problem in road networks. Recently, Agarwal et al. [1] studied the expected NN-queries, where the location of each input point and/or query point is specified as a probability density function. Wang and Zhang [22] improved the result of Agarwal et al. on the expected NN-queries for the case of an uncertain query point.

A brute-force way handles a group k-NN query in $O(nm)$ time by computing the sum-of-distances for all data points in $O(nm)$ time and then using a selection algorithm to find the k closest points in P in $O(n)$ time. It is unlikely that we achieve a $o(n)$ time algorithm without any preprocessing because of the lower bound for the selection problem [4]. To avoid $\Omega(n)$ query time, especially when k is much smaller than n, we may consider the following approach: compute the region that has the minimum sum-of-distance with respect to Q, and then expand this region by increasing the sum-of-distance value. During the expansion, we report the data points hit by the expanding region until we have k points reported. In this case, we may not need to consider all the points in P.

On the other hand, in the Euclidean space, this approach seems not easy to use because of the following reasons.

1. Computing a point that has the minimum sum-of-distance (known as the Fermat-Weber point [9]) is known to be hard because the equations to compute the point cannot be solved into closed form for $m > 2$ [18].

2. In the plane, the region that has the same sum-of-distance value from m points in the Euclidean space forms an m-ellipse, which is an algebraic curve with very high degree $O(2^m)$ [16].

Our Results. We study the group k-NN query problem in the L_1 plane, that is, the distance of a data point is determined to be the sum of the L_1 distances, and present two efficient algorithms that solve the problem: RNGALGO and SGMTALGO. We assume that the maximum number m of query points is known in advance and $m \leq n$. In many applications, this is a reasonable restriction.

– RNGALGO answers a group k-NN query in $O(m^2 \log n + k(\log \log n + \log m))$ time after preprocessing P into a data structure in $O(m^2 n \log^2 n)$ time and space.

– SGMTALGO answers a group k-NN query in $O(m^2 \log n + (k+m) \log^2 n)$ time after preprocessing P into a data structure in $O(m^2 n \log n)$ time and $O(m^2 n)$ space.

Note that both of our query algorithms spend $o(n)$ time to answer a query when k and m are reasonably smaller than n. Moreover, our approach can easily be used for the problem in the L_∞ metric, and farthest neighbor queries in the same setting.

2   Observations on the Sum-of-Distance Function

In this section we investigate properties of the sum-of-distance function sumdist_Q: $\mathbb{R}^2 \to \mathbb{R}$ with respect to a query set Q, defined to be $\text{sumdist}_Q(p) = \sum_{q \in Q} \text{dist}(p, q)$, where $\text{dist}(p, q)$ is the L_1 distance between two points p and q.

We first observe that sumdist_Q is a convex function. Note that the L_1 distance function dist is convex and piecewise linear. Since sumdist_Q is the sum of those functions, it is also a convex function; the sum of convex functions is also convex [20]. For any real number $c \in \mathbb{R}$, let $A_Q(c) := \{x \in \mathbb{R}^2 \mid \text{sumdist}_Q(p) \leq c\}$ be the sublevel set of function sumdist_Q. The sublevel set of a convex function is convex by definition [20]. We thus get the following lemma.

Lemma 1. *The sum-of-distance function* sumdist_Q *is convex, and therefore its sublevel set* $A_Q(c)$ *is convex for any* $c \in \mathbb{R}$.

2.1   The Set of Points Minimizing sumdist_Q

Consider the problem of identifying the region of points that minimize the function sumdist_Q. This problem is well known as the Fermat-Weber problem [9], and a result in the L_p metric was given in 1964 [23]. We present a simple proof for the problem in the L_1 plane in the following.

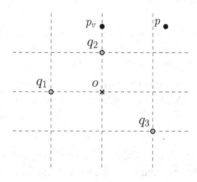

Fig. 1. The grid constructed from $Q = \{q_1, q_2, q_3\}$ and two data points p_v and p. The point o minimizes sumdist_Q.

For each query point, draw a horizontal line and a vertical line passing through the point. See Fig. 1. We denote by $G(Q)$ the grid of Q constructed in this way. Let x_v be the median among all the x-coordinates of the query points when $|Q|$ is odd.

Lemma 2. *For any data points $p = (x, y)$ and $p_v = (x_v, y)$, we have* $\text{sumdist}_Q(p) > \text{sumdist}_Q(p_v)$, *where $x \neq x_v$ when $|Q|$ is odd.*

Proof. See Fig. 1 for an illustration to the proof. Without loss of generality, we assume that $x > x_v$. Let Q_1 denote the set of query points whose x-coordinates are at most x_v, and $Q_2 = Q \setminus Q_1$. So we have $|Q_1| > |Q_2|$. For any $q \in Q_1$ and $q' \in Q_2$, we have the followings.

$$\text{dist}(p, q) = \text{dist}(p_v, q) + |x - x_v|, \quad \text{dist}(p, q') \geq \text{dist}(p_v, q') - |x - x_v|.$$

Because $|Q_1| > |Q_2|$, we have

$$\begin{aligned}
\text{sumdist}_Q(p) &= \sum_{q \in Q_1} \text{dist}(p, q) + \sum_{q' \in Q_2} \text{dist}(p, q') \\
&\geq \sum_{q \in Q_1} (\text{dist}(p_v, q) + |x - x_v|) + \sum_{q' \in Q_2} (\text{dist}(p_v, q') - |x - x_v|) \\
&> \text{sumdist}_Q(p_v).
\end{aligned}$$

$\square$

Let y_v be the median among all the y-coordinates of the query points when $|Q|$ is odd. Then, the point (x_v, y_v) minimizes sumdist_Q over all points in $\mathbb{R}^2$ when $|Q|$ is odd. When $|Q|$ is even, the *median cell* of $G(Q)$ is the cell bounded by the two consecutive

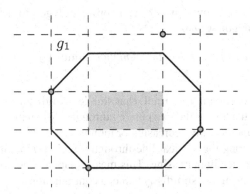

Fig. 2. The grid $G(Q)$ (dashed lines), the median cell (gray region) that minimizes function sumdist_Q and the boundary of the sublevel set $A(c)$ that has the same sumdist_Q value (solid line segments)

vertical lines through the query points whose x-coordinates are the medians among all the x-coordinates of Q, and by the two consecutive horizontal lines through the query points whose y-coordinates are the medians among all the y-coordinates of Q. See Fig. 2. The following lemma can be proved in a similar way to Lemma 2.

Lemma 3. *When $|Q|$ is odd, only the point (x_v, y_v) minimizes the function sumdist_Q, and when $|Q|$ is even, the median cell of $G(Q)$ does.*

2.2 Properties of Cells of $G(Q)$

Let g be any cell of $G(Q)$. Because of the way in which $G(Q)$ is constructed, every interior point of g has the same number of query points lying to the left of it; the same claim holds on the query points lying to the right, above, and below of the point. Therefore we denote by $m_l(g), m_r(g), m_t(g)$ and $m_b(g)$ the numbers of query points that are to the left, right, above, and below of any point in g, respectively.

Lemma 4. *For an interior point r of any cell g of $G(Q)$, there is a line ℓ passing through r such that for every point r' of $g \cap \ell$, sumdist$_Q(r') = $ sumdist$_Q(r)$. Moreover, this property holds for any line parallel to ℓ.*

Proof. Let $r = (x, y)$ be a interior point of a cell g. For a point $r' = (x + \delta, y)$ in g, sumdist$_Q(r') = $ sumdist$_Q(r) + \delta(m_l(g) - m_r(g))$. For a point $r'' = (x, y + \delta)$ in g sumdist$_Q(r'') = $ sumdist$_Q(r) + \delta(m_b(g) - m_t(g))$.

Let $\text{slp}(g) = \frac{m_r(g) - m_l(g)}{m_b(g) - m_t(g)}$. For any line ℓ with slope $\text{slp}(g)$ passing through r, let r' be a point in $g \cap \ell$. Then $r' = (x + \delta_x, y + \delta_y)$ with $\delta_y / \delta_x = \text{slp}(g)$.

$$\text{sumdist}_Q(r') = \text{sumdist}_Q(r) + \delta_x(m_l(g) - m_r(g)) + \delta_y(m_b(g) - m_t(g))$$
$$= \text{sumdist}_Q(r) + \delta_x(m_l(g) - m_r(g)) + \delta_x \cdot \text{slp}(g)(m_b(g) - m_t(g))$$
$$= \text{sumdist}_Q(r).$$

$\square$

By Lemmas 1 and 4, we know that $A(c)$ is a convex polygon for any fixed c, and the complexity of the polygon is $O(m)$.

Lemma 5. *One corner of a cell has the minimum* sumdist$_Q$ *value over all the points in the cell.*

Proof. Assume that some point r of a cell g has smaller sumdist$_Q$ value than that of any corner of g. Let h be the horizontal line passing through r. Then the function sumdist$_Q$ on $h \cap g$ is either monotonically decreasing or monotonically increasing. Similarly, the function sumdist$_Q$ along the vertical line through r within g is either monotonically decreasing or monotonically increasing. This means that we can move r to one of the corners without increasing the sumdist$_Q$ value, a contradiction. $\square$

Let ℓ^+ and ℓ^- be the right and the left sides of ℓ, respectively. Then, one of $g \cap \ell^+$ and $g \cap \ell^-$ contains a corner s that minimizes sumdist$_Q$ over g. By Lemma 1, every point of the side containing s has sumdist$_Q$ value at most sumdist$_Q(r)$, and every point in the side not containing s has sumdist$_Q$ value at least sumdist$_Q(r)$.

3 Algorithm

In this section, we present a query algorithm to compute the k nearest neighbors from P for a given query Q and k. The basic idea of our algorithm is as follows. Given a query with Q and k, we first compute the set $A(\min_{x \in \mathbb{R}^2} \text{sumdist}_Q(x))$, and then expand $A(c)$ by increasing c. During the expansion, we report each data point in P hit by $A(c)$ until we report k data points. To do this efficiently, we preprocess P such that we consider $A(c)$ for only $O(k + m^2)$ distinct values of c in the increasing order of sumdist$_Q$ value and report the k nearest neighbors.

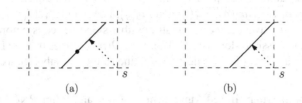

(a)   (b)

Fig. 3. Two types of events : (a) point event (b) corner event

We construct $G(Q)$ and then compute $A(c)$ setting c to be $\min_{x \in \mathbb{R}^2}$ sumdist$_Q(x)$ using Lemma 3. If there are more than k data points in $A(c)$, then we report k points among them. In the general case, we expand $A(c)$ by increasing c. During the expansion, we encounter the following events.

- *Point event*: $A(c)$ hits a data point (Fig. 3(a)).
- *Corner event*: $A(c)$ hits a corner of a cell (Fig. 3(b)).

We keep these events in an *event queue* Q, which is a priority queue indexed by the sumdist$_Q$ value of its associated point. After initializing the event queue, we insert the events of the cells adjacent to the median cell. We then process events one by one in the order from Q as follows. For a point event e, we report the data point p associated with the event. If this point is the k-th point to be reported, we are done. In the other case, we find the next event in the cell containing p, and then insert it to Q. For a corner event e, we consider each cell having e as a corner and disjoint in its interior from $A(\text{sumdist}_Q(e))$, and find the next event e' in each such cell. We also find the next event in the cell that reports e. We then insert them to Q.

The correctness of our algorithm immediately follows from the convexity of $A(c)$ and the fact that the events are processed in the increasing order of sumdist$_Q$ value. That is, when we process a point event e, the data point of e has the minimum sumdist$_Q$ value among all unreported data points, so our algorithm correctly reports the k nearest neighbors in P with respect to Q.

Let us now analyze the complexity of our algorithm. We preprocess P to construct the orthogonal range counting query structure, which takes $O(n \log n)$ time [8]. We then construct the grid $G(Q)$ in $O(m^2)$ time. In the worst case, we need to test all the cells for emptiness, which takes $O(m^2 \log n)$ time. Since $A(c)$ is convex, the number of cells intersected by the boundary of $A(c)$ is $O(m)$ for any fixed c. There can be $O(m^2)$ corner events and k point events until we report k nearest neighbors, so it takes $O((m^2 + k) \log m)$ time to insert and delete them from the event queue. The only remaining part of the algorithm is to process the data points contained in a cell to support the operation of finding next point events in the cell efficiently. We will describe two methods for this and analyze their time complexities.

3.1 Detecting Events in a Cell

We propose two methods to find events in a cell in the increasing order of their sumdist$_Q$ values. There are four corners in a cell, so finding them is easy. To find events in a cell g, we sweep g by a line with slope slp(g) from the corner with the minimum sumdist$_Q$ value over all points in the cell g. As aforementioned, the slope slp(g) is determined by $m_l(g)$, $m_r(g)$, $m_t(g)$ and $m_b(g)$. For example, slp$(g_1) = 1$ in Fig. 2. Since m is given in advance, we know all possible slopes of cells before Q is given. Let a set of all the possible slopes be S. There are $O(m^2)$ slopes in S by Lemma 4. We preprocess P for all slopes in S as described in the following subsections.

Orthogonal Range Query Based Algorithm. The first algorithm, RNGALGO, is based on an orthogonal range query structure, namely, the range tree [8]. Rahul et al. [19] introduced a data structure based on the orthogonal range query structure such that the first k points from n weighted points in $\mathbb{R}^d$ can be reported in sorted order in $O(\log^{d-1} n + k \log \log n)$ query time using the data structure. This structure can be constructed in $O(n \log^d n)$ time and space in the preprocssing phase.

Consider a slope $s \in S$. For each data point p, we set the weight of p to the y-intercept value of the line of slope s passing through p. Then we construct the data structure of Rahul et al. for the weighted data points. We do this for every slope of S and maintain one data structure of Rahul et al. for each slope. Let $\mathcal{R}$ denote the set of these data structures.

By using $\mathcal{R}$, we can get the first point event from each cell in $O(\log n)$ time, and then we spend $O(\log \log n)$ time for finding the next point event in each cell. Therefore, it takes $O(m^2 \log n + k \log \log n)$ time to find k nearest neighbors in $O(m^2 n \log^2 n)$ preprocessing time and space.

Theorem 1. *The algorithm* RNGALGO *reports the k nearest neighbors in* $O(m^2 \log n + k(\log \log n + \log m))$ *time after preprocessing P into a data structure in* $O(m^2 n \log^2 n)$ *time and space.*

Segment Dragging Query Based Algorithm. The second algorithm, SGMTALGO, uses the segment dragging query [3,6,15] for preprocessing P with respect to slp(g). The segment dragging query, informally speaking, is to determine the next point hit by the given query segment of orientation θ when it is dragged along two tracks. There are three types of segment dragging queries:(a) dragging by parallel tracks, (b) dragging out of a corner, and (c) dragging into a corner (See Fig. 4).

Chazelle [6] and Mitchell [15] showed that one can preprocess a set P of n points into a data structure of size $O(n)$ in $O(n \log n)$ time that answers the segment dragging queries of type (a) and (b) in $O(\log n)$ time. Bae et al. [3] proposed a data structure that answers segment dragging queries of type (c) in $O(\log^2 n)$ time. Those data structures require $O(n)$ space and $O(n \log n)$ preprocessing time.

We sweep a cell using a segment in three different types (See Fig. 5). From the corner s with the minimum sumdist$_Q$ value, we sweep the cell using a segment dragging of type (b) 'dragging out of a corner' until the segment hits another corner. We then apply a segment dragging of type (a) 'dragging by parallel tracks' from the corner until it hits

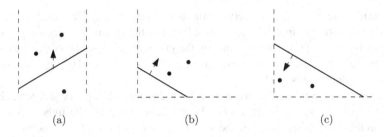

Fig. 4. Three types of segment dragging : (a) dragging by parallel tracks (b) dragging out of a corner (c) dragging into a corner

the third corner. Afterwards, we apply a segment dragging of type (c) 'dragging into a corner' until it hits the last corner.

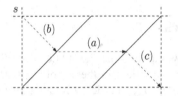

Fig. 5. We sweep a cell using a segment in three different types of dragging

If we adopt the segment dragging query structure, we need $O(\log^2 n)$ time for each point event. Therefore, it takes $O((k+m)\log^2 n)$ time in total to report $k+O(m)$ point events with $O(m^2 n \log n)$ preprocessing time and $O(m^2 n)$ space if we only consider the cells containing a data point.

Theorem 2. *The algorithm* SGMTALGO *computes the* k *nearest neighbors in* $O(m^2 \log n + (k + m)\log^2 n)$ *time after preprocessing* P *into a data structure in* $O(m^2 n \log n)$ *time using* $O(m^2 n)$ *space.*

4   Group Farthest Neighbor Queries

The farthest neighbor query problem is also one of the fundamental problems in computer science. The group farthest neighbor (GFN) query problem is a generalization of the farthest neighbor query problem where more than one query point are given at the same time, and the distance of a data point is measured with respect to the query points. The group k-farthest neighbor (k-FN) query problem is an analogue to the group k-NN query problem.

There have been a few previous works on the k-FN query problem. Cheong et al. [7] studies the farthest neighbor query problem for a set of points on a convex polytope. Katoh and Iwano [13] proposed an algorithm for finding k farthest pairs. Gao et al. [10] presented heuristic algorithms to solve the problem in the Euclidean space.

We consider the group k-FN query problem in the L_1 plane. The basic idea is the same as done for the group k-NN query problem. For the group k-NN query problem, we expand $A(c)$ until it hits k data points. In the group k-FN query problem, we sweep the plane in the opposite direction; we shrink $A(c)$ from $c = \infty$ until the region contains only the $n - k$ data points in P.

We preprocess P in the same way as done for the group k-NN query problem. Given a query with Q and k, we construct $G(Q)$ and then sweep all the unbounded cells of $G(Q)$ in the descending order of the sumdist$_Q$ value. To handle the point and the corner events, we construct a max-heap structure instead. We can handle each event in the same way as done for the group k-NN query problem in the previous sections. We repeat this process until we find the k farthest neighbor points in P. Since the boundary of $A(\infty)$ intersects only with the unbounded cells of $G(Q)$, and we handle the events in order, the algorithm reports the k farthest neighbors correctly. The running time and space complexity of this algorithm are the same as those of the algorithms for the group k-NN query problem.

5   Concluding Remarks

In this paper, we propose two algorithms, RNGALGO and SGMTALGO, to solve the group nearest neighbor query problem in the L_1 plane. Our approach can be easily extended for the problem in the L_∞ metric by rotating all the data points and query points by $\pi/4$. We also show that the group farthest neighbor query problem in the L_1 metric can be solved by the similar approach.

As aforementioned, we can solve this problem in $O(nm)$ time in a straightforward way. Without any preprocessing we cannot avoid $\Omega(n)$ time because of the lower bound for the selection problem [4]. For RNGALGO, if $m = o(\sqrt{\frac{n}{\log n}})$ and $k = o(\frac{n}{\log \log n + \log m})$, then the query time of RNGALGO is $o(n)$. For SGMTALGO, if $m = o(\sqrt{\frac{n}{\log n}})$ and $k = o(\frac{n}{\log^2 n})$, then the query time of SGMTALGO is $o(n)$.

The algorithm SGMTALGO requires smaller time and space for preprocessing than the other algorithm RNGALGO, and for the query time RNGALGO outperforms SGMTALGO.

References

1. Agarwal, P.K., Efrat, A., Sankararaman, S., Zhang, W.: Nearest-neighbor searching under uncertainty. In: Proceedings of the 31st Symposium on Principles of Database Systems, PODS 2012, pp. 225–236. ACM, New York (2012)
2. Arya, S., Mount, D.M., Netanyahu, N.S., Silverman, R., Wu, A.Y.: An optimal algorithm for approximate nearest neighbor searching fixed dimensions. J. ACM 45(6), 891–923 (1998)
3. Bae, S.W., Korman, M., Tokuyama, T.: All farthest neighbors in the presence of highways and obstacles. In: Das, S., Uehara, R. (eds.) WALCOM 2009. LNCS, vol. 5431, pp. 71–82. Springer, Heidelberg (2009)
4. Bent, S.W., John, J.W.: Finding the median requires 2n comparisons. In: Proceedings of the Seventeenth Annual ACM Symposium on Theory of Computing, STOC 1985, pp. 213–216. ACM, New York (1985)

5. Beygelzimer, A., Kakade, S., Langford, J.: Cover trees for nearest neighbor. In: Proceedings of the 23rd International Conference on Machine Learning, ICML 2006, pp. 97–104. ACM, New York (2006)

6. Chazelle, B.: An algorithm for segment-dragging and its implementation. Algorithmica 3(1), 205–221 (1988)

7. Cheong, O., Shin, C.-S., Vigneron, A.: Computing farthest neighbors on a convex polytope. Theoretical Computer Science 296(1), 47–58 (2003)

8. de Berg, M., Cheong, O., van Kreveld, M., Overmars, M.: Computational Geometry: Algorithms and Applications, 3rd edn. Springer (2008)

9. Durier, R., Michelot, C.: Geometrical properties of the fermat-weber problem. European Journal of Operational Research 20(3), 332–343 (1985)

10. Gao, Y., Shou, L., Chen, K., Chen, G.: Aggregate farthest-neighbor queries over spatial data. In: Yu, J.X., Kim, M.H., Unland, R. (eds.) DASFAA 2011, Part II. LNCS, vol. 6588, pp. 149–163. Springer, Heidelberg (2011)

11. Indyk, P., Motwani, R.: Approximate nearest neighbors: towards removing the curse of dimensionality. In: Proceedings of the Thirteeth Annual ACM Symposium on Theory of Computing, STOC 1998, pp. 604–613. ACM, New York (1998)

12. Jagadish, H.V., Ooi, B.C., Tan, K.-L., Yu, C., Zhang, R.: idistance: An adaptive b+-tree based indexing method for nearest neighbor search. ACM Trans. Database Syst. 30(2), 364–397 (2005)

13. Katoh, N., Iwano, K.: Finding k farthest pairs and k closest/farthest bichromatic pairs for points in the plane. International Journal of Computational Geometry and Applications 05(01n02), 37–51 (1995)

14. Li, Y., Li, F., Yi, K., Yao, B., Wang, M.: Flexible aggregate similarity search. In: Proceedings of the 2011 ACM SIGMOD International Conference on Management of Data, SIGMOD 2011, pp. 1009–1020. ACM, New York (2011)

15. Mitchell, J.: L_1 shortest paths among polygonal obstacles in the plane. Algorithmica 8(1), 55–88 (1992)

16. Nie, J., Parrilo, P., Sturmfels, B.: Semidefinite representation of the k-ellipse. In: Algorithms in Algebraic Geometry. The IMA Volumes in Mathematics and its Applications, vol. 146, pp. 117–132. Springer, New York (2008)

17. Papadias, D., Shen, Q., Tao, Y., Mouratidis, K.: Group nearest neighbor queries. In: Proceedings of the 20th International Conference on Data Engineering, March-April 2, pp. 301–312 (2004)

18. Papadias, D., Tao, Y., Mouratidis, K., Hui, C.K.: Aggregate nearest neighbor queries in spatial databases. ACM Trans. Database Syst. 30(2), 529–576 (2005)

19. Rahul, S., Gupta, P., Janardan, R., Rajan, K.S.: Efficient top-k queries for orthogonal ranges. In: Katoh, N., Kumar, A. (eds.) WALCOM 2011. LNCS, vol. 6552, pp. 110–121. Springer, Heidelberg (2011)

20. Rockafellar, R.T.: Convex Analysis. Princeton University Press (1996)

21. Sproull, R.: Refinements to nearest-neighbor searching in k-dimensional trees. Algorithmica 6, 579–589 (1991)

22. Wang, H., Zhang, W.: The L1 Nearest Neighbor Searching with Uncertain Queries. ArXiv e-prints (November 2012)

23. Witzgall, C.: Optimal location of a central facility: mathematical models and concepts. National Bureau of Standards (1964)

24. Yiu, M., Mamoulis, N., Papadias, D.: Aggregate nearest neighbor queries in road networks. IEEE Transactions on Knowledge and Data Engineering 17(6), 820–833 (2005)

Modelling the Power Supply Network – Hardness and Approximation

Alexandru Popa

Department of Communications & Networking, Aalto University School
of Electrical Engineering, Aalto, Finland
alexandru.popa@aalto.fi

Abstract. In this paper we study a problem named *graph partitioning with supply and demand (GPSD)*, motivated by applications in energy transmission. The input consists of an undirected graph G with the nodes partitioned into two sets: suppliers and consumers. Each supply node has associated a capacity and each consumer node has associated a demand. The goal is to find a subgraph of G and to partition it into trees, such that in each tree: (i) there is precisely one supplier and (ii) the total demand of the consumers is less than or equal to the capacity of the supplier. Moreover, we want to maximize the demand of all the consumers in such a partition.

We also study a variation of the GPSD, termed *energy delivery (ED)*. In this paper we show the following results:

1. A $2k$-approximation algorithm for the GPSD problem, where k is the number of suppliers. This is the first approximation algorithm proposed for the general case.
2. A 2-approximation for the GPSD in the case of two suppliers implies a polynomial time algorithm for the famous *minimum sum 2-disjoint paths problem*, which is not known if it is in **P** or **NP**-hard.
3. The ED problem in the case of two or more suppliers is hard to approximate within any factor, assuming $\mathbf{P} \neq \mathbf{NP}$.

1 Introduction

Motivation. In the recent years, the competition among the electricity providers has increased rapidly and the companies struggle to improve the market value of the services provided. Thus, a lot of effort is put into the optimization of the energy distribution (e.g. lowering the cost of maintenance, construction, delivery). Since there are a lot of factors which have to be taken into consideration, such as consumer loads and characteristics of the geographical area, the problem of finding the optimal distribution system is very complex. Many approaches to model the power supply network have been presented in the literature using algorithmic and mathematical techniques (e.g. [1,10,3,12,4,11]).

The distribution network is an important part of the total electrical supply system since reports show that 80% of customer service interruptions are due to problems in the distribution network [8]. To improve the service, in many

T-H.H. Chan, L.C. Lau, and L. Trevisan (Eds.): TAMC 2013, LNCS 7876, pp. 62–71, 2013.
© Springer-Verlag Berlin Heidelberg 2013

countries automation is applied to the distribution networks [8,9]. Due to the complexity of the problem and the large number of variables involved, restricted models, which address only some of the challenges of the system, are studied. In this paper we consider the following two problems, the first one proposed by Ito et al. [6] and the second one newly introduced in this paper, which are an attempt to model the power distribution system.

In the first problem, termed *graph partitioning with supply and demand (GPSD)* the network is represented as an undirected graph where some of the nodes are suppliers and the others are consumers. Each supplier and consumer has associated a capacity and, respectively, a demand. The goal is to provide as much electricity as possible. In graph theoretic terms, this task is equivalent to finding a subnetwork (as not all the customers may be satisfied) and to partition it into trees such that:

1. Each tree has precisely one supply vertex (the others are consumers).
2. The total demand of all the consumers in a tree does not exceed the capacity of the supplier.

The goal is to provide as much electricity as possible and, thus, the total demand of the consumers in this subgraph has to be maximized.

The second problem, named *energy delivery (ED)*, is a variation of the GPSD problem. The input is the same as in the GPSD problem, but the difference is that a supplier s can power a consumer c even if not all the consumers on the path from s to c are powered by s (still, those consumers have to be powered by some other supplier). This is a natural extension of the GPSD problem and models more precisely the real-life problems (the network defined by a supplier and the consumers powered by it may not necessarily be a subtree of the general network). Also, notice that we require that there exists a path from s to c where all the nodes are powered (without this restriction the problem is simply the multiple knapsack problem as any supplier can power any consumer).

Related Results. Ito et al. consider first the GPSD problem in the case when the distribution network is a tree [6]. In the same paper, they show that the decision problem of GPSD (i.e., if there exists a partition such that each consumer is satisfied) is solvable in polynomial time. They also prove a pseudo-polynomial time algorithm and a fully polynomial time approximation scheme (FPTAS) for GPSD on trees. A pseudo-polynomial time algorithm for series-parallel graphs and partial k-trees (i.e., graphs with bounded treewidth) is given in [7]. Series parallel graphs are reconsidered in [5] where a PTAS in the case of a network with exactly one supplier is presented. The GPSD problem is known to be **APX**-hard [5] and, thus, it is unlikely that a PTAS for this problem exists.

Our Results. Despite the large amount of attention that GPSD received, approximation algorithms are known only for restricted classes of graphs (trees, series-parallel graphs). In this paper we present the first approximation algorithm for GPSD for arbitrary graphs. The approximation factor achieved is $2k$, where k is the number of supply vertices in the graph.

Then, we show that a $(2 - \epsilon)$-approximation algorithm for the GPSD problem in the case of two suppliers, implies a polynomial time algorithm for the *minimum sum 2-disjoint paths problem*, whose complexity is still unknown.

The third result concerns the newly defined ED problem. We show that ED is hard to approximate within any factor, unless $\mathbf{P} = \mathbf{NP}$. This is surprising since the ED problem seems easier than the GPSD problem.

The rest of the paper is organized as follows. In Section 2 we formally introduce the three problems and give preliminary definitions. Then, in Section 3 we present the approximation algorithm and the hardness result for GPSD and in Section 4 the hardness result for the ED problem. Section 5 is reserved for conclusions and open problems.

2   Preliminaries

In this section we introduce notation and preliminary definitions. First we give the formal definitions of the two problems studied.

Problem 1 (Graph Partitioning with Supply and Demand). The input consists of an undirected graph $G = (V, E)$, where $V = S \cup D$, $S \cap D = \emptyset$ (i.e., the vertices of V are partitioned into two sets S and D) and two functions $c : S \to \mathbb{R}_+$ and $d : D \to \mathbb{R}_+$. The vertices in S are named *supply* vertices (or suppliers) and the ones in D are termed *demand* vertices (or consumers).

The goal of the problem is to find a subgraph of G and partition it into trees $T_1 = (V_1, E_1), T_2 = (V_2, E_2), \ldots, T_k = (V_k, E_k)$, where $k = |S|$ such that:

1. Each tree T_i contains exactly one supply vertex s_i and $\sum_{v \in V_i \setminus \{s_i\}} d(v) \leq c(s_i)$.
2.
$$\sum_{i=1}^{k} \sum_{v \in V_i \setminus \{s_i\}} d(v)$$

is maximized.

Problem 2 (Energy Delivery). The input of the ED problem is identical to the input of the GPSD problem. The goal is to find a subset of consumers and to partition it into k sets $S_1, S_2, \ldots, S_k$ (one for each supplier) such that:

1. For each supplier i and each consumer c in S_i, there is a path from i to c such that each vertex on this path is either powered by i or is powered by another supplier.
2.
$$\sum_{v \in S_i} d(v) \leq c(i), \forall \text{ supplier } i$$

3.
$$\sum_{i=1}^{k} \sum_{v \in S_i} d(v)$$

is maximized.

We now present the *minimum sum 2-disjoint paths* problem which we use to prove the hardness result for the GPSD.

Definition 1 (Minimum sum 2-disjoint paths problem). *The input consists of a graph G and two pairs of vertices (s_1, t_1) and (s_2, t_2). Find two disjoint paths from s_1 to t_1 and s_2 to t_2 whose total length is minimized.*

The complexity of the *minimum sum 2-disjoint paths* problem is open, although a similar version in which $s_1 = s_2$ and $t_1 = t_2$ is in **P**. We use the *subset sum* problem, defined next, to prove the hardness of the ED problem.

Definition 2 (Subset sum). *The input consists of a set $S = \{s_1, \ldots, s_n\}$ of integers and another integer k. Is there a subset $S' \subset S$ whose elements sum precisely to k?*

The following theorem and its proof can be found in [2].

Theorem 1. *The subset sum problem is* **NP***-complete.*

3   Graph Partitioning with Supply and Demand

In this section we present a $2k$-approximation algorithm for the GPSD problem. First, we show a 2-approximation algorithm for the graphs which contain exactly one supply vertex. Then, we apply this algorithm to general graphs to obtain the $2k$-approximation.

3.1   A 2-Approximation Algorithm for Graphs with One Supply Vertex

There are several ideas which one can try in order to obtain an approximation algorithm. We first present some of them and show why they do not lead to a good solution. Finally, we present the correct algorithm.

Largest Demand First. The first idea is to select nodes greedily as follows: at each step select the vertex with the largest demand which is adjacent to the vertices selected so far or to the supply vertex. A similar algorithm leads to a 2-approximation for the *knapsack problem* and one might think that it gives a good approximation in this case also. The problem is that we might have different branches in the graph (unlike the knapsack where all the elements are available from the beginning) and the algorithm might go on the wrong branch. The following counterexample presents this situation.

Counterexample 1. *Let $G = (V, E)$ with $V = \{1, 2, 3, 4\}$. There are 3 edges in G: $(1, 2), (1, 3), (3, 4)$. The supply vertex is 1 and has capacity n. The demands are: $d(2) = 2$, $d(3) = 1$, $d(4) = n - 1$. The optimal solution has value n (selecting nodes 3 and 4). The algorithm presented above selects first vertex 2 (as it has higher demand) and then vertex 3, thus satisfying a demand of 3.*

Smallest Demand First. To fix the previous algorithm one may try to select first the nodes with the smallest demand. However, a similar counterexample can be constructed.

Counterexample 2. *Let* $G = (V, E)$ *with* $V = \{1, 2, 3, 4\}$. *There are 3 edges in* G: $(1, 2), (1, 3), (3, 4)$. *The supply vertex is 1 and has capacity* n. *The demands are:* $d(2) = 1, d(3) = 2, d(4) = n - 2$. *The optimal solution has value* n *(selecting nodes 3 and 4). The algorithm presented above selects first vertex 2 (as it has smaller demand) and then vertex 3, thus satisfying a demand of 3.*

Using Paths Instead of Single Vertices. We observe that the algorithms which select only one node at one step do not work as they can be tricked by an adversary which points them on the wrong direction. Therefore, the next idea is to select greedily an entire path, rather than a single node. The procedure is presented in Algorithm 1. In the following, we define the *length* of the path in the graph as the sum of the demands of the consumers on the path.

Input: An instance of the GPSD problem with one supply vertex v.

1. Let $SP(i)$ be the shortest path from v to a demand vertex i and let $|SP(i)|$ be the sum of the demands of its vertices.
2. $Cost \leftarrow 0$, $Sol \leftarrow \emptyset$.
3. While there exists a node i with $|SP(i)| > 0$ and $Cost + |SP(i)| \leq c(v)$ do:
   (a) Select the node i with the maximum value of $|SP(i)|$ such that $Cost + |SP(i)| \leq c(v)$.
   (b) $Sol \leftarrow Sol \cup SP(i)$; $Cost \leftarrow Cost + |SP(i)|$
   (c) Remove from G the nodes in $SP(i)$.
   (d) Recompute $SP(i)$ for each node in the new graph.

Output: Sol.

Algorithm 1. Algorithm using shortest paths (unbounded approximation ratio)

Algorithm 1 fails to give a good approximation if the optimal solution consists of a tree with multiple branches (as the algorithm uses a path and then removes it from the graph). The following counterexample presents this situation.

Counterexample 3. *Consider a star with* n *branches (i.e., one center vertex connected with* n *end-points) where the supply vertex is one of the end-points and has capacity* n. *All the other nodes have demand 1. The solution returned by the algorithm has value 2 (after one round,* $n - 1$ *endpoints remain disconnected) while the optimal solution has value* n.

Final Algorithm. We modify Algorithm 1 to allow overlapping paths. At step 3(c), instead of removing all the nodes on $SP(i)$ from G, we set their demands to 0.

Input: An instance of the GPSD problem with one supply vertex v.

1. Let $SP(i)$ be the shortest path from v to a demand vertex i and let $|SP(i)|$ be the sum of the demands of its vertices.
2. $Cost \leftarrow 0$, $Sol \leftarrow \emptyset$.
3. While there exists a node i with $|SP(i)| > 0$ and $Cost + |SP(i)| \leq c(v)$ do:
    (a) Select the node i with the maximum value of $|SP(i)|$ such that $Cost + |SP(i)| \leq c(v)$.
    (b) $Sol \leftarrow Sol \cup SP(i)$; $Cost \leftarrow Cost + |SP(i)|$
    (c) Set $d(x) \leftarrow 0$ for all $x \in SP(i)$.
    (d) Recompute $SP(i)$ for each node in the new graph.

Output: Sol.

Algorithm 2. A 2-approximation algorithm for graphs with one supply vertex

Theorem 2. *Algorithm 2 is a 2-approximation for the GPSD problem with one supply vertex.*

Proof. If at the end $Cost \geq c(v)/2$, then the algorithm is a 2-approximation as $c(v)$ is an upper bound on the value of the optimal solution.

Now consider the case when the algorithm stops (i.e., there are no nodes that can be supplied with power) and $Cost < c(v)/2$. We show that the algorithm has reached the optimal solution. Assume by contradiction that there exists a vertex q which is powered in the optimal solution and is not selected by Algorithm 2. The shortest path from v to q has total demand less than $c(v)/2$ (and in fact less than $Cost$). Otherwise, the path from v to q would have been selected at a previous step of the algorithm (since it has larger demand than any other path selected by Algorithm 2). Thus we can add the vertex q to our solution, together with all the vertices from the path from v to q (contradiction).   □

The approximation factor of Algorithm 2 is asymptotically tight as we show in the following example.

Example 1. Let $G = (V, E)$ with $V = \{1, 2, 3, 4\}$. There are 3 edges in G: $(1, 2), (1, 3), (3, 4)$. The supply vertex is 1 and has capacity $2n$. The demands are: $d(2) = n$, $d(3) = n$, $d(4) = 1$. The optimal solution has value $2n$ (selecting nodes 2 and 3). The algorithm presented above selects the vertices 3 and 4, thus satisfying a demand of $n + 1$.

3.2   An Approximation Algorithm for General Graphs

In this subsection we generalize Algorithm 2 to obtain a $2k$-approximation for the GPSD problem. The idea is to try Algorithm 2 for each supply vertex in turn and select the best solution.

Theorem 3. *Algorithm 3 is a $2k$-approximation for the GPSD problem.*

Input: An instance of the GPSD problem with k supply vertices $v_1, v_2, \ldots, v_k$.

1. $Sol' \leftarrow \emptyset$,
2. For $i = 1$ to k do:
   (a) Let G' the graph G with the vertices $v_1, v_2, \ldots, v_{i-1}, v_{i+1}, \ldots, v_k$ removed.
   (b) Apply Algorithm 2 on graph G' with supply vertex v_i and let Sol be the output.
   (c) If the value of Sol is greater than the value of Sol' then $Sol' \leftarrow Sol$.

Output: Sol'.

Algorithm 3. A $2k$-approximation algorithm for GPSD problem

Proof. Denote by OPT the value of the optimal solution and by $|Sol'|$ the value of the solution returned by Algorithm 3. Let

$$\text{OPT} = \sum_{i=1}^{k} \text{OPT}_i$$

where OPT_i is the demand satisfied by the i'th supply vertex in an optimal solution. We know that: $|Sol'| \geq \max_{i=1}^{k} \text{OPT}_i/2$. Thus,

$$|Sol'| \geq \text{OPT}/2k$$

and the theorem follows.   □

The approximation algorithm presented above is straightforward and may seem extremely easy to improve. However, in the next section we give evidence that an algorithm with a significantly better ratio might be difficult to find.

3.3   Hardness of the GPSD problem

First, we present a couple of natural greedy algorithms which, intuitively, might lead to a better approximation ratio. Unfortunately, all of them fail to give an approximation ratio better than $2k$ (we invite the reader to find counterexamples).

Wrong algorithm 1. *Consider an ordering of the suppliers (e.g., sort the suppliers in decreasing order of their capacity). Apply the 2-approximation algorithm on the first supplier, remove all the edges of the solution, then apply the 2-approximation algorithm on the second supplier and so on.*

One might be tempted to think that a fixed ordering is the reason for which the above algorithm fails. However, this is not the case since the following algorithm fails as well.

Wrong algorithm 2. *Consider all the possible orderings of the suppliers. For each such ordering we apply Algorithm 1 and output the best solution of all orderings.*

Then, another idea is to generalize the Algorithm 3 from Subsection 3.1.

Wrong algorithm 3. *First, let $SP_i(j)$ be the shortest path from supplier i to a demand vertex j and let $|SP_i(j)|$ be the sum of the demands of its vertices. At each step, we select the path $SP_i(j)$ with the highest demand (and, of course, which does not exceed the capacity of supplier i and does not intersect the paths of the other suppliers).*

The problem with all the algorithms above is that, when we select a path from one of the suppliers, we can block the other suppliers from reaching the consumers they want.

Next, we show a reduction which shows that a $(2 - \epsilon)$-approximation for the GPSD problem, implies a polynomial time algorithm for the minimum sum 2-disjoint paths problem.

Given a graph G with n vertices and two pairs (s_1, t_1), (s_2, t_2), we construct the following instance of the GPSD problem with the same underlying graph. The two vertices s_1 and s_2 are suppliers with weights $X + a$ and $X + n + b$, respectively, where X is very large and a and b are between 1 and n. All the other vertices are consumers. Vertex t_1 has demand X, vertex t_2 has demand $X + n$ and all the other consumers have demand 1.

The result is stated in the following theorem.

Theorem 4. *There exist two disjoint paths between s_1 and t_1, and s_2 and t_2, first of length at most a and the second of length at most b if and only if both consumers t_1 and t_2 are powered.*

Proof. If we have two disjoint paths between s_1 and t_1 and s_2 and t_2, of length a and, respectively, b, then we can power the two consumers via this path and thus, the first implication, follows.

We prove the reverse implication. First notice that t_2 can be powered only by s_2 since s_1 does not have enough energy (since a and b are between 1 and n). Then, if both t_1 and t_2 are powered, then there has to exist a path from s_1 to t_1 of length at most a and a path from s_2 to t_2 of length at most b.

If we choose X large enough, then the hardness result follows.

Theorem 5. *A polynomial time $(2 - \epsilon)$ approximation algorithm for the GPSD problem implies a polynomial time exact algorithm for the minimum sum 2-disjoint paths problem.*

Intuitively, the hardness of GPSD problem with more suppliers comes from two directions. First, each supplier has to power as much energy as possible (this is the difficulty in the one-supplier case). Secondly, the trees generated by the suppliers must be disjoint and an algorithm has to construct those trees simultaneously. Otherwise, if the trees are constructed sequentially, one of the trees, can "cut" the trees of other suppliers.

In order to understand better the problem, we formulate a relaxed variant in which all the consumers have unit demand. This variant becomes trivial in the one supplier case, but, in the case of two suppliers there is not an obvious polynomial time algorithm. We leave this as an open problem.

Open problem 1. *The input is a graph $G = (V, E)$, two vertices a and b and two integers A and B. Find two disjoint and connected subsets $S_1, S_2 \subset V$ (if possible), with $|S_1| \geq A$ and $|S_2| \geq B$ and $a \in S_1$ and $b \in S_2$.*

4   Inapproximability of the ED Problem

In this section we show that the ED problem cannot be approximated within any factor, even for two suppliers, unless $\mathbf{P} = \mathbf{NP}$. We present a reduction from the *subset sum* problem.

Given a set of numbers $S = \{a_1, a_2, \ldots a_n\}$ and an integer k, we create the following instance of the ED problem.

- There are $n + 1$ consumers and 2 suppliers, A and B.
- The $n + 1$ consumers $v_1, \ldots, v_{n+1}$ form a line (i.e. v_1 is connected with v_2, v_2 with v_3 and so on) and have demands $a_1, a_2, \ldots, a_n$ and respectively C (where C is very large).
- Supplier A is connected with all the vertices v_1 up to v_n and has capacity k. The other supplier, B, has capacity $C + \sum_{i=1}^{n} a_n - k$ and is connected only to v_1.

The hardness is stated in the following theorem.

Theorem 6. *The ED problem cannot be approximated within a factor c, for any $c \in R_+$, unless $\mathbf{P} = \mathbf{NP}$.*

Proof. Since C is very large, the node v_{n+1} can be powered only by the supplier B. However, this can happen only if A uses all its capacity (i.e. the nodes powered by A have total demand precisely k), since the only path from X to v_{n+1} is $Xv_1v_2, \ldots v_n v_{n+1}$. In turn, A uses fully its capacity, if and only if there exists a subset of $\{a_1, \ldots, a_n\}$ which sums to k. Thus, if we can approximate the ED within any factor c we can see if the node v_{n+1} is powered and, respectively, we can solve the subset sum in polynomial time. Therefore, the theorem follows.

5   Conclusions and Open Problems

In this paper we study two problems, namely *graph partitioning with supply and demand (GPSD)* and *energy delivery (ED)*, which attempt to model the power supply network. First, we show a $2k$-approximation algorithm for the GPSD problem. Then, we show that a $(2 - \epsilon)$- approximation algorithm for GPSD in the case of two suppliers, implies a polynomial time algorithm for the minimum sum 2-disjoint paths problem. Finally, we introduce the ED problem and prove that it cannot be approximated within any factor, unless $\mathbf{P} = \mathbf{NP}$.

A natural open problem is to close the gap between the approximation upper and lower bounds for the GPSD problem. Another open problem is to design exact or fixed parameter algorithms. As the problems seem hard to solve on arbitrary instances, it is interesting to find restricted versions of the problems which are polynomial time solvable.

It is also intriguing what is the complexity of Problem 1 since at a first glance it seems very easy to solve in polynomial time. However, at a more careful look, no simple algorithm seems to work.

Acknowledgements. The author would like to thank the anonymous reviewers for their useful comments.

References

1. Adams, R.N., Laughton, M.A.: Optimal planning of power networks using mixed-integer programming. part 1: Static and time-phased network synthesis. Proceedings of the Institution of Electrical Engineers 121(2), 139–147 (1974)
2. Cormen, T.H., Stein, C., Rivest, R.L., Leiserson, C.E.: Introduction to Algorithms, 2nd edn. McGraw-Hill Higher Education (2001)
3. Crawford, D.M., Holt Jr., S.B.: A mathematical optimization technique for locating and sizing distribution substations, and deriving their optimal service areas. IEEE Transactions on Power Apparatus and Systems 94(2), 230–235 (1975)
4. El-Kady, M.A.: Computer-aided planning of distribution substation and primary feeders. IEEE Transactions on Power Apparatus and Systems PAS-103(6), 1183–1189 (1984)
5. Ito, T., Demaine, E.D., Zhou, X., Nishizeki, T.: Approximability of partitioning graphs with supply and demand. Journal of Discrete Algorithms 6(4), 627–650 (2008)
6. Ito, T., Zhou, X., Nishizeki, T.: Partitioning trees of supply and demand. In: Bose, P., Morin, P. (eds.) ISAAC 2002. LNCS, vol. 2518, pp. 612–623. Springer, Heidelberg (2002)
7. Ito, T., Zhou, X., Nishizeki, T.: Partitioning graphs of supply and demand. In: ISCAS (1), pp. 160–163 (2005)
8. Teng, J.H., Lu, C.-N.: Feeder-switch relocation for customer interruption cost minimization. IEEE Transactions on Power Delivery 17(1), 254–259 (2002)
9. Kersting, W.H., Phillips, W.H., Doyle, R.C.: Distribution feeder reliability studies. In: Rural Electric Power Conference, pp. B4-1–7 (April 1997)
10. Masud, E.: An interactive procedure for sizing and timing distribution substations using optimization techniques. IEEE Transactions on Power Apparatus and Systems PAS-93(5), 1281–1286 (1974)
11. Peponis, G.J., Papadopoulos, M.P.: New dynamic, branch exchange method for optimal distribution system planning. IEE Proceedings-Generation, Transmission and Distribution 144(3), 333–339 (1997)
12. Wall, D.L., Thompson, G.L., Northcote-Green, J.E.D.: An optimization model for planning radial distribution networks. IEEE Transactions on Power Apparatus and Systems PAS-98(3), 1061–1068 (1979)

Approximation Algorithms for a Combined Facility Location Buy-at-Bulk Network Design Problem

Andreas Bley[1,*], S. Mehdi Hashemi[2], and Mohsen Rezapour[1,**]

[1] Institute for Mathematics, TU Berlin, Straße des 17. Juni 136, 10623 Berlin, Germany
{bley,rezapour}@math.tu-berlin.de
[2] Department of Computer Science, Amirkabir University of Technology, No. 424, Hafez Ave., Tehran, Iran
hashemi@aut.ac.ir

Abstract. We consider a generalization of the connected facility location problem where the clients must be connected to the open facilities via shared capacitated (tree) networks instead of independent shortest paths. This problem arises in the planning of fiber optic telecommunication access networks, for example. Given a set of clients with positive demands, a set of potential facilities with opening costs, a set of capacitated access cable types, and a core cable type of infinite capacity, one has to decide which facilities to open, how to interconnect them using a Steiner tree of infinite capacity core cables, and which access cable types to install on which potential edges such that these edges form a forest and the installed capacities suffice to simultaneously route the client demands to the open facilities via single paths. The objective is to minimize the total cost of opening facilities, building the core Steiner tree among them, and installing the access cables. In this paper, we devise a constant-factor approximation algorithm for problem instances where the access cable types obey economies of scale. In the special case where only multiples of a single cable type can be installed on the access edges, a variant of our algorithm achieves a performance guarantee of 6.72.

1 Introduction

We study a generalization of the *Connected Facility Location* (ConFL) problem where not only direct connections between clients and open facilities, but also shared access trees connecting multiple clients to an open facility are allowed. Accordingly, also more realistic capacity and cost structures with flow-dependent buy-at-bulk costs for the access edges are considered. The resulting *Connected Facility Location with Buy-at-Bulk edge costs* (BBConFL) problem captures the central aspects of both the buy-at-bulk network design problem and the ConFL problem. In this paper, we study the approximability of the BBConFL problem. Although both the ConFL and the buy-at-bulk network design problem have

---

[*] Supported by the DFG research center MATHEON 'Mathematics for key technologies'.
[**] Supported by the DFG research training group 'Methods for Discrete Structures'.

T-H.H. Chan, L.C. Lau, and L. Trevisan (Eds.): TAMC 2013, LNCS 7876, pp. 72–83, 2013.

been well studied in the past, the combination of them has not been considered in the literature, to the best of our knowledge.

A typical telecommunication network consists of a backbone network with (almost) unlimited capacity on the links and several local access networks. In such a network, the traffic originating from the clients is sent through access networks to gateway or core nodes, which provide routing functionalities and access to the backbone network. The backbone then provides the connectivity among the core nodes, which is necessary to route the traffic further towards its destination. Designing such a network involves selecting the core nodes, connecting them with each other, and choosing and dimensioning the links that are used to route the traffic from the clients to the selected core nodes.

We model this planning problem as the BBConFL problem. We are given an undirected graph $G = (V, E)$ with nonnegative edge lengths $c_e \in \mathbb{Z}_{\geq 0}$, $e \in E$, obeying the triangle inequality, a set $F \subset V$ of potential facilities with opening costs $f_i \in \mathbb{Z}_{\geq 0}$, $i \in F$, and a set of clients $D \subset V$ with demands $d_j \in \mathbb{Z}_{>0}$, $j \in D$. We are also given K types of access cables that may be used to connect clients to open facilities. A cable of type i has capacity $u_i \in \mathbb{Z}_{>0}$ and cost (per unit length) $\sigma_i \in \mathbb{Z}_{\geq 0}$. Core cables, which are used to connect the open facilities, have a cost (per unit length) of $M \in \mathbb{Z}_{\geq 0}$ and infinite capacity. The task is to find a subset $I \subseteq F$ of facilities to open, a Steiner tree $S \subseteq E$ connecting the open facilities, and a forest $E' \subseteq E$ with a cable installation on its edges, such that E' connects each client to exactly one open facility and the installed capacities suffice to route all clients' demands to the open facilities. We are allowed to install multiple copies and types of access cables on each edge of E'. The objective is to minimize the total cost, where the cost for using edge e in the core Steiner tree is Mc_e and the cost for installing a single access cable of type i on edge e is $\sigma_i c_e$. We also consider the variant with only a single cable type, which we denote by *Single-Cable Connected facility location problem* (Single-Cable-ConFL).

The classical ConFL problem is special case of the BBConFL problem with only one cable type of unit capacity. This problem is well-studied in the literature. Gupta et al. [10] obtain a 10.66-approximation for this problem, based on LP rounding. Swamy and Kumar [15] later improved the approximation ratio to 8.55, using a primal-dual algorithm. Using sampling techniques, the guarantee was later reduced to 4 by Eisenbrand et al. [4], and to 3.19 by Grandoni et al. [7].

The (unsplittable) *Single-Sink Buy-at-Bulk problem (u-SSBB)* can be viewed as a special case of the BBConFL problem where the set of interconnected facilities are given in advance. Several approximation algorithms for u-SSBB have been proposed in the literature. Using LP rounding techniques, Garg et al. [5] developed a $O(k)$ approximation, where k is the number of cable types. Hassin et al. [11] provide a constant factor approximation for the single cable version of the problem. The first constant factor approximation for the problem with multiple cable types is due to Guha et al. [9]. Talwar [16] showed that an IP formulation of this problem has a constant integrality gap and provided a factor 216 approximation algorithm. Using sampling techniques, the approximation was reduced to 145.6 by Jothi et al. [12], and later to 40.82 by Grandoni et al. [6].

If we omit the requirement to connect the open facilities by a core Steiner tree, then the BBConFL problem reduces to a *k-cable facility location problem*. For this problem, Ravi et al. [13] provide an $O(k)$ approximation algorithm.

The rest of the paper is structured as follows. In Section 2, we describe our constant factor approximation algorithm for BBConFL extending the algorithm of Guha et al. [9] to incorporate also the selection of facilities to open as well as the Steiner tree (of infinite capacity) interconnecting them. In Section 3, we study the single cable version of the problem and present a factor 6.72 approximation algorithm for this problem.

2 Approximating BBConFL

In this section, we present a constant factor approximation algorithm for the BBConFL, which uses the ideas of Guha's algorithm [9] for the single sink buy-at-bulk network design problem to design the access trees of the solution.

First, we define another problem similar to the BBConFL with slightly different cost function, called *modified-BBConFL*. In this problem, each access cable has a fixed cost of σ_i, a flow dependent incremental cost of $\delta_i = \frac{\sigma_i}{u_i}$, and unbounded capacity. That is, for using one copy of cable type i on edge e and transporting D flow unit on e, a cost of $(\sigma_i + D\delta_i)c_e$ is incurred.

It is not hard to see that any ρ-approximation to the modified problem gives a 2ρ-approximation to the corresponding original buy-at-bulk ConFL. Furthermore, we will show later that there exist near optimal solutions of the modified problem that have a nice tree-like structure with each cable type being installed in a corresponding layer. We will exploit this special structure in our algorithm to compute approximate solutions for the modified problem and, thereby, also approximate solutions for the original buy-at-bulk ConFL.

In the modified-BBConFL, we may assume w.l.o.g. that $\sigma_1 < ... < \sigma_K$ and that $\delta_1 > ... > \delta_K$. In addition, we assume that $2\sigma_K < M$. Note that in our and many other applications, it is natural to assume that $\sigma_K << M$.

First, we prune the set of cable types such that all cables are considerably different. As shown in [9], this can be done without increasing the cost of the optimal solution too much.

Theorem 1. *For a predefined constant $\alpha \in (0, \frac{1}{2})$, we can prune the set of cables such that, for any i, we have $\sigma_{i+1} > \frac{1}{\alpha} \cdot \sigma_i$ and $\delta_{i+1} < \alpha \cdot \delta_i$ hold and the cost of the optimal solution increases by at most $\frac{1}{\alpha}$.*

We observe that, as demand along an edge increases, there are break-points at which it becomes cheaper to use the next larger cable type. For $1 \leq i < K$, we define b_i such that $\sigma_{i+1} + b_i\delta_{i+1} = 2\alpha(\sigma_i + b_i\delta_i)$. Intuitively, b_i is the demand at which it becomes considerably cheaper to use a cable type $i + 1$ rather than a cable type i. It has been shown in [9] that the break-points and modified cable cost functions satisfy the following properties.

Lemma 2. *For all i, we have $u_i \leq b_i \leq u_{i+1}$. For any i and $D \geq b_i$, we have $\sigma_{i+1} + D\delta_{i+1} \leq 2 \cdot \alpha(\sigma_i + D\delta_i)$.*

Let $b_K = \frac{M-\sigma_K}{\delta_K}$ be the edge flow at which the cost of using cable type K and a core link are the same. Suppose we install cable type i whenever the edge flow is in the range $[b_{i-1}, b_i]$, $1 \le i \le K$, where $b_0 = 0$. It can be shown that, if the edge flow is in the range $[b_{i-1}, u_i]$, then considering only the fixed cost σ_i (times the edge length) for using cable type i on the edge and ignoring the flow dependent incremental cost will underestimate the true edge cost only by a factor 2. Similarly, if the edge flow is in $[u_i, b_i]$, then considering only the flow dependent cost δ_i times the flow and ignoring the fixed cost underestimates the cost by only a factor 2. This means that any solution can be converted to a layered solution, loosing at most a factor 2 in cost, where layer i consists of (i) a Steiner forest using cable type i and carrying a flow of at least b_{i-1} on each edge, and (ii) a shortest path forest with each edge carrying a flow of at least u_i. In the following theorem, we define the structural properties of such layered solutions more formally. As in [9], for sake of simplicity, we assume that there are extra loop-edges such that property (iii) can be enforced for any solution.

Theorem 3. *Modified-BBConFL has a solution with the following properties:*

  (i) *The incoming demand of each open facility is at least b_K.*
  (ii) *Cable $i + 1$ is used on edge e only if at least b_i demand is routed across e.*
  (iii) *All demand which enters a node, except an open facility, using cable i, leaves that node using cables i or $i + 1$.*
  (iv) *The solution's cost is at most $2(\frac{1}{\alpha} + 1)$ times the optimum cost.*

Proof. Consider an optimum solution of the modified-BBConFL. Let T^* be the tree connecting the open facilities in the optimum solution. Consider those open facilities whose incoming demand is less than b_K. We can find an unsplittable flow on the edges of T^* sending the aggregated demand from these facilities to some other open facilities such that the resulting solution obeys property (i) and the total flow on any edge of the Steiner tree is at most b_K. Therefore the cost of closing these facilities and sending the corresponding demands to some other open facility using access links can be bounded by the core Steiner tree cost of the optimal solution, so we close these facilities and reroute demands. Now identify the set of remaining open facilities to a single sink, and update the edge length metric appropriately. The resulting solution is now a (possibly sub-optimal) single-sink network design solution. Results in [9] imply that there is a near-optimal solution to this single-sink instance which obeys the properties (ii) and (iii), with a factor $(\frac{2}{\alpha} + 1)$ loss in the total access cable cost. Hence, we can transform our modified-BBConFL solution to a solution which satisfies properties (ii)–(iv), too.   □

Our algorithm constructs a layered solution with the properties described in Theorem 3 in a bottom-up fashion, aggregating the client demands repeatedly and alternating via Steiner trees and via direct assignments (or, equivalently, via shortest path trees) to values exceeding u_i and b_i. In phase i, we first aggregate the (already pre-aggregated) demands of value at least b_{i-1} to values of at least u_i using cable type i on the edges of an (approximate) Steiner tree connecting these demands. Then we further aggregate these aggregates to values of at least (a constant fraction of) b_i solving a corresponding *Lower Bounded Facility Location* (LBFL) problem [1,8,14], where all clients may serve as facilities to aggregate demand (except for the last phase, where only real facilities are eligible). The LBFL problem is a generalization of the facility location problem where each open facility is required to serve a certain minimum amount of demand.

Let D_i be the set of demand points we have at the i-th stage. Initially $D_1 = D$. Algorithm 1 describes the steps of the algorithm in more detail.

Algorithm 1.

---

1. Guess a facility r from the optimum solution.
2. *For cable type $i = 1, 2, ..., K - 1$ Do*
   - *Steiner Trees:* Construct a ρ_{ST}-approximate Steiner tree T_i on terminals $D_i \cup \{r\}$ for edge costs σ_i per unit length. Install a cable of type i on each edge of this tree. Root this tree at r. Transport the demands from D_i upwards along the tree. Walking upwards along this tree, identify edges whose demand is larger than u_i and cut the tree at these edges.
   - *Consolidate:* For every tree in the forest created in the preceding step, transfer the total demand in the root of tree, which is at least u_i, back to one of its sources using a shortest path of cable type i. Choose this source with probability proportional to the demand at the source.
   - *Shortest Path:* Solve the LBFL problem with clients D_1, facility opening cost 0 at all nodes, facility lower bound b_i, and edge costs δ_i per unit length. The solution is a forest of shortest path trees. Then route the current demands along these trees to their roots, installing cables of type i.
   - *Consolidate:* For every root in the forest created in the preceding step, transfer the total demand in the root of tree, which is at least b_i, back to one of its sources with probability proportional to the demand at that source using a shortest path with cables of type i. Let D_{i+1} be the resulting demand locations.
3. *For cable type K Do*
   - Construct a ρ_{ST}-approximate Steiner tree T_K on terminals $D_K \cup \{r\}$ for edge costs σ_K per unit length. Install a cable of type K on each edge of this tree. Root this tree at r. Transport the demands from D_K upwards along the tree. Walking along this tree, identify edges whose demand is larger than u_K and cut the tree at these edges. For every tree in the created forest, transfer the total demand in the root of tree back to one of its sources with probability proportional to the demand at that source via a shortest path, using cables of type K.
   - Solve the LBFL problem with clients D_1, facility set F, opening costs f_i, facility lower bound b_K, and edge costs δ_K per unit length. We obtain a forest of shortest path trees. Then route the current demands along these trees to their roots, installing cables of type K. Let F' be the set of open facilities.
4. Compute a ρ_{ST}-approximate Steiner tree T_{core} on terminals $F' \cup \{r\}$ for edge costs M per unit length. Install the core link on the edges of T_{core}.

To solve LBFL, we employ the bicriteria $\mu\rho_{FL}$-approximation algorithm devised by Guha et al. [8], which relaxes the lower bound on the minimum demand served by a facility by a factor $\beta = \frac{\mu-1}{\mu+1}$. Here ρ_{FL} is the best known approximation for the facility location problem.

It remains to show that the computed solution is an approximate solution. Let C_i^*, S^*, and O^* be the amount paid for cables of type i, for the core Steiner tree, and for opening facilities in the near-optimal solution, respectively. We define C_i to be the total cost paid for cables of type i in the returned solution. Let D_j^i be the demand of node j at stage i of the algorithm. Let T_i, P_i and N_i be cost incurred in the Steiner tree step, the shortest path step, and the consolidation steps of iteration i, respectively. Also, let T_i^I and T_i^F denote the incremental and the fixed cost components of the Steiner tree step at iteration i. Analogously, P_i^I and P_i^F denote the incremental and the fixed costs incurred in the shortest path step. Recall that the set of access cable types has been reduced depending on the constant parameter $\alpha \in (0, \frac{1}{2})$. How to choose this parameter appropriately will be discussed later.

The following Lemma carries over from the single sink buy-at-bulk problem studied in [9] to our problem in a straightforward way.

Lemma 4.

(i) At the end of each consolidation step, every node has $E[D_j^i] = d_j$.
(ii) $E[N_i] \leq T_i + P_i$ for each i.
(iii) $P_i^F \leq P_i^I$ and $T_i^I \leq T_i^F$ for each i.

The following lemma bounds the fixed costs of the cables installed in the Steiner tree phase i of our algorithm.

Lemma 5. $E[T_i^F] \leq \rho_{ST}\left(\sum_{j=1}^{i-1} \frac{1}{\beta}(2\alpha)^{i-j}C_j^* + \sum_{j=i}^{K} \alpha^{j-i}C_j^* + \frac{1}{2}\alpha^{K-i}S^*\right)$ for each i.

Proof. We construct a feasible Steiner tree for stage i as follows. Consider the near-optimum solution, and consider only those nodes which are candidate terminals in stage i of our algorithm. We remove all the cables if the total demand flowing across it is zero. Otherwise we replace the cable with a cable of type i. Note that, being in stage i, the expected demand on each cable $j < i$ is at least βb_i. Hence, by Lemma 2, the expected cost of all replacement cables for cables of type $j < i$ is bounded by $\frac{1}{\beta}(2\alpha)^{i-j}C_j^*$.

Similarly, the expected cost of the replacement cables for the cables $j > i$ are bounded by $\alpha^{j-i}C_j^*$, using the fixed costs scale. Finally, the cost on a core link used to connect candidate terminals to r is reduced at least by $\frac{1}{2}\alpha^{K-i}S^*$. Altogether, the expected fixed cost of this Steiner tree, which is a possible solution to the Steiner tree problem in stage i, is bounded by

$$\sum_{j=1}^{i-1} \frac{1}{\beta}(2\alpha)^{i-j}C_j^* + \sum_{j=i}^{K} \alpha^{j-i}C_j^* + \frac{1}{2}\alpha^{K-i}S^* .$$

As we use a ρ_{ST}-approximation algorithm to solve this Steiner tree problem in our algorithm, the claim follows. $\qquad\square$

In a similar way, we can also bound the incremental costs of the cables installed in the shortest path phase i of our algorithm.

Lemma 6. $E[P_i^I] \leq \mu \cdot \rho_{FL} \sum_{j=1}^{i} \alpha^{i-j} \cdot C_j^*$ for each i.

Proof. Consider the forest defined by the edges with cable types 1 to i in the near-optimum solution and replace all cables of type less than i by cables of type i. The cost of replacing all cables of type $j < i$ is bounded by $\alpha^{i-j} \cdot C_j^*$, using the incremental costs scale. The resulting tree provides a feasible solution for the shortest path stage i. As our algorithm applies a bicriteria $\mu \cdot \rho_{FL}$-approximation algorithm to solve the lower bounded facility location problem in this stage, the claim follows.   □

The opening costs and the incremental shortest path costs in the final stage of our algorithm can be bounded as follows.

Lemma 7. $E[P_K^I + f(F')] \leq \mu \cdot \rho_{FL}(\sum_{i=1}^{K} \alpha^{K-i} \cdot C_i^* + O^*)$

Proof. Now, consider the forest given by all access edges of the near-optimum solution and replace all cables (of type less than K) by cables of type K. For each $i < K$, the incremental cost of the new solution is a fraction α^{K-i} of the incremental cost of the optimal solution's cable i portion. The set of facilities opened in the solution, combined with the cables, constitutes a feasible solution for the LBFL problem solved in the final stage, and its cost is no more than $\sum_{i=1}^{K} \alpha^{K-i} C_i^* + O^*$. Using the bicriteria $\mu \cdot \rho_{FL}$-approximation algorithm, the claim follows.   □

Finally, the cost of the core Steiner tree have to be bounded.

Lemma 8. $E[T_{core}] \leq \rho_{ST}(S^* + \frac{1}{\beta} \sum_{j=1}^{K} (C_j^* + C_j))$

Proof. Let F^*, T_{core}^* and T_{access}^* be the set of open facilities, the tree connecting them, and the forest connecting clients to open facilities in the near-optimum solution, respectively. Let T_{access} be the forest connecting clients to open facilities in the solution returned by the algorithm. We construct a feasible Steiner tree on $F' \cup \{r\}$, whose expected cost is $S^* + \frac{1}{\beta} \sum_{j=1}^{K} (C_j^* + C_j)$. In the algorithm's solution, each facility $l \in F'$ serves at least a total demand of βb_K. This demand is also served by the set of optimal facilities in the near-optimum solution. Therefore, at least βb_K demand can be routed between each facility $l \in F'$ and the facilities of F^* along edges of $T_{access}^* \cup T_{access}$ (using the access links). Hence, we obtain a feasible Steiner tree on $F' \cup F^*$, using core links, whose cost is at most $S^* + \frac{1}{\beta} \sum_{j=1}^{K} (C_j^* + C_j)$.   □

Together, Lemmas 4–8 imply our main result.

Theorem 9. *Algorithm 1 is a constant factor approximation for BBConFL.*

Proof. By Lemmas 4–6, the total expected cost of access links is bounded by

$$4\sum_{i=1}^{K}\left[\mu\rho_{FL}\sum_{j=1}^{i}\alpha^{i-j}C_j^* + \rho_{ST}\left(\sum_{j=i}^{K}\alpha^{j-i}C_j^* + \sum_{j=1}^{i-1}\frac{1}{\beta}(2\alpha)^{i-j}C_j^* + \frac{1}{2}\alpha^{K-i}S^*\right)\right]$$

$$\leq 4\left(\frac{\mu\cdot\rho_{FL}}{1-\alpha} + \frac{\rho_{ST}}{1-\alpha} + \frac{\rho_{ST}}{\beta(1-2\alpha)}\right)\sum_{i=1}^{K}C_i^* + \frac{2\cdot\rho_{ST}}{1-\alpha}S^*$$

Additionally, using Lemmas 7 and 8, the total cost of installing core links and opening facilities is bounded by

$$\mu\rho_{FL}O^* + \rho_{ST}S^* + \frac{\rho_{ST}}{\beta}\left(\sum_{i=1}^{K}C_i^* + \sum_{i=1}^{K}C_i\right).$$

Altogether, we obtain a bound of

$$\mu\rho_{FL}O^* + \left[\frac{\rho_{ST}}{\beta} + 4\left(1+\frac{\rho_{ST}}{\beta}\right)\left(\frac{\mu\rho_{FL}+\rho_{ST}}{1-\alpha} + \frac{\rho_{ST}}{\beta(1-2\alpha)}\right)\right]\sum_{i=1}^{K}C_i^*$$

$$+ \left[\left(1+\frac{\rho_{ST}}{\beta}\right)\left(\frac{2\rho_{ST}}{1-\alpha}\right) + \rho_{ST}\right]S^*$$

for the worst case ratio between the algorithm's solution and a near optimal solution, restricted according to Theorem 3, of the modified-BBConFL. With Theorems 1 and 3, this yields a worst case approximation guarantee of $\frac{2}{\alpha}(\frac{1}{\alpha}+1)$ times the above ratio against an unrestricted optimal solution of the modified-BBConFL.

Finally, we lose another factor of 2 in the approximation guarantee when evaluating the approximate solution for the modified-BBConFL with respect to the original BBConFL problem. For appropriately chosen fixed parameters α, β, and μ, we nevertheless obtain a constant factor approximation algorithm for BBConFL. □

3   Approximating Single-Cable-ConFL

In this section, we consider a simpler version of the problem, where only multiples of a single cable type can be installed. Let $u > 0$ be the capacity of the only cable type available. We may assume that the cost of this cable is one. The algorithm presented in this paper can easily be adapted for $\sigma > 1$.

We obtain an approximation algorithm for this problem by modifying the algorithmic framework proposed in [7] as shown in Algorithm 2 on the next page. In this Algorithm, $c(v,u)$ denotes the distance between u and v, and $c(v,U) = \min_{u\in U} c(v,u)$. Again, the algorithm uses a constant parameter $\alpha \in (0,1]$, whose setting will be discussed later.

One easily verifies that Algorithm 2 computes a feasible solution. Clearly, T' is a Steiner tree connecting the open facilities F'. The existence of (and a

---

Algorithm 2.

---

1. Guess a facility r from the optimum solution.
   Mark each client $j \in D$ with probability $\frac{\alpha \cdot d_j}{M \cdot u}$. Let D' be the set of marked clients.
2. Compute a ρ_{ST}-approximate Steiner tree T_1 on terminals $D' \cup \{r\}$.
3. Define a FL instance with clients D, facilities F, costs $c'_{ij} := \frac{d_j}{u} c(i, j)$, $j \in D$ and $i \in F$, and opening costs $f'_i := f_i + M \cdot c(i, D' \cup \{r\})$, $i \in F$.
   Compute a (λ_F, λ_C)-bifactor-approximate solution $U = (F', \sigma)$ to this instance, where $\sigma(j) \in F'$ indicates the facility serving $j \in D$ in U.
4. Augment T_1 with shortest paths from each $i \in F'$ to T_1.
   Let T' be the augmented tree.
   Output F' and T' as open facilities and core Steiner tree, respectively.
5. Compute a ρ_{ST}-approximate Steiner tree T_2 on terminals $D \cup \{r\}$.
6. // *Using the results in [11,13], we now install capacities to route the clients' demands to open facilities F'.*
   - For each $j \in D$ with $d_j > u/2$, install $\lceil d_j/u \rceil$ cables from j to its closest open facility in F'.
   - Considering only clients with $d_j \leq u/2$, partition T_2 into disjoint subtrees such that the total demand of each subtree not containing r is in $[u/2, u]$ and the total demand of the subtree containing r is at most u; see [11].
   - Install one cable on each edge contained in any subtree.
   - For each subtree not containing r, install one cable from the client closest to an open facility to this facility.

---

polynomial time algorithm to find) a partition of the tree T_2 into subtrees of total demand between $u/2$ and u each, except for the subtree containing r, has been shown in [11], given that each individual demand is at most u. From that, it follows immediately that all clients j with $d_j \leq u/2$ can be routed within their respective subtree towards the client closest to an open facility and then further on to this facility without exceeding the capacity u on these edges.

It remains to show that the computed solution is an approximate solution. Let O'_U and C'_U be the (modified) opening and connection costs of the solution U of the facility location problem solved in Step 3. Furthermore, let I^*, S^*, and F^* be the set of open facilities, the Steiner tree connecting them, and the forest connecting the clients to the open facilities in the optimal solution, respectively. Also let $\sigma^*(j) \in I^*$ be the facility serving $j \in D$ in the optimal solution. The opening cost, cable installation costs, and core Steiner tree costs of the algorithm's solution and of the optimal solution are denoted by O, C, T and O^*, C^*, T^*, respectively. Let $c(E') := \sum_{e \in E'} c_e$ for any $E' \subseteq E$,

Lemma 10. *The cable cost induced in Step 6 is at most $c(T_2) + 2 \cdot C'_U$.*

Proof. Using the result in [11], the total flow on any edge of the Steiner tree T_2 induced by grouping the demands into disjoint subtrees is at most u. Thus, one copy of the cable on all edges in T_2 is sufficient to accommodate the flow on the edges of T_2, which contributes $c(T_2)$ to the total cable installation cost.

Let $C_1, C_2, ..., C_T$ be the sets of clients in each subtree and for each C_t let $j_t \in C_t$ be the client which is closest to an open facility in F'. The modified connection costs in U are

$$C_U' = \sum_t \sum_{j \in C_t} \frac{d_j}{u} c(j, \sigma(j)) + \sum_{j \in D : d_j > \frac{u}{2}} \frac{d_j}{u} c(j, \sigma(j))$$

$$\geq \sum_t \sum_{j \in C_t} \frac{d_j}{u} c(j, \sigma(j)) + \sum_{j \in D : d_j > \frac{u}{2}} \frac{1}{2} c(j, \sigma(j)) \ .$$

Since the algorithm sends the total demand of C_t via j_t, we have

$$C_U' \geq \sum_t \frac{\sum_{j \in C_t} d_j}{u} c(j_t, \sigma(j_t)) + \sum_{j \in D : d_j > \frac{u}{2}} \frac{1}{2} c(j, \sigma(j)) \geq \frac{1}{2} C_{AC} \ ,$$

where C_{AC} is the cost of the cables installed by the algorithm between the subtrees and the closest open facilities and between the large demand clients and the open facilities. Altogether the total cost of buying cables to route the traffic is at most $c(T_2) + 2 \cdot C_U'$. $\qquad \square$

Lemma 11. *The opening and core connection cost of the computed solution satisfy $O + T \leq O_U' + M \cdot c(T_1)$.*

Proof. Algorithm 2 opens the facilities chosen in the FL solution and connects these facilities by the tree T'. Since the modified opening costs f' in Step 3 include both the original cost for opening F' and the cost for augmenting T_1 to T', the sum of the opening cost and core connection cost of the final solution are at most $O_U' + M \cdot c(T_1)$. $\qquad \square$

Lemma 12. *The expected cost of T_1 is at most $\frac{\rho_{ST}}{M}(T^* + \alpha C^*)$.*

Proof. We obtain a feasible Steiner tree on $D' \cup \{r\}$ by joining the optimal solution's Steiner tree S^* and the paths connecting each client in D' with its corresponding open facility in I^* in the optimal solution. The expected cost of the resulting subgraph is at most

$$\sum_{e \in S^*} c(e) + \frac{\alpha}{M} \sum_{j \in D} \frac{d_j}{u} \cdot l(j, I^*) \leq \frac{T^*}{M} + \frac{\alpha}{M} C^* \ ,$$

where $l(j, I^*)$ denotes the length of the path connecting j to its open facility in I^* using edges of F^*. The last inequality holds since in $\sum_{j \in D} \frac{d_j}{u} \cdot l(j, I^*)$ instead of installing an integral number of cables on every edge, we install multiples of $\frac{1}{u}$ on every edge, which is a lower bound for C^*. Thus the expected cost of the ρ_{ST}-approximate Steiner tree on $D' \cup \{r\}$ is at most $\frac{\rho_{ST}}{M}(T^* + \alpha C^*)$. $\qquad \square$

Lemma 13. *The cost of T_2 is at most $\rho_{ST}(T^* + C^*)$.*

Proof. Clearly $S^* \cup F^*$ defines a feasible Steiner tree on $D \cup \{r\}$. $\qquad \square$

Lemma 14. $E[O'_U + C'_U] \leq \lambda_F(O^* + \alpha C^*) + \lambda_C(C^* + \frac{0.807}{\alpha}T^*)$.

Proof. We provide a feasible solution for the facility location problem, whose expected opening cost is $O^* + \alpha C^*$ and whose expected connection cost is $C^* + \frac{0.807}{\alpha}T^*$. Choose facilities $\sigma^*(D') \cup \{r\}$. The expected opening cost is at most

$$\sum_{i \in I^*} f_i + M \cdot \frac{\alpha}{M} \sum_{j \in D} \frac{d_j}{u} \cdot l(j, \sigma^*(j)) \leq O^* + \alpha C^* .$$

Now, replace j by several copies of co-located unit-demand clients. In order to bound the expected connection cost, we apply the core connection game described in [4] (see also Lemma 2 in [7] for ConFL with unit-demand clients, probability $\frac{\alpha}{M \cdot u}$ (which is the same to mark each client $j \in D$ with probability $\frac{\alpha \cdot d_j}{M \cdot u}$), core S^*, mapping $\sigma = \sigma^*$, and $w(e) = \frac{c(e)}{u}$ which yields

$$E[\sum_{j \in D} c'(j, \sigma^*(D') \cup \{r\})] \leq \sum_{j \in D} c'(j, I^*) + \frac{0.807}{\frac{\alpha}{M \cdot u}} \cdot \frac{w(T^*)}{M}$$

$$\leq \sum_{j \in D} \frac{d_j}{u} l(j, I^*) + \frac{0.807}{\frac{\alpha}{M \cdot u}} \cdot \frac{T^*}{M \cdot u} \leq C^* + \frac{0.807}{\alpha}T^* .$$
□

Theorem 15. *For a proper choice of α, Algorithm 2 is an 6.72-approximation algorithm for Single-Cable-ConFL.*

Proof. By Lemmas 10–13, we have

$$E[O + T + C] \leq O'_U + 2 \cdot C'_U + \rho_{ST}(2T^* + (\alpha + 1)C^*) .$$

Applying Lemma 14, we can bound the first two terms, which yields

$$E[O + T + C] \leq \rho_{ST}(2T^* + (\alpha+1)C^*) + 2[\lambda_F(O^* + \alpha C^*) + \lambda_C(C^* + \frac{0.807}{\alpha}T^*)]$$

$$= (2\lambda_F)O^* + 2(\lambda_C \frac{0.807}{\alpha} + \rho_{ST})T^* + (\rho_{ST}(\alpha + 1) + 2(\lambda_F \alpha + \lambda_C))C^* . \quad (1)$$

Applying Byrka's $(\lambda_F, 1 + 2 \cdot e^{-\lambda_F})$-bifactor approximation algorithm [2] for the facility location subproblem and the (currently best known) $ln(4)$-approximation algorithm for the Steiner tree problem [3] and setting $\alpha = 0.5043$ and $\lambda_F = 2.1488$, inequality (1) implies $E[O + T + C] \leq 6.72(O^* + T^* + C^*)$. □

For unit demands, one can derive a stronger bound of $c(T_2) + C_U$ for the cable installation costs using the techniques proposed in [11] for the single sink network design problem. Adapting Step 6 of the algorithm and adjusting the parameters α and λ_F accordingly, one easily obtains a 4.57-approximation algorithm for the Single-Cable-ConFL problem with unit demands.

References

1. Ahmadian, S., Swamy, C.: Improved Approximation Guarantees for Lower-Bounded Facility Location. In: Proc. of WAOA (2012)
2. Byrka, J.: An Optimal Bifactor Approximation Algorithm for the Metric Uncapacitated Facility Location Problem. In: Charikar, M., Jansen, K., Reingold, O., Rolim, J.D.P. (eds.) APPROX and RANDOM 2007. LNCS, vol. 4627, pp. 29–43. Springer, Heidelberg (2007)
3. Byrka, J., Grandoni, F., Rothvoß, T., Sanità, L.: An improved LP-based approximation for Steiner tree. In: Proc. of STOC 2010, pp. 583–592 (2010)
4. Eisenbrand, F., Grandoni, F., Rothvoß, T., Schäfer, G.: Connected facility location via random facility sampling and core detouring. Journal of Computer and System Sciences 76, 709–726 (2010)
5. Garg, N., Khandekar, R., Konjevod, G., Ravi, R., Salman, F.S., Sinha, A.: On the integrality gap of a natural formulation of the single-sink buy-at-bulk network design problem. In: Aardal, K., Gerards, B. (eds.) IPCO 2001. LNCS, vol. 2081, pp. 170–184. Springer, Heidelberg (2001)
6. Grandoni, F., Rothvoß, T.: Network design via core detouring for problems without a core. In: Abramsky, S., Gavoille, C., Kirchner, C., Meyer auf der Heide, F., Spirakis, P.G. (eds.) ICALP 2010. LNCS, vol. 6198, pp. 490–502. Springer, Heidelberg (2010)
7. Grandoni, F., Rothvoß, T.: Approximation algorithms for single and multicommodity connected facility location. In: Günlük, O., Woeginger, G.J. (eds.) IPCO 2011. LNCS, vol. 6655, pp. 248–260. Springer, Heidelberg (2011)
8. Guha, S., Meyerson, A., Munagala, K.: Hierarchical placement and network design problems. In: Proc. of FOCS 2000, pp. 603–612 (2000)
9. Guha, S., Meyerson, A., Munagala, K.: A constant factor approximation for the single sink edge installation problems. In: Proc. of STOC 2001, pp. 383–388 (2001)
10. Gupta, A., Kleinberg, J., Kumar, A., Rastogi, R., Yener, B.: Provisioning a virtual private network: A network design problem for multicommodity flow. In: Proc. of STOC 2001, pp. 389–398 (2001)
11. Hassin, R., Ravi, R., Salman, F.S.: Approximation algorithms for a capacitated network design problem. In: Jansen, K., Khuller, S. (eds.) APPROX 2000. LNCS, vol. 1913, pp. 167–176. Springer, Heidelberg (2000)
12. Jothi, R., Raghavachari, B.: Improved approximation algorithms for the single-sink buy-at-bulk network design problems. In: Hagerup, T., Katajainen, J. (eds.) SWAT 2004. LNCS, vol. 3111, pp. 336–348. Springer, Heidelberg (2004)
13. Ravi, R., Sinha, A.: Integrated logistics: Approximation algorithms combining facility location and network design. In: Cook, W.J., Schulz, A.S. (eds.) IPCO 2002. LNCS, vol. 2337, pp. 212–229. Springer, Heidelberg (2002)
14. Svitkina, Z.: Lower-bounded facility location. Trans. on Algorithms 6(4) (2010)
15. Swamy, C., Kumar, A.: Primal-dual algorithms for connected facility location problems. Algorithmica 40, 245–269 (2004)
16. Talwar, K.: The single-sink buy-at-bulk LP has constant integrality gap. In: Cook, W.J., Schulz, A.S. (eds.) IPCO 2002. LNCS, vol. 2337, pp. 475–480. Springer, Heidelberg (2002)

k-means++ under Approximation Stability

Manu Agarwal[1,*], Ragesh Jaiswal[2], and Arindam Pal[3,**]

[1] IIT Rajasthan
[2] IIT Delhi
[3] TCS Innovations Lab

Abstract. The Lloyd's algorithm, also known as the k-means algorithm, is one of the most popular algorithms for solving the k-means clustering problem in practice. However, it does not give any performance guarantees. This means that there are datasets on which this algorithm can behave very badly. One reason for poor performance on certain datasets is bad initialization. The following simple sampling based seeding algorithm tends to fix this problem: pick the first center randomly from among the given points and then for $i \geq 2$, pick a point to be the ith center with probability proportional to the squared distance of this point from the previously chosen centers. This algorithm is more popularly known as the k-means++ seeding algorithm and is known to exhibit some nice properties. These have been studied in a number of previous works [AV07, AJM09, ADK09, BR11]. The algorithm tends to perform well when the optimal clusters are *separated* in some sense. This is because the algorithm gives preference to further away points when picking centers. Ostrovsky et al.[ORSS06] discuss one such separation condition on the data. Jaiswal and Garg [JG12] show that if the dataset satisfies the separation condition of [ORSS06], then the sampling algorithm gives a constant approximation with probability $\Omega(1/k)$. Another separation condition that is strictly weaker than [ORSS06] is the approximation stability condition discussed by Balcan et al.[BBG09]. In this work, we show that the sampling algorithm gives a constant approximation with probability $\Omega(1/k)$ if the dataset satisfies the separation condition of [BBG09] and the optimal clusters are not too small. We give a negative result for datasets that have small optimal clusters.

1 Introduction

The k-means clustering problem is defined as follows:

Given n points $\mathcal{X} = \{x_1, ..., x_n\} \in \mathbb{R}^d$, find k points $\{c_1, ..., c_k\} \in \mathbb{R}^d$ (these are called centers) such that the following objective function is minimized:

$$\phi_{\{c_1,...,c_k\}}(\mathcal{X}) = \sum_{x \in \mathcal{X}} \min_{c \in \{c_1,...,c_k\}} D(x, c)$$

* This work was done while the author was visiting IIT Delhi and was supported by the Summer Research Fellowship programme at IIT Delhi.
** Major part of this work was done while the author was at IIT Delhi.

T-H.H. Chan, L.C. Lau, and L. Trevisan (Eds.): TAMC 2013, LNCS 7876, pp. 84–95, 2013.
© Springer-Verlag Berlin Heidelberg 2013

where $D(x,c)$ denotes the square of the Euclidean distance between points x and c.

Note that the k centers define an implicit clustering of the points in $\mathcal{X}$ as all the points that have the same closest center are in the same cluster. This problem is known to be an NP-hard problem when $k \geq 2$. We can generalize the problem for any distance measure by defining the distance function D accordingly. Such generalized version of the problem is known as the k-median problem with respect to a given distance measure. Here, we will talk about the k-means problem and then generalize our results for the k-median problem with respect to distance measures that are metrics in an approximate sense.

As discussed in the abstract, the most popular algorithm for solving the k-means problem is the Lloyd's algorithm that can be described as follows: (i) Pick k centers arbitrarily (ii) consider the implicit clustering induced by these centers (iii) move the centers to the respective centroids of these induced clusters and then repeat (ii) and (iii) until the solution does not improve. Even though this algorithm works extremely well in practice, it does not have any performance guarantees, the main problem being arbitrary initialization. This means that the algorithm takes a very long time to converge or the final solution is arbitrarily bad compared to the optimal. The following simple sampling algorithm that is more popularly known as the k-means++ seeding algorithm seems to fix the problem to some extent:

(**SampAlg**) Pick the first center uniformly at random from $\mathcal{X}$. Choose a point $x \in \mathcal{X}$ to be the ith center for $i \geq 2$ with probability proportional to the squared distance of x from the nearest previously chosen centers, i.e., with probability $\frac{\min_{c \in \{c_1,\ldots,c_{i-1}\}} D(x,c)}{\phi_{\{c_1,\ldots,c_{i-1}\}}(\mathcal{X})}$.

In this work, we study some properties of this simple sampling algorithm. First, let us look at the previous works.

Previous Work. The above algorithm, apart from being simple, easy-to-implement, and quick, exhibits some very nice theoretical properties. Arthur and Vassilvitskii [AV07] show that **SampAlg** gives $O(\log k)$ approximation in expectation. They also give an example where the algorithm gives solution with approximation factor $\Omega(\log k)$ in expectation. Ailon et al. [AJM09] and Aggarwal et al. [ADK09] show that this algorithm is a constant factor pseudo-approximation algorithm. This means that **SampAlg** gives a solution that is within a constant factor of the optimal (w.r.t. k centers) if it is allowed to output more than k centers. Brunsch and Röglin [BR11] gave an example where **SampAlg** gives an approximation factor of $(2/3 - \epsilon) \log k$ with probability exponentially small in k thus closing the open question regarding whether the sampling algorithm gives a constant approximation with not-too-small probability. Jaiswal and Garg [JG12] observe that **SampAlg** behaves well for datasets that satisfy the separation condition $\frac{\Delta_{k-1}(\mathcal{X})}{\Delta_k(\mathcal{X})} \geq c$, where $\Delta_i(\mathcal{X})$ denotes the optimal value of the cost for the i-means problem on data $\mathcal{X}$. They show that under this separation condition, the algorithm gives a constant approximation factor with probability $\Omega(1/k)$. This separation condition was discussed by Ostrovsky et al. [ORSS06] who also observe that **SampAlg** behaves well under such separation and construct a PTAS for the k-means problem using a variant of **SampAlg** in their algorithm. Balcan et al. discuss a strictly weaker separation condition than [ORSS06]. This separation condition has gained prominence and a number of followup works has been done. In this work, we show that **SampAlg** behaves well even under this weaker separation property. Next, we discuss our results in more detail.

Our Results. Let us first discuss the [BBG09] separation condition. This is known as the $(1+\alpha, \epsilon)$-approximation stability condition.

Definition 1 ($(1+\alpha, \epsilon)$-approximation stability). *Let $\alpha > 0, 1 \geq \epsilon > 0$. Let $\mathcal{X} \in \mathbb{R}^d$ be a point set and let $C_1^*, ..., C_k^*$ denote the optimal k clusters of $\mathcal{X}$ with respect to the k-means objective. $\mathcal{X}$ is said to satisfy $(1+\alpha, \epsilon)$-approximation stability if any $(1+\alpha)$-approximate clustering $C_1, ..., C_k$ is ϵ-close to $C_1^*, ..., C_k^*$. ϵ-closeness means that at most ϵ fraction of points have to be reassigned in $C_1, ..., C_k$ to be able to match $C_1^*, ..., C_k^*$.*

Note that for a fixed value of ϵ, the larger the value of α the stronger is the separation between the optimal clusters. Our techniques easily generalize for large values of α. The above condition captures how stable the optimal clustering is under approximate clustering solutions. This separation condition has been shown to be strictly weaker than the [ORSS06] separation condition. More specifically, it has been shown (see Section 6 in [BBG09] and Lemma 5.1 in [ORSS06]) if a dataset $\mathcal{X}$ satisfies the separation condition $\frac{\Delta_k(\mathcal{X})}{\Delta_{k-1}(\mathcal{X})} \leq \epsilon$, then any near-optimal k-means solution is ϵ-close to the optimal k-means solution. They also give an example that shows that the other direction does not hold.

Main Theorem for k-means The next theorem gives our main result for the k-means problem. Here the distance measure is square of the Euclidean distance.

Theorem 1 (Main Theorem). *Let $0 < \epsilon, \alpha \leq 1$. Let $\mathcal{X} \in \mathbb{R}^d$ be a dataset that satisfies the $(1+\alpha, \epsilon)$-approximation stability and each optimal cluster has size at least $(60\epsilon n/\alpha^2)$. Then the sampling algorithm **SampAlg** gives an 8-approximation to the k-means objective with probability $\Omega(1/k)$.*

When $\alpha > 1$, we get the following result.

Theorem 2 (Main Theorem, large α). *Let $0 < \epsilon \leq 1$ and $\alpha > 1$. Let $\mathcal{X} \in \mathbb{R}^d$ be a dataset that satisfies the $(1+\alpha, \epsilon)$-approximation stability and each optimal cluster has size at least $70\epsilon n$. Then the sampling algorithm **SampAlg** gives an 8-approximation to the k-means objective with probability $\Omega(1/k)$.*

Generalization to k-median w.r.t. Approximate Metrics. The above result can be generalized for the k-median problem with respect to distance measures that are approximately metric. This means that the distance measure D satisfies the following two properties:

Definition 2 (γ-approximate symmetry). *Let $0 < \gamma \leq 1$. Let $\mathcal{X}$ be some data domain and D be a distance measure with respect to $\mathcal{X}$. D is said to satisfy the γ-approximate symmetry property if the following holds:*

$$\forall x, y \in \mathcal{X}, \gamma \cdot D(y, x) \leq D(x, y) \leq (1/\gamma) \cdot D(y, x). \tag{1}$$

Definition 3 (δ-approximate triangle inequality). *Let $0 < \delta \leq 1$. Let $\mathcal{X}$ be some data domain and D be a distance measure with respect to $\mathcal{X}$. D is said to satisfy the δ-approximate triangle inequality if the following holds:*

$$\forall x, y, z \in \mathcal{X}, D(x, z) \leq (1/\delta) \cdot (D(x, y) + D(y, z)). \tag{2}$$

Here is our main theorem for the general k-median problem.

Theorem 3 (k-median). *Let $0 < \epsilon, \gamma, \delta, \alpha \leq 1$. Consider the k-median problem with respect to a distance measure that satisfies γ-symmetry and δ-approximate triangle inequality. Let $\mathcal{X} \in \mathbb{R}^d$ be a dataset that satisfies the $(1 + \alpha, \epsilon)$-approximation stability and each optimal cluster has size at least $(20\epsilon n/\delta^2\alpha^2)$. Then the sampling algorithm **SampAlg** gives an $\frac{8}{(\gamma\delta)^2}$-approximation to the k-median objective with probability $\Omega(1/k)$.*

When $\alpha > 1$, we get the following result.

Theorem 4 (k-median, large α). *Let $0 < \epsilon, \gamma, \delta \leq 1$ and $\alpha > 1$. Consider the k-median problem with respect to a distance measure that satisfies γ-symmetry and δ-approximate triangle inequality. Let $\mathcal{X} \in \mathbb{R}^d$ be a dataset that satisfies the $(1 + \alpha, \epsilon)$-approximation stability and each optimal cluster has size at least $(20\epsilon n/\delta^2)$. Then the sampling algorithm **SampAlg** gives an $\frac{8}{(\gamma\delta)^2}$-approximation to the k-median objective with probability $\Omega(1/k)$.*

Negative result for small clusters The above two Theorems show that the sampling algorithm behaves well when the data satisfies the Approximation-stability property and the optimal clusters are large. This leaves open the question as to what happens when the clusters are small. The next Theorem shows a negative result if the clusters are small. We show that if the clusters are small, then in the worst case, **SampAlg** gives $O(\log k)$ approximation with probability exponentially small in k.

Theorem 5. *Let $0 < \epsilon, \alpha \leq 1$. Consider the k-means problem. There exists a dataset $\mathcal{X} \in \mathbb{R}^d$ such that the following holds:*

- *$\mathcal{X}$ satisfies the $(1 + \alpha, \epsilon)$ approximation stability property, and*
- ***SampAlg** achieves an approximation factor of $\left(\frac{1}{2} \cdot \log k\right)$ with probability at most $e^{-\sqrt{k} - o(1)}$.*

2   Proof of Theorems 1 and 2

We follow the framework of Jaiswal and Garg [JG12]. We denote the dataset by
$\mathcal{X} = \{x_1, ..., x_n\} \in \mathbb{R}^d$. Let $C_1^*, ..., C_k^*$ denote the optimal k clusters with respect
to the k-means objective function and let $c_1^*, ..., c_k^*$ denote the centroids of these
optimal clusters. We denote the optimal cost with OPT, i.e.,

$$OPT = \sum_{x \in \mathcal{X}} \min_{c \in \{c_1^*, ..., c_k^*\}} D^2(c, x),$$

where $D(.,.)$ denotes the Euclidean distance between any pair of points. For
any point $x \in \mathcal{X}$, we denote the distance of this point to the closest center in
$\{c_1^*, ..., c_k^*\}$ with $w(x)$ and the distance of this point to the second closest center
with $w_2(x)$.

The following Lemma from [BBG09] will be crucial in our analysis.

Lemma 1 (Lemma 4.1 in [BBG09]). *If the dataset satisfies $(1 + \alpha, \epsilon)$-approximation-stability for the k-means objective, then*

(a) If $\forall i, |C_i^*| \geq 2\epsilon n$, then less than ϵn points have $w_2^2(x) - w^2(x) \leq \frac{\alpha \cdot OPT}{\epsilon n}$.

(b) For any $t > 0$, at most $t\epsilon n$ points have $w^2(x) \geq \frac{OPT}{t\epsilon n}$.

Let $c_1, ..., c_i$ denote the centers that are chosen by the first i iterations of
SampAlg and let $j_1, ..., j_i$ denote the indices of the optimal clusters to which
these centers belong, i.e., if $c_p \in C_q^*$, then $j_p = q$. Let $J_i = \{j_1\} \cup ... \cup \{j_i\}$ and let
$\bar{J}_i = \{1, ..., k\} \setminus J_i$. So, J_i denotes the clusters that are *covered* and $\bar{J}_i$ denotes the
clusters that are not covered by the end of the ith iteration. An optimal cluster
being *covered* means that a point has been chosen as a center from the cluster.
Let $\mathcal{X}_i = \cup_{j \in J_i} C_j^*$ and let $\bar{\mathcal{X}}_i = \cup_{j \in \bar{J}_i} C_j^*$.

Let B_1 be the subset of points in $\bar{\mathcal{X}}_i$ such that for any point $x \in B_1$, $w_2^2(x) -$
$w^2(x) \leq \frac{\alpha \cdot OPT}{\epsilon n}$. Let B_2 denote the subset of points in $\bar{\mathcal{X}}_i$ such that for every
point $x \in B_2$, $w^2(x) \geq \frac{\alpha^2 \cdot OPT}{6\epsilon n}$. Note that from Lemma 1, we have that $|B_1| \leq$
ϵn and $|B_2| \leq 6\epsilon n/\alpha^2$. Let $B = B_1 \cup B_2$ and $\bar{B} = \bar{\mathcal{X}}_i \setminus B$. We have $|B| \leq 7\epsilon n/\alpha^2$.

Lemma 2. *Let $\beta = \frac{1 - \alpha/2}{6 + \alpha}$. For any $x \in \bar{B}$ we have, we have $D^2(x, c_t) \geq$
$\beta \cdot D^2(x, c_{j_t}^*)$.*

Proof. Let j be the index of the optimal cluster to which x belongs. Note that
$w^2(x) = D^2(x, c_j^*)$ and $w_2^2(x) \leq D^2(x, c_{j_t}^*)$. Figure 1 shows this arrangement. For
any $x \in \bar{B}$, we have:

$$w_2^2(x) - w^2(x) \geq \frac{\alpha \cdot OPT}{\epsilon n} \geq 6 \cdot w^2(x)/\alpha$$
$$\Rightarrow w_2^2(x) \geq (1 + 6/\alpha) \cdot w^2(x) \qquad (3)$$

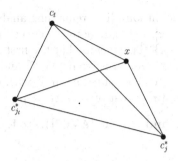

Fig. 1. x belongs to the uncovered cluster j

We will now argue that $D^2(x, c_t) \geq \beta \cdot D^2(x, c^*_{j_t})$. For the sake of contradiction, assume that $D^2(x, c_t) < \beta \cdot D^2(x, c^*_{j_t})$. Then we observe the following inequalities.

$$2 \cdot D^2(x, c^*_j) + 2 \cdot D^2(x, c_t) \geq D^2(c_t, c^*_j) \quad \text{(triangle inequality)}$$

$$\Rightarrow 2 \cdot D^2(x, c^*_j) + 2 \cdot D^2(x, c_t) \geq D^2(c_t, c^*_{j_t}) \quad \text{(since } D^2(c_t, c^*_j) \geq D^2(c_t, c^*_{j_t}))$$

$$\Rightarrow 2 \cdot D^2(x, c^*_j) + 2 \cdot D^2(x, c_t) \geq \frac{1}{2} \cdot D^2(x, c^*_{j_t}) - D^2(x, c_t) \quad \text{(triangle inequality)}$$

$$\Rightarrow 3 \cdot D^2(x, c_t) \geq \frac{1}{2} \cdot D^2(x, c^*_{j_t}) - 2 \cdot D^2(x, c^*_j)$$

$$\Rightarrow 3\beta \cdot D^2(x, c^*_{j_t}) > \frac{1}{2} \cdot D^2(x, c^*_{j_t}) - 2 \cdot D^2(x, c^*_j)$$

$$\text{(using assumption } D^2(x, c_t) < \beta \cdot D^2(x, c^*_{j_t}))$$

$$\Rightarrow D^2(x, c^*_j) > \frac{(1 - 6\beta)}{4} \cdot D^2(x, c^*_{j_t})$$

$$\Rightarrow w^2(x) > \frac{1}{1 + 6/\alpha} \cdot w_2^2(x) \quad \text{(since } D^2(x, c^*_{j_t}) \geq w_2^2(x) \text{ and } \beta = \frac{1-\alpha/2}{6+\alpha})$$

This contradicts with Equation (3). Hence, we get that for any $x \in \bar{B}$ and any $t \in \{1, ..., i\}$, we have $D^2(x, c_t) \geq \beta \cdot D^2(x, c^*_{j_t})$. This proves the Lemma.

Let $W_{min} = \min_{t \in [k]} \left(\sum_{x \in C^*_t, x \in \bar{B}} w_2^2(x) \right)$. Let C_i denote the set of centers $\{c_1, ..., c_i\}$ that are chosen in the first i iterations of **SampAlg**. Let $\mathcal{X}_i = \cup_{t \in J_i} C^*_t$ and $\bar{\mathcal{X}}_i = \mathcal{X} \setminus \mathcal{X}_i$. So, in some sense, $\mathcal{X}_i$ denote the points that are covered by the algorithm after step i and $\bar{\mathcal{X}}_i$ are the uncovered points. For any subset of points $Y \in \mathcal{X}$, $\phi_{C_i}(Y)$ is the cost of the points in Y with respect to the centers C_i, i.e., $\phi_{C_i}(Y) = \sum_{x \in Y} \min_{c \in C_i} D^2(x, c)$. We can now present our next useful lemma which says that the cost of the uncovered points is significant. Note that this implies that the probability of a point being picked from an uncovered clusters in step $(i + 1)$ is significant.

Lemma 3. *Let* $\beta = \frac{1-\alpha/2}{6+\alpha}$. $\phi_{\{c_1, ..., c_i\}}(\bar{\mathcal{X}}_i) \geq \beta \cdot (k - i) \cdot W_{min}$.

Proof. This Lemma follows from the definition of W_{min} and Lemma 2.

We will need a few more definitions. The remaining analysis will be on the lines of a similar analysis in [JG12]. Let E_i denote the event that the set J_i contains i distinct indices from $\{1, ..., k\}$. This means that the first i sampled centers cover i optimal clusters. The next Lemma is from [AV07] and shows that given that event E_i happens, the expected cost of points in $\mathcal{X}_i$ with respect to C_i is at most some constant times the optimal cost of $\mathcal{X}_i$ with respect to $\{c_1^*, ..., c_k^*\}$.

Lemma 4 (Lemma 3.1 and 3.2 in [AV07]). $\forall i, \mathbf{E}[\phi_{\{c_1,...,c_i\}}(\mathcal{X}_i)|E_i] \leq 4 \cdot \phi_{\{c_1^*,...,c_k^*\}}(\mathcal{X}_i)$.

The next Lemma (this is Lemma 4 in [JG12]) shows that the probability that **SampAlg** returns a good solution depends on the probability of the event E_k, i.e., the event that all the clusters get covered.

Lemma 5. $\Pr\left[\phi_{\{c_1,...,c_k\}}(\mathcal{X}) \leq 8 \cdot \phi_{\{c_1^*,...,c_k^*\}}(\mathcal{X})\right] \geq (1/2) \cdot \Pr[E_k]$

Proof. From the previous Lemma, we know that $\mathbf{E}[\phi_{\{c_1,...,c_k\}}(\mathcal{X})|E_k] \leq 4 \cdot \phi_{\{c_1^*,...,c_k^*\}}(\mathcal{X})$. Using Markov, we get that $\Pr[\phi_{\{c_1,...,c_k\}}(\mathcal{X}) > 8 \cdot \phi_{\{c_1^*,...,c_k^*\}}(\mathcal{X})|E_k] \leq 1/2$. Removing the conditioning on E_k, we get the desired Lemma.

We will now argue in the remaining discussion that $\Pr[E_k] \geq 1/k$. This follows from the next Lemma that shows that $\Pr[E_{i+1}|E_i] \geq \frac{k-i}{k-i+1}$.

Lemma 6. $\Pr[E_{i+1}|E_i] \geq \frac{k-i}{k-i+1}$.

Proof. $\mathbf{Pr}[E_{i+1} \mid E_i]$ is just the conditional probability that the $(i+1)^{th}$ center is chosen from the set $\bar{\mathcal{X}}_i$ given that the first i centers are chosen from i different optimal clusters. This probability can be expressed as

$$\mathbf{Pr}[E_{i+1} \mid E_i] = \mathbf{E}\left[\frac{\phi_{\{c_1,...,c_i\}}(\bar{\mathcal{X}}_i)}{\phi_{\{c_1,...,c_i\}}(\mathcal{X})} \mid E_i\right] \tag{4}$$

For the sake of contradiction, let us assume that

$$\mathbf{E}\left[\frac{\phi_{\{c_1,...,c_i\}}(\bar{\mathcal{X}}_i)}{\phi_{\{c_1,...,c_i\}}(\mathcal{X})} \mid E_i\right] = \mathbf{Pr}[E_{i+1} \mid E_i] < \frac{k-i}{k-i+1} \tag{5}$$

Applying Jensen's inequality, we get the following:

$$\frac{1}{\mathbf{E}\left[\frac{\phi_{\{c_1,...,c_i\}}(\mathcal{X})}{\phi_{\{c_1,...,c_i\}}(\bar{\mathcal{X}}_i)} \mid E_i\right]} \leq \mathbf{E}\left[\frac{\phi_{\{c_1,...,c_i\}}(\bar{\mathcal{X}}_i)}{\phi_{\{c_1,...,c_i\}}(\mathcal{X})} \mid E_i\right] < \frac{k-i}{k-i+1}$$

This gives the following:

$$1 + \frac{1}{k-i} < \mathbf{E}\left[\frac{\phi_{\{c_1,...,c_i\}}(\mathcal{X})}{\phi_{\{c_1,...,c_i\}}(\bar{\mathcal{X}}_i)} \mid E_i\right]$$

$$= \mathbf{E}\left[\frac{\phi_{\{c_1,...,c_i\}}(\mathcal{X}_i) + \phi_{\{c_1,...,c_i\}}(\bar{\mathcal{X}}_i)}{\phi_{\{c_1,...,c_i\}}(\bar{\mathcal{X}}_i)} \mid E_i\right]$$

$$= 1 + \mathbf{E}\left[\frac{\phi_{\{c_1,\dots,c_i\}}(\mathcal{X}_i)}{\phi_{\{c_1,\dots,c_i\}}(\bar{\mathcal{X}_i})} \mid E_i\right]$$

$$\Rightarrow \frac{1}{k-i} \leq \mathbf{E}\left[\frac{\phi_{\{c_1,\dots,c_i\}}(\mathcal{X}_i)}{\beta \cdot (k-i) \cdot W_{min}} \mid E_i\right] \quad \text{(using Lemma 3)}$$

$$\leq \frac{\mathbf{E}[\phi_{\{c_1,\dots,c_i\}}(X_i) \mid E_i]}{\beta \cdot (k-i) \cdot W_{min}}$$

$$\leq \frac{4 \cdot \phi_{\{c_1^*,\dots,c_k^*\}}(\mathcal{X})}{\beta \cdot (k-i) \cdot W_{min}} \quad \text{(using Lemma 4)}$$

$$\Rightarrow \frac{W_{min}}{\text{OPT}} \leq \frac{4}{\beta} = 4 \cdot \frac{6+\alpha}{1-\alpha/2} \tag{6}$$

The above gives us an upper bound on W_{min}. Next, we get a lower bound on W_{min} that contradicts with the above bound. Let j be the index of the optimal cluster such that $\sum_{x \in C_j^*, x \in \bar{B}} w_2^2(x)$ is minimized. Note that $W_{min} = \sum_{x \in C_j^*, x \in \bar{B}} w_2^2(x)$. We note that for any $x \notin B_1$, we have $w_2^2(x) - w^2(x) \geq \frac{\alpha \cdot \text{OPT}}{\epsilon n}$. This gives us the following:

$$\forall x \notin B_1, x \in C_j^*, w_2^2(x) \geq \frac{\alpha \cdot \text{OPT}}{\epsilon n}$$

$$\Rightarrow W_{min} = \sum_{x \in C_j^*, x \in \bar{B}} w_2^2(x) \geq \frac{\alpha \cdot \text{OPT}}{\epsilon n} \cdot \frac{52\epsilon n}{\alpha^2} = \frac{52}{\alpha} \cdot \text{OPT} \tag{7}$$

The above being true since all clusters are of size at least $\frac{60\epsilon n}{\alpha^2}$. Note that this contradicts with equation (6) since $\alpha \leq 1$.

This concludes the proof of Theorem 1.

Proof (Proof of Theorem 2). We run through the same proof as discussed above with the following quantities redefined as follows: Let B_1 be the subset of points in $\bar{\mathcal{X}_i}$ such that for any point $x \in B_1$, $w_2^2(x) - w^2(x) \leq \frac{\alpha \cdot \text{OPT}}{\epsilon n}$. Let B_2 denote the subset of points in $\bar{\mathcal{X}_i}$ such that for every point $x \in B_2$, $w^2(x) \geq \frac{\text{OPT}}{6\epsilon n}$. Note that from Lemma 1, we have that $|B_1| \leq \epsilon n$ and $|B_2| \leq 6\epsilon n$. Let $B = B_1 \cup B_2$ and $\bar{B} = \bar{\mathcal{X}_i} \setminus B$. We have $|B| \leq 7\epsilon n$. Now, we note that Lemma 3 works for $\beta = \frac{\alpha - 1/2}{6+\alpha}$. This changes equation (6) as follows:

$$\frac{W_{min}}{\text{OPT}} \leq \frac{4}{\beta} = 4 \cdot \frac{6+\alpha}{\alpha - 1/2} \tag{8}$$

Furthermore, equation (7) gets modified to the following:

$$W_{min} = \sum_{x \in C_j^*, x \in \bar{B}} w_2^2(x) \geq \frac{\alpha \cdot \text{OPT}}{\epsilon n} \cdot (56\epsilon n) = 56\alpha \cdot \text{OPT} \tag{9}$$

The above being true since all clusters are of size at least $70\epsilon n$. Note that this contradicts with equation (8) since $\alpha > 1$.

3   Small Cluster

In the previous section, we saw a positive result on datasets that have large optimal clusters. In this section, we show that if the dataset have optimal clusters that are small in size, then **SampAlg** may have a bad behavior. More formally, we will prove Theorem 5 in this Section. We will need the following result from [BR11] for proving this Theorem.

Theorem 6 (Theorem 1 from [BR11]). *Let* $r : \mathbb{N} \to \mathbb{R}$ *be a real function. If* $r(k) = \delta^* \log k$ *for a fixed real* $\delta^* \in (0, 2/3)$, *then there is a class of instances on which* **SampAlg** *achieves an* $r(k)$-*approximation with probability at most* $e^{1-(3/2)\delta^* - o(1)}$.

Let $\mathcal{X}_{BR}$ denote the dataset on which **SampAlg** gives an approximation factor of $((1/3) \log k')$ with probability at most $e^{-\sqrt{k'} - o(1)}$ when solving the k'-means problem. We will construct another dataset using $\mathcal{X}_{BR}$ and show that **SampAlg** behaves poorly on this dataset. We will need the following fact from [BR11] for our analysis:

Fact 1 ([BR11]). $OPT(k', \mathcal{X}_{BR}) = \frac{k'(k'-1)}{2}$.

Consider the dataset $\mathcal{X} = \mathcal{X}_{far} \cup \mathcal{X}_{BR}$ where $\mathcal{X}_{far}$ has the following properties:

1. $\mathcal{X}_{BR} \cap \mathcal{X}_{far} = \phi$,
2. $|\mathcal{X}_{far}| = |\mathcal{X}_{BR}| \cdot \left(\frac{1}{\epsilon} - 1\right)$.
3. All points in $\mathcal{X}_{far}$ are located at a point c such that the distance of every point $x \in \mathcal{X}_{BR}$ from c is at least $4 \cdot \sqrt{\frac{(1+\alpha)(k-1)(k-2)}{2 \cdot |\mathcal{X}_{far}|}}$.

We solve the k-means problem for $k = k'+1$ on the dataset $\mathcal{X}$ that has $n = \frac{|\mathcal{X}_{BR}|}{\epsilon}$ points. Note that the size of the smallest optimal cluster for this dataset is of size $\epsilon n/k$. We first observe cost of the optimal solution of $\mathcal{X}$.

Lemma 7. $OPT(k, \mathcal{X}) = k'(k'-1)/2$.

Proof. This is simple using the Fact 1.

We now show that $\mathcal{X}$ has the $(1 + \alpha, \epsilon)$-approximation stability property.

Lemma 8. $\mathcal{X}$ *satisfies the* $(1 + \alpha, \epsilon)$-*approximation stability property.*

Proof. Consider any $(1+\alpha)$-approximate solution for the dataset $\mathcal{X}$. Let $c_1, ..., c_k$ be the centers with respect to this approximate solution. We have $\phi_{\{c_1,...,c_k\}}(\mathcal{X}) \leq (1+\alpha) \cdot (k-1)(k-2)/2$. Consider the center in $\{c_1, ..., c_k\}$ that is closest to the point c. Let this center be c_j. Then we note that:

$$D^2(c, c_j) \leq \frac{(1+\alpha)(k-1)(k-2)}{2 \cdot |\mathcal{X}_{far}|}$$

Since the distance of every point in $\mathcal{X}_{BR}$ from point c is at least $4 \cdot \sqrt{\frac{(1+\alpha)(k-1)(k-2)}{2 \cdot |\mathcal{X}_{far}|}}$, we get that all points in $\mathcal{X}_{far}$ are correctly classified. Furthermore, since the number of points in $\mathcal{X}_{BR}$ is at most ϵ fraction of total points, we get that the total number of mis-classified points cannot be more than ϵn and hence the data $\mathcal{X}$ satisfies the $(1 + \alpha, \epsilon)$ approximation stability property.

4   Proof of Theorems 3 and 4

Consider the k-median problem with respect to a distance measure $D(.,.)$ that satisfies the γ-symmetry and δ-approximate triangle inequality. The following Lemma is a generalized version of the Lemma in [BBG09] for any given distance measure. The proof remains the same as the proof of Lemma 3.1 in [BBG09].

Lemma 9 (Generalization of Lemma 3.1 in [BBG09]). *If the dataset satisfies* $(1 + \epsilon, \alpha)$-*approximation-stability for the k-median objective, then*

(a) If $\forall i, |C_i^*| \geq 2\epsilon n$, *then less than* ϵn *points have* $w_2(x) - w(x) \leq \frac{\alpha \cdot OPT}{\epsilon n}$.
(b) For any $t > 0$, *at most* $t\epsilon n$ *points have* $w(x) \geq \frac{OPT}{t\epsilon n}$.

where $w(x)$ *denotes the distance of the point* x *to the closest optimal center as per the distance measure* D *and* $w_2(x)$ *is the distance to the second closest center.*

We now prove a generalized version of Lemma 2 for distance measures that satisfy γ-symmetry and δ-approximate triangle inequality. We can redefine some of the previous quantities for this case. Let B_1 be the subset of points in $\bar{\mathcal{X}}_i$ such that for any point $x \in B_1$, $w_2(x) - w(x) \leq \frac{\alpha \cdot OPT}{\epsilon n}$. Let B_2 denote the subset of points in $\bar{\mathcal{X}}_i$ such that for every point $x \in B_2$, $w(x) \geq \frac{\delta^2 \alpha^2 \cdot OPT}{\epsilon n}$. Note that from Lemma 9, we have that $|B_1| \leq \epsilon n$ and $|B_2| \leq \frac{\epsilon n}{\delta^2 \alpha^2}$. Let $B = B_1 \cup B_2$ and we have $|B| \leq \frac{2\epsilon n}{\delta^2 \alpha^2}$. Let $\bar{B} = \bar{\mathcal{X}}_i$.

Lemma 10. *Let* $\beta = \frac{\delta^2 + \frac{1}{\alpha} - 1}{(1 + \frac{1}{\delta^2 \alpha})(1 + \delta)}$. *For any* $x \in \bar{B}$, *we have* $D(x, c_t) \geq \beta \cdot D(x, c_{j_t}^*)$.

Proof. Consider any point $x \in \bar{B}$. Let $x \in C_j^*$. In other words, j is the index of the optimal cluster to which x belongs. Note that $w(x) = D(x, c_j^*)$ and $w_2(x) \leq D(x, c_{j_t}^*)$. Please refer Figure 1 that shows this arrangement. For any $x \in \bar{B}$, we have:

$$w_2(x) - w(x) \geq \frac{\alpha \cdot OPT}{\epsilon n} \geq \frac{1}{\delta^2 \alpha} \cdot w(x)$$

$$\Rightarrow w_2(x) \geq \left(1 + \frac{1}{\delta^2 \alpha}\right) \cdot w(x) \tag{10}$$

We will now argue that $D(x, c_t) \geq \beta \cdot D(x, c_{j_t}^*)$. For the sake of contradiction, assume that $D(x, c_t) < \beta \cdot D(x, c_{j_t}^*)$. Then we observe the following inequalities.

$$D(x, c_t) + D(x, c_j^*) \geq \delta \cdot D(c_t, c_j^*)$$
$$(\delta\text{-approximate triangle inequality})$$
$$\Rightarrow D(x, c_t) + D(x, c_j^*) \geq \delta \cdot D(c_t, c_{j_t}^*)$$
$$(\text{since } D(c_t, c_j^*) \geq D(c_t, c_{j_t}^*))$$
$$\Rightarrow D(x, c_t) + D(x, c_j^*) \geq \delta \cdot (\delta \cdot D(x, c_{j_t}^*) - D(x, c_t))$$
$$(\delta\text{-approximate triangle inequality})$$

$$\Rightarrow (1+\delta) \cdot D(x, c_t) \geq \delta^2 \cdot D(x, c_{j_t}^*) - D(x, c_j^*)$$
$$\Rightarrow (1+\delta) \cdot \beta \cdot D(x, c_{j_t}^*) > \delta^2 \cdot D(x, c_{j_t}^*) - D(x, c_j^*)$$
$$\text{(using assumption } D(x, c_t) < \beta \cdot D(x, c_{j_t}^*))$$
$$\Rightarrow D(x, c_j^*) > (\delta^2 - \beta(1+\delta)) \cdot D(x, c_{j_t}^*)$$
$$\Rightarrow w(x) > \frac{1}{\left(1 + \frac{1}{\delta^2 \alpha}\right)} \cdot w_2(x)$$
$$\left(\text{since } D(x, c_{j_t}^*) \geq w_2(x) \text{ and } \beta = \frac{\delta^2 + \frac{1}{\alpha} - 1}{(1 + \frac{1}{\delta^2 \alpha})(1+\delta)}\right)$$

This contradicts with Equation (10). Hence, we get that for any $x \in \bar{B}$ and any $t \in \{1, ..., i\}$, we have $D(x, c_t) \geq \beta \cdot D(x, c_{j_t}^*)$. This proves the Lemma.

The rest of the proof remains the same as that for the k-means problem of the previous section. The main difference that arises due to the generalization is that instead of using Lemma 4 we will have to use the following generalized version. This is Lemma 3 in [JG12].

Lemma 11. $\forall i, \mathbb{E}[\phi_{\{c_1,...,c_i\}}(\mathcal{X}_i)|E_i] \leq \frac{4}{(\gamma\delta)^2} \cdot \phi_{\{c_1^*,...,c_k^*\}}(\mathcal{X}_i).$

So the approximation factor changes from 8 to $8/(\gamma\delta)^2$ due to this generalization. Finally, equation (6) changes as follows:

$$\frac{W_{min}}{\text{OPT}} \leq \frac{4}{\beta} = 4 \cdot \frac{(1 + \frac{1}{\delta^2\alpha})(1+\delta)}{\delta^2 + \frac{1}{\alpha} - 1} \tag{11}$$

Furthermore, equation (7) gets modified to the following:

$$W_{min} = \sum_{x \in C_j^*, x \in \bar{B}} w_2^2(x) \geq \frac{\alpha \cdot \text{OPT}}{\epsilon n} \cdot \frac{18\epsilon n}{\delta^2 \alpha^2} = \frac{18}{\delta^2 \alpha} \cdot \text{OPT} \tag{12}$$

The above being true since all clusters are of size at least $\frac{20\epsilon n}{\delta^2 \alpha^2}$. Note that this contradicts with equation (11) since $\alpha \leq 1$.

Proof (Proof of Theorem 4). We run through the same proof as discussed above with the following quantities redefined as follows: Let B_1 be the subset of points in $\bar{\mathcal{X}}_i$ such that for any point $x \in B_1$, $w_2^2(x) - w^2(x) \leq \frac{\alpha \cdot \text{OPT}}{\epsilon n}$. Let B_2 denote the subset of points in $\bar{\mathcal{X}}_i$ such that for every point $x \in B_2$, $w^2(x) \geq \frac{\delta^2 \cdot \text{OPT}}{\epsilon n}$. Note that from Lemma 1, we have that $|B_1| \leq \epsilon n$ and $|B_2| \leq \epsilon n/\delta^2$. Let $B = B_1 \cup B_2$ and $\bar{B} = \bar{\mathcal{X}}_i$. We have $|B| \leq 2\epsilon n/\delta^2$. Now, we note that Lemma 3 works for $\beta = \frac{\delta^2 + \alpha - 1}{(1 + \alpha/\delta^2)(1+\delta)}$. This changes equation (6) as follows:

$$\frac{W_{min}}{\text{OPT}} \leq \frac{4}{\beta} = 4 \cdot \frac{(1 + \alpha/\delta^2)(1+\delta)}{\delta^2 + \alpha - 1} \tag{13}$$

Furthermore, equation (7) gets modified to the following:

$$W_{min} = \sum_{x \in C_j^*, x \in \bar{B}} w_2^2(x) \geq \frac{\alpha \cdot \text{OPT}}{\epsilon n} \cdot (18\epsilon n/\delta^2) = \frac{18\alpha}{\delta^2} \cdot \text{OPT} \tag{14}$$

The above being true since all clusters are of size at least $20\epsilon n/\delta^2$. Note that this contradicts with equation (13) since $\alpha > 1$.

Acknowledgements. Ragesh Jaiswal would like to thank the anonymous referee of [JG12] for initiating the questions discussed in this paper.

References

[ADK09]   Aggarwal, A., Deshpande, A., Kannan, R.: Adaptive sampling for k-means clustering. In: Dinur, I., Jansen, K., Naor, J., Rolim, J. (eds.) APPROX and RANDOM 2009. LNCS, vol. 5687, pp. 15–28. Springer, Heidelberg (2009)

[AJM09]   Ailon, N., Jaiswal, R., Monteleoni, C.: Streaming k-means approximation. In: Advances in Neural Information Processing Systems (NIPS 2009), pp. 10–18 (2009)

[AV07]   Arthur, D., Vassilvitskii, S.: k-means++: the advantages of careful seeding. In: In Proceedings of the 18th Annual ACM-SIAM Symposium on Discrete Algorithms (SODA 2007), pp. 1027–1035 (2007)

[ABS10]   Awasthi, P., Blum, A., Sheffet, O.: Stability yields a PTAS for k-median and k-means clustering. In: Proceedings of the 51st Annual IEEE Symposium on Foundations of Computer Science (FOCS 2010), pp. 309–318 (2010)

[BBG09]   Balcan, M.-F., Blum, A., Gupta, A.: Approximate clustering without the approximation. In: Proceedings of the 20th Annual ACM-SIAM Symposium on Discrete Algorithms (SODA 2009), pp. 1068–1077 (2009)

[BR11]   Brunsch, T., Röglin, H.: A bad instance for k-means++. In: Ogihara, M., Tarui, J. (eds.) TAMC 2011. LNCS, vol. 6648, pp. 344–352. Springer, Heidelberg (2011)

[JG12]   Jaiswal, R., Garg, N.: Analysis of k-means++ for separable data. In: Proceedings of the 16th International Workshop on Randomization and Computation, pp. 591–602 (2012)

[JKS12]   Jaiswal, R., Kumar, A., Sen, S.: A Simple D^2-sampling based PTAS for k-means and other Clustering Problems. In: Gudmundsson, J., Mestre, J., Viglas, T. (eds.) COCOON 2012. LNCS, vol. 7434, pp. 13–24. Springer, Heidelberg (2012)

[ORSS06] Ostrovsky, R., Rabani, Y., Schulman, L.J., Swamy, C.: The effectiveness of lloyd-type methods for the k-means problem. In: Proceedings of the 47th IEEE FOCS, pp. 165–176 (2006)

An Exact Algorithm for TSP in Degree-3 Graphs via Circuit Procedure and Amortization on Connectivity Structure*

Mingyu Xiao[1] and Hiroshi Nagamochi[2]

[1] School of Computer Science and Engineering,
University of Electronic Science and Technology of China, China
myxiao@gmail.com
[2] Department of Applied Mathematics and Physics, Graduate School of Informatics,
Kyoto University, Japan
nag@amp.i.kyoto-u.ac.jp

Abstract. The paper presents an $O^*(1.2312^n)$-time and polynomial-space algorithm for the traveling salesman problem in an n-vertex graph with maximum degree 3. This improves the previous time bound for this problem. Our algorithm is a simple branch-and-search algorithm. The only branch rule is designed on a cut-circuit structure of a graph induced by unprocessed edges. To improve a time bound by a simple analysis on measure and conquer, we introduce an amortization scheme over the cut-circuit structure by defining the measure of an instance to be the sum of not only weights of vertices but also weights of connected components of the induced graph.

Keywords: Traveling Salesman Problem, Exact Exponential Algorithms, Cubic Graphs, Connectivity, Measure and Conquer.

1 Introduction

The traveling salesman problem (TSP) is one of the most famous and intensively studied problems in computational mathematics. Many algorithmic methods have been investigated to beat this challenge of finding the shortest route visiting each member of a collection of n locations and returning to the starting point. The first $O^*(2^n)$-time dynamic programming algorithm for TSP is back to early 1960s. However, in the last half of a century no one can break the barrier of 2 in the base of the running time. To make steps toward the long-standing and major open problem in exact exponential algorithms, TSP in special classes of graphs, especially degree bounded graphs, has also been intensively studied. Eppstein [5] showed that TSP in degree-3 graphs (a graph with maximum degree i is called a degree-i graph) can be solved in $O^*(1.260^n)$ time

---

* Supported by National Natural Science Foundation of China under the Grant 60903007 and Fundamental Research Funds for the Central Universities under the Grant ZYGX2012J069.

T.-H.H. Chan, L.C. Lau, and L. Trevisan (Eds.): TAMC 2013, LNCS 7876, pp. 96–107, 2013.

and polynomial space, and TSP in degree-4 graphs can be solved in $O^*(1.890^n)$ time and polynomial space. Iwama and Nakashima [9] refined Eppstein's algorithm for degree-3 graphs by showing that the worst case in Eppstein's algorithm will not always happen and claimed an improved bound $O^*(1.251^n)$. Gebauer [8] designed an $O^*(1.733^n)$-time exponential-space algorithm for TSP in degree-4 graphs, which is improved to $O^*(1.716^n)$ time and polynomial space by Xiao and Nagamochi [13]. Bjorklund et al. [2] also showed TSP in degree bounded graph can be solved in $O^*((2-\varepsilon)^n)$ time, where $\varepsilon > 0$ depends on the degree bound only. There is a Monte Carlo algorithm to decide whether a graph is Hamiltonian or not in $O^*(1.657^n)$ time [1]. For planar TSP and Euclidean TSP, there are sub-exponential algorithms based on small separators [3].

In this paper, we present an improved deterministic algorithm for TSP in degree-3 graphs, which runs in $O^*(2^{\frac{3}{10}n}) = O^*(1.2312^n)$ time and polynomial space. The algorithm is simple and contains only one branch rule that is designed on a cut-circuit structure of a graph induced by unprocessed edges. We will apply the measure and conquer method to analyze the running time. Note that our algorithm for TSP in degree-4 graphs in [13] is obtained by successfully applying the measure and conquer method to TSP for the first time. However, direct application of measure and conquer to TSP in degree-3 graphs may only lead to an $O^*(1.260^n)$-time algorithm. To effectively analyze our algorithm, we use an amortization scheme over the cut-circuit structures by setting weights to both vertices and connected components of the induced graph.

Due to the limited space, some proofs of lemmas are not included in the extended abstract. Readers are referred to [14] for a full version of this paper.

2   Preliminaries

In this paper, a graph $G = (V, E)$ stands for an undirected edge-weighted graph with maximum degree 3, which possibly has multiple edges, but no self-loops. For a subset $V' \subseteq V$ of vertices and a subset $E' \subseteq E$ of edges, the subgraphs induced by V' and E' are denoted by $G[V']$ and $G[E']$ respectively. We also use cost(E') to denote the total weight of edges in E'. For any graph G', the sets of vertices and edges in G' are denoted as $V(G')$ and $E(G')$ respectively. Two vertices in a graph are *k-edge-connected* if there are k-edge-disjoint paths between them. A graph is *k-edge-connected* if every pair of vertices in it are k-edge-connected. Given a graph with an edge weight, the *traveling salesman problem* (TSP) is to find a Hamiltonian cycle of minimum total edge weight.

Forced TSP. We introduce the *forced traveling salesman problem* as follows. An instance is a pair (G, F) of a graph $G = (V, E)$ and a subset $F \subseteq E$ of edges, called *forced edges*. A Hamiltonian cycle of G is called a *tour* if it passes though all the forced edges in F. The objective of the problem is to compute a tour of minimum weight in the given instance (G, F). An instance is called *infeasible* if no tour exists. A vertex is called *forced* if there is a forced edge incident on it. For convenience, we say that the *sign* of an edge e is 1 if e is a forced edge and 0 if e is an unforced edge. We use sign(e) to denote the sign of e.

U-graphs and U-components. We consider an instance (G, F). Let $U = E(G) - F$ denote the set of unforced edges. A subgraph H of G is called a *U-graph* if H is a trivial graph or H is induced by a subset $U' \subseteq U$ of unforced edges (i.e., $H = G[U']$). A maximal connected *U*-graph is called a *U-component*. Note that each connected component in the graph $(V(G), U)$ is a *U*-component.

For a vertex subset X (or a subgraph X) of G, let cut(X) denote the set of edges in $E = F \cup U$ that join a vertex in X and a vertex not in X, and denote cut$_F(X) = \text{cut}(X) \cap F$ and cut$_U(X) = \text{cut}(X) \cap U$. Edge set cut$(X)$ is also called a *cut* of the graph. We say that an edge is *incident* on X if the edge is in cut(X). The *degree* $d(v)$ of a vertex v is defined to be $|\text{cut}(\{v\})|$. We also denote $d_F(v) = |\text{cut}_F(\{v\})|$ and $d_U(v) = |\text{cut}_U(\{v\})|$. A *U*-graph H is *k-pendent* if $|\text{cut}_U(H)| = k$. A *U*-graph H is called *even* (resp., *odd*) if $|\text{cut}_F(H)|$ is even (resp., odd). A *U*-component is 0-pendent.

For simplicity, we may regard a maximal path of forced edges between two vertices u and v as a single forced edge uv in an instance (G, F), since we can assume that $d_F(v) = 2$ always implies $d(v) = 2$ for any vertex v.

An *extension* of a 6-cycle is obtained from a 6-cycle $v_1v_2v_3v_4v_5v_6$ and a 2-clique ab by joining them with two independent edges av_i and bv_j $(i \neq j)$. An extension of a 6-cycle always has exactly eight vertices. Fig. 1(a) and (b) illustrate two examples of extensions of a 6-cycle. A chord of an extension of a 6-cycle is an edge joining two vertices in it but different from the eight edges $v_1v_2, v_2v_3, v_3v_4, v_4v_5, v_5v_6, v_6v_1, av_i, bv_j$ and ab. A subgraph H of a *U*-component in an instance (G, F) is *k-pendent critical*, if it is a 6-cycle or an extension of a 6-cycle with $|\text{cut}_U(H)| = k$ and $|\text{cut}_F(H)| = 6 - k$ (i.e., H has no chord of unforced/forced edge). A 0-pendent critical *U*-component is also simply called a *critical graph* or *critical U-component*.

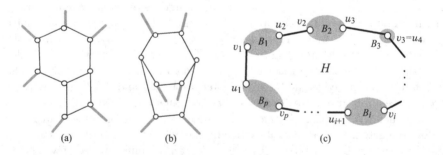

(a)   (b)   (c)

Fig. 1. (a), (b) Extensions of a 6-cycle; (c) A circuit in a 2-edge-connected graph H

Circuits and Blocks. We consider a nontrivial 2-edge-connected *U*-component H in an instance (G, F). A *circuit* C in H is a maximal sequence $e_1, e_2, \ldots, e_p$ of edges $e_i = u_iv_i \in E(H)$ $(1 \leq i \leq p)$ such that for each $e_i \in C$ $(i \neq p)$, the next edge $e_{i+1} \in C$ is given by a subgraph B_i of H such that cut$_U(B_i) = \{e_i, e_{i+1}\}$. Note that cut$_U(B_p) = \{e_p, e_1\}$. See Fig. 1(c) for an illustration of a circuit C.

We say that each subgraph B_i is a *block* along C and vertices v_i and u_{i+1} are the *endpoints* of block B_i. By the maximality of C, we know that any two vertices in each block B_i are 2-edge-connected in the induced subgraph $G[B_i]$. It is possible that a circuit in a 2-edge-connected graph H may contain only one edge $e = u_1v_1$. For this case, vertices u_1 and v_1 are connected by three edge-disjoint paths in H and the circuit is called *trivial*, where the unique block is the U-component H. Each nontrivial circuit contains at least two blocks, each of which is a 2-pendent subgraph of H. In our algorithm, we will consider only nontrivial circuits C. A block B_i is called *trivial* if $|V(B_i)| = 1$ and $d_F(v) = 1$ for the only vertex v in it (v is of degree 3 in G). A block B_i is called *reducible* if $|V(B_i)| = 1$ and $d_F(v) = 0$ for the only vertex v in it (v is of degree 2 in G). A block B_i with $V(B_i) = \{v_i = u_{i+1}\}$ is either trivial or reducible in a 2-edge-connected graph.

For convenience, we call a maximal sequence $P = \{e_1, e_2, \ldots, e_p\}$ of edges $e_i = u_i u_{i+1} \in E(H)$ ($1 \le i \le p-1$) a *chain* if all vertices u_j ($j = 2, 3, \ldots, p-1$) are forced vertices. In the definition of chains, we allow $u_1 = u_p$. Observe that each chain is contained in the same circuit.

We easily observe the following properties on circuits and blocks (e.g., see [10] and [11]).

Lemma 1. *Each edge in a 2-edge-connected U-component H of a degree-3 graph is contained in exactly one circuit. A partition of $E(H)$ into circuits can be obtained in polynomial time.*

Lemma 2. *An instance (G, F) is infeasible if G is not 2-edge-connected or it violates the parity condition: (i) every U-component is even; and (ii) the number of odd blocks along every circuit is even.*

On the other hand, a polynomially solvable case is found by Eppstein [5].

Lemma 3. [5] *A minimum cost tour of an instance (G, F) such that every U-component is trivial or a component of a 4-cycle can be found in polynomial time.*

3   Branch-and-search Algorithms

Our algorithm is a branch-and-search algorithm: we search the solution by iteratively branching on the current instance to generate several smaller instances until the current instance becomes polynomially solvable. In this paradigm, we will get a search tree whose root and leaves represent an input instance and polynomially solvable instances, respectively. The size of the search tree is the exponential part of the running time of the algorithm. For a measure μ of the instance, let $C(\mu)$ denote the maximum number of leaves in the search tree generated by the algorithm for any instance with measure μ. When we branch on an instance (G, F) with k branches such that the i-th branch decreases the measure μ of (G, F) by at least a_i, we obtain the following recurrence $C(\mu) \le C(\mu - a_1) + C(\mu - a_2) + \cdots + C(\mu - a_k)$. Solving this recurrence, we get $C(\mu) = [\alpha(a_1, a_2, \ldots, a_k)]^\mu$, where $\alpha(a_1, a_2, \ldots, a_k)$ is the largest root of the function $f(x) = 1 - \sum_{i=1}^{k} x^{-a_i}$. We need to find out the worst recurrence in the algorithm to evaluate the size of the search tree. In this paper, we represent the above recurrence by a vector $(a_1; a_2; \cdots; a_k)$ of measure decreases, called a *branch vector* (cf. [7]). In particular, when $a_i = a_{i+1} = \cdots = a_j$ for some $i \le j$, it may be written as $(a_1; a_2; \cdots a_{i-1}; [a_i]_{j-i+1}; a_{j+1}; \cdots; a_k)$, and a vector $([a]_k)$ is simply written as $[a]_k$. When we compare two branch vectors $\mathbf{b} = (a_1; a_2)$ $(a_1 \le a_2)$ and $\mathbf{b}' = (a_1'; a_2')$ such that "$a_i \le a_i'$ $(i = 1, 2)$" or "$a_1' = a_1 + \varepsilon$ and $a_2' = a_2 - \varepsilon$ for some $0 \le \varepsilon \le (a_2 - a_1)/2$," we only consider branch vector $\mathbf{b}$ in analysis, since a solution α from $\mathbf{b}$ is not smaller than that from $\mathbf{b}'$ (cf. [7]). We say that $\mathbf{b}$ *covers* $\mathbf{b}'$ in this case.

4   Reduction Operations

A reduction operation reduces an instance into a smaller instance without branching. This section shows our reduction operations designed based on the structures of edge-cuts with size at most 4 in the graph.

The unique unforced edge incident on a 1-pendent U-graph is *eliminable*. For any subgraph H of G with $|\text{cut}(H)| = 2$, we call the unforced edges in $\text{cut}(H)$ *reducible*. Eliminable edges can be deleted from the graph and reducible edges need to be included into F.

We also have the following lemmas to deal with edge-cuts of size 3 and 4.

Lemma 4. *Let (G, F) be an instance where G is a graph with maximum degree 3. For any subgraph X with $|\text{cut}(X)| = 3$, we can replace X with a single vertex x and update the cost on the three edges incident on x preserving the optimality of the instance.*

Similar to Lemma 4, we simplify the following subgraphs X with $|\text{cut}(X)| = 4$. We consider a subgraph X with $|\text{cut}_F(X)| = 4$ and $|\text{cut}_U(X)| = 0$. Denote $\text{cut}(X)$ by $\{y_1x_1, y_2x_2, y_3x_3, y_4x_4\}$ with $x_i \in V(X)$ and $y_i \in V - V(X)$, where $x_i \ne x_j$ $(1 \le i < j \le 4)$. We define I_i $(i = 1, 2, 3)$ to be instances of the problem of finding two disjoint paths P and P' of minimum total cost in X such that all vertices and forced edges in X appear in exactly one of the two paths, and one of the two paths is from x_i to x_4 and the other one is from x_{j_1} to x_{j_2} $(\{j_1, j_2\} = \{1, 2, 3\} - \{i\})$. We say that I_i *infeasible* if it has no solution.

A subgraph X is *4-cut reducible* if $|\text{cut}_F(X)| = 4$, $|\text{cut}_U(X)| = 0$, and at least one of the three problems I_1, I_2 and I_3 defined above is infeasible. We have the following lemma to reduce the 4-cut reducible subgraph.

Lemma 5. *Let (G, F) be an instance where G is a graph with maximum degree 3. A 4-cut reducible subgraph X can be replaced with one of the following subgraphs X' with four vertices and $|\text{cut}_F(X')| = 4$ so that the optimality of the instance is preserved:*

(i) *four single vertices (i.e., there is no solution to this instance);*
(ii) *a pair of forced edges; and*
(iii) *a 4-cycle with four unforced edges.*

Lemma 6. *Let X be an induced subgraph of a degree-3 graph G such that X contains at most eight vertices of degree 3 in G. Then X is 4-cut reducible if $|\mathrm{cut}_F(X)| = 4$, $|\mathrm{cut}_U(X)| = 0$, and X contains at most two unforced vertices.*

Lemma 4 and Lemma 5 imply a way of simplifying some local structures of an instance. However, it is not easy to find solutions to problems I_i in the above two lemmas. In our algorithm, we only do this replacement for X containing no more than 10 vertices and then the corresponding problems I_i can be solved in constant time by a brute force search.

We define the operation of *3/4-cut reduction*: If there a subgraph X of G with $|V(X)| \leq 10$ such that $|\mathrm{cut}(X)| = 3$ or X is 4-cut reducible, then we simplify the graph by replacing X with a graph according to Lemma 4 or Lemma 5. Note that a 3/4-cut can be found in polynomial time if it exists and then this reduction operation can be implemented in polynomial time.

We can also easily reduce all multiple edges. An instance (G, F) is called a *reduced* instance if G is 2-edge-connected, (G, F) satisfies the parity condition, and has none of reducible edges, eliminable edges, multiple edges, 3-cut or 4-cut reducible subgraphs. Note that a reduced instance has no triangle, otherwise there would be a 3-cut reducible subgraph. An instance is called *2-edge-connected* if every U-component in it is 2-edge-connected. The initial instance $(G, F = \emptyset)$ is assumed to be 2-edge-connected, otherwise it is infeasible by Lemma 2. In our algorithm, we will always keep instances 2-edge-connected after applying reduction/branching operations.

5   Algorithms Based on Circuit Procedures

The *circuit procedure* is one of the most important operations in our algorithm. The procedure will determinate each edge in a circuit to be included to F or to be deleted from the graph. It will be widely used as the only branching operation in our algorithm.

5.1   Circuit Procedure

Processing circuits. Determining an unforced edge means either including it to F or deleting it from the graph. When an edge is determined, the other edges in the same circuit containing this edge can also be determined directly by reducing eliminable edges. We call the series of procedures applied to all edges in a circuit together as a *circuit procedure*. Thus, in the circuit procedure, after we start to process a circuit C either by including an edge $e_1 \in C$ to F or by deleting e_1 from the graph, the next edge e_{i+1} of e_i becomes an eliminable edge and we continue to determine e_{i+1} either by deleting it from the graph if block B_i is odd and $e_i = u_i v_i$ is included to F (or B_i is even and e_i is deleted); or by including it to F otherwise. Circuit procedure is a fundamental operation to build up our proposed algorithm. Note that a circuit procedure determines only the edges in the circuit. During the procedure, some unforced edges outside the circuit may become reducible and so on, but we do not determine them in this execution.

Lemma 7. *Let H be a 2-edge-connected U-component in an instance (G, F) and C be a circuit in H. Let (G', F') be the resulting instance after applying circuit procedure on C. Then*

(i) *each block B_i of C becomes a 2-edge-connected U-component in (G', F'); and*
(ii) *any other U-component H' than H in (G, F) remains unchanged in (G', F').*

We call a circuit *reducible* if it contains at least one reducible edge. We can apply the circuit procedure on a reducible circuit directly starting by including a reducible edge to F. In our algorithm, we will deal with reducible edges by processing a reducible circuit. When the instance becomes a reduced instance, we may not be able to reduce the instance directly. Then we search the solution by "branching on a circuit." *Branching on a circuit C at edge $e \in C$* means branching on the current instance to generate two instances by applying the circuit procedure to C after including e to F and deleting e from the graph respectively. Branching on a circuit is the only branching operation used in our algorithm. However, a naive branch-and-search algorithm using circuit procedures yields only an $O^*(1.260^n)$ time algorithm (see [14] for the details).

5.2 The Algorithm

A block is called a *normal block* if it is none of trivial, reducible and 2-pendent critical. A normal block is *minimal* if no subgraph of it is a normal block along any circuit. Note that when F is not empty, each U-component has at least one nontrivial circuit. Our recursive algorithm for forced TSP only consists of the following two main steps:

1. First apply the reduction rules to a given instance until it becomes a reduced one; and
2. Then take any U-component H that is neither trivial nor a 4-cycle (if no such U-component H, then the instance is polynomially solvable by Lemma 3), and branch on a nontrivial circuit C in H, where C is chosen so that
(1) no normal block appears along C (i.e., C has only trivial and 2-pendent critical blocks) if this kind of circuits exist; and
(2) a minimal normal block B_1 in H appears along C otherwise.

It is easy to see that after applying the reduction rules on a 2-edge-connected instance, the resulting instance remains 2-edge-connected. By this observation and Lemma 7, we can guarantee that an input instance is always 2-edge-connected.

6   Analysis

We analyze our algorithm by the measure and conquer method [6]. In the measure and conquer method, a measure μ for instance size should satisfy the *measure condition*: (i) when $\mu \le 0$ the instance can be solved in polynomial time; (ii) the measure w will never increase in each operation in the algorithm; and (iii) the measure will decrease in each of the subinstances generated by applying a branching rule. We introduce *vertex-weight* as follows. For each vertex v, we set its vertex-weight $w(v)$ to be

$$w(v) = \begin{cases} w_3 = 1 & \text{if } d_U(v) = 3 \\ w_{3'} & \text{if } d_U(v) = 2 \text{ and } d_F(v) = 1 \\ 0 & \text{otherwise.} \end{cases}$$

We will determine the best value of $w_{3'}$ such that the worst recurrence in our algorithm is best. Let $\Delta_3 = w_3 - w_{3'}$. For a subset of vertices (or a subgraph) X, we also use $w(X)$ to denote the total vertex-weight in X.

6.1   Amortization on Connectivity Structures

By using the above vertex weight setting, we cannot improve the running time bound of the naive algorithm easily. But we can also decrease the number of U-components by one in the bottleneck cases. This observation suggests us a new idea of an amortization scheme over the cut-circuit structure by setting a weight on each U-component in the graph. We also set a weight (which is possibly negative, and bounded from above by a constant $c \geq 0$) to each U-component. Let μ be the sum of all vertex weight and U-component weight. We will use μ to measure the size of the search tree generated by our algorithm. The measure μ will also satisfy the measure condition. Initially there is only one U-component and $\mu < n+c$ holds, which yields a time bound of $O^*(\alpha^\mu) = O^*(\alpha^{n+c}) = O^*(\alpha^n)$ for the maximum branch factor α.

A simple idea is to set the same weight to each nontrivial U-component. It is possible to improve the previous best result by using this simple idea. However, to get further improvement, in this paper, we set several different component-weights. Our purpose is to distinguish some "bad" U-components, which will be characterized as "critical" U-components. Branching on a critical U-component may lead to a bottleneck recurrence in our algorithm. So we set a different component-weight to this kind of components to get improvement. For each U-component H, we set its component-weight $c(H)$ to be

$$c(H) = \begin{cases} 0 & \text{if } H \text{ is trivial} \\ -4w_{3'} & \text{if } H \text{ is a 4-cycle} \\ \gamma & \text{if } H \text{ is a critical } U\text{-component} \\ \delta & \text{otherwise,} \end{cases}$$

where we set $c(H) = -4w_{3'}$ so that $c(H) + w(H) = 0$ holds for every 4-cycle U-component H.

We also require that the vertex-weight and component-weight satisfy the following requirements

$$2\Delta_3 \geq \gamma \geq \delta \geq \Delta_3 \geq \frac{1}{2}w_3, \ w_{3'} \geq \frac{1}{5}w_3 \text{ and } \gamma - \delta \leq w_{3'}. \tag{1}$$

6.2   Decrease of Measure after Reduction Operations

We show that the measure will not increase after applying any reduction operation in an 2-edge-connected instance. Since an input instance is 2-edge-connected,
there is no eliminable edge. In fact, we always deal with eliminable edges in circuit procedures. For reducible edges, we deal with them during a process of a
reducible circuit (including the reducible edges to F and dealing with the resulting eliminable edges). We will show that μ never increases after processing a
circuit. The measure μ will not increase after deleting any unforced parallel edge.
The following lemma also shows that reducing a 3/4-cut reducible subgraph does
not increase μ.

Lemma 8. *For a given instance,*
(i) *reducing a 3-cut reducible subgraph does not increase the measure μ; and*
(ii) *reducing a 4-cut reducible subgraph X decreases the measure μ by $w(X) + c(X)$.*

6.3   Decrease of Measure after Circuit Procedures

Next we consider how much amount of measure decreases by processing a circuit.
We consider that the measure μ becomes zero whenever we find an instance
infeasible by Lemma 2. After processing a circuit $\mathcal{C} = \{e_i = u_i v_i \mid 1 \leq i \leq p\}$ in
a U-component H, each block B_i along $\mathcal{C}$ becomes a new U-component, which
we denote by $\bar{B}_i$. We define the *direct benefit* $\beta'(B_i)$ from B_i to be the decrease
in vertex-weight of the endpoints v_i and u_{i+1} of B_i minus the component-weight
$c(\bar{B}_i)$ in the new instance after the circuit procedure. Immediately after the
procedure, the measure μ decreases by $w(H) + c(H) - \sum_i (w(\bar{B}_i) + c(\bar{B}_i)) =
c(H) + \sum_i \beta'(B_i)$. After the circuit procedure, we see that the vertex-weights
of endpoints of each non-reducible and nontrivial block B_i decreases by Δ_3 and
Δ_3 (or w_3 and w_3) respectively if B_i is even, and by Δ_3 and w_3 (or w_3 and
Δ_3) respectively if B_i is odd. Summarizing these, the direct benefit $\beta'(B)$ from
a block B is given by

$$\beta'(B) = \begin{cases} 0 & \text{if } B \text{ is reducible,} \\ w_{3'} & \text{if } B \text{ is trivial,} \\ w_3 + \Delta_3 - \delta & \text{if } B \text{ is odd and nontrivial,} \\ 2w_3 - \delta & \text{if } B \text{ is even and non-reducible, and } \text{cut}_U(B) \text{ is deleted,} \\ 2\Delta_3 - \gamma & \text{if } B \text{ is 2-pendent critical, and } \text{cut}_U(B) \text{ is included in } F, \\ w(B) & \text{if } B \text{ is a 2-pendent 4-cycle, and } \text{cut}_U(B) \text{ is included in } F, \\ 2\Delta_3 - \delta & \text{otherwise (i.e., } B \text{ is even, non-reducible but not} \\ & \quad \text{a 2-pendent critical } U\text{-graph or a 2-pendent 4-cycle,} \\ & \quad \text{and } \text{cut}_U(B) \text{ is included to } F). \end{cases}$$
(2)

By (1) and (2), we have that $\beta'(B_i) \geq 0$ for any type of block B_i, which implies
that the decrease $c(H) + \sum_i \beta'(B_i) \geq c(H) \geq 0$ (where H is not a 4-cycle) is in
fact nonnegative, i.e., the measure μ never increases by processing a circuit.

After processing a circuit $\mathcal{C}$, a reduction operation may be applicable to some U-components $\bar{B}_i$ and we can decrease μ more by reducing them. The *indirect benefit* $\beta''(B)$ from a block B is defined as the amount of μ decreased by applying reduction rules on the U-component $\bar{B}$ after processing the circuit. Since we have shown that μ never increases by applying reduction rules, we know that $\beta''(B)$ is always nonnegative. The *total benefit* (*benefit*, for short) from a block B is $\beta(B) = \beta'(B) + \beta''(B)$.

Lemma 9. *After processing a circuit $\mathcal{C}$ in a 2-edge-connected U-component H (not necessary being reduced) and applying reduction rules until the instance becomes a reduced one, the measure μ decreases by $c(H) + \sum_i \beta(B_i)$, where B_i are the blocks along circuit $\mathcal{C}$.*

The indirect benefit from a block depends on the structure of the block. In our algorithm, we hope that the indirect benefit is as large as possible. Here we prove some lower bounds on it for some special cases.

Lemma 10. *Let H be a U-component containing no induced triangle and $\mathcal{C}'$ be a reducible circuit in it such that there is exactly one reducible block along $\mathcal{C}'$. The measure μ decreases by at least $2\Delta_3$ by processing the reducible circuit $\mathcal{C}'$ and applying reduction rules.*

Lemma 11. *In the circuit procedure for a circuit $\mathcal{C}$ in a reduced instance, the indirect benefit from a block B along $\mathcal{C}$ satisfies*

$$\beta''(B) \geq \begin{cases} 2\Delta_3 & \text{if } B \text{ is odd and nontrivial,} & \text{(i)} \\ w(B) - \beta'(B) & \text{if } B \text{ is a 2-pendent cycle or critical graph,} \\ & \text{and } \mathrm{cut}_U(B) \text{ is deleted,} & \text{(ii)} \\ \delta & \text{if } B \text{ is even but not reducible or a 2-pendent} \\ & \text{cycle, and } \mathrm{cut}_U(B) \text{ is deleted,} & \text{(iii)} \\ 0 & \text{otherwise.} & \text{(iv)} \end{cases}$$

6.4 Branch Vectors of Branching on Circuits

In the algorithm, branching on a circuit generates two instances (G_1, F_1) and (G_2, F_2). By Lemma 9, we get branch vector

$$(c(H) + \sum_i \beta_1(B_i); c(H) + \sum_i \beta_2(B_i)),$$

where $\beta_j(B)$, $\beta_j'(B)$ and $\beta_j''(B)$ denote the functions $\beta(B)$, $\beta'(B)$ and $\beta''(B)$ evaluated in (G_j, F_j), $j = 1, 2$ for clarifying how branch vectors are derived in the subsequent analysis. We have the following branch vectors for two different choices of circuits $\mathcal{C}$ on which our algorithm branches.

Lemma 12. *Assume that a circuit $\mathcal{C}$ in a U-component H (not a 4-cycle) chosen in Step 2 of the algorithm has only trivial and 2-pendent critical blocks. Then we can branch on the circuit $\mathcal{C}$ with one of three branch vectors $[6w_{3'} + \gamma]_2$, $(\gamma + 2w_3 + 6w_{3'}; \delta + 2w_3 - \gamma)$ and $(\delta + 2(2\Delta_3 - \gamma); \delta + 2(2w_3 + 4w_{3'}))$.*

Lemma 13. *Assume that a circuit C in a U-component H (not a 4-cycle) chosen in Step 2 of the algorithm has a minimal normal block B_1. Then we can branch on the circuit C with one of eight branch vectors*

$[4w_3 - 2w_{3'}]_2$, $(2w_3; 6w_3 - 2w_{3'})$, $(4\Delta_3 - \delta; 4w_3 + 4\Delta_3)$, $(4\Delta_3 - \gamma; 8w_3)$,

$(\delta + 6w_3 - 2w_{3'} - 2\gamma; \delta + 6w_3 + 4w_{3'} - \gamma; \delta + 4w_3 + 6w_{3'})$,

$(\delta + 4w_3 + 2w_{3'} - \gamma; \delta + 4w_3 + 8w_{3'}; \delta + 2w_3 + 4w_{3'})$,

$(\delta + 8w_3 - 6w_{3'} - 3\gamma; \delta + 8w_3 + 6w_{3'} - \gamma; \delta + 4w_3 + 4w_{3'})$ *and*

$(\delta + 6w_3 - 2w_{3'} - 2\gamma; \delta + 6w_3 + 10w_{3'}; \delta + 2w_3 + 3w_{3'})$.

A quasiconvex program is obtained from (1) and the 13 branch vectors from Lemma 12 and Lemma 13 in our analysis. There is a general method to solve quasiconvex programs [4]. We look at branch vectors $[6w_{3'} + \gamma]_2$ in Lemma 12, and $[4w_3 - 2w_{3'}]_2$ in Lemma 13. Note that $\min\{6w_{3'} + \gamma, 4w_3 - 2w_{3'}\}$ under the constraint $2\Delta_3 \geq \gamma$ gets the maximum value at the time when $6w_{3'} + \gamma = 4w_3 - 2w_{3'}$ and $2\Delta_3 = \gamma$. We get $w_{3'} = \frac{1}{3}$ and $\gamma = \frac{4}{3}$. With this setting and $\delta \in [1.2584, 1.2832]$, all other branch vectors will not be the bottleneck in our quasiconvex program. This gives a time bound $O^*(\alpha^\mu)$ with $\alpha = 2^{\frac{3}{10}} < 1.2312$.

Theorem 1. *TSP in an n-vertex graph G with maximum degree 3 can be solved in $O^*(1.2312^n)$ time and polynomial space.*

The bottlenecks in the analysis are attained by branch vectors $[6w_{3'} + \gamma]_2$ in Lemma 12, $[4w_3 - 2w_{3'}]_2$ in Lemma 13 and $2\Delta_3 \geq \gamma$ in (1).

7   Concluding Remarks

In this paper, we have presented an improved exact algorithm for TSP in degree-3 graphs. The basic operation in the algorithm is to process the edges in a circuit by either including an edge in the circuit to the solution or excluding it from the solution. The algorithm is analyzed by using the measure and conquer method and an amortization scheme over the cut-circuit structure of graphs, wherein we introduce not only weights of vertices but also weights of U-components to define the measure of an instance.

The idea of amortization schemes introducing weights on components may yield better bounds for other exact algorithms for graph problems if how reduction/branching procedures change the system of components is successfully analyzed.

References

1. Bjorklund, A.: Determinant sums for undirected Hamiltonicity. In: Proc. 51st Annual IEEE Symp. on Foundations of Computer Science, pp. 173–182 (2010)
2. Bjorklund, A., Husfeldt, T., Kaski, P., Koivisto, M.: The travelling salesman problem in bounded degree graphs. ACM Transactions on Algorithms 8(2), 18 (2012)
3. Dorn, F., Penninkx, E., Bodlaender, H.L., Fomin, F.V.: Efficient exact algorithms on planar graphs: Exploiting sphere cut decompositions. Algorithmica 58(3), 790–810 (2010)

4. Eppstein, D.: Quasiconvex analysis of multivariate recurrence equations for backtracking algorithms. ACM Trans. on Algorithms 2(4), 492–509 (2006)
5. Eppstein, D.: The traveling salesman problem for cubic graphs. J. Graph Algorithms and Applications 11(1), 61–81 (2007)
6. Fomin, F.V., Grandoni, F., Kratsch, D.: Measure and conquer: Domination – a case study. In: Caires, L., Italiano, G.F., Monteiro, L., Palamidessi, C., Yung, M. (eds.) ICALP 2005. LNCS, vol. 3580, pp. 191–203. Springer, Heidelberg (2005)
7. Fomin, F.V., Kratsch, D.: Exact Exponential Algorithms. Springer (2010)
8. Gebauer, H.: Finding and enumerating Hamilton cycles in 4-regular graphs. Theoretical Computer Science 412(35), 4579–4591 (2011)
9. Iwama, K., Nakashima, T.: An improved exact algorithm for cubic graph TSP. In: Lin, G. (ed.) COCOON 2007. LNCS, vol. 4598, pp. 108–117. Springer, Heidelberg (2007)
10. Nagamochi, H., Ibaraki, T.: A linear time algorithm for computing 3-edge-connected components in multigraphs. J. of Japan Society for Industrial and Applied Mathematics 9(2), 163–180 (1992)
11. Nagamochi, H., Ibaraki, T.: Algorithmic Aspects of Graph Connectivities, Encyclopedia of Mathematics and Its Applications. Cambridge University Press (2008)
12. Woeginger, G.J.: Exact algorithms for NP-hard problems: A survey. In: Jünger, M., Reinelt, G., Rinaldi, G. (eds.) Combinatorial Optimization - Eureka, You Shrink! LNCS, vol. 2570, pp. 185–207. Springer, Heidelberg (2003)
13. Xiao, M., Nagamochi, H.: An improved exact algorithm for TSP in degree-4 graphs. In: Gudmundsson, J., Mestre, J., Viglas, T. (eds.) COCOON 2012. LNCS, vol. 7434, pp. 74–85. Springer, Heidelberg (2012)
14. Xiao, M., Nagamochi, H.: An exact algorithm for TSP in degree-3 graphs via circuit procedure and amortization on connectivity structure. CoRR abs/1212.6831 (2012)

Non-crossing Connectors in the Plane*

Jan Kratochvíl and Torsten Ueckerdt

Department of Applied Mathematics, Faculty of Mathematics and Physics,
Charles University in Prague, Czech Republic
{honza,torsten}@kam.mff.cuni.cz

Abstract. We consider the non-crossing connectors problem, which is stated as follows: Given n regions $R_1, \ldots, R_n$ in the plane and finite point sets $P_i \subset R_i$ for $i = 1, \ldots, n$, are there non-crossing connectors γ_i for (R_i, P_i), i.e., arc-connected sets γ_i with $P_i \subset \gamma_i \subset R_i$ for every $i = 1, \ldots, n$, such that $\gamma_i \cap \gamma_j = \emptyset$ for all $i \neq j$?

We prove that non-crossing connectors do always exist if the regions form a collection of pseudo-disks, i.e., the boundaries of every pair of regions intersect at most twice. We provide a simple polynomial-time algorithm if each region is the convex hull of the corresponding point set, or if all regions are axis-aligned rectangles. We prove that the general problem is NP-hard, even if the regions are convex, the boundaries of every pair of regions intersect at most four times and P_i consists of only two points on the boundary of R_i for $i = 1, \ldots, n$.

Finally, we prove that the non-crossing connectors problem lies in NP, i.e., is NP-complete, by a reduction to a non-trivial problem, and that there indeed are problem instances in which every solution has exponential complexity, even when all regions are convex pseudo-disks.

1 Introduction

Connecting points in a non-crossing way is one of the most basic algorithmic problems in discrete mathematics. It has been considered in various settings with most diverse motivations. For example, connecting vertices in a graph via vertex-disjoint or edge-disjoint paths is a fundamental problem in graph theory, the latter being one of Karp's original NP-complete problems [8]. Both problems remain NP-complete even when restricted to planar graphs [15]. Besides having numerous applications, e.g., in network routing and VLSI design, the disjoint path problem in planar graphs also plays a key-role in various theoretical contexts, such as in the seminal work of Robertson and Seymour on the graph minor project [18].

The *homotopic routing problem*, sometimes called river routing, is the most important geometric version of the disjoint path problem. Here a set of points in the plane have to be connected via non-crossing continuous curves that have the same homotopy type as a set of given curves, called the sketch. The sketch also

---

* Research was supported by GraDR EUROGIGA project No. GIG/11/E023. The authors thank Maria Saumell, Stefan Felsner and Irina Mustata for fruitful discussions.

T.-H.H. Chan, L.C. Lau, and L. Trevisan (Eds.): TAMC 2013, LNCS 7876, pp. 108–120, 2013.

prescribes the position of the points as well as some obstacles, and the computed curves have to be non-crossing and disjoint from the obstacles. For example, in the field of map schematization one draws highly simplified metro maps or road networks in such a way that the position of all important points (metro stations, cities, landmarks, . . .) is fixed and the curves (metro lines, roads, rivers, . . .) are drawn with the same homotopy type as given by geographic data [24]. Another application of the homotopic routing problem comes from VLSI layout, where modules (predesigned circuit components) have to be interconnected correctly, satisfying certain design rules [16,2,19]. If the sketch is already non-crossing, then homotopic shortest paths can be computed efficiently [4]. Furthermore, the disjoint path problem in planar graphs can be solved via the homotopic routing problem in case the end points of all paths lie on a bounded number of faces [17]. We refer to the survey of Schrijver for more on homotopic routing [21].

In most of the above applications the homotopy type is fixed to ensure that the final curves are "close" to the sketch. However, there is no ad-hoc guarantee on the distance between a computed curve and the curve in the sketch. In this paper, we pursue a different approach by demanding that each computed curve is contained in a prescribed region. Given n finite sets of points and a region for each point set, the task is to connect all points in each set by a curve completely contained in the corresponding region, such that no two curves intersect.

This problem, called the *non-crossing connectors problem*, has been posed in the field of imprecise points, where one is given a set $P_v \subset \mathbb{R}^2$ for each vertex v of a planar graph G and wants to find a plane embedding of G such that each vertex v lies in its set P_v. Deciding whether such an embedding with straight edges exists is known to be NP-complete for cycles, even if all P_v are vertical line segments or all P_v are disks [14]. If the graph is a matching, NP-completeness has been shown if $|P_v| \leq 3$ [1], or P_v is a vertical line segment of unit length [23]. The latter case remains NP-complete if every edge $\{u, v\}$ is allowed to be a monotone curve within the convex hull of $P_u \cup P_v$ [22]. At the Eurogiga GraDR Kick-off meeting in Prague July 7–8, 2011 Bettina Speckman has asked for an efficient algorithm to decide whether pairs of unit cubes can be connected within their convex hull by non-crossing arbitrary continuous curves. In this paper we investigate general simply-connected regions for the curves, under the assumption that the position of points is already fixed.

The non-crossing connectors problem is also closely related to the *clustered planarity problem*, c-planarity problem for short [6]. Consider a clustered graph, i.e., a planar graph G together with a set of subsets of vertices, called clusters. Any two clusters are either disjoint or one contains the other. The c-planarity problem asks to embed G in a planar way and identify a simply-connected region for each cluster containing precisely the vertices of that cluster and whose boundaries are disjoint. Moreover, every edge shall intersect a region boundary at most once. The complexity of the c-planarity problem is open, even when the embedding of G is fixed [3]. In this case it remains to identify the region for each cluster, or equivalently finding a continuous curve that connects exactly the vertices of that cluster and is disjoint from the interior of every edge. Moreover, these curves shall be non-crossing [7]. This is exactly the non-crossing connectors problem. Hence the c-planarity problem with fixed embedding, whose complexity status is open, can be reduced to the non-crossing connectors problem.

A more entertaining motivation for non-crossing connectors can be found in recreational mathematics. Sam Loyds Cyclopedia of 5000 puzzles, tricks and conundrums published in 1914 refers to a puzzle of connecting houses to gates by non-crossing paths. This is nothing else than an instance of the non-crossing connectors problem. And so is also the recently booming Lab Mice Puzzle game.[1]

Our Results. We consider the non-crossing connectors problem, that is we want to connect several point sets via non-crossing continuous curves, each contained in a corresponding region. W.l.o.g. the curves can be thought of as polylines with a finite number of bends. Given an N-element point set P, each curves γ_i is asked to go through (connect) a fixed subset P_i of P of two or more points. Any two such curves, called *connectors*, shall be non-crossing, i.e., have empty intersection. In particular, w.l.o.g. P is partitioned into subsets $P_1, \ldots, P_n$ and we ask for a set of n non-crossing connectors $\gamma_1, \ldots, \gamma_n$ with $P_i \subset \gamma_i$ for $i = 1, \ldots, n$. It is easily seen that the order in which γ_i visits the points P_i may be fixed arbitrarily. Indeed, we could embed any planar graph G_i with $|P_i|$ vertices and curved edges onto P_i, even while prescribing the position of every vertex in G_i.

Non-crossing connectors as described above do always exist. But the situation gets non-trivial if we fix subsets $R_1, \ldots, R_n$ in the plane, called *regions*, and impose $\gamma_i \subset R_i$ for every $i = 1, \ldots, n$.

In Section 3 we prove that non-crossing connectors do always exist if the given regions form a collection of pseudo-disks, i.e., the boundaries of every pair of regions intersect at most twice. In Section 4 we show that deciding whether or not non-crossing connectors exist is polynomial time solvable if the regions are axis-aligned rectangles or every region is the convex hull of the corresponding point set, while in Section 5 we prove that the problem is NP-complete, even if the regions are convex, the boundaries of every pair of regions intersect at most four times, and $|P_i| = 2$ for every $i = 1, \ldots, n$. We start with some notation in Section 2 and show that the non-crossing connectors problem is in NP.

Some proofs are omitted or only sketched here. All proofs in full detail can be found in the forthcoming journal version of the paper, and on arxiv [12].

2   The Non-crossing Connectors Problem

The non-crossing connectors problem is formally defined as follows.

NON-CROSSING CONNECTORS
Given: Collection $R_1, \ldots, R_n$ of subsets of the plane and a finite point set $P_i \subset R_i$, for $i = 1, \ldots, n$ with $P_i \cap P_j = \emptyset$ for $i \neq j$.
Question: Is there a collection $\gamma_1, \ldots, \gamma_n$ of curves, such that $P_i \subset \gamma_i \subset R_i$ for $i = 1, \ldots, n$ and $\gamma_i \cap \gamma_j = \emptyset$ for $i \neq j$?

---

[1] Both the examples have been pointed out by Marcus Schaefer.

Throughout this paper we consider simply-connected (hence path-connected) closed regions which are equal to the closure of their interior only. Then the boundary of every region R_i is a simple closed curve, denoted by ∂R_i. We assume here and for the rest of the paper that $\partial R_i \cap \partial R_j$ is a finite point set. We may think of $\bigcup_{i=1}^{n} \partial R_i$ as an embedded planar graph $G = (V, E)$ with vertex set $V = \{p \in \mathbb{R}^2 \mid p \in \partial R_i \cap \partial R_j, i \neq j\}$ and edge set $E = \{e \subset \mathbb{R}^2 \mid e$ is a connected component of $\bigcup \partial R_i \setminus V\}$. A point $p \in \partial R_i \cap \partial R_j$ is either a *crossing point* or a *touching point*, depending on whether the cyclic order of edges in ∂R_i and ∂R_j around p is alternating or not. We allow points in P_i to lie on some boundary ∂R_j, although such points can always be moved off the boundary without affecting the existence of non- crossing connectors. We say that two regions R_i, R_j are *k-intersecting* for $k \geq 0$ if $|\partial R_i \cap \partial R_j| \leq k$ and all these points are crossing points, i.e., w.l.o.g. k is even. A set $R_1, \ldots, R_n$ of regions is *k-intersecting* if this is the case for any two of them. For example, $R_1, \ldots, R_n$ are 0-intersecting if and only if they form a nesting family, i.e., $R_i \cap R_j \in \{\emptyset, R_i, R_j\}$ for $i \neq j$.

Regions $R_1, \ldots, R_n$ are a called a *collection of pseudo-disks* if they are 2-intersecting. Two pseudo-disks may also have one touching point. However, this can be locally modified into two crossing points without affecting the existence of non-crossing connectors. Pseudo-disks for example include homothetic copies of a fixed convex point set, but they are not convex in general. A collection of axis-aligned rectangles is always 4-intersecting, but not necessarily 2-intersecting. Finally, a family of convex polygons with at most k corners each is $2k$-intersecting.

Consider a set of non-crossing connectors to a given set of point sets and regions, i.e., a particular solution of a particular instance of the non-crossing connectors problem. Connectors must not cross each other, but we do not bound the number of intersections between a connector and the region boundaries. One may ask whether there is always a solution (if any) in which the total number of such intersections is bounded by a polynomial in the number of crossings between region boundaries. In particular, if the instance is feasible, is there a solution whose complexity is polynomial in the size of the input? We answer this question in the negative even when regions are convex and pseudo-disks and point sets are of size two only. The idea is based on a construction from [11].

Theorem 1. *For every positive integer n, there exists a collection of $n+1$ convex pseudo-disks $R_0, R_1, \ldots, R_n$ and pairs of points $P_i = \{A_i, B_i\} \subset R_i$ such that in every solution to the non-crossing connectors problem the connecting curve $A_n B_n$ crosses the boundary of R_0 in at least 2^{n-1} crossing points.*

Proof. Figure 1 depicts the example for $n = 4$. The regions can easily be deformed to be convex. We refer to the full paper for a complete proof.   □

Theorem 1 shows that the most natural guess-and-verify-a-solution approach requires exponential time and fails to prove NP-membership for the non-crossing connectors problem. However, the problem does belong to NP.

Proposition 1. NON-CROSSING CONNECTORS *is in NP.*

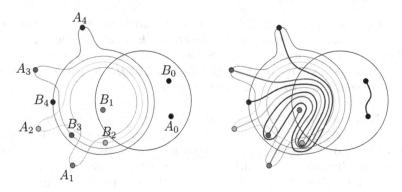

Fig. 1. A schematic illustration of an instance of the non-crossing connectors problem in which every solution has exponential complexity

Proof. We reduce our problem to WEAK REALIZABILITY OF ABSTRACT TOPO-LOGICAL GRAPHS. Abstract topological graphs (AT-graphs, for short) have been introduced in [10] as triples (V, E, R) where (V, E) is a graph and R is a set of pairs of edges of E. The AT-graph (V, E, R) is weakly realizable if (V, E) has a drawing (not necessarily non-crossing) in the plane such that $ef \in R$ whenever the edges e, f cross in the drawing. WEAK REALIZABILITY OF AT-GRAPHS is NP-complete. The NP-hardness was shown in [9], and the NP-membership was shown relatively recently in [20].

We assume the input of the problem is described as a plane graph G (the boundaries of the regions) with the incidence structure of the points of P to the faces of the graph. (W.l.o.g. no point in P lies on the boundary of any region.) We refer to Fig. 2 for an illustrative example.

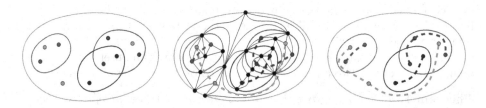

Fig. 2. From NON-CROSSING CONNECTORS to WEAK REALIZABILITY OF AT-GRAPHS

Add vertices and edges to $G \cup P$ to create a planar vertex 3-connected super-graph G'' of G. Such a graph G'' has a topologically unique non-crossing drawing in the plane (up to the choice of the outer face and its orientation) which fixes the location of the points of P inside their regions. To create $\widetilde{G}$, for every i, add *connecting edges* that connect the points of P_i by a path.

Now define an AT-graph with underlying graph $\widetilde{G}$ by allowing the connecting edges of points P_i to cross anything but other connecting edges and the edges resulting from the boundary of R_i. No other edges are allowed to cross, in particular, no edges of G'' may cross each other. It is straightforward to see that a weak realization of this AT-graph is a collection of non-crossing connectors, and vice versa. Thus the NP-membership follows by the result of Schaefer et al. [20] and the fact that our construction of $\widetilde{G}$ is polynomial. $\qquad\square$

3   Pseudo-disks

In this section we prove that non-crossing connectors do always exist if the given regions form a collection of pseudo-disks. We begin with an auxiliary lemma.

Lemma 1. *Let R, R' be pseudo-disks, $p \in \partial R \backslash R'$ be a point, and $\gamma \subset R \cap R'$ a curve that intersects ∂R exactly twice. Then the connected component of $R \backslash \gamma$ not containing p is completely contained in the interior of R'.*

Proof. Let C be the connected component of $R \setminus \gamma$ not containing p. Let q, r be the intersections of γ with ∂R. Then $\partial R \setminus \{p, q, r\}$ is a set of three disjoint curves. The curve between p and q as well as between p and r contains a point in $\partial R'$, since $p \notin R'$ and $q, r \in \gamma \subset R'$. Because R, R' are pseudo-disks the points from $\partial R'$ between p and q and between p and r are the only points in $\partial R \cap \partial R'$ and hence the third curve $\delta = C \cap \partial R$ between q and r is completely contained in R'. Since the closed curve $\delta \cup \gamma \subset R'$ is the boundary of C and $\delta \cap \partial R' = \emptyset$, we conclude that C is completely contained in the interior of R'. $\qquad\square$

Theorem 2. *If $R_1, \ldots, R_n$ is a collection of pseudo-disks, then non-crossing connectors exist for any finite point sets $P_i \subset R_i$ $(i = 1, \ldots, n)$ with $P_i \cap P_j = \emptyset$ for $i \neq j$.*

Proof. The proof is constructive. Let $R_1, \ldots, R_n$ be a collection of pseudo-disks and P_i a finite subset of R_i for $i = 1, \ldots, n$. We assume w.l.o.g. that every ∂R_i is a closed polygonal curve of finite complexity. Moreover, we assume that the regions are labeled from $1, \ldots, n$, such that for every $i = 2, \ldots, n$ the set $\partial R_i \backslash (R_1 \cup \cdots \cup R_{i-1})$ is non-empty, i.e., contains some point p_i. For example, we may order the regions by non-decreasing x-coordinate of their rightmost point. Note that rightmost points of pseudo-disks do not coincide. For simplicity we add p_i to P_i for every $i = 1, \ldots, n$ (and denote the resulting point set again by P_i). Clearly, every collection of non-crossing connectors for the new point sets is good for the original point sets, too.

We start by defining a connector γ_1 for (R_1, P_1) arbitrarily, such that $P_1 \subset \gamma_1 \subset R_1$, $\gamma_1 \cap P_i = \emptyset$ for every $i \geq 2$, and γ_1 is a polyline of finite complexity. To keep the number of operations in the upcoming construction finite we consider polylines of finite complexity only. That is, whenever we define a curve we mean a polyline of finite complexity even if we do not explicitly say so.

For $i = 2, \ldots, n$ assume that we have non-crossing connectors $\gamma_1, \ldots, \gamma_{i-1}$, such that $(\bigcup_{j<i} \gamma_j) \cap (\bigcup_{k \geq i} P_k) = \emptyset$. We want to define a connector γ_i for (R_i, P_i). The set $R_i \backslash (\bigcup_{j<i} \gamma_j)$ has finitely many connected components $\{C_k\}_{k \in K}$

with $|K| < \infty$. Every point in P_i is contained in exactly one C_k. Let C_0 be the component containing the additional point $p_i \in P_i$. The informal idea is the following. We reroute some of the existing connectors until P_i is completely contained in C_0. Then, we define a connector γ_i for (R_i, P_i) arbitrarily, such that $P_i \subset \gamma_i \subset C_0$, as well as $\gamma_i \cap P_j = \emptyset$ for $j > i$ and $\gamma_i \cap \gamma_j = \emptyset$ for $j < i$. The reader may consider Fig. 3 for an illustration of the upcoming operation. For better readability the parts of connectors are omitted in Fig. 3, which have an endpoint in the interior of R_i. However, those, as well as the point sets P_j with $j > i$, will be circumnavigated by the curve δ.

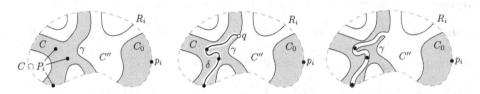

Fig. 3. Rerouting the curve γ bordering the components C and C'', such that the subset of P_i formerly contained in C is contained in C'' afterwards

Every connector γ_j for $j < i$ is a simple curve. Thus every connected component of $R_i \setminus (\bigcup_{j<i} \gamma_j)$ contains a point from the boundary of R_i, i.e., the adjacency graph between the components is a tree T on vertex set $\{C_k\}_{k \in K}$, which we consider to be rooted at C_0. Let $C \neq C_0$ be a component, such that $C \cap P_i \neq \emptyset$ but $C' \cap P_i = \emptyset$ for every descendant C' of C in T. Let γ be the curve in R_i that forms the border between C and its father C'' in T, i.e., γ intersects ∂R_i only at its endpoints and is a subset of some connector γ_{j^*} for (R_{j^*}, P_{j^*}). In particular, $j^* < i$ and hence $p_i \notin R_{j^*}$. Applying Lemma 1 with $p = p_i$ we get that C is contained in the interior of R_{j^*}.

Let $q \notin P_{j^*}$ be any interior point of γ and δ be any curve with endpoint q, such that $(P_i \cap C) \subset \delta \subset (C \cup \{q\}) \subset R_{j^*}$, as well as $\delta \cap (\bigcup_{j<i} \gamma_j) = \{q\}$ and $\delta \cap (\bigcup_{j \neq i} P_j) = \emptyset$. We reroute γ_{j^*} within a small distance around δ. More formally, define a simply-connected set $D \supset \delta$ to be a thickening of the curve δ by some small $\varepsilon > 0$, such that D is still contained in $R_{j^*} \setminus (\bigcup_{j \neq i} P_j \cup \bigcup_{j<i} \gamma_j)$. Note that $D \not\subset C$ if P_i (and hence δ) contains points on the boundary of R_i. However, we can ensure that $D \subset R_{j^*}$ since C lies in the *interior* of R_{j^*}. Moreover, we can choose ε small enough, such that ∂D intersects γ only in two points q_1 and q_2, which are ε-close to q.

Next, the part of γ between q_1 and q_2 is replaced by the part of ∂D between q_1 and q_2 that runs through C. This rerouting of γ (and implicitly the connector γ_{j^*}) may (or may not) change the subtree of T rooted at C''. But it does not affect any component of $R_i \setminus \bigcup_{j<i} \gamma_j$ that is not in this subtree. Moreover, C'' is extended by $D \cap C$, which contains all points in $P_i \cap C$. Hence, the so-to-speak total distance of the points in P_i from C_0 in T is decreased. After finitely many steps we have $P_i \subset C_0$ and thus can define the connector γ_i for (R_i, P_i).   □

4   Polynomially Decidable Cases

We provide an obviously necessary condition for the existence of non-crossing connectors for convex regions. We show that this condition is also necessary when every region is the convex hull of the corresponding point set, as well as when every region is an axis-aligned rectangle. We conclude that in both cases the existence of non-crossing connectors can be tested in polynomial time.

Consider two convex regions, a white region W and a black region B, with non-empty intersection $W \cap B$, which we call the *center* and consider to be gray. The closure of a connected component of the symmetric difference $W \triangle B$ is called a *leaf* and colored white if it is a subset of the white region and black otherwise. Since $\partial W \cap \partial B$ consists of crossing points only, the leaves appear alternatingly in black and white around the gray center. We say that a pair of white leaves and a pair of black leaves form a *cross* if they appear around the center in the cyclic order white–black–white–black. Moreover, a cross is called a *filled cross* if each of the four leaves contains at least one point from the corresponding set.

Observation 1. *Non-crossing connectors do not exist if some pair of regions contain a filled cross. In other words, the absence of filled crosses is a necessary condition for the existence of non-crossing connectors.*

Note that as long as the union and the intersection of any two regions is simply-connected, filled crosses are well-defined and Observation 1 holds.

Proposition 2. *If every region is the convex hull of the corresponding point set, i.e., $R_i = \mathrm{conv}(P_i)$ for all $i = 1, \ldots, n$, then non-crossing connectors exist if and only if the regions form a collection of pseudo-disks.*

Proof. The "if"-part is Theorem 2.

If the regions do not form a collection of pseudo-disks then some pair of regions contains a cross. Since the regions are convex polygons, each leaf of such a cross contains a corner of the corresponding region. From $R_i = \mathrm{conv}(P_i)$ for all $i = 1, \ldots, n$ follows that every corner of R_i is an element of P_i, i.e., every cross is a filled cross. Hence non-crossing connectors do not exist by Observation 1. □

Next assume that every region is an axis-aligned rectangle. We show that then the obviously necessary condition in Observation 1 is also sufficient.

Theorem 3. *A set of axis-aligned rectangles admits a set of non-crossing connectors if and only if it does not contain a filled cross.*

Proof. The "only if"-part is given by Observation 1. We prove the "if"-part by applying Theorem 2. To this end we consider axis-aligned rectangles $R_1, \ldots, R_n$ no two of which form a filled cross. If there is no cross at all, then the rectangles are pseudo-disks and non-crossing connectors exist by Theorem 2. Assume some pair of rectangles is a cross, but *not* a filled cross. W.l.o.g. $\{R_1 = [x_1^1, x_2^1] \times [y_1^1, y_2^1], R_2 = [x_1^2, x_2^2] \times [y_1^2, y_2^2]\}$ is a cross where $R_1 \cap R_2 = [x_1^1, x_2^1] \times [y_1^2, y_2^2]$ is inclusion-minimal among all crosses, and the leaf $C = [x_1^1, x_2^1] \times [y_2^2, y_2^1] \subset R_1$ contains no point from P_1. The situation is illustrated in Fig. 4. Figuratively speaking we chop off C (actually a slight superset C' of C) from R_1 in order to reduce the total number of crosses of all rectangles. More precisely, choose $\varepsilon > 0$ small enough that $C' := [x_1^1, x_2^1] \times [y_2^2 - \varepsilon, y_2^1]$ contains no point from P_1, and that the y-coordinate of no corner of a rectangle $\neq R_2$ lies between $y_2^2 - \varepsilon$ and y_2^2. We replace R_1 by $\tilde{R}_1 := R_1 \setminus C'$.

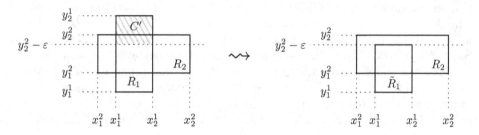

Fig. 4. Chopping off C' with $C' \cap P_1 = \emptyset$ from the rectangle R_1 to obtain $\tilde{R}_1$

From the inclusion-minimality of $R_1 \cap R_2$ follows that every rectangle R_k crosses $\tilde{R}_1$ only if it crosses R_1, too. Hence, the total number of crosses has decreased by at least one and the rectangles $\tilde{R}_1, R_2, \ldots, R_n$ contain no *filled* cross. Repeating the procedure at most $\binom{n}{2}$ times finally results in a collection of axis-aligned rectangles, which are subsets of the original rectangles, and contain no cross at all. By Theorem 2 non-crossing connectors exist for the smaller rectangles, which are good for the original rectangles, too.   □

Corollary 1. *If the regions are n axis-aligned rectangles, or $R_i = \text{conv}(P_i)$ for $i = 1, \ldots, n$, then it can be tested in $\mathcal{O}(n^2)$ whether or not non-crossing connectors exist.*

Proof. By Proposition 2 and Theorem 3 we check whether some pair of regions forms a filled cross. If so the answer is 'No', and if not the answer is 'Yes'.   □

5 NP-Completeness

In this section we prove NP-completeness of the non-crossing connectors problem. By Proposition 1 the problem is in NP. We prove NP-hardness, even if the regions and their point sets are very restricted.

Theorem 4. *The non-crossing connectors problem is NP-complete, even if the regions are 4-intersecting convex polygons with at most 8 corners and for every $i = 1, \ldots, n$ the set P_i consists of only two points on the boundary of R_i.*

The proof of Theorem 4 is rather technical and pretty long. Most of its techni-calities, including the use of zones and segment gadgets, are due to the fact that we use *convex* regions only. Dropping convexity but keeping all the other restric-tions allows for a much shorter and less technical proof. Due to space limitations we present here only the proof sketch with non-convex, but still 4-intersecting, regions.

We do a polynomial reduction from PLANAR 3-SAT, i.e., we are given a formula ψ in conjunctive normal form where each clause has at most 3 literals, that is positive or negated variables. Moreover, the *formula graph* G_ψ, namely the bipartite graph whose vertices are the clauses and variables of ψ, and whose edges are given by $\{x, c\}$ with variable x contained in clause c, is planar. It is known [13] that PLANAR 3-SAT is NP-complete, even if every variable appears in at most 3 clauses, i.e., G_ψ has maximum degree 3 [5].

Clause Gadget. We define the clause gadget, which consists of 5 regions as depicted in Fig. 5. For every clause c we define a black region R_c and a blue region $\bar{R}_c$, which are 4-intersecting. The colors are added just for better readability of the figures. We color the 2-element point set corresponding to every region in the same color as the region. Assume c has size 3. We define 3 pairwise disjoint red regions R_{xc}, one for each variable x in c, such that the regions appear in the same clockwise order as the edges $\{x, c\}$ around c in G_ψ. For every pair $\{x, c\} \in E(G_\psi)$ the red region R_{xc} has one component inside the black region R_c containing both red points, and one outside R_c containing no red point.

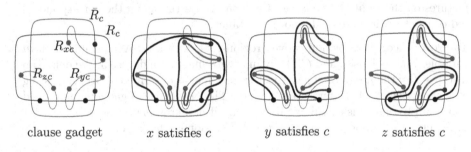

clause gadget x satisfies c y satisfies c z satisfies c

Fig. 5. Clause gadget for clause c consisting of literals x, y, z

If the connector γ_{xc} for R_{xc} is completely contained in R_c, we say that *variable x satisfies clause c*. If the clause has size 2, then only two of the red regions are associated with the variables. Moreover we put the point of the third red region, which is contained in the blue region $\bar{R}_c$, anywhere outside the black region R_c instead of inside R_c. Hence this "artificial variable" can not satisfy the clause.

It is not difficult to see that for any non-crossing connectors of a clause gadget at least one variable satisfies the clause. Moreover, as verified by Fig. 5, non-crossing connectors do exist as soon as one variable satisfies the clause.

Variable Gadget. We assume w.l.o.g. that every variable x appears in three clauses, positive in c_1 and negated in c_2, c_3, or vice versa. Now x is associated with a red region in the clause gadgets for c_1, c_2 and c_3, each containing a part outside the corresponding black region. We bring together these three parts (together with the corresponding black and blue regions) and overlap them as shown in Fig. 6.

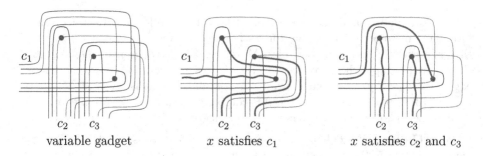

| variable gadget | x satisfies c_1 | x satisfies c_2 and c_3 |

Fig. 6. Variable gadget for variable x contained in clauses c_1, c_2, c_3

Note that if $\gamma_{xc_1} \subset R_{c_1}$, then $\gamma_{xc_i} \not\subset R_{c_i}$, for $i = 2, 3$. In other words, if x satisfies c_1 then x satisfies neither c_2 nor c_3. Similarly, if x satisfies c_2 or c_3 then x does not satisfy c_1. Moreover, Fig. 6 shows that non-crossing connectors do exist as long as x does not satisfy c_1 and c_2 or c_3 at the same time. For better readability blue regions and black and blue connectors are omitted in both pictures on the right. These connectors can always run along the corresponding red connector within a small enough distance.

To prove Theorem 4 for non-convex regions, let ψ be a 3-SAT formula with m variables and n clauses, and G_ψ be planar with maximum degree 3. We define an instance $\mathcal{I}$ of the non-crossing connectors problem as described above, consisting of at most $3(m + n)$ 4-intersecting regions. The properties of gadgets mentioned above imply that ψ is satisfiable if and only if non-crossing connectors exist for $\mathcal{I}$. A detailed argument is given in the full version of the paper.

6   Conclusion

We have shown that there are instances of the non-crossing connectors problem in which every solution is exponentially complex in the input size, even if all regions are convex pseudo-disks, and hence though the non-crossing connectors always exist in such a case, one cannot hope for a polynomial-time construction. But we conjecture that whenever non-crossing connectors exist for axis-aligned rectangles, then there is a solution with polynomially many crossing points between connectors and region boundaries.

References

1. Aloupis, G., Cardinal, J., Collette, S., Demaine, E., Demaine, M., Dulieu, M., Fabila-Monroy, R., Hart, V., Hurtado, F., Langerman, S., Saumell, M., Seara, C., Taslakian, P.: Non-crossing matchings of points with geometric objects. Computational Geometry (2012), http://dx.doi.org/10.1016/j.comgeo.2012.04.005
2. Cole, R., Siegel, A.: River routing every which way, but loose. In: 25th Annual Symposium on Foundations of Computer Science, pp. 65–73 (1984)
3. Di Battista, G., Frati, F.: Efficient c-planarity testing for embedded flat clustered graphs with small faces. In: Hong, S.-H., Nishizeki, T., Quan, W. (eds.) GD 2007. LNCS, vol. 4875, pp. 291–302. Springer, Heidelberg (2008)
4. Efrat, A., Kobourov, S.G., Lubiw, A.: Computing homotopic shortest paths efficiently. In: Möhring, R.H., Raman, R. (eds.) ESA 2002. LNCS, vol. 2461, pp. 411–423. Springer, Heidelberg (2002)
5. Fellows, M.R., Kratochvíl, J., Middendorf, M., Pfeiffer, F.: The complexity of induced minors and related problems. Algorithmica 13, 266–282 (1995)
6. Feng, Q., Cohen, R., Eades, P.: How to draw a planar clustered graph. In: Li, M., Du, D.-Z. (eds.) COCOON 1995. LNCS, vol. 959, pp. 21–30. Springer, Heidelberg (1995)
7. Feng, Q., Cohen, R., Eades, P.: Planarity for clustered graphs. In: Spirakis, P.G. (ed.) ESA 1995. LNCS, vol. 979, pp. 213–226. Springer, Heidelberg (1995)
8. Karp, R.M.: Reducibility among combinatorial problems. Complexity of Computer Computations (1972)
9. Kratochvíl, J.: String graphs II. Recognizing string graphs is NP-hard. Journal of Combinatorial Theory, Series B 52(1), 67–78 (1991)
10. Kratochvíl, J., Lubiw, A., Nešetřil, J.: Noncrossing subgraphs in topological layouts. SIAM Journal on Discrete Mathematics 4(2), 223–244 (1991)
11. Kratochvíl, J., Matoušek, J.: String graphs requiring exponential representations. Journal of Combinatorial Theory, Series B 53(1), 1–4 (1991)
12. Kratochvíl, J., Ueckerdt, T.: Non-crossing connectors in the plane. CoRR, abs/1201.0917 (2012)
13. Lichtenstein, D.: Planar formulae and their uses. SIAM Journal on Computing 11(2), 329–343 (1982)
14. Löffler, M.: Existence of simple tours of imprecise points. In: 23rd European Workshop on Computational Geometry, EuroCG (2007)
15. Lynch, J.F.: The equivalence of theorem proving and the interconnection problem. SIGDA Newsl. 5(3), 31–36 (1975)
16. Pinter, R.Y.: River routing: Methodology and analysis. In: 3rd CalTech Conf. on Very Large-Scale Integration, pp. 141–163 (1983)
17. Robertson, N., Seymour, P.D.: Graph minors. VII. disjoint paths on a surface. J. Comb. Theory Ser. B 45(2), 212–254 (1988)
18. Robertson, N., Seymour, P.D.: Graph minors. XIII: the disjoint paths problem. J. Comb. Theory Ser. B 63(1), 65–110 (1995)
19. Sarrafzadeh, M., Liao, K.F., Wong, C.K.: Single-layer global routing. IEEE Transactions on Computer-Aided Design of Integrated Circuits and Systems 13(1), 38–47 (1994)
20. Schaefer, M., Sedgwick, E., Štefankovič, D.: Recognizing string graphs in NP. Journal of Computer and System Sciences 67(2), 365–380 (2003)
21. Schrijver, A.: Homotopic routing methods, pp. 329–371. Springer, Berlin (1990)
22. Speckmann, B.: Personal communication (2011)

23. Verbeek, K.: Non-crossing paths with fixed endpoints. Master's Thesis, Eindhoven (2008)
24. Verbeek, K.: Homotopic C-oriented routing. In: Proceedings of 28th European Workshop on Computational Geometry, EuroCG 2012, pp. 173–176 (2012)

Minimax Regret 1-Sink Location Problems in Dynamic Path Networks

Siu-Wing Cheng[1], Yuya Higashikawa[2], Naoki Katoh[2,*], Guanqun Ni[3],
Bing Su[4,6], and Yinfeng Xu[3,5,6,**]

[1] Department of Computer Science and Engineering, The Hong Kong University
of Science and Technology, Hong Kong
scheng@cse.ust.hk

[2] Department of Architecture and Architectural Engineering, Kyoto University,
Japan
{as.higashikawa,naoki}@archi.kyoto-u.ac.jp

[3] Business School, Sichuan University, Chengdu 610065, China
{gqni,yfxu}@scu.edu.cn

[4] School of Economics and Management, Xi'an Technological University, Xi'an
710032, China
subing684@sohu.com

[5] School of Management, Xi'an Jiaotong University, Xi'an 710049, China

[6] State Key Lab for Manufacturing Systems Engineering, Xi'an 710049, China

Abstract. This paper considers minimax regret 1-sink location problems in *dynamic path networks*. A dynamic path network consists of an undirected path with positive edge lengths and constant edge capacity and the vertex supply which is nonnegative value, called weight, is unknown but only the interval of weight is known. A particular assignment of weight to each vertex is called a *scenario*. Under any scenario, the cost of a sink is defined as the minimum time to complete evacuation for all weights (evacuees), and the *regret* of a sink location x is defined as the cost of x minus the cost of an optimal sink. Then, the problem is to find a point as a sink such that the maximum regret for all possible scenarios is minimized. We present an $O(n \log^2 n)$ time algorithm for minimax regret 1-sink location problems in dynamic path networks, where n is the number of vertices in the network.

Keywords: minimax regret, sink location, dynamic flow, path networks, evacuation problem.

1 Introduction

The Tohoku-Pacific Ocean Earthquake happened in Japan on March 11, 2011, and many people failed to evacuate and lost their lives due to severe attack by tsunamis. From the viewpoint of disaster prevention from city planning and

* Supported by JSPS Grant-in-Aid for Scientific Research(B)(21300003).
** Supported by National Natural Science Foundation of China under Grants 71071123, 61221063 and Program for Changjiang Scholars and Innovative Research Team in University under Grant IRT1173.

T.-H.H. Chan, L.C. Lau, and L. Trevisan (Eds.): TAMC 2013, LNCS 7876, pp. 121–132, 2013.

evacuation planning, it has now become extremely important to establish effective evacuation planning systems against large scale disasters. In particular, arrangements of tsunami evacuation buildings in large Japanese cities near the coast has become an urgent issue. To determine appropriate tsunami evacuation buildings, we need to consider where evacuation buildings are assigned and how to partition a large area into small regions so that one evacuation building is designated in each region. This produces several theoretical issues to be considered. Among them, this paper focuses on the location problem of the evacuation building assuming that we fix the region such that all evacuees in the region are planned to evacuate to this building. In this paper, we consider the simplest case for which the region consists of a single road.

The evaluation criterion of the building location is the time required to complete the evacuation. This is a kind of facility location problem which has been studied by Mamada et al [11] in which the region is modeled as a tree network such that a nonnegative weight that represents the number of evacuees at each vertex is known, and an $O(n \log^2 n)$ time algorithm was proposed to find an optimal location of a sink (the location of an evacuation building). However, the vertex weight (the number of evacuees at a vertex) varies depending on the time (e.g., in an office area in a big city there are many people during the daytime on weekdays while there are much less people on weekends or during the night time). So, in order to take into account the uncertainty of the vertex weights, we consider a minimax regret criterion assuming that for each vertex, we only know the interval of the vertex weight. We will treat such uncertainty in this paper by formulating the problem as the minimax regret 1-sink location problem in dynamic path networks. A particular realization (assignment of a weight to each vertex) is called a scenario. The problem can be understood as a 2-person Stackelberg game as follows. The first player picks a location x of a sink and the second player chooses a scenario s that maximizes the regret which is defined as the cost of x (the minimum time to complete evacuation) minus the cost of an optimal sink under the scenario s. The objective of the first player is to choose x that minimizes the regret.

Recently several researchers studied the minimax regret 1-median problem and efficient algorithms have been proposed [2, 3, 5, 14]. See also [1, 4–7, 10, 13] for related minimax regret location problems.

In this paper, we propose an $O(n \log^2 n)$ time algorithm for the minimax regret 1-sink location problem on a path assuming that a path is considered as a network consisting of a vertex set and an edge set in which an interval of the vertex weight is associated with each vertex, and the travel time and the capacity are associated with each edge that represent the time required to traverse the edge and the upper bound on the number of evacuees that can enter the edge per unit time, respectively.

2   Preliminaries

2.1   Definition

Let $P = (V, E)$ be a path where $V = \{v_0, v_1, ..., v_n\}$ and $E = \{e_1, e_2, ..., e_n\}$ such that v_{i-1} and v_i are endpoints of e_i for $1 \le i \le n$. Let $\mathcal{N} = (P, l, W, c, \tau)$ be a dynamic flow network with the underlying undirected graph being a path P, where l is a function that associates each edge e_j with positive length $l(e_j)$, W is also a function that associates each vertex $v_i \in V$ with an interval of weight (the number of the evacuees) $W(v_i) = [\underline{w}_i, \overline{w}_i]$ with $0 < \underline{w}_i \leq \overline{w}_i$, c is a constant representing the capacity of each edge: the least upper bound for the number of the evacuees passing a point in an edge per unit time, and τ is also a constant representing the time required for traversing the unit distance of each evacuee. We call such networks with path structures *dynamic path networks*. Let S denote the Cartesian product of all $W(v_i)$ for $1 \leq i \leq n$ (i.e., a set of scenarios):

$$S = \prod_{1 \leq i \leq n} [\underline{w}_i, \overline{w}_i]. \tag{1}$$

When a scenario $s \in S$ is given, we use the notation $w_i(s)$ to denote the weight of each vertex $v_i \in V$ under the scenario s.

In the following, suppose that a path P is embedded on a real line and each vertex $v_i \in V$ is associated with the line coordinate x_i such that $x_i = x_0 + \sum_{1 \leq j \leq i} l(e_j)$ for $1 \leq i \leq n$. We also use a notation P to denote the set of all points x such that $x_0 \leq x \leq x_n$. For a point $x \in P$, we also use a notation x to denote the line coordinate of the point, and the *left side* of x (resp. the *right side* of x) to denote the part of P consisting of all points $t \in P$ such that $t < x$ (resp. $t > x$). Suppose that a sink (evacuation center) is located at a point $x \in P$. Let $\Theta_L(x, s)$ (resp. $\Theta_R(x, s)$) denote the minimum time required for all evacuees on the left side (resp. the right side) of x to complete evacuation to x under a scenario $s \in S$. Note that we assume that the capacity of the entrance of an evacuation building is infinite, and thus, if we place a sink in a vertex v_i, all evacuees of v_i can finish their evacuation in no time. Then, by [9], $\Theta_L(x, s)$ and $\Theta_R(x, s)$ are expressed as follows:

$$\Theta_L(x, s) = \max_{0 \leq i \leq n-1} \left\{ (x - x_i)\tau + \left\lceil \frac{\sum_{0 \leq j \leq i} w_j(s)}{c} \right\rceil - 1 \;\middle|\; x > x_i \right\},$$

$$\Theta_R(x, s) = \max_{1 \leq i \leq n} \left\{ (x_i - x)\tau + \left\lceil \frac{\sum_{i \leq j \leq n} w_j(s)}{c} \right\rceil - 1 \;\middle|\; x < x_i \right\}.$$

For the ease of exposition, we assume that $c = 1$ (the case of $c > 1$ can be treated in essentially the same manner), and also omit the constant part (i.e., -1) from these equations in the following discussion. Thus, we redefine $\Theta_L(x, s)$ and $\Theta_R(x, s)$ as

$$\Theta_L(x, s) = \max_{0 \leq i \leq n-1} \left\{ (x - x_i)\tau + \sum_{0 \leq j \leq i} w_j(s) \;\middle|\; x > x_i \right\}, \tag{2}$$

$$\Theta_R(x, s) = \max_{1 \leq i \leq n} \left\{ (x_i - x)\tau + \sum_{i \leq j \leq n} w_j(s) \;\middle|\; x < x_i \right\}. \tag{3}$$

Additionally, we regard $\Theta_L(x_0, s)$ and $\Theta_R(x_n, s)$ as 0 in the subsequent discussion. Now, under $s \in S$, the minimum time required for the evacuation to $x \in P$ of all evacuees is defined by

$$\Theta(x, s) = \max\left\{\Theta_L(x, s), \Theta_R(x, s)\right\}. \tag{4}$$

Let $f_L^i(t, s)$ and $f_R^i(t, s)$ denote functions defined as follows: for $0 \le i \le n - 1$,

$$f_L^i(t, s) = (t - x_i)\tau + \sum_{0 \le j \le i} w_j(s) \ (t > x_i), \tag{5}$$

and for $1 \le i \le n$,

$$f_R^i(t, s) = (x_i - t)\tau + \sum_{i \le j \le n} w_j(s) \ (t < x_i). \tag{6}$$

Then, $\Theta_L(t, s)$ and $\Theta_R(t, s)$ are expressed as follows:

$$\Theta_L(t, s) = \max_{0 \le i \le n-1}\left\{f_L^i(t, s)\,\middle|\, t > x_i\right\}, \tag{7}$$

$$\Theta_R(t, s) = \max_{1 \le i \le n}\left\{f_R^i(t, s)\,\middle|\, t < x_i\right\}. \tag{8}$$

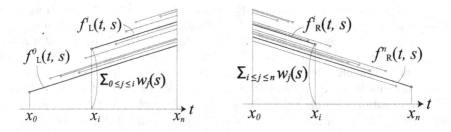

Fig. 1. Functions $f_L^i(t, s)$ for $0 \le i \le n - 1$ **Fig. 2.** Functions $f_R^i(t, s)$ for $1 \le i \le n$

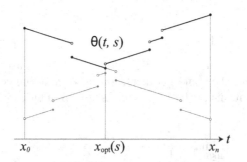

Fig. 3. A function $\Theta(t, s)$

The function $f_L^i(t, s)$ is drawn as a left-open half line with a positive slope starting from $(x_i, \sum_{0 \le j \le i} w_j(s))$ (see Fig. 1) while $f_R^i(t, s)$ is drawn as a right-open half line with a negative slope ending at $(x_i, \sum_{i \le j \le n} w_j(s))$ (see Fig. 2).

Thus $\Theta_L(t, s)$ is the upper envelope of these n half lines, and so $\Theta_L(t, s)$ is a strictly monotone increasing function of t. Symmetrically, $\Theta_R(t, s)$ is a strictly monotone decreasing function of t. Therefore, $\Theta(t, s)$ is a unimodal function, so there is a unique point in P which minimizes $\Theta(t, s)$ (see Fig. 3). In the following, let $x_{\mathrm{opt}}(s)$ denote such a point in P: $x_{\mathrm{opt}}(s) = \mathrm{argmin}_{x_0 \le t \le x_n} \Theta(t, s)$. We have the following propositions.

Proposition 1. *Under a scenario $s \in \mathcal{S}$, (i) $x_{\mathrm{opt}}(s)$ is unique, (ii) for $x < x_{\mathrm{opt}}(s)$, $\Theta_L(x, s) < \Theta_R(x, s)$ holds, and (iii) for $x > x_{\mathrm{opt}}(s)$, $\Theta_L(x, s) > \Theta_R(x, s)$ holds.*

Note that Proposition 1(ii)(iii) implies that $\Theta(x, s) = \Theta_R(x, s)$ holds for $x < x_{\mathrm{opt}}(s)$ and $\Theta(x, s) = \Theta_L(x, s)$ holds for $x > x_{\mathrm{opt}}(s)$.

We define the *regret* for x under s as

$$R(x, s) = \Theta(x, s) - \Theta(x_{\mathrm{opt}}(s), s). \tag{9}$$

Moreover, we also define the *maximum regret* for x as

$$R_{\max}(x) = \max\{R(x, s) \mid s \in \mathcal{S}\}. \tag{10}$$

If $R_{\max}(x) = R(x, s^*)$ for a scenario s^*, we call s^* the *worst case scenario* for x. The goal is to find a point $x^* \in P$, called the *minimax regret sink*, which minimizes $R_{\max}(x)$ over $x \in P$, i.e., the objective is

$$\text{minimize } \{R_{\max}(x) \mid x \in P\}. \tag{11}$$

2.2   Properties

For a scenario $s \in \mathcal{S}$ and an integer p such that $0 \le p \le n$, let s_p^+ denote a scenario such that $w_p(s_p^+) = \overline{w}_p$ and $w_i(s_p^+) = w_i(s)$ for $i \ne p$ and s_p^- denote a scenario such that $w_p(s_p^-) = \underline{w}_p$ and $w_i(s_p^-) = w_i(s)$ for $i \ne p$.

By (5), $f_L^i(t, s)$ is defined on $x_i < t \le x_n$ for $i = 0, 1, \ldots, n - 1$. Thus, for a point x such that $x_i < x \le x_n$, we have $f_L^i(x, s) \le f_L^i(x, s_p^+)$ and $f_L^i(x, s_p^-) \le f_L^i(x, s)$. Moreover, by these facts and (7), we also have $\Theta_L(x, s) \le \Theta_L(x, s_p^+)$ and $\Theta_L(x, s_p^-) \le \Theta_L(x, s)$. Generally we have the following claim.

Claim 1. *For a scenario $s \in \mathcal{S}$, a point $x \in P$ and an integer p such that $x_0 \le x_p \le x$ (resp. $x \le x_p \le x_n$),*
(i) $\Theta_L(x, s) \le \Theta_L(x, s_p^+)$ (resp. $\Theta_R(x, s) \le \Theta_R(x, s_p^+)$) holds, and
(ii) $\Theta_L(x, s_p^-) \le \Theta_L(x, s)$ (resp. $\Theta_R(x, s_p^-) \le \Theta_R(x, s)$) holds.

From Claim 1, we obtain the following lemma.

Lemma 1. *For a scenario $s \in \mathcal{S}$ and an integer p such that $x_0 \le x_p \le x_{\mathrm{opt}}(s)$ (resp. $x_{\mathrm{opt}}(s) \le x_p \le x_n$),*
(i) $x_{\mathrm{opt}}(s_p^+) \le x_{\mathrm{opt}}(s)$ (resp. $x_{\mathrm{opt}}(s) \le x_{\mathrm{opt}}(s_p^+)$) holds, and
(ii) $x_{\mathrm{opt}}(s) \le x_{\mathrm{opt}}(s_p^-)$ (resp. $x_{\mathrm{opt}}(s_p^-) \le x_{\mathrm{opt}}(s)$) holds.

Proof. We will prove Lemma 1(i) by contradiction: suppose that $x_{\mathrm{opt}}(s_p^+) > x_{\mathrm{opt}}(s)$ for p such that $x_0 \leq x_p \leq x_{\mathrm{opt}}(s)$. Let x_{mid} be the mid point of $x_{\mathrm{opt}}(s_p^+)$ and $x_{\mathrm{opt}}(s)$: $x_{\mathrm{mid}} = (x_{\mathrm{opt}}(s_p^+) + x_{\mathrm{opt}}(s))/2$. Then, by Proposition 1(ii)(iii), we have

$$\Theta_L(x_{\mathrm{mid}}, s_p^+) < \Theta_R(x_{\mathrm{mid}}, s_p^+) \text{ and } \Theta_R(x_{\mathrm{mid}}, s) < \Theta_L(x_{\mathrm{mid}}, s). \tag{12}$$

Note that by $x_{\mathrm{opt}}(s) < x_{\mathrm{mid}}$ and the assumption of $x_p \leq x_{\mathrm{opt}}(s)$, we have $x_p < x_{\mathrm{mid}}$. Thus, by (8), we also have

$$\Theta_R(x_{\mathrm{mid}}, s_p^+) = \Theta_R(x_{\mathrm{mid}}, s). \tag{13}$$

Thus, by (12) and (13), we obtain $\Theta_L(x_{\mathrm{mid}}, s_p^+) < \Theta_L(x_{\mathrm{mid}}, s)$ which contradicts Claim 1(i). Other cases can be similarly treated. $\square$

Corollary 1. *For a scenario $s \in \mathcal{S}$ and an integer p such that $x_0 \leq x_p \leq x_{\mathrm{opt}}(s)$ (resp. $x_{\mathrm{opt}}(s) \leq x_p \leq x_n$), $x_p \leq x_{\mathrm{opt}}(s_p^+)$ (resp. $x_{\mathrm{opt}}(s_p^+) \leq x_p$) holds.*

Proof. Assume that $x_p > x_{\mathrm{opt}}(s_p^+)$ for p such that $x_0 \leq x_p \leq x_{\mathrm{opt}}(s)$, then $x_{\mathrm{opt}}(s_p^+) < x_{\mathrm{opt}}(s)$. By Lemma 1(ii), we have $x_{\mathrm{opt}}(s) \leq x_{\mathrm{opt}}(s_p^-)$, and by applying Lemma 1(ii) with s replaced by s_p^+, we also have $x_{\mathrm{opt}}(s_p^-) \leq x_{\mathrm{opt}}(s_p^+)$, implying $x_{\mathrm{opt}}(s_p^+) \geq x_{\mathrm{opt}}(s)$, contradiction. The other case can also be proved in the same manner. $\square$

Lemma 2. *For a scenario $s \in \mathcal{S}$, a point $x \in P$ and an integer p such that $x_0 \leq x_p \leq x$ (resp. $x \leq x_p \leq x_n$), suppose that q is a maximum integer such that $\Theta_L(x, s) = f_L^q(x, s)$ (resp. q is a minimum integer such that $\Theta_R(x, s) = f_R^q(x, s)$). Then*
(i) let r be a maximum integer such that $\Theta_L(x, s_p^+) = f_L^r(x, s_p^+)$ (resp. let r be a minimum integer such that $\Theta_R(x, s_p^+) = f_R^r(x, s_p^+)$), then $x_q \leq x_r$ (resp. $x_q \geq x_r$) holds,
(ii) let r be a maximum integer such that $\Theta_L(x, s_p^-) = f_L^r(x, s_p^-)$ (resp. let r be a minimum integer such that $\Theta_R(x, s_p^-) = f_R^r(x, s_p^-)$), then $x_q \geq x_r$ (resp. $x_q \leq x_r$) holds.

Proof. We only prove (i). Suppose otherwise, i.e., $x_q > x_r$. Then by the maximality of r and $f_L^r(x, s_p^+)$,

$$f_L^q(x, s_p^+) < f_L^r(x, s_p^+) \tag{14}$$

holds. Also by the maximality of $f_L^q(x, s)$,

$$f_L^q(x, s) \geq f_L^r(x, s) \tag{15}$$

holds. Thus by (14) and (15), we have $f_L^q(x, s_p^+) - f_L^q(x, s) < f_L^r(x, s_p^+) - f_L^r(x, s)$, namely, $\sum_{0 \leq i \leq q}(w_i(s_p^+) - w_i(s)) < \sum_{0 \leq i \leq r}(w_i(s_p^+) - w_i(s))$, which contradicts $x_q > x_r$. $\square$

A scenario $s \in S$ is said to be *left-dominant* (resp. *right-dominant*) if for some i with $0 \leq i \leq n$, $w_j(s) = \overline{w}_j$ for $0 \leq j < i$ and $w_j(s) = \underline{w}_j$ for $i \leq j \leq n$ hold (resp. $w_j(s) = \underline{w}_j$ for $0 \leq j < i$ and $w_j(s) = \overline{w}_j$ for $i \leq j \leq n$ hold). Let S_L (resp. S_R) denote the set of all left-dominant (resp. right-dominant) scenarios. S_L consists of the following $n+1$ scenarios:

$$s_L^i = (\overline{w}_0, \ldots, \overline{w}_i, \underline{w}_{i+1}, \ldots, \underline{w}_n) \quad \text{for } i = 0, 1, \ldots, n-1,$$
$$s_L^n = (\overline{w}_0, \overline{w}_1, \ldots, \overline{w}_n), \tag{16}$$

and S_R consists of the following $n+1$ scenarios:

$$s_R^i = (\underline{w}_0, \ldots, \underline{w}_i, \overline{w}_{i+1}, \ldots, \overline{w}_n) \quad \text{for } i = 0, 1, \ldots, n-1,$$
$$s_R^n = (\underline{w}_0, \underline{w}_1, \ldots, \underline{w}_n). \tag{17}$$

The following is a key theorem.

Theorem 1. *For any point $x \in P$, there exists a worst case scenario for x which belongs to $S_L \cup S_R$.*

Proof. Suppose that s is a worst case scenario for x. We prove that if $x_{\mathrm{opt}}(s) < x$, $R(x, s^*) \geq R(x, s)$ holds for some left-dominant scenario s^* while otherwise, $R(x, s^*) \geq R(x, s)$ holds for some right-dominant scenario s^*. We only consider the case of $x_{\mathrm{opt}}(s) < x$ since the other case can be similarly treated. Then, by Proposition 1, $\Theta(x, s) = \Theta_L(x, s)$ holds: for some integer k such that $x_k < x$,

$$\Theta(x, s) = f_L^k(x, s). \tag{18}$$

We now show that $R(x, s_L^k) \geq R(x, s)$ holds, i.e., s_L^k is also a worst case scenario for x. If s is not equal to s_L^k, there exists an integer p such that [Case 1] $x_k < x_p \leq x_n$ and $w_p(s) > \underline{w}_p$ or [Case 2] $x_0 \leq x_p \leq x_k$ and $w_p(s) < \overline{w}_p$. If we can show that $R(x, s_p^-) \geq R(x, s)$ holds for [Case 1] and $R(x, s_p^+) \geq R(x, s)$ holds for [Case 2], we will eventually obtain $R(x, s_L^k) \geq R(x, s)$ by repeatedly applying the same discussion as long as there exists such an integer p. In the following, we consider two subcases: (I) $x_{\mathrm{opt}}(s) < x_k$ and (II) $x_k \leq x_{\mathrm{opt}}(s)$. We only give the proof for Case 1.

[**Case 1**]: In this case, we consider a scenario s_p^-. We consider two subcases.
(I) See Fig. 4. By (5), $f_L^k(x, s_p^-) = (x - x_k)\tau + \sum_{0 \leq j \leq k} w_j(s_p^-)$, and by $x_k < x_p$, $\sum_{0 \leq j \leq k} w_j(s_p^-) = \sum_{0 \leq j \leq k} w_j(s)$ holds, thus we have $f_L^k(x, s_p^-) = f_L^k(x, s)$. By Lemma 1(ii), $x_{\mathrm{opt}}(s_p^-) \leq x_{\mathrm{opt}}(s)$ holds, thus $\Theta(x, s_p^-) = f_L^k(x, s_p^-)$. By these facts and (18), we have

$$\Theta(x, s_p^-) = \Theta(x, s) \tag{19}$$

By applying Claim 1(ii) with x replaced by $x_{\mathrm{opt}}(s)$, we have $\Theta_R(x_{\mathrm{opt}}(s), s_p^-) \leq \Theta_R(x_{\mathrm{opt}}(s), s)$. Also, by $x_{\mathrm{opt}}(s) < x_p$, we have $\Theta_L(x_{\mathrm{opt}}(s), s_p^-) = \Theta_L(x_{\mathrm{opt}}(s), s)$ (by the same reason for $f_L^k(x, s_p^-) = f_L^k(x, s)$ above). Thus

$$\Theta(x_{\mathrm{opt}}(s), s_p^-) \leq \Theta(x_{\mathrm{opt}}(s), s) \tag{20}$$

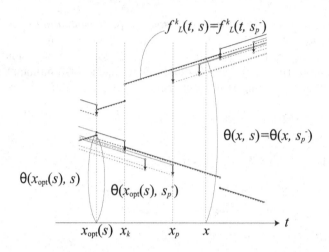

Fig. 4. Illustration of Case 1(I)

holds. Also, by the optimality of $x_{\mathrm{opt}}(s_p^-)$ under s_p^-,

$$\Theta(x_{\mathrm{opt}}(s_p^-), s_p^-) \le \Theta(x_{\mathrm{opt}}(s), s_p^-) \tag{21}$$

holds. Therefore, by (19), (20) and (21), we obtain $\Theta(x, s_p^-) - \Theta(x_{\mathrm{opt}}(s_p^-), s_p^-) \ge \Theta(x, s) - \Theta(x_{\mathrm{opt}}(s), s)$, i.e., $R(x, s_p^-) \ge R(x, s)$.

(**II**) We will show that (19), (20) and (21) also hold in this subcase. Because of $x_k < x_p$, we have $f_L^k(x, s_p^-) = f_L^k(x, s)$ in the same manner as in Case 1(I), thus (19) holds. In order to show that (20) holds, we consider the following two cases (a) and (b).

(a) Case of $x_{\mathrm{opt}}(s) < x_p$. Then by applying Claim 1(ii) with x replaced by $x_{\mathrm{opt}}(s)$, we have $\Theta_R(x_{\mathrm{opt}}(s), s_p^-) \le \Theta_R(x_{\mathrm{opt}}(s), s)$. Because of $x_{\mathrm{opt}}(s) < x_p$, we also have $\Theta_L(x_{\mathrm{opt}}(s), s_p^-) = \Theta_L(x_{\mathrm{opt}}(s), s)$, thus (20) holds.

(b) Case of $x_{\mathrm{opt}}(s) \ge x_p$. Then by applying Claim 1(ii) with x replaced by $x_{\mathrm{opt}}(s)$, we have $\Theta_L(x_{\mathrm{opt}}(s), s_p^-) \le \Theta_L(x_{\mathrm{opt}}(s), s)$. Because of $x_{\mathrm{opt}}(s) \ge x_p$, we also have $\Theta_R(x_{\mathrm{opt}}(s), s_p^-) = \Theta_R(x_{\mathrm{opt}}(s), s)$, thus (20) holds.

Also, by the optimality of $x_{\mathrm{opt}}(s_p^-)$ under s_p^-, (21) clearly holds. Therefore we also obtain $R(x, s_p^-) \ge R(x, s)$ in this subcase.   □

3   Algorithm

We will show an $O(n \log^2 n)$ time algorithm that computes x^* which minimizes a function $R_{\max}(t)$. By Theorem 1, we have

$$R_{\max}(t) = \max_{s \in \mathcal{S}_L \cup \mathcal{S}_R} R(t, s). \tag{22}$$

Thus, we consider $2n + 2$ left and right-dominant scenarios.

We now show how to efficiently compute $R_{\max}(x_i)$ for $i = 0, 1, \ldots, n$, and then how to compute $R_{\max}(x^*)$. In order to evaluate $R(x_i, s)$ for $s \in \mathcal{S}_L \cup \mathcal{S}_R$, we need to compute $\Theta(x_{\mathrm{opt}}(s), s)$ in advance. We then explain how we efficiently evaluate $R(x_i, s)$ for all dominant scenarios s and obtain $R_{\max}(x_i)$.

First, we show how to compute $\Theta(x_{\mathrm{opt}}(s_L^k), s_L^k)$ for $k = 0, 1, \ldots, n$. Computing $\Theta(x_{\mathrm{opt}}(s_R^k), s_R^k)$ can be done similarly, and thus omitted. In order to compute $\Theta(x_{\mathrm{opt}}(s_L^k), s_L^k)$ for a given k, we are required to evaluate $\Theta_L(x_i, s_L^k)$ and $\Theta_R(x_i, s_L^k)$ for $i = 0, 1, \ldots, n$. We now consider constructing *partial persistent priority search trees* [8] T_L and T_R for all left-dominant scenarios. In the following, we show only how to construct T_L, however, T_R can be constructed similarly. T_L consists of a priority search tree T_L^0 and path data structures $P_L^0, P_L^1, \ldots, P_L^n$. We first construct T_L^0 which has $n + 1$ leaves $l_0, l_1, \ldots, l_n$ corresponding to vertices $v_0, v_1, \ldots, v_n$ and internal nodes such that each internal node v has pointers to left and right children, each leaf l_i for $i = 1, 2, \ldots, n$ has the value $\sum_{j \in [1,i]} (\overline{w}_j - \underline{w}_j)$, and each node (including each leaf) v has an interval $[i_{\min}(v), i_{\max}(v)]$ where $i_{\min}(v)$ and $i_{\max}(v)$ are the indices of a minimum and maximum leaves of a subtree rooted at v, the value

$$\max\{-x_i\tau + \sum_{j \in [0,i]} w_j(s_L^0) \mid i_{\min}(v) \le i \le i_{\max}(v)\} \tag{23}$$

and the corresponding index of the leaf that attains the maximum. Note that for a leaf l_i, $i_{\min}(l_i) = i_{\max}(l_i) = i$ holds. By computing values of nodes in decreasing order of depth, T_L^0 can be constructed in $O(n)$ time and $O(n)$ space. Subsequently, we construct path data structures P_L^k for $k = 0, 1, \ldots, n$ along the path in T_L^0 from the leaf l_k to the root (see Fig. 5) such that each node v on the path P_L^k has the value $V_L^k(v)$ defined as

$$\max\{-x_i\tau + \sum_{j \in [0,i]} w_j(s_L^k) \mid i_{\min}(v) \le i \le i_{\max}(v)\}. \tag{24}$$

Note that the value of (23) can be represented as $V_L^0(v)$, and thus P_L^0 can be constructed in $O(\log n)$ time and $O(\log n)$ space by using T_L^0. In practice, we construct P_L^k for $k = 1, 2, \ldots, n$ in the following manner. Suppose that $P_L^0, \ldots, P_L^{k-1}$ have been constructed. We then follow the path P_L^k from the leaf l_k to the root and store $V_L^k(v)$ at each node v on P_L^k, which takes $O(\log n)$ time and $O(\log n)$ space. At the leaf l_k, we set the value $V_L^k(l_k) = -x_k\tau + \sum_{j \in [0,k]} w_j(s_L^k)$ by computing $V_L^0(l_k)$ plus $\sum_{j \in [1,k]} (\overline{w}_j - \underline{w}_j)$ (recall that these values are stored at the leaf l_k in T_L^0). For an internal node v on P_L^k, let c_l^v and c_r^v be the left and right children of v. If c_r^v is not on P_L^k, we compute its value $V_L^k(c_r^v)$ as $V_L^0(c_r^v)$ plus $\sum_{j \in [1,k]} (\overline{w}_j - \underline{w}_j)$, and set $V_L^k(v)$ as the maximum of $V_L^k(c_r^v)$ and $V_L^k(c_l^v)$. If c_l^v is not on P_L^k, since $V_L^k(c_l^v) = V_L^{k-1}(c_l^v)$ holds and $V_L^{k-1}(c_l^v)$ is already computed in the previous step, we only set $V_L^k(v)$ as the maximum of $V_L^k(c_r^v)$ and $V_L^k(c_l^v)$. Thus we can construct $P_L^0, P_L^1, \ldots, P_L^n$ in $O(n \log n)$ time and $O(n \log n)$ space. We have the following claim.

Claim 2. *For all left-dominant scenarios, partial persistent priority search trees T_L and T_R can be constructed in $O(n \log n)$ time and $O(n \log n)$ space.*

We now show how to compute $\Theta_L(x_i, s_L^k)$ for some integers $i \in [1, n]$ and $k \in [0, n]$ by using T_L (recall that we assume $\Theta_L(x_0, s) = 0$ and $\Theta_R(x_n, s) = 0$ for any scenario s). By the definition of (5), functions $f_L^0(t, s_L^k), f_L^1(t, s_L^k), \ldots, f_L^{i-1}(t, s_L^k)$ are defined at $t = x_i$ while $f_L^i(t, s_L^k), f_L^{i+1}(t, s_L^k), \ldots, f_L^n(t, s_L^k)$ are not, thus we are required to compute the maximum of $V_L^k(l_j)$ for $j \in [0, i-1]$, i.e., the maximum of $V_L^k(c_l^v)$ for each node v on P_L^i. There are two cases [Case 1] $i \le k$ (see Fig. 6) and [Case 2] $i > k$ (see Fig. 7).

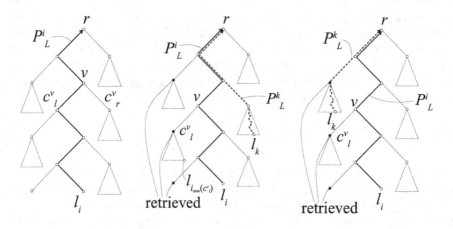

Fig. 5. Illustration of P_L^i **Fig. 6.** Illustration of Case 1 **Fig. 7.** Illustration of Case 2

[**Case 1**]: We follow the path P_L^i from leaf l_i to the root. Every time we visit an internal node v, we examine whether its left child c_l^v is on P_L^i or not. If not, we get the rightmost leaf $i_{\max}(c_l^v)$ in the subtree rooted at c_l^v, retrieve the value $V_L^{i_{\max}(c_l^v)}(c_l^v)$ stored in $P_L^{i_{\max}(c_l^v)}$ since $V_L^k(c_l^v) = V_L^{i_{\max}(c_l^v)}(c_l^v)$ holds. We continue to do this computation and to take the maximum value among those retrieved, which takes $O(\log n)$ time.

[**Case 2**]: The task we do is similar to Case 1. Every time we visit an internal node v before P_L^i encounters the node on P_L^k, we examine whether its left child c_l^v is on P_L^i or not. If not, we retrieve the value $V_L^0(c_l^v)$ stored in T_L^0 and add $\sum_{j \in [1,k]} (\overline{w}_j - \underline{w}_j)$ to the retrieved value since $V_L^k(c_l^v) = V_L^0(c_l^v) + \sum_{j \in [1,k]} (\overline{w}_j - \underline{w}_j)$ holds. We continue to do this computation and to take the maximum value among those retrieved before encountering the node on P_L^k, and after that, we do the same computation as in Case 1, which takes $O(\log n)$ time.

Similarly, we can also compute $\Theta_R(x_i, s_L^k)$ in $O(\log n)$ time once we have constructed T_R. We have the following claim.

Claim 3. *For any integers $i \in [0, n]$ and $k \in [0, n]$, $\Theta(x_i, s_L^k)$ can be computed in $O(\log n)$ time once T_L and T_R have been constructed.*

By Claim 3, we have the following lemma.

Lemma 3. *For any integer $k \in [0, n]$, $\Theta(x_{\mathrm{opt}}(s_L^k), s_L^k)$ can be computed in $O(\log^2 n)$ time once T_L and T_R have been constructed.*

Proof. By the unimodality of $\Theta(t, s_L^k)$ and Claim 3, we can compute by binary search in $O(\log^2 n)$ time

$$x_{i_L} = \min\{x_h \mid \Theta_L(x_h, s_L^k) \geq \Theta_R(x_h, s_L^k)\}, \tag{25}$$

$$x_{i_R} = \max\{x_h \mid \Theta_R(x_h, s_L^k) \geq \Theta_L(x_h, s_L^k)\}. \tag{26}$$

Note that $i_L - i_R \leq 1$. If $i_L = i_R$, then $x_{\mathrm{opt}}(s_L^k) = x_{i_L}$ holds. Otherwise, let $\Theta_L(x_{i_L}, s_L^k) = f_L^{h_L}(x_{i_L}, s_L^k)$ and $\Theta_R(x_{i_R}, s_L^k) = f_R^{h_R}(x_{i_R}, s_L^k)$ for some integers h_L and h_R such that $h_L \leq i_R$ and $h_R \geq i_L$. If two line segments $f_L^{h_L}(t, s_L^k)$ and $f_R^{h_R}(t, s_L^k)$ intersect at a point whose x-coordinate is at least x_{i_R} and at most x_{i_L}, i.e., if $f_L^{h_L}(x_{i_R}, s_L^k) < f_R^{h_R}(x_{i_R}, s_L^k)$ and $f_L^{h_L}(x_{i_L}, s_L^k) > f_R^{h_R}(x_{i_L}, s_L^k)$ hold, then x-coordinate of the intersection point is $x_{\mathrm{opt}}(s_L^k)$. If $f_L^{h_L}(x_{i_R}, s_L^k) \geq f_R^{h_R}(x_{i_R}, s_L^k)$ holds, then $x_{\mathrm{opt}}(s_L^k) = x_{i_L}$ holds. If $f_L^{h_L}(x_{i_L}, s_L^k) \leq f_R^{h_R}(x_{i_L}, s_L^k)$ holds, then $x_{\mathrm{opt}}(s_L^k) = x_{i_R}$ holds. $\qquad\square$

Note that $f_L^i(t, s_L^k)$ (resp. $f_R^i(t, s_L^k)$) is not defined at $t = x_i$ for any i, nevertheless in the proof of Lemma 3, we use the notation $f_L^{h_L}(x_{i_R}, s_L^k)$ even for $h_L = i_R$ (resp. $f_R^{h_R}(x_{i_L}, s_L^k)$ even for $h_R = i_L$) to represent the value $\sum_{j \in [0, i_R]} w_j(s_L^k)$ (resp. $\sum_{j \in [i_L, n]} w_j(s_L^k)$). By Lemma 3, we obtain the following lemma and corollary.

Lemma 4. *$\Theta(x_{\mathrm{opt}}(s_L^k), s_L^k)$ for $k = 0, 1, \ldots, n$ can be computed in $O(n \log^2 n)$ time.*

Proof. By Claim 2, we can construct T_L and T_R in $O(n \log n)$ time. By this and Lemma 3, we can compute $\Theta(x_{\mathrm{opt}}(s_L^k), s_L^k)$ for $k = 0, 1, \ldots, n$ in $O(n \log n + n \log^2 n) = O(n \log^2 n)$ time. $\qquad\square$

Corollary 2. *$\Theta(x_{\mathrm{opt}}(s_L^k), s_L^k)$ and $\Theta(x_{\mathrm{opt}}(s_R^k), s_R^k)$ for $k = 0, 1, \ldots, n$ can be computed in $O(n \log^2 n)$ time.*

Now we turn to the problem of how to compute $R_{\max}(x^*)$. For a given integer $i \in [0, n]$, since we need to compute $\Theta_L(x_i, s_L^k) - \Theta(x_{\mathrm{opt}}(s_L^k), s_L^k)$ and $\Theta_R(x_i, s_L^k) - \Theta(x_{\mathrm{opt}}(s_L^k), s_L^k)$ in order to evaluate $R(x_i, s_L^k)$ for $k = 0, 1, \ldots, n$, we prepare partial persistent priority search trees $\hat{T}_L$ and $\hat{T}_R$ in the same manner as constructing T_L and T_R. Here $\hat{T}_L$ consists of $\hat{T}_L^0$ and $\hat{P}_L^k$ for $k = 0, 1, \ldots, n$ where T_L^0 has $n + 1$ leaves $l_0, l_1, \ldots, l_n$ and internal nodes such that each node v has the value $V_L^0(v) - \Theta(x_{\mathrm{opt}}(s_L^0), s_L^0)$ and each leaf l_i for $i = 1, 2, \ldots, n$ has the value $\sum_{j \in [1, i]}(\overline{w}_j - \underline{w}_j) + \Theta(x_{\mathrm{opt}}(s_L^{i-1}), s_L^{i-1}) - \Theta(x_{\mathrm{opt}}(s_L^i), s_L^i)$, and $\hat{P}_L^k$ for $k = 0, 1, \ldots, n$ is the path in $\hat{T}_L^0$ from the leaf l_k to the root such that each node v on $\hat{P}_L^k$ has the value $V_L^k(v) - \Theta(x_{\mathrm{opt}}(s_L^k), s_L^k)$. The only difference between $\hat{T}_L$ and T_L is the existence of some offset values in $\hat{T}_L^0$ and $\hat{P}_L^k$ at each node, and the same thing can be said for $\hat{T}_R$ and T_R. Thus, by Claim 2, we can construct $\hat{T}_L$ and $\hat{T}_R$ in $O(n \log n)$ time and $O(n \log n)$ space once we obtain $\Theta(x_{\mathrm{opt}}(s_L^k), s_L^k)$ for $k = 0, 1, \ldots, n$. By this and Lemma 4, $\hat{T}_L$ and $\hat{T}_R$ for all left-dominant sce-narios are constructed in $O(n \log^2 n)$ total time, and the same thing can be said for all right-dominant scenarios. Using these data structures, we can compute $\max_{s \in \mathcal{S}_L} R(x_i, s)$ and $\max_{s \in \mathcal{S}_R} R(x_i, s)$ in $O(n \log n)$, respectively. Thus, we can also compute $R_{\max}(x_i)$ in $O(n \log n)$. By (22), $R_{\max}(t)$ is an upper envelope of $2n + 2$ functions of $R(t, s_L^k)$ and $R(t, s_R^k)$ for $k = 0, 1, \ldots, n$. Since $\Theta(t, s)$ and $R(t, s)$ are unimodal in t by (9), $R_{\max}(t)$ is clearly unimodal. Therefore, we can compute x^* which minimizes $R_{\max}(t)$ in $O(n \log^2 n)$ time.

Theorem 2. *The minimax regret sink x^* can be computed in $O(n \log^2 n)$ time and $O(n \log n)$ space.*

References

1. Averbakh, I., Berman, O.: Algorithms for the robust 1-center problem on a tree. European Journal of Operational Research 123(2), 292–302 (2000)
2. Bhattacharya, B., Kameda, T.: A linear time algorithm for computing minmax regret 1-median on a tree. In: Gudmundsson, J., Mestre, J., Viglas, T. (eds.) COCOON 2012. LNCS, vol. 7434, pp. 1–12. Springer, Heidelberg (2012)
3. Brodal, G.S., Georgiadis, L., Katriel, I.: An $O(n \log n)$ version of the Averbakh-Berman algorithm for the robust median of a tree. Operations Research Letters 36(1), 14–18 (2008)
4. Chen, D., Chen, R.: A relaxation-based algorithm for solving the conditional p-center problem. Operations Research Letters 38(3), 215–217 (2010)
5. Chen, B., Lin, C.: Minmax-regret robust 1-median location on a tree. Networks 31(2), 93–103 (1998)
6. Conde, E.: Minmax regret location-allocation problem on a network under uncertainty. European Journal of Operational Research 179(3), 1025–1039 (2007)
7. Conde, E.: A note on the minmax regret centdian location on trees. Operations Research Letters 36(2), 271–275 (2008)
8. Driscoll, J.R., Sarnak, N., Sleator, D.D., Tarjan, R.E.: Making data structures persistent. Journal of Computer and System Sciences 38(1), 86–124 (1989)
9. Kamiyama, N., Katoh, N., Takizawa, A.: An efficient algorithm for evacuation problem in dynamic network flows with uniform arc capacity. IEICE Transactions 89-D(8), 2372–2379 (2006)
10. Kouvelis, P., Yu, G.: Robust discrete optimization and its applications. Kluwer Academic Publishers, Dordrecht (1997)
11. Mamada, S., Uno, T., Makino, K., Fujishige, S.: An $O(n \log^2 n)$ Algorithm for the Optimal Sink Location Problem in Dynamic Tree Networks. Discrete Applied Mathematics 154(16), 2387–2401 (2006)
12. McCreight, E.M.: Priority Search Trees. SIAM Journal on Computing 14(2), 257–276 (1985)
13. Ogryczak, W.: Conditional median as a robust solution concept for uncapacitated location problems. TOP 18(1), 271–285 (2010)
14. Puerto, J., Rodríguez-Chía, A.M., Tamir, A.: Minimax regret single-facility ordered median location problems on networks. Informs Journal on Computing 21(1), 77–87 (2009)

A Notion of a Computational Step
for Partial Combinatory Algebras

Nathanael L. Ackerman[1] and Cameron E. Freer[2]

[1] Department of Mathematics
Harvard University
nate@math.harvard.edu
[2] Computer Science and Artificial Intelligence Laboratory
Massachusetts Institute of Technology
freer@math.mit.edu

Abstract. Working within the general formalism of a partial combinatory algebra (or PCA), we introduce and develop the notion of a *step algebra*, which enables us to work with individual computational steps, even in very general and abstract computational settings. We show that every partial applicative structure is the closure of a step algebra obtained by repeated application, and identify conditions under which this closure yields a PCA.

Keywords: partial combinatory algebra, step algebra, computational step.

1   Introduction

A key feature of a robust notion of computation is having analogous notions of *efficient* computations. More precisely, given a definition of objects that are computable (given unlimited resources such as time, space, or entropy), one would like corresponding definitions of those that are efficiently computable (given bounds on such resources).

As the notion of computation has been generalized in many directions, the related notions of efficiency have not always followed. One important generalization of computation is that given by partial combinatory algebras. Here our goal is to provide one such corresponding notion of *a single step of a computation*, by introducing *step algebras*. Step algebras attempt to describe, using a similar formalism, what it means to carry out one step of a computation; we believe that they may be useful in the analysis of efficient computation in this general context, as well as in other applications, as we will describe.

1.1   Partial Combinatory Algebras

The class of *partial combinatory algebras* (PCAs) provides a fundamental formulation of one notion of abstract computation. PCAs generalize the combinatorial

T-H.H. Chan, L.C. Lau, and L. Trevisan (Eds.): TAMC 2013, LNCS 7876, pp. 133–143, 2013.
© Springer-Verlag Berlin Heidelberg 2013

calculi of Schönfinkel [Sch24] and Curry [Cur29], and have connections to realizability, topos theory, and higher-type computation. For an introduction to
PCAs, see van Oosten [vO08] or Longley [Lon95].

PCAs are flexible and general enough to support many of the standard operations and techniques of (generalized) computation, and despite the austerity
of their definition, each PCA will contain a realization of every partial computable function. As has been shown over the years, many of the natural models
of computation are PCAs. For example, partial computable functions relative
to any fixed oracle, partial continuous functions $\mathbb{N}^{\mathbb{N}} \to \mathbb{N}$ as in Kleene's higher-
type computability, nonstandard models of Peano arithmetic, and certain Scott
domains all form PCAs (see, e.g., [vO08, §1.4]), as do the partial α-recursive
functions for a given admissible ordinal α.

Furthermore, by work of van Oosten [vO06], there is a notion of reducibility
between PCAs extending the notion of Turing reducibility arising from ordinary
computation with oracles. Also, by augmenting PCAs with limiting operations,
Akama [Aka04] has shown how they can interpret infinitary λ-calculi.

Realizability was initially used as a tool to study concrete models of intuitionistic theories. As a result, the use of PCAs has been expanded to give models
of intuitionistic set theory, as in work on realizability toposes (generalizing the
effective topos; see, e.g, [vO08]) and, more recently, in work by Rathjen [Rat06].

In all these facets of computation, PCAs provide a clean and powerful generalization of the notion of computation. However, the elements of a PCA represent
an entire computation, and PCAs generally lack a notion of a single step of a
computation. In this paper we aim to provide one such notion.

1.2   Other Approaches to Abstract Algorithmic Computations

One approach, other than PCAs, to abstract algorithmic computation is the
finite algorithmic procedure of H. Friedman [Fri71]; for details and related approaches, including its relationship to recursion in higher types, see Fenstad
[Fen80, §0.1]. In such procedures, there is a natural notion of a step provided by
the ordered list of instructions.

Another, more abstract, attempt to capture the notion of computational step
is suggested by Moschovakis' *recursors* [Mos01]. In this setting, the result of
an abstract computation is the least fixed point of a continuous operator on
a complete lattice. As this least fixed point can be obtained as the supremum
of iteratively applying this continuous operator to the minimal element of the
lattice, one might consider a computational step to be a single application of
this continuous operator.

Still another approach is that of Gurevich's *abstract state machines* [Gur00].
Within this formalism, the notion of a single step of an algorithm has also been
an important concept [BDG09].

These and other approaches do provide useful analyses of the notion of a single
computation step. Here our goal is analogous but is in the more general setting of
PCAs, where one is never explicitly handed a list of fine-grained computational

instructions, and must instead treat each potentially powerful and long (even nonhalting) subcomputation as a black box.

1.3   Outline of the Extended Abstract

We begin by defining PCAs and recalling their key properties. In the following section we introduce *step algebras*, the main notion of this abstract. A PCA always gives rise to a step algebra (by considering application to be a single step), but additional hypotheses on a step algebra are needed to ensure that the closure under repeated application is itself a PCA.

In the most general cases, a step algebra gives us little handle on its closure. Therefore we consider additional computable operations such as pairing, the use of registers, and serial application. These operations lead us to conditions under which a step algebra yields a PCA (by closure under repeated application). On the other hand, we conjecture that every PCA (up to isomorphism) comes from a suitable step algebra in this way.

Finally, we briefly discuss potential extensions of this work, including a generalization of computational complexity to the setting of PCAs, and an analysis of reversible computations in PCAs.

2   Preliminaries

Before proceeding to step algebras, we briefly recall the definition and key properties of partial applicative structures and partial combinatory algebras. For many more details, see [vO08] or [Lon95].

Definition 1. *Suppose A is a nonempty set and $\circ : A \times A \to A$ is a partial map. We say that $\mathbb{A} = (A, \circ)$ is a* **partial applicative structure** *(PAS). We write PAS to denote the class of partial applicative structures.*

This map $\circ$ is often called application. When the map $\circ$ is total, we say that $\mathbb{A}$ is a *total PAS*. When there is no risk of confusion (e.g., from application in another PAS), we will write ab to denote $a \circ b$. Furthermore, we adopt the standard convention of association to the left, whereby abc denotes $(ab)c$ (but not $a(bc)$ in general).

Given an infinite set of variables, the set of terms over a PAS $\mathbb{A} = (A, \circ)$ is the least set containing these variables and all elements of A that is closed under application. For a closed term t (i.e., without variables) and element $a \in A$, we write $t \downarrow a$, and say that term t denotes element a, when a is the result of repeated reduction of subterms of t. We write $t\downarrow$, and say that t denotes, when there is some a such that $t \downarrow a$. For closed terms t, s, the expression $t = s$ means that they denote the same value, and $t \simeq s$ means that if either t or s denotes, then $t = s$. This notation extends to non-closed terms (and means the corresponding expression of closed terms for every substitution instance).

Definition 2. *Let* $\mathbb{A} = (A, \circ)$ *be a PAS. We say that* $\mathbb{A}$ *is combinatorially complete when for every* $n \in \mathbb{N}$ *and any term* $t(x_1, \ldots, x_{n+1})$, *there is an element* $a \in A$ *such that for all* $a_1, \ldots, a_{n+1} \in A$, *we have* $aa_1 \cdots a_n{\downarrow}$ *and*

$$aa_1 \cdots a_{n+1} \simeq t(a_1, \ldots, a_{n+1}).$$

A **partial combinatory algebra** *(PCA) is a combinatorially complete PAS.*

The following lemma is standard.

Lemma 1. *Let* $\mathbb{A} = (A, \circ)$ *be a PAS. Then* $\mathbb{A}$ *is a PCA if and only if there are elements* $S, K \in A$ *satisfying*

- $Kab = a$,
- $Sab{\downarrow}$, *and*
- $Sabc \simeq ac(bc)$

for all $a, b, c \in A$.

Using the S and K combinators, many convenient objects or methods can be obtained in arbitrary PCAs, including pairing and projection operators, all natural numbers (via Church numerals), and definition by cases. Note that we use $\langle \cdot, \cdot \rangle$ to denote elements of a cartesian product (and not, as in other texts, for pairing in a PCA or for lambda abstraction).

Furthermore, by combinatory completeness, in every PCA each (ordinary) partial computable function corresponds to some element, and a wide range of computational facts about ordinary partial computable functions translate to arbitrary PCAs, including a form of lambda abstraction and the existence of fixed point operators (giving a version of the recursion theorem). However, one key feature that arbitrary PCAs do *not* admit is a sort of induction whereby one is able to iterate over all programs in sequence.

2.1 Examples

For concreteness, we present two of the most canonical examples of PCAs, although PCAs admit many more powerful or exotic models of computation, as described in the introduction.

Example 1. Kleene's first model is $\mathcal{K}_1 = (\mathbb{N}, \circ)$ *where* $a \circ b = \varphi_a(b)$, *i.e., the application of the partial computable function with index* a *to the input natural number* b.

In $\mathcal{K}_1$ there is a natural notion of a computational step, provided by, e.g., a single operation (i.e., overwriting the current cell, making a state transition, and moving the read/write head) on a Turing machine.

Example 2. Kleene's second model is $\mathcal{K}_2 = (\mathbb{N}^{\mathbb{N}}, \circ)$ *where application* $a \circ b$ *involves treating the element* a *of Baire space as (code for) a partial continuous map* $\mathbb{N}^{\mathbb{N}} \to \mathbb{N}^{\mathbb{N}}$, *applied to* $b \in \mathbb{N}^{\mathbb{N}}$.

While one may typically think of an *element* of $\mathbb{N}^{\mathbb{N}}$ as naturally having a step-by-step representation (e.g., whereby a real is represented as a sequence of nested intervals), application itself does not admit an obvious decomposition in terms of steps, and so even here step algebras may provide additional granularity.

3   Step Algebras

We now proceed to consider a framework for computations where the fundamental "steps" are abstract functions. To do so, we define step algebras, the key notion of this extended abstract. A step algebra is a PAS with certain extra structure and properties, whose elements are an object paired with a state; as with PCAs, these elements can be thought of as either data or code. When an element is treated as code, its state will be irrelevant, but when an element is treated as data, the state of the data can inform what is done to it.

Specifically, if $\mathbb{A} = (A \times I, \circledast)$ is a step algebra, then for $\mathbf{a}, \mathbf{b} \in \mathbb{A}$ we will think of $\mathbf{a} \circledast \mathbf{b}$ as the result of applying code $\mathbf{a}$ to data $\mathbf{b}$ for *one time step*. An intuition that does not translate exactly, but is nonetheless useful, is to imagine the PAS as a "generalized Turing machine". In this context, the code for a Turing machine can be thought of as a function telling the Turing machine how to increment its state and tape by one time step. The result of running the Turing machine is then the result of iterating this fixed code. Another (imperfect) analogy is with lambda calculi, where a single step naturally corresponds with a single β-reduction.

With this intuition in mind, we will be interested in the operation of "iterating" the application of our functions until they halt. In order to make this precise, we will need a notion capturing when the result of a single operation has halted. We will achieve this by requiring one state, $\Downarrow$, to be a "halting state". In particular, we will require that if an element is in the halting state, then any code applied to it does not change it.

Definition 3. *Suppose* $\mathbb{A} = (A \times I, \circledast)$ *is a total PAS such that I has two (distinct) distinguished elements* $0, \Downarrow$*, and let* $\langle \cdot, \cdot \rangle$ *denote elements of the cartesian product* $A \times I$*. We say that $\mathbb{A}$ is a* **step algebra** *when*

1. $\langle a, i \rangle \circledast \mathbf{b} = \langle a, j \rangle \circledast \mathbf{b}$ *for all $a \in A$, $i, j \in I$, and $\mathbf{b} \in \mathbb{A}$, and*
2. $\mathbf{a} \circledast \langle b, \Downarrow \rangle = \langle b, \Downarrow \rangle$ *for all $\mathbf{a} \in \mathbb{A}$ and $b \in A$.*

We write Step *to denote the class of step algebras.*

As with PCAs, in step algebras we will use the convention that when $\mathbf{a}, \mathbf{b} \in \mathbb{A}$, then $\mathbf{ab}$ represents the element $\mathbf{a} \circledast \mathbf{b}$, and will associate to the left.

The elements of a step algebra are meant to each describe a single step of a computation. Under this intuition, an entire (partial) computation arises from repeated application. Specifically, suppose $\mathbf{a} = \langle a, 0 \rangle$ and $\mathbf{b} = \langle b, 0 \rangle$ are elements of a step algebra $\mathbb{A} = (A \times I, \circledast)$. Then the computation describing a applied to b corresponds to the sequence

$$\mathbf{b}, \ \mathbf{ab}, \ \mathbf{a(ab)}, \ \mathbf{a(a(ab))}, \ \dots,$$

where if the sequence ever reaches a halting state $(c, \Downarrow)$, the computation is deemed to have finished with output c. Note that because of Definition 3.2, at most one such halting state can ever be reached by repeated application of $\mathbf{a}$ to $\mathbf{b}$. We now make precise this intuition of a partial computation arising from steps.

Definition 4. *We define the* **closure** *map* $\mathcal{P}$: Step $\to$ PAS *as follows. Suppose* $\mathbb{A} = (A \times I, \circledast)$ *is a step algebra. Let* $\mathcal{P}(\mathbb{A})$ *be the PAS* $(A, \circ)$ *such that*

$$a \circ b = c$$

if and only if there is a sequence of elements $\mathbf{c}_0, \dots, \mathbf{c}_m \in \mathbb{A}$ *satisfying*

- $\langle b, 0 \rangle = \mathbf{c}_0$,
- $\langle a, 0 \rangle \circledast \mathbf{c}_i = \mathbf{c}_{i+1}$ *for* $0 \le i < m$, *and*
- $\langle c, \Downarrow \rangle = \mathbf{c}_m$.

We now show that in fact every PAS is the closure of some step algebra.

Lemma 2. *The map* $\mathcal{P}$ *is surjective. Namely, for every PAS* $\mathbb{B} = (B, \circ)$ *there is a step algebra* $\mathbb{A} = (B \times \{0, \Downarrow\}, \circledast)$ *such that* $\mathcal{P}(\mathbb{A}) = \mathbb{B}$.

Proof. Let $\circledast$ be such that

- $\langle a, i \rangle \circledast \langle b, 0 \rangle = \langle ab, \Downarrow \rangle$ for all $a, b \in B$ such that $ab{\downarrow}$ and $i \in \{0, \Downarrow\}$,
- $\langle a, i \rangle \circledast \langle b, 0 \rangle = \langle b, 0 \rangle$ for all $a, b \in B$ such that $ab{\uparrow}$ and $i \in \{0, \Downarrow\}$, and
- $\mathbf{a} \circledast \langle b, \Downarrow \rangle = \langle b, \Downarrow \rangle$ for all $\mathbf{a} \in \mathbb{A}$ and $b \in B$.

Note that $\mathbb{A}$ is a step algebra, as Definition 3.1 holds because the first and second points in the definition of $\circledast$ here are independent of i. We have $\mathcal{P}(\mathbb{A}) = \mathbb{B}$ because for all $a, b \in B$, if $ab{\downarrow}$ then $\langle a, 0 \rangle \circledast \langle b, 0 \rangle = \langle ab, \Downarrow \rangle$, and if $ab{\uparrow}$ then $\langle a, 0 \rangle$ acts as the constant function on $\langle b, 0 \rangle$. $\qquad \square$

As we have just seen, every PAS comes from a step algebra where each computation occurs as a single step. Note that the collection of *terms* in a PAS can be made into a PAS itself by repeated reductions that consist of a single application of elements in the underlying PAS. Associated to such a PAS of terms, there is a natural step algebra, in which each step corresponds to a single term reduction.

Example 3. Suppose $\mathbb{A} = (A, \circ)$ is a PAS. Let A^* be the collection of closed terms of $\mathbb{A}$. Suppose the term a contains a leftmost subterm of the form $b \circ c$, for $b, c \in A$. Let a^+ be the result of replacing this subterm with the value of $b \circ c$ in A, if $b \circ c{\downarrow}$, and let a^+ be any value not in A otherwise. If a contains no such subterm, let $a^+ = a$. We define the step algebra $(A^* \times \{0, 1, \Uparrow, \Downarrow\}, \circledast^*)$ by the following equations, for all $a, b \in A^*$ and $i \in \{0, 1, \Uparrow, \Downarrow\}$.

- $\langle a, i \rangle \circledast^* \langle b, 0 \rangle = \langle (a) \circ (b), 1 \rangle$;
- $\langle a, i \rangle \circledast^* \langle b, 1 \rangle = \langle b^+, 1 \rangle$, when $b^+ \in A^*$ and $b^+ \ne b$;
- $\langle a, i \rangle \circledast^* \langle b, 1 \rangle = \langle b, \Uparrow \rangle$ when $b^+ \notin A^*$;
- $\langle a, i \rangle \circledast^* \langle b, 1 \rangle = \langle b, \Downarrow \rangle$ when $b^+ = b$; and
- $\langle a, i \rangle \circledast^* \langle b, \Uparrow \rangle = \langle b, \Uparrow \rangle$.

Now let $(A^*, \circ^*) = \mathcal{P}((A^* \times \{0, 1, \Uparrow, \Downarrow\}, \circledast^*))$. It is then easily checked that for any sequences $a, b \in A^*$ and $c \in A$, we have $a \circ^* b = c$ if and only if $(a) \circ (b)$ evaluates to c in $\mathbb{A}$.

We now turn to a context where we are guaranteed to have more concrete tools at our disposal, with the goal of finding conditions that ensure that the closure of a step algebra is a PCA.

4   Complete Step Algebras and PCAs

The notion of a step algebra is rather abstract, and provides relatively little structure for us to manipulate. We now introduce some basic computational operations that will ensure that the closure of a step algebra is a PCA. These are modeled after the standard techniques for programming on a Turing machine (or other ordinary model of computation), but make use of our abstract notion of computation for the basic steps.

Specifically, there are four types of operations that we will consider. First, there is a very abstract notion of "hidden variables"; this allows us to read in and keep track of two elements, for future use. Second, there is the notion of an iterative step algebra; given two pieces of code, this provides code that runs the first until it halts, then runs the second until it halts on the output of the first, and finally returns the result of the second. We also allow for passing the hidden variables from the code running the pair to each of the individual pieces of code it runs. Third, we require code that returns the first hidden variable. Fourth, we require a pairing operation that allows us to either run the first hidden variable on the first element of the pair, or run the second hidden variable on the second element of the pair, or run the first element of the pair on the second.

We will show that the closure of a step algebra having such operations contains S and K combinators. In particular, by Lemma 1, this will show that having such operations ensures that the closure is a PCA.

We now introduce the notion of hidden variables; while this definition is quite general, we will make use of it later in the specific ways we have just sketched.

Definition 5. *Suppose* $\mathbb{A} = (A \times I, \circledast, v_0, v_1, r)$ *is such that*

- $(A \times I, \circledast)$ *is a step algebra,*
- $r : A \to A$ *is total, and*
- $v_0, v_1 : A \to A \cup \{\emptyset\}$ *are total.*

We say $\mathbb{A}$ *has* **hidden variables** *when for all* $b \in A$ *there is a (necessarily unique)* $a^b \in A$ *satisfying*

- $\langle r(a), 0 \rangle \circledast \langle b, 0 \rangle = \langle a^b, \Downarrow \rangle$ *and*
- $v_1(a^b) = b$ *and* $v_0(a^b) = v_1(a)$.

We will use the notation $a^{b,c}$ *to mean the element* $(a^b)^c$.

The rough idea is to require a stack containing at least two elements (which we sometimes refer to as the *registers*). The code $r(a)$ reads in the next element, b, and returns the code for "a with b pushed on the stack". In this view, $v_1(a)$ is the element most recently read by the code.

Before proceeding to see how this formalism is used, we make a few observations. First, we have not required that the states are preserved (although some step algebras may nonetheless keep track of their state); this is because we will mainly treat the objects we read in only as code, not data. Second, note that we have only assumed that the stack has two elements. (Likewise, some step algebras may happen to keep track of the entire stack — e.g., as part of a reversible computation.)

Definition 6. *Suppose* $\mathbb{A} = (A \times I, \circledast, (v_0, v_1, r), (\pi_0, \pi_1, t))$ *is such that*

- $(A \times I, \circledast, v_0, v_1, r)$ *is a step algebra with hidden variables,*
- $\pi_0, \pi_1 : A \times A \times A \times A \times I \to I$ *with* π_0, π_1 *both total and injective in the last coordinate (i.e.,* $\pi_i(a_0, a_1, l_0, l_1, \cdot)$ *is injective for each* $a_0, a_1, l_0, l_1 \in A$), and
- $t : A \times A \to A$ *is total.*

We then say that $\mathbb{A}$ *is an* **iterative step algebra** *if whenever* $a_0, a_1, l_0, l_1, b \in A$ *with* $\pi_0^*(\cdot) = \pi_0(a_0, a_1, l_0, l_1, \cdot)$, $\pi_1^*(\cdot) = \pi_1(a_0, a_1, l_0, l_1, \cdot)$, *and* $t = t(a_0, a_1)^{l_0, l_1}$, *we have*

- $\langle t, 0 \rangle \circledast \langle b, 0 \rangle = \langle b, \pi_0^*(0) \rangle$,

- $\langle t, 0 \rangle \circledast \langle b, \pi_0^*(j) \rangle = \langle b', \pi_0^*(j') \rangle$ *when* $\langle a_0^{l_0, l_1}, 0 \rangle \circledast \langle b, j \rangle = \langle b', j' \rangle$ *and* $j' \neq \Downarrow$,
- $\langle t, 0 \rangle \circledast \langle b, \pi_0^*(j) \rangle = \langle b', \pi_1^*(0) \rangle$ *when* $\langle a_0^{l_0, l_1}, 0 \rangle \circledast \langle b, j \rangle = \langle b', \Downarrow \rangle$,

- $\langle t, 0 \rangle \circledast \langle b, \pi_1^*(j) \rangle = \langle b', \pi_1^*(j') \rangle$ *when* $\langle a_1^{l_0, l_1}, 0 \rangle \circledast \langle b, j \rangle = \langle b', j' \rangle$ *and* $j' \neq \Downarrow$, *and*
- $\langle t, 0 \rangle \circledast \langle b, \pi_1(j) \rangle = \langle b', \Downarrow \rangle$ *when* $\langle a_1^{l_0, l_1}, 0 \rangle \circledast \langle b, j \rangle = \langle b', \Downarrow \rangle$.

Intuitively, a step algebra is iterative when for every pair of code a_0 and a_1, there is code such that when it has l_0 and l_1 as its stack variables and is given a piece of data in state 0, it first runs a_0 with stack values (l_0, l_1) until it halts, then resets the state to 0 and runs a_1 with stack values (l_0, l_1) until it halts, and finally returns the result.

Note that while we have only defined this for pairs of code, the definition implies that elements can be found which iteratively run sequences of code of any finite length. We write $t_{a_0, \ldots, a_m}$ for code that first runs a_0 (with appropriate stack values) until it halts, then runs a_1 (with the same stack values) until it halts, and so on.

There are two subtleties worth mentioning about π_0^* and π_1^*. First, these take as input the states of t as well as the code that t is following. This is because we want it to be possible, in some cases, for π_0^*, π_1^* to keep track of the operations being performed.

Second, while we have assumed that π_0^* and π_1^* are always injective, we have not assumed that they have disjoint images (even outside of $\{\Downarrow\}$). One example that might be helpful to keep in mind is the case of $I = \mathbb{N} \cup \{\Downarrow, \Uparrow\}$ where each element of our step algebra is constant on elements whose state is in $\{\Downarrow, \Uparrow\}$, where π_0^*, π_1^* are constant on $\{\Downarrow, \Uparrow\}$, and where $\pi_i^*(n) = 2 \cdot n + i + 1$. In this case we can think of the state $\Uparrow$ as "diverges", i.e., a state that if reached will never halt, and we can think of of the maps π_i as using the natural bijections between even and odd natural numbers to "keep track" of what state we are in as we apply multiple pieces of code.

We are now able to give the two conditions that guarantee the desired combinators.

Definition 7. *Suppose* $\mathbb{A} = (A \times I, \circledast, (v_0, v_1, r))$ *is a step algebra with hidden variables. We say* $\mathbb{A}$ *that has* **constant functions** *when there is some* $c \in A$ *such that for all* $x, y \in A$, *we have* $\langle c^x, 0 \rangle \circ \langle y, 0 \rangle = \langle x, \Downarrow \rangle$.

We can think of c as code that simply returns the value in its first register. In this case, c^x is then code that already has x in its first register and that returns the value in its first register. In particular, we have the following easy lemma.

Lemma 3. *Suppose* $\mathbb{A} = (A \times I, \circledast, (v_0, v_1, r))$ *is an iterative step algebra with a constant function* c. *Let* $(A, \circ) = \mathcal{P}\big((A \times I, \circledast)\big)$ *be the closure of* $\mathbb{A}$. *Then for all* $x, y \in A$, *we have* $(r(c) \circ x) \circ y = x$.

Proof. The code $r(c)$ first reads x into its first register, and then returns c^x, which itself is code that returns what is in the first register (i.e., x).   □

In particular, if $\mathbb{A}$ is an iterative step algebra with a constant function, then the closure of $\mathbb{A}$ has a K combinator.

Definition 8. *Suppose* $\mathbb{A} = (A \times I, \circledast, (v_0, v_1, r), ([\cdot, \cdot], p, p_0, p_1, p_2))$ *is such that*

- $(A \times I, \circledast, v_0, v_1, r)$ *is a step algebra with hidden variables,*
- $[\cdot, \cdot] : A \times A \to A$ *is total, and*
- $p, p_0, p_1, p_2 \in A$.

We then say that $\mathbb{A}$ *has* **pairing** *when for all* $a_0, a_1, b_0, b_1 \in A$ *and* $j \in I$,

- $\langle p^{a_0, a_1}, 0 \rangle \circledast \langle b_0, 0 \rangle = \langle [b_0, b_0], \Downarrow \rangle$,
- $\langle p_0^{a_0, a_1}, 0 \rangle \circledast \langle [b_0, b_1], j \rangle = \langle [b', b_1], j' \rangle$ *when* $\langle a_0, 0 \rangle \circledast \langle b_0, j \rangle = \langle b', j' \rangle$,
- $\langle p_1^{a_0, a_1}, 0 \rangle \circledast \langle [b_0, b_1], j \rangle = \langle [b_0, b'], j' \rangle$ *when* $\langle a_1, 0 \rangle \circledast \langle b_1, j \rangle = \langle b', j' \rangle$,
- $\langle p_2^{a_0, a_1}, 0 \rangle \circledast \langle [b_0, b_1], j \rangle = \langle [b_0, b'], j' \rangle$ *when* $\langle b_0, 0 \rangle \circledast \langle b_1, j \rangle = \langle b', j' \rangle$ *and* $j' \neq \Downarrow$, *and*
- $\langle p_2^{a_0, a_1}, 0 \rangle \circledast \langle [b_0, b_1], j \rangle = \langle b', j' \rangle$ *when* $\langle b_0, 0 \rangle \circledast \langle b_1, j \rangle = \langle b', \Downarrow \rangle$.

We say that $\mathbb{A} = (A \times I, \circledast, (v_0, v_1, r), (\pi_0, \pi_1, t), ([\cdot, \cdot], p, p_0, p_1, p_2))$ *is an iterative step algebra with pairing when* $(A \times I, \circledast, (v_0, v_1, r), (\pi_0, \pi_1, t))$ *is an iterative step algebra and* $(A \times I, \circledast, (v_0, v_1, r), ([\cdot, \cdot], p, p_0, p_1, p_2))$ *is a step algebra with pairing. We will sometimes abuse notation and speak of the closure of* $\mathbb{A}$ *to mean* $\mathcal{P}\big((A \times I, \circledast)\big)$.

Intuitively, we say that $\mathbb{A}$ has pairing when there is an external pairing function $[\cdot, \cdot]$ along with an element of $\mathbb{A}$ that takes an element and pairs it with itself; an element that applies what is in the first register to the first element of the pair; an element that applies what is in the second register to an element of the pair; and an element that applies the first element of the pair to the second element, returning the answer if it halts.

Lemma 4. *Suppose* $\mathbb{A} = \left(A \times I, \circledast, (v_0, v_1, r), (\pi_0, \pi_1, t), ([\,\cdot\,,\cdot\,], p, p_0, p_1, p_2) \right)$ *is an iterative step algebra with pairing. Then the closure of* $\mathbb{A}$ *has an* S *combinator.*

Proof sketch. Suppose $(A, \circ)$ is the closure of $\mathbb{A}$. Let $S_2 = t_{p,p_0,p_1,p_2}$ and let $S = r(r(S_2))$. Intuitively, S_2 takes an argument d and then runs the following subroutines in succession:

- Return $[d, d]$.
- Return $[v_0(S_2) \circ d, d]$.
- Return $[v_0(S_2) \circ d, v_1(S_2) \circ d]$.
- Return $(v_0(S_2) \circ d) \circ (v_1(S_2) \circ d)$.

But then Sab is code that first reads in a, then reads in b (and moves a to the 0th register), and then performs the above. Hence S is the desired combinator. □

Definition 9. *Let* $\mathbb{A} = (A \times I, \circledast)$ *be a step algebra. We say that* $\mathbb{A}$ *is a* **complete** *step algebra when it can be extended to an iterative step algebra with pairing and with constant functions.*

Theorem 1. *If* $\mathbb{A}$ *is a complete step algebra, its closure is a PCA.*

Proof. This follows immediately from Lemma 1, Lemma 3, and Lemma 4. □

Conjecture 1. Every PCA is isomorphic to a PCA that arises as the closure of a complete step algebra (for a suitable notion of isomorphism).

5   Future Work

Here we have begun developing a notion of a single step of a computation, in the setting of PCAs. Having done so, we can now begin the project of developing robust notions of efficient computation in this general setting. For example, we aim to use step algebras to extend a notion of computational complexity to arbitrary PCAs (e.g., by considering suitably parametrized families of step algebras).

Many questions also remain about the class of step algebras whose closures yield the same PCA. In particular, there are many natural options one might consider within the partition on step algebras induced in this way. For example, the relationship between a step algebra and the one obtained by uniformly collapsing every n-step sequence into a single element, or those obtained by many other transformations, remains unexplored.

Finally, we plan to use step algebras to develop a notion of reversible computation in the general context of PCAs. The fine-grained analysis of computational steps might be used to ensure that each step is injective (whether by requiring that a complete step algebra keep track of its entire stack, or obtained by other means). Under an appropriate formulation of reversibility, one might explore whether, for every PCA, there is an essentially equivalent one in which computation is fully reversible.

Acknowledgements. The authors thank Bob Lubarsky for helpful conversations, and thank Rehana Patel, Dan Roy, and the anonymous referees for comments on a draft. This publication was made possible through the support of grants from the John Templeton Foundation and Google. The opinions expressed in this publication are those of the authors and do not necessarily reflect the views of the John Templeton Foundation.

References

[Aka04]   Akama, Y.: Limiting partial combinatory algebras. Theoret. Comput. Sci. 311(1-3), 199–220 (2004)

[BDG09] Blass, A., Dershowitz, N., Gurevich, Y.: When are two algorithms the same? Bull. Symbolic Logic 15(2), 145–168 (2009)

[Cur29]   Curry, H.B.: An analysis of logical substitution. Amer. J. Math. 51(3), 363–384 (1929)

[Fen80]   Fenstad, J.E.: General recursion theory: an axiomatic approach. Perspectives in Mathematical Logic. Springer, Berlin (1980)

[Fri71]   Friedman, H.: Algorithmic procedures, generalized Turing algorithms, and elementary recursion theory. In: Logic Colloquium 1969 (Proc. Summer School and Colloq., Manchester, 1969), pp. 361–389. North-Holland, Amsterdam (1971)

[Gur00]   Gurevich, Y.: Sequential abstract-state machines capture sequential algorithms. ACM Trans. Comput. Log. 1(1), 77–111 (2000)

[Lon95]   Longley, J.: Realizability toposes and language semantics. PhD thesis, University of Edinburgh, College of Science and Engineering, School of Informatics (1995)

[Mos01] Moschovakis, Y.N.: What is an algorithm? In: Mathematics Unlimited—2001 and Beyond, pp. 919–936. Springer, Berlin (2001)

[Rat06]   Rathjen, M.: Models of intuitionistic set theories over partial combinatory algebras. In: Cai, J.-Y., Cooper, S.B., Li, A. (eds.) TAMC 2006. LNCS, vol. 3959, pp. 68–78. Springer, Heidelberg (2006)

[Sch24]   Schönfinkel, M.: Über die Bausteine der mathematischen Logik. Math. Ann. 92(3-4), 305–316 (1924)

[vO06]   van Oosten, J.: A general form of relative recursion. Notre Dame J. Formal Logic 47(3), 311–318 (2006)

[vO08]   van Oosten, J.: Realizability: an introduction to its categorical side. Studies in Logic and the Foundations of Mathematics, vol. 152. Elsevier B. V., Amsterdam (2008)

Selection by Recursively Enumerable Sets[*]

Wolfgang Merkle[1], Frank Stephan[2], Jason Teutsch[3],
Wei Wang[4], and Yue Yang[2]

[1] Institut für Informatik, Universität Heidelberg, 69120 Heidelberg, Germany
merkle@math.uni-heidelberg.de
[2] Department of Mathematics, National University of Singapore,
Singapore 119076, Republic of Singapore
fstephan@comp.nus.edu.sg and matyangy@nus.edu.sg
[3] Computer Science and Engineering, Pennsylvania State University
teutsch@cse.psu.edu
[4] Department of Philosophy, Sun Yat-Sen University,
135 Xingang Xi Road, Guangzhou 510275, Guangdong, P.R. of China
wwang.cn@gmail.com

Abstract. For given sets A, B and Z of natural numbers where the members of Z are $z_0, z_1, \ldots$ in ascending order, one says that A is selected from B by Z if $A(i) = B(z_i)$ for all i. Furthermore, say that A is selected from B if A is selected from B by some recursively enumerable set, and that A is selected from B in n steps iff there are sets $E_0, E_1, \ldots, E_n$ such that $E_0 = A$, $E_n = B$, and E_i is selected from E_{i+1} for each $i < n$.

The following results on selections are obtained in the present paper. A set is ω-r.e. if and only if it can be selected from a recursive set in finitely many steps if and only if it can be selected from a recursive set in two steps. There is some Martin-Löf random set from which any ω-r.e. set can be selected in at most two steps, whereas no recursive set can be selected from a Martin-Löf random set in one step. Moreover, all sets selected from Chaitin's Ω in finitely many steps are Martin-Löf random.

1 Introduction

Post [12] introduced various important reducibilities in recursion theory among which the one-one reducibility is the strictest one; here A is one-one reducible to B iff there is a one-one recursive function F such that $A(x) = B(F(x))$ for all x. In a setting of closed left-r.e. sets, Jain, Stephan and Teutsch [2] investigated a strengthening of one-one reductions were it is required in addition that F is strictly increasing or, equivalently, that F is the principal function of an infinite recursive set. The present paper relaxes the latter notion of reducibility and considers reductions given by principal functions of infinite sets that are recursively enumerable (r.e., for short).

[*] F. Stephan is supported in part by NUS grant R252-000-420-112. Part of the work was done while W. Merkle, F. Stephan and Y. Yang visited W. Wang at the Sun Yat-Sen University.

T.-H.H. Chan, L.C. Lau, and L. Trevisan (Eds.): TAMC 2013, LNCS 7876, pp. 144–155, 2013.
© Springer-Verlag Berlin Heidelberg 2013

Recall that the principal function of an infinite set Z is the strictly increasing function F such that Z can be written as $\{F(0), F(1), \ldots\}$. In case for such Z and F some set A is reduced to some set B in the sense that $A(x) = B(F(x))$ for all x, this reduction could also be viewed as retrieving A from the asymmetric join B where the two "halves" of the join are not coded at the even and odd positions, respectively, of the join as usual but are coded into the positions that correspond to members and nonmembers, respectively, of the set Z.

Definition 1. A set A is selected from a set B by a set Z if Z is infinite and for the principal function F of Z it holds that $A(i) = B(F(i))$ for all i. A set A is selected from a set B, $A \sqsubset B$ for short, if A is selected from B by some r.e. set. Furthermore, say that A is selected from B in n steps iff there are sets $E_0, E_1, \ldots, E_n$ such that $E_0 = A$, $E_n = B$ and E_i is selected from E_{i+1} for each $i < n$.

The set B has selection rank n, if n is the maximum number such that some set A can be selected from B in n steps but not in $n - 1$ steps.

It makes sense to consider selection in several steps as the selection relation is not transitive: it follows by Theorems 14 and 15 that there is a Martin-Löf random set from which every recursive set can be selected in two steps but not in one step.

Note that for any infinite set Z, the principal function F of Z depends uniquely on Z. Furthermore, for a selection of A from B in n steps as in Definition 1, where F_m selects E_m from E_{m+1}, one can easily see that the function $\widetilde{F}$ given by

$$\widetilde{F}(y) = F_{n-1}(F_{n-2}(\ldots F_2(F_1(F_0(y)))\ldots))$$

satisfies that $A(i) = B(\widetilde{F}(i))$. However, since the selection relation is not transitive, in general, the range of $\widetilde{F}$ is not recursively enumerable and one cannot use the function $\widetilde{F}$ to select A from B in one step.

Early research in algorithmic randomness was formulated in terms of admissible selection rules. More precisely, given a certain way of selecting a subsequence from the characteristic sequence of a set, a set is called random iff all of its subsequences selected this way satisfy the condition in the law of large numbers that in the limit the frequencies of the symbols 0 and 1 are both equal to $1/2$ [6]. Furthermore, van Lambalgen's Theorem [5] states that if one decomposes a set A by selecting along a recursive set and its complement into two infinite halves B_0 and B_1 then A is Martin-Löf random iff B_0 and B_1 are Martin-Löf random relative to each other. From this viewpoint it is natural to ask whether the choice of the selection along a recursive set can be generalised here to the choice along an r.e. set and how this is compatible with randomness notions. For notions whose definition involves the halting problem, in particular for Kurtz random relative to K, Schnorr random relative to K and Martin-Löf random relative to K, it can easily be shown that if B has one of these randomness properties and A is selected from B then A has also the same randomness property and that in the case of Martin-Löf randomness relative to K, one even gets the full equivalent of van Lambalgen's Theorem.

These initial and obvious connections ask for deeper investigation in order to see how far these correspondences go and thus, one of the central questions investigated in this paper is when random sets can be selected from extremely nonrandom ones and vice versa. Furthermore, the notion of ω-r.e. sets — which also play an important role in algorithmic randomness — fits well with the notion of selection by r.e. sets and strong connections are found. Hence, the present work aims at establishing some basic properties of the selection relation and at clarifying its interplay with other established recursion-theoretic notions like Martin-Löf randomness, immunity, enumeration-properties and initial segment complexity.

In the sequel it is shown that a set is ω-r.e. if and only if it can be selected from an infinite and coinfinite recursive set in two steps. Every recursive set E has selection rank of at most 2, where the selection rank is 2 if and only if the set is infinite and coinfinite. Furthermore, the truth-table cylinder of the halting problem has selection rank 1. Every set selected from Ω in finitely many steps is Martin-Löf random (but differs from Ω in case at least one of the selections is nontrivial). There are Martin-Löf random sets which behave differently, for example, all ω-r.e. sets can be selected from some Martin-Löf random set.

2 Selection and ω-r.e. Sets

Recall that a set A is ω-r.e. iff there is a recursive function f and a sequence of sets $A_0, A_1, \ldots$ such that the A_s form a recursive approximation to A where the number of mind changes is bounded by f, that is,

- the sets A_s are uniformly recursive, that is, the mapping $(x, s) \mapsto A_s(x)$ is a recursive function of two arguments;
- for all x and all sufficiently large stages s, $A(x) = A_s(x)$;
- $A_0 = \emptyset$ and for every x there at most $f(x)$ stages s with $A_s(x) \neq A_{s+1}(x)$.

Note that a set A is r.e. if and only if it is ω-r.e. with a bounding function f as above that satisfies the additional constraint that $f(x) = 1$ for all x. The r.e. sets and ω-r.e. sets have been well-studied in recursion theory [6, 9–11, 14].

Our first result is that the ω-r.e. sets are closed downwards under the selection relation.

Theorem 2. *Assume that A is selected from B and B is an ω-r.e. set. Then A is an ω-r.e. set, too.*

Proof. Let the recursive approximation $B_0, B_1, \ldots$ and the recursive function f witness that B is an ω-r.e. set. Furthermore, let W be an r.e. set selecting A from B. There is a strictly increasing recursive function g such that its range is a recursive subset W_0 of W. Fix some recursive approximation $W_0, W_1, \ldots$ of W with $W_0 \subseteq W_1 \subseteq \ldots$ and let A_s be the set selected from B_s by W_s. Then $A_0 = \emptyset$ because $B_0 = \emptyset$. Furthermore, it can easily be seen that $A_s(n) \neq A_{s+1}(n)$ requires that there is an $x \leq g(n)$ with $W_s(x) \neq W_{s+1}(x)$ or $B_s(x) \neq B_{s+1}(x)$. Since for each such x these two conditions can be true for at most 1 and for at most $f(x)$ stages s, respectively, the total number of stages s where $A_s(n) \neq A_{s+1}(n)$ holds is at most $g(n) + 1 + f(0) + \ldots + f(g(n))$. Hence A is an ω-r.e. set. $\qquad\square$

Theorem 3. *Let O be the set of odd numbers and A be an ω-r.e. set. Then A can be selected from O in two steps, that is, there is a set B such that $A \sqsubseteq B$ and $B \sqsubseteq O$.*

Proof. Let the recursive approximation $A_0, A_1, \ldots$ and the recursive function f witness that A is an ω-r.e. set and let $h(n) = n^2 \cdot (1 + f(0) + f(1) + \ldots + f(n))$. It suffices to construct B such that, first, $A(n) = B(h(n))$, that is, the range of h selects A from B and, second, the set B is selected from O by some r.e. set W.

Split the natural numbers into consecutive intervals $I_0, J_0, I_1, J_1, I_2, J_2, \ldots$ where the length of I_n is $1 + f(0) + f(1) + \ldots + f(n)$, and the length of J_n is $h(0) + 1$ for $n = 0$ and is $h(n) - h(n-1)$ for $n \geq 1$. Let W_0 be the union of all J_n and let $k_0 = 0$. During stages $s = 0, 1, \ldots$, one applies the following updates:

1. Let B_s be the set selected by W_s from O;
2. if $A_s(k_s) \neq B_s(h(k_s))$ then let $W_{s+1} = W_s \cup \{\max(I_{k_s} \setminus W_s)\}$ else let $W_{s+1} = W_s$;
3. let $k_{s+1} = \min(\{k_s + 1\} \cup \{j : A_s(j) \neq A_{s+1}(j)\})$.

Say a stage s is enumerating in case on reaching its second step the condition of the if-clause is satisfied. First it is shown that for every enumerating stage s, the set $I_{k_s} \setminus W_s$ is nonempty, hence indeed the maximum member of this set is enumerated into W during stage s. Fix n and consider the enumerations of members of I_n into W in step 2. After each such enumeration at some stage s, any further such enumeration requires that at some stage $t > s$, one has $k_{t+1} \leq n < k_t$, which by step 3 in turn requires that the approximation to A has a mind change of the form $A_t(j) \neq A_{t+1}(j)$ where $j \leq n$ and $s < t$. Since for enumerations of distinct members of I_n there must be distinct such pairs (j, t), by choice of f at most $1 + f(0) + f(1) + \ldots + f(n) = |I_n|$ members of I_n are enumerated into W.

Next let $z_0, z_1, \ldots$ be the members of W in ascending order and for all s let $z_0^s, z_1^s, \ldots$ be the members of W_s in ascending order. By choice of the lengths of the intervals J_n and since W_0 was chosen as the union of these intervals, one has $z_{h(n)}^0 = \max J_n$ for all n. Furthermore, at most $|I_0| + \cdots + |I_n| \leq |J_n|$ times at some stage s a number smaller than $z_{h(n)}^s$ is enumerated into W_{s+1}. Hence for all such stages, one has

$$z_{h(n)-1}^s = z_{h(n)}^s - 1, \quad \text{hence} \quad z_{h(n)}^{s+1} = z_{h(n)}^s - 1 \quad \text{and} \quad O(z_{h(n)}^{s+1}) = 1 - O(z_{h(n)}^s).$$

In particular, during each enumerating stage s the value of the previous approximation to $B(h(k_s))$ is flipped from $O(z_{h(k_s)}^s)$ to $O(z_{h(k_s)}^{s+1})$, and after step 2 of each stage s, one has $A_s(k_s) = B_{s+1}(h(k_s))$. By induction on stages one can then show as an invariant of the construction that during each stage s at the end of step 2 it holds that

$$A_s(j) = B_{s+1}(h(j)) = O(z_{h(j)}^{s+1}) \quad \text{for all } s \text{ and all } j \leq k_s.$$

This concludes the verification of the construction because k_s goes to infinity by step 3 and because the sets A_s, B_s and W_s converge pointwise to A, B and W, respectively. □

The two preceding theorems give rise to the following corollary.

Corollary 4. *For any set A the following assertions are equivalent:*

1. *A is ω-r.e.;*
2. *A can be selected from O in two steps;*
3. *A can be selected from O in finitely many steps.*

As a further consequence of Theorem 3, there is a set B that is selected from a recursive set but has a logarithmic lower bound on the plain Kolmogorov complexity $C(\sigma)$ of its initial segments $\sigma = B(0)B(1)\ldots B(n)$, hence, in particular, the set B is complex [3]. Here the plain Kolmogorov complexity $C(\sigma)$ of a string σ is the length of the shortest program p such that $U(p) = \sigma$ for some fixed underlying universal machine U, see the textbook of Li and Vitányi [6] for more details. Note that the bound in Corollary 5 is optimal up to a constant factor by the proof of Theorem 11 below, which yields as a special case that every set selected from a recursive set has infinitely many initial segments of at most logarithmic complexity.

Corollary 5. *There is a set B selected from O such that for almost all n it holds that $C(B(0)B(1)\ldots B(n)) \geq 0.5 \cdot \log(n)$.*

Proof. Section 3 provides a closer look at Chaitin's Ω, which is the standard example of a Martin-Löf random left-r.e. set. From these properties it is immediate that for almost all n it holds that $C(\Omega(0)\Omega(1)\ldots\Omega(n)) > n - 3\log n$ and that Ω is ω-r.e. with bounding function $f(n) = 2^{n+1} - 1$. Applying the construction in the proof of Theorem 3 with A equal to Ω, one has $1 + f(0) + f(1) + \ldots + f(n) \leq 2^{n+2}$, hence $h(n) \leq 3^n$ for almost all n. So one can retrieve $\Omega(0)\Omega(1)\ldots\Omega(n)$ from $B(0)B(1)\ldots B(3^n)$. The corollary now follows by some elementary rearrangements.   □

Proposition 7 determines the rank of certain sets with rank 0, 1 or 2. The corresponding arguments use again Theorems 3 and 11, together with the following absorption principle for selections by recursive sets.

Proposition 6. *Let A be selected from E by the r.e. set W and let E be selected from B by the recursive set V. Then A is selected from B.*

Proof. Let $v_0, v_1, \ldots$ and $w_0, w_1, \ldots$ be the members of V and W, respectively, in ascending order. Note that $n \in A$ iff $w_n \in E$ iff $v_{w_n} \in B$, hence A is selected from B by the r.e. set $\{v_{w_0}, v_{w_1}, \ldots\}$.   □

Proposition 7. *Exactly the sets $\emptyset$ and $\mathbb{N}$ have selection rank 0. Every finite and every cofinite set that differs from $\emptyset$ and $\mathbb{N}$ has selection rank 1. Every recursive set that is infinite and coinfinite has selection rank 2.*

Proof. The only set that can be selected from the empty set is the empty set itself, hence the empty set has rank 0; the same argument works for $\mathbb{N}$. Next consider a finite set B. In case B differs from $\emptyset$ and $\mathbb{N}$, some set different from B

can be selected from A, thus the selection rank of B is at least 1. However, in case a set A is selected from B in several steps, then A and all the intermediate sets are finite and all selecting sets can be taken to be recursive. Then B can be selected from A in a single step according to Proposition 6, hence the selection rank of A is at most 1. Again, an almost identical argument works for the symmetric case of a coinfinite set.

The selection rank of O is at most 2 by Corollary 2, and is at least 2 because by Theorems 3 and 11 the ω-r.e set Ω can be selected from O in two steps but not in one step. Given any infinite and coinfinite recursive set B, the set B can be selected from O by a recursive set and vice versa. By the absorption principle in Proposition 6, from B and O exactly the same sets can be selected in exactly the same number of steps, hence B and O share the same selection rank.   □

Proposition 8 shows that also nonrecursive sets can have a low selection rank. The proof of the proposition is based on the fact that every set weakly truth-table reducible to the halting problem K is also one-one reducible to its truth-table cylinder by a strictly increasing reduction function; however, due to lack of space, details are omitted. Recall that by definition A is truth-table reducible to B if there are recursive functions f and g where f maps pairs of numbers and strings to bits and the reduction is given by $A(x) = f(x, B(0), B(1), \ldots, B(g(x)))$ for every x. Recall further that by definition a set B is a truth-table cylinder if there are three recursive functions pad, and, neg such that for all x and y, $\mathrm{pad}(x) > x$, $B(\mathrm{pad}(x)) = B(x)$, $B(\mathrm{neg}(x)) = 1 - B(x)$ and $B(\mathrm{and}(x,y)) = B(x) \cdot B(y)$. Furthermore, for any set X, one can choose a truth-table cylinder in the truth-table degree of X and by appropriately restricting this choice to a specific truth-table cylinder obtain *the* truth-table cylinder X^{tt} of X.

Proposition 8. *The truth-table cylinder K^{tt} of the halting problem has selection rank 1 and every ω-r.e. set can be selected from it by a recursive set W.*

3   Selection and Ω

Chaitin's Ω is a standard example for a Martin-Löf random set which is in addition also an ω-r.e. set [1]. It will turn out that Ω has various special properties and some but not all of them are shared by Martin-Löf random sets in general. The following gives an overview about Martin-Löf randomness.

Using a characterisation of Schnorr [13], one can say that a set A is Martin-Löf random [7] iff no r.e. martingale M succeeds on A. In this context, a martingale is a function from binary strings to nonnegative real numbers such that $M(\sigma) = (M(\sigma 0) + M(\sigma 1))/2$. M succeeds on a set A iff the set $\{M(\sigma) : \sigma \preceq A\}$ has the supremum ∞, where $\sigma \preceq A$ means that $\sigma(x) = A(x)$ for all x in the domain of σ; similarly one can compare strings with respect to $\preceq$. Furthermore, M is called r.e. iff $\{(\sigma, q) : \sigma \in \{0,1\}^*, q \in \mathbb{Q}, M(\sigma) > q\}$ is recursively enumerable; M is recursive iff the just defined set is recursive. Without loss of generality, one can take a recursive martingale to be $\mathbb{Q}$-valued and can show that whenever some recursive martingale succeeds on A then some $\mathbb{Q}$-valued recursive martingale succeeds on A where in addition the function $\sigma \mapsto M(\sigma)$ is a recursive mapping which returns on input σ the canonical representation (as a pair of numerator and denominator) of $M(\sigma)$. This also holds relativised to oracles. Furthermore, one can say that a partial-recursive martingale M succeeds on A iff for every $\sigma \preceq A$, $M(\sigma), M(\sigma 0), M(\sigma 1)$ are all defined, for every $\sigma \preceq A$ the relation $M(\sigma) = (M(\sigma 0) + M(\sigma 1))/2$ holds and the supremum of $\{M(\sigma) : \sigma \preceq A\}$ is ∞. It is known that if a recursive martingale succeeds on A, then also a partial-recursive martingale succeeds on A; furthermore, if a partial-recursive martingale succeeds on A then a r.e. martingale succeeds on A. A further characterisation by Zvonkin and Levin [15] is that A is Martin-Löf random iff there is no partial-recursive function G compressing A. Here G compresses A iff G maps strings to strings, the domain of G is prefix free – that is whenever $G(p)$ is defined then $G(pq)$ is undefined for all $p, q \in \{0,1\}^*$ with $q \neq \varepsilon$ – and there are infinitely many n for which there is a p of length at most n with $G(p) = A(0)A(1) \ldots A(n)$. The interested reader is referred to standard textbooks on algorithmic randomness for more information [6, 9].

In this section, the relations between Ω and $\sqsubset$ are investigated. First, Theorem 9 shows every set selected from Ω is Martin-Löf random. Second, the next result shows that one cannot select Ω nontrivially in several steps from itself, that is, there are no sets $E_0, E_1, \ldots, E_n$ such that $E_m \sqsubset E_{m+1}$ via an $W_{e_m} \neq \mathbb{N}$ for all $m < n$ and $E_0 = E_n = \Omega$.

Note that for this section, for an r.e. set W with recursive enumeration $W_0, W_1, \ldots$ (that is, the W_s are uniformly recursive, $W = \bigcup_s W_s$ and $W_0 \subseteq W_1 \subseteq \ldots$), one defines the convergence module $c_W(x)$ is the first stage $s \geq x$ such that $W_s(y) = W(y)$ for all $y \leq x$. Furthermore, one fixes an approximation $\Omega_0, \Omega_1, \ldots$ from the left for Ω, that is, this approximation satisfies the following three conditions:

- the Ω_s are uniformly recursive;
- for all x and almost all s, $\Omega(x) = \Omega_s(x)$;
- whenever $\Omega_{s+1} \neq \Omega_s$, then the least element x in the symmetric difference satisfies $x \in \Omega_{s+1} - \Omega_s$.

Now one defines the convergence module of Ω at x as $c_\Omega(x) = \min\{s \geq x : \forall y \leq x \, [\Omega_s(y) = \Omega(y)]\}$. Note that c_Ω, due to Ω being Martin-Löf random, grows much faster than c_W for any given r.e. set W; in particular there is a constant c such that, for all $x > 0$, $c_\Omega(x-1) + c \geq c_W(x)$. This is used in several of the proofs below, in particular as martingales working on Ω and currently having the task to predict $\Omega(x)$, can from the already known values $\Omega(0) \ldots \Omega(x-1)$ figure out which $y \leq x$ are in finitely many fixed r.e. sets and therefore reconstruct the nature of reductions up to x.

Theorem 9 answers an open question by Kjos-Hanssen, Stephan and Teutsch [4, Question 6.1] on whether a set selected from Ω by an r.e. set is Martin-Löf random; the corresponding question with respect to selections by co-r.e. sets also mentioned there is still open.

Theorem 9. *If A is selected from Ω in finitely many steps then A is Martin-Löf random.*

Proof. Assume that there are a number n and sets $E_0, E_1, \ldots, E_n$, $A = E_0$, $\Omega = E_n$ and for all $m < n$ there is an increasing function F_m with $x \in E_m \Leftrightarrow F(x) \in E_{m+1}$ and the range of F_m being an r.e. set. Note that one knows $F_m(0), F_m(1), \ldots, F_m(y)$ at time s iff all elements of the range of F_m below $F_m(y)$ are enumerated within s time steps (with respect to some given recursive enumeration of the range of F_m). As the convergence module of Ω dominates the convergence module of every r.e. set there is a constant c such that one can, for every $x > 0$ and every $m < n$, compute $F_m(y)$ for all y with $F_m(y) \leq x$ within time $c_\Omega(x-1) + c$.

Now assume by way of contradiction that A is not Martin-Löf random. Miller [8] showed that there is an oracle B which is low for Ω and PA-complete; that is, B satisfies that Ω is Martin-Löf random relative to B and that every partial-recursive $\{0,1\}$-valued function has a total B-recursive extension. It is known that every set which is not Martin-Löf random is not recursively random relative to such an oracle B; hence there is a B-recursive martingale M which succeeds on A.

Now it is shown that M can be transformed into a partial B-recursive martin-gale N succeeding on Ω in contradiction to the choice of B; this N will be defined inductively and the $N(\sigma a)$ will be defined for all $\sigma \preceq \Omega$ and all $a \in \{0,1\}$. This is done by inductively defining sequences $\Phi(\sigma)$ from σ for some partial-recursive function Φ and then letting $N(\sigma) = M(\Phi(\sigma))$. As a starting point, let $\Phi(\varepsilon) = \varepsilon$ and hence $N(\varepsilon) = M(\varepsilon)$. Inductively, $\Phi(\sigma a)$ is defined from $\Phi(\sigma)$ and hence $N(\sigma a)$ from $N(\sigma)$.

Now for any given σ where $\Phi(\sigma)$ and $N(\sigma)$ are defined, one does for $a = 0, 1$ the following: Let $s = t + c$ for the first time $t \geq |\sigma|$ with $\sigma \preceq \Omega_t$; if this time t does not exist then $N(\sigma 0)$ and $N(\sigma 1)$ are undefined. Having s, one computes approximations $F_{m,0}, F_{m,1}, \ldots$ to F_m where $F_{m,s}(y)$ is the y-th element of the set of strings enumerated into the range of F_m within s steps with respect to some recursive enumeration. Let

$$\widetilde{F}_s(y) = F_{n-1,s}(F_{n-2,s}(\ldots(F_{1,s}(F_{0,s}(y)))\ldots))$$

and $\widetilde{F}(y)$ be the limit of $\widetilde{F}_s(y)$. Note that when $\sigma \preceq \Omega$ then $\widetilde{F}_s(y) = \widetilde{F}(y)$ for all y with $\widetilde{F}(y) \leq |\sigma| + 1$ because of the above domination properties; note that the t there would be $c_\Omega(|\sigma|)$. Furthermore, for all y, either $\widetilde{F}_s(y)$ is undefined or $\widetilde{F}_s(y) \geq \widetilde{F}(y)$.

If there is a y such that $\widetilde{F}_s(y) = |\sigma|$ then let $\Phi(\sigma a) = \Phi(\sigma)a$ else let $\Phi(\sigma a) = \Phi(\sigma)$. Furthermore, $N(\sigma a) = M(\Phi(\sigma a))$.

Now one analyses the behaviour of N on Ω. Note that whenever $\sigma a \preceq \Omega$ and $\widetilde{F}_s(y) \in \mathrm{dom}(\sigma a)$ then $\widetilde{F}_s(y) = \widetilde{F}(y)$ where the s is as above. As a consequence, one has for the maximal y with $\widetilde{F}_s(y) \in \mathrm{dom}(\sigma a)$ that $\Phi(\sigma a) = \Omega(\widetilde{F}(0))\Omega(\widetilde{F}(1))\ldots\Omega(\widetilde{F}(y))$ and hence $\Phi(\sigma a) \preceq A$. It follows that N works on Ω like a delayed version of M on A; in particular as M takes on A arbitrarily large values, so does N on Ω. This would mean that N succeeds on Ω in contradiction to the assumption that Ω is Martin-Löf random relative to the oracle B. Thus, against the assumption, the set A has to be Martin-Löf random. $\square$

A similar proof (which is omitted due to page constraints) shows the following result.

Theorem 10. *One cannot select Ω nontrivially in several steps from itself, that is, there are no $n > 0$ and no sets $E_0, E_1, \ldots, E_n$ such that $E_m \sqsubset E_{m+1}$ via an $W_{e_m} \neq \mathbb{N}$ for all $m < n$ and $E_0 = E_n = \Omega$.*

Furthermore, one can also show the following: interchange even and odd positions by letting $\widetilde{\Omega}(2n) = \Omega(2n{+}1)$ and $\widetilde{\Omega}(2n{+}1) = \Omega(2n)$; the set $\widetilde{\Omega}$ cannot be selected from Ω in any number of steps.

4   Selection and Martin-Löf Random Sets in General

After having investigated relations between selection and the special Martin-Löf random set Ω, the focus is now on relations between selection and Martin-Löf random sets in general. First, Theorems 11 and 12 exhibit classes of sets from which no Martin-Löf random set can be selected in one step. Furthermore, Theorems 14 and 15 assert that there is a Martin-Löf random set from which one can select all ω-r.e. sets in up to two steps, whereas no recursive set can be selected from any Martin-Löf random set in one step.

Theorem 11. *Assume that B is Turing reducible to a Turing-incomplete r.e. set. Then no set selected from B is Martin-Löf random.*

Proof. Let B be stated as in the theorem. Recall that a sufficient criterion for a set A to be not Martin-Löf random is that there are infinitely many n such that the plain Kolmogorov complexity of $A(0)A(1)\ldots A(n)$ is bounded proportionally to $\log(n)$. Indeed, in the following it is shown that there are a constant c and infinitely many n such that $C(A(0)A(1)\ldots A(n)) \leq 2 \cdot \log(n) + c$.

Consider any $A \sqsubset B$ and let W be the r.e. set with $A(n) = B(w_n)$ for the n-th element w_n of W in ascending order. Let $b_0, b_1, b_2, \ldots$ be a recursive one-one enumeration of W and let $e_0 = 0$ and e_{n+1} be the first number $d > e_n$ such that $b_{e_n} < b_d$. Note that the mapping $m \mapsto b_{e_m}$ is recursive. Now given any m, let n be the number with $b_{e_m} = w_n$, note that $m \leq n$. Knowing m and n, one can compute $w_0, w_1, \ldots, w_n$.

There is a recursive approximation $B_0, B_1, \ldots$ to B such that the convergence module g of this approximation does not permit to compute the diagonal halting problem K. In particular there are infinitely many $m \in K$ such that m is enumerated into K at a stage s larger than $g(b_{e_m})$ and all w_k with $k \leq n$ satisfy $B_s(w_k) = B(w_k)$. Hence, for these m and the corresponding n, $A(0)A(1)\ldots A(n)$ can be described by m and n using the time s when m is enumerated into K and the members $w_0, w_1, \ldots, w_n$ of W obtained from m, n and conjecturing that $A(k) = B_s(w_k)$ for $k = 0, 1, \ldots, n$. For the right parameters, the s exists and the corresponding data can be computed and the resulting string is correct. As one can describe m and n by two numbers of $\log(n)$ binary digits (the number of digits must be the same for permitting to separate out the digits from m from those for n), $C(A(0)A(1)\ldots A(n)) \leq 2 \cdot \log(n) + c$ for some constant c and infinitely many n. It follows that A is not Martin-Löf random. $\square$

Theorem 12. *Every truth-table degree contains a set B such that no set selected from B is Martin-Löf random.*

Proof. This proof is mainly based on the fact that every truth-table degree contains a retraceable set; here a set B is retraceable iff there is a partial-recursive function ψ which returns for every $x \in B$ a canonical index of the set $\{y \leq x : y \in B\}$; on $x \notin B$, ψ can either be undefined or return any information, either wrong or right. For example, if E is a given set then the set $B = \{x_0, x_1, \ldots\}$ with $x_0 = 1$ and $x_{n+1} = 2x_n + E(n)$ for all n is a retraceable set of the same truth-table degree as E. So fix such B and ψ with B being inside the given truth-table degree. The proof follows now in general the proof of Theorem 11 with the adjustment that it is shown that for each $A \sqsubseteq B$ there are a constant c and infinitely many n such that the plain Kolmogorov complexity of $A(0)A(1) \ldots A(n)$ is bounded by $3 \cdot \log(n) + c$, which then gives that A is not Martin-Löf random.

Consider any $A \sqsubseteq B$ and let W be the r.e. set with $A(n) = B(w_n)$ for the n-th element w_n of W in ascending order. Without loss of generality, $0 \in A$. Let $b_0, b_1, b_2, \ldots$ be a recursive one-one enumeration of W and let $e_0 = 0$ and e_{n+1} be the first number $d > e_n$ such that $b_{e_n} < b_d$. Note that the mapping $m \mapsto b_{e_m}$ is recursive. Now given any m, let n be the number with $b_{e_m} = w_n$, note that $m \leq n$. Knowing m and n, one can compute $w_0, w_1, \ldots, w_n$. Furthermore, let k be such that w_k is the maximal of the $w_0, w_1, w_2, \ldots, w_n$ with $w_k \in B$. Note that $k \leq n$ and k exists as $0 \in A \wedge w_0 \in B$.

Hence, for each n and the corresponding $m, k \leq n$, one can compute $w_0, w_1, \ldots, w_n$ from m, n and use $\psi(w_k)$ to find out which of these numbers are in B. Hence $A(0)A(1) \ldots A(n)$ can be computed from n, m, k. One can code m, n, k as 3 binary numbers of $\log(n)$ digits each and gets therefore that there are a constant c and infinitely many n such that $C(A(0)A(1) \ldots A(n)) \leq 3 \cdot \log(n) + c$. Hence the set A is not Martin-Löf random. $\square$

If one would start with a hyperimmune set B then every $A \sqsubseteq B$ is also hyperimmune and therefore not Martin-Löf random. Hence one has the following result similar to the previous one.

Proposition 13. *There are uncountably many sets B such that no set that is selected from B in one or several steps is Martin-Löf random.*

The following result stands in contrast to Theorem 9, which says that one cannot select any nonrandom set from Ω in arbitrarily many steps. Note that the resulting set B is like Ω also an ω-r.e. Martin-Löf random set. The lengthy proof is omitted due to page constraints.

Theorem 14. *There is a Martin-Löf random set B such that some set selected from B is not Martin-Löf random and every ω-r.e. set can be selected from B in two steps.*

Theorem 15 below shows that the above bound of two steps cannot be brought down to one; indeed, recursive sets can be selected from the above B in exactly two steps. Note that the proof of Theorem 15 indeed shows that it is not possible to select from a Martin-Löf random set any set that obeys certain upper bounds on the complexity of its initial segments.

Theorem 15. *It is not possible to select a recursive set from a Martin-Löf random set.*

Proof. Assume that $A \sqsubseteq B$ via an r.e. set W and A is recursive and F is the function which lists W in ascending order (F is not recursive). So $A(x) = B(F(x))$ for all x and $W = \{F(0), F(1), \ldots\}$. Let $u_0, u_1, \ldots$ be an ascending recursive enumeration of a recursive subset of W which is selected such that W has at least 3^n elements below a given u_n. Now one shows that there is a partial-recursive function G with prefix-free domain which compresses B, that is, for which there are infinitely many $p \in \mathrm{dom}(G)$ with $G(p)$ being a prefix of B which is longer than $|p|$; this would then be an alternative way to prove that B is not Martin-Löf random.

On input $p = 0^n 10^m 1 b_0 b_1 \ldots b_m c_0 c_1 \ldots c_k$, $G(p)$ first checks whether $k + 1 = u_n - d$ where d is the binary value of $b_0 b_1 \ldots b_m$. In the case that this is true, $G(p)$ enumerates the W until d many elements at places $\widetilde{F}(0), \widetilde{F}(1), \ldots, \widetilde{F}(d)$ have been enumerated into W with $\widetilde{F}(0) < \widetilde{F}(1) < \ldots < \widetilde{F}(d) = u_n$. If this is eventually achieved and if $d \geq n$, then G outputs a string $\sigma \in \{0,1\}^{u_n+1}$ which is obtained by letting $\sigma(\widetilde{F}(d')) = A(d')$ for all $d' \leq d$ and by filling the remaining missing $k + 1$ values in σ below the position u_n according to the string $c_0 c_1 \ldots c_k$. This results in a string of length u_n which is computed from a p of length $n + 2m + k + 4$; by taking m as small as possible, one has that $m \leq \log(d) + 1$ and $n \leq \log(d)$, thus one has a length bounded by $u_n + 3 \log(d) - d$ which is, for all sufficiently large n and d (as $d \geq n$) smaller than u_n.

One has now to show that one can always choose d, m, $b_0 b_1 \ldots b_m$ and $c_0 c_1 \ldots c_k$ such that the corresponding output $G(p)$ is $B(0) B(1) \ldots B(u_n)$. To see this, let d be the number of strings in W up to u_n (which is larger than n) and $m = \log(d)$ and $b_0 b_1 \ldots b_m$ be the binary representation of d. Furthermore, let $k = u_n - d - 1$. One gets that $\widetilde{F}(d') = F(d')$ for all $d' \leq d$. Now one chooses $c_0 c_1 \ldots c_k$ such that the missing positions in σ which are not covered by $F(0), F(1), \ldots, F(d)$ are covered with the corresponding bits of B. Hence one has that for the so selected p that $G(p)$ equals $B(0) B(1) \ldots B(u_n)$. It is furthermore easy to verify that the domain of G is prefix-free. □

5   Conclusion

The present paper focussed on the question when a set A is one-one reducible to B via the principal function of an r.e. set and generalised this notion also to reductions in several steps, as this reducibility is not transitive. The investigations show that there is a rich relation between this type of reducibility and ω-r.e. sets and Martin-Löf random sets. Future work might in particular address the question for which numbers $n \in \{0, 1, 2, \ldots, \infty\}$ there are sets A of selection rank n; for $n = 0, 1, 2$, examples are given within this paper and all of these examples are ω-r.e. sets. As the current investigations centered on ω-r.e. sets, subsequent research might also aim for more insights concerning the selection relation among sets that are not ω-r.e. or even not Δ_2^0. For example, one might ask whether every set selected from a strongly random set in finitely many steps is again strongly random; this closure property holds for 2-randomness and also for 2-genericity but not for Martin-Löf randomness and also not for 1-genericity.

References

1. Calude, C.S.: Chaitin Ω numbers, Solovay machines and incompleteness. Theoretical Computer Science 284, 269–277 (2002)
2. Jain, S., Stephan, F., Teutsch, J.: Closed left-r.e. sets. In: Ogihara, M., Tarui, J. (eds.) TAMC 2011. LNCS, vol. 6648, pp. 218–229. Springer, Heidelberg (2011)
3. Kjos-Hanssen, B., Merkle, W., Stephan, F.: Kolmogorov complexity and the recursion theorem. Transactions of the American Mathematical Society 363, 5465–5480 (2011)
4. Kjos-Hanssen, B., Stephan, F., Teutsch, J.: Arithmetic complexity via effective names for random sequences. ACM Transactions on Computational Logic 13(3), 24:1–24:18 (2012)
5. van Lambalgen, M.: The axiomatization of randomness. The Journal of Symbolic Logic 55(3), 1143–1167 (1990)
6. Li, M., Vitányi, P.: An Introduction to Kolmogorov Complexity and its Applications, 3rd edn. Springer (2008)
7. Martin-Löf, P.: The definition of random sequences. Information and Control 9, 602–619 (1966)
8. Miller, J.S.: The K-degrees, low for K degrees, and weakly low for K sets. Notre Dame Journal of Formal Logic 50(4), 381–391 (2010)
9. Nies, A.: Computability and Randomness. Oxford Logic Guides, vol. 51. Oxford University Press, Oxford (2009)
10. Odifreddi, P.: Classical Recursion Theory. Studies in Logic and the Foundations of Mathematics, vol. 125. North-Holland (1989)
11. Odifreddi, P.: Classical Recursion Theory, Volume II. Studies in Logic and the Foundations of Mathematics, vol. 143. Elsevier (1999)
12. Post, E.L.: Recursively enumerable sets of positive integers and their decision problems. Bulletin of the American Mathematical Society 50, 284–316 (1944)
13. Schnorr, C.P.: Zufälligkeit und Wahrscheinlichkeit. Springer Lecture Notes in Mathematics (1971)
14. Soare, R.I.: Recursively Enumerable Sets and Degrees. Springer, Heidelberg (1987)
15. Zvonkin, A.K., Levin, L.A.: The complexity of finite objects and the development of the concepts of information and randomness by means of the theory of algorithms. Russian Mathematical Surveys 25, 83–124 (1970)

On the Boundedness Property of Semilinear Sets

Oscar H. Ibarra[1,*] and Shinnosuke Seki[2,3,**]

[1] Department of Computer Science
University of California, Santa Barbara, CA 93106, USA
ibarra@cs.ucsb.edu
[2] Helsinki Institute of Information Technology (HIIT)
[3] Department of Information and Computer Science
Aalto University, P.O.Box 15400, FI-00076, Aalto, Finland
shinnosuke.seki@aalto.fi

Abstract. An additive system to generate a semilinear set is k-bounded if it can generate any element of the set by repeatedly adding vectors according to its rules so that pairwise differences between components in any intermediate vector are bounded by k except for those that have achieved their final target value. We look at two (equivalent) representations of semilinear sets as additive systems: one without states (the usual representation) and the other with states, and investigate their properties concerning boundedness: decidability questions, hierarchies (in terms of k), characterizations, etc.

Keywords: semilinear set, generator without states, generator with states, bounded, multitape NFA, decidable, undecidable.

1 Introduction

Semilinear sets have been extensively investigated because of their connection to context-free grammars [14] and their many decidable properties that have found applications in various fields such as complexity and computational theory [9,13], formal verification [15], and DNA self-assembly [1]. Nevertheless, there are still interesting problems that remain unresolved, for example, the long-standing open question of S. Ginsburg [3] of whether or not it is decidable if an arbitrary semilinear set is a finite union of stratified linear sets. The purpose of this paper is to examine the "boundedness" properties of additive systems that generate semilinear sets.

A linear set Q is a subset of $\mathbb{N}^n$ (the set of n-dimensional nonnegative integer vectors) that can be specified by a *linear generator* (c, V) as $Q = \{c + i_1 v_1 + \cdots + i_r v_r \mid i_1, \ldots, i_r \in \mathbb{N}\}$, where $c \in \mathbb{N}^n$ is a constant vector and $V = \{v_1, \ldots, v_r\} \subseteq \mathbb{N}^n$ is a *finite* set of periodic vectors. A process for (c, V) to generate a vector $v \in Q$ can be described as a sequence of intermediate vectors $u_0, u_1, \ldots, u_k$, where $u_0 = c$, $u_k = v$, and for $1 \le j \le k$, $u_j - u_{j-1} \in V$. We say that the linear

---

* Supported in part by NSF Grants CCF-1143892 and CCF-1117708.
** Supported in part by HIIT Pump Priming Project Grants 902184/T30606.

T-H.H. Chan, L.C. Lau, and L. Trevisan (Eds.): TAMC 2013, LNCS 7876, pp. 156–168, 2013.

generator (c, V) is *k-bounded* if any vector v in Q admits a generating process where each intermediate vector has the property that the difference in any of its two components neither of which has reached its final value is at most k. (Actually, this definition is valid for any additive vector generating system.) A semilinear set $Q \subseteq \mathbb{N}^n$ is a finite union of linear sets so that a collection of linear generators of the linear sets comprising Q is a conventional way to specify Q and called a *generator* of Q.

The aim of this paper is to compare the conventional generator with another automata-like system to generate semilinear sets, which we call a *generator with states*. We investigate their properties concerning boundedness: decidability questions, hierarchies (in terms of k), and characterizations in terms of synchronized multitape NFAs (which has been recently studied in [2,10,11,12,16]). We show that for any $k \geq 1$, every k-bounded generator with states can be converted into an equivalent $(k - 1)$-bounded one (Proposition 2). Thus, the hierarchy among bounded generators with states with respect to k collapses. This is in marked contrast with the existence of an infinite hierarchy among bounded generators *without* states (Theorem 3). As for the decidability problems, we first show that it is decidable whether a given generator with states is k-bounded for a given $k \geq 0$, and then show the decidability of the problem of determining the existence of such k (Lemma 1). We also show that it is decidable whether a given semilinear set can be generated by a k-bounded (stateless, i.e., conventional) generator for some $k \geq 0$. This is a corollary of our characterization result that a unary n-tuple language L is accepted by a 0-synchronized n-tape NFA if and only if there exists a $k \geq 0$ such that the semilinear set $Q(L) = \{(i_1, i_2, \ldots, i_n) \mid (a^{i_1}, \ldots, a^{i_n}) \in L\}$ can be generated by a k-bounded *stateless* generator (Corollary 3).

Our motivation for studying bounded semilinear sets is that if we know that a semilinear set is bounded, then it can be defined in a simple way in terms of a synchronized multitape NFA, which in turn can be reduced to an ordinary one-tape NFA (as we will see later). Hence decision questions concerning bounded semilinear sets (e.g., disjointness, containment, equivalence, etc.) and their analysis can be reduced to similar questions concerning finite automata.

The paper is organized as follows. After the preliminary section (Sect. 2), we introduce the notion of bounded generator in Sect. 3. We prove the main characterization results in Sect. 4. In Sect. 5, we show that if a linear set Q admits one stateless unbounded linear generator, then all stateless linear generators of Q are also unbounded. We conjecture that this generalizes to semilinear sets. Sect. 6 is an appendix.

2 Preliminaries

For the set $\mathbb{N}$ of natural numbers and $n \geq 1$, $\mathbb{N}^n$ denotes the set of (n-dimensional nonnegative integer) vectors including the zero vector $\bar{0} = (0, 0, \ldots, 0)$. For a vector $v = (i_1, \ldots, i_n) \in \mathbb{N}^n$, $v[j]$ denotes its j-th component, that is, $v[j] = i_j$. A set of vectors $Q \subseteq \mathbb{N}^n$ is called a *linear set* if there is a vector $c \in \mathbb{N}^n$ (constant vector) and a finite (possibly-empty) set $V = \{v_1, \ldots, v_r\} \subseteq \mathbb{N}^n \setminus \{\bar{0}\}$ of nonzero

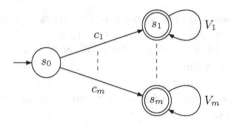

Fig. 1. A generator with states that simulates the vector generation process by a (stateless) generator $G = \{(c_1, V_1), \ldots, (c_m, V_m)\}$

vectors (periodic vectors) such that $Q = \{c + i_1 v_1 + \cdots + i_r v_r \mid i_1, \ldots, i_r \in \mathbb{N}\}$. We denote Q also as $c + V^*$. We call the pair (c, V) a *linear generator* of Q. A set of vectors is called a *semilinear set* if it is a finite union of linear sets. The set of linear generators of linear sets comprising Q is called a *generator* of Q.

Let Σ be an alphabet, and Σ^* be the set of words over Σ. For $w \in \Sigma^*$, let $|w|$ be the number of letters (symbols) in w. For an n-letter alphabet $\Sigma = \{a_1, a_2, \ldots, a_n\}$, the *Parikh map* of a word $w \in \Sigma^*$, denoted by $\psi(w)$, is the vector $(|w|_{a_1}, \ldots, |w|_{a_n})$, where $|w|_{a_i}$ denotes the number of occurrences of the letter a_i in w. The Parikh map (or image) of a language $L \subseteq \Sigma^*$ is defined as $\psi(L) = \{\psi(w) \mid w \in L\}$.

A language $L \subseteq \Sigma^*$ is *bounded* if it is a subset of $w_1^* \cdots w_n^*$ for some nonempty words $w_1, \ldots, w_n \in \Sigma^*$. If all of $w_1, \ldots, w_n$ are pairwise-distinct letters, then L is especially called *letter-bounded*. A bounded language $L \subseteq w_1^* \cdots w_n^*$ is *semilinear* if the set $Q(L) = \{(i_1, \ldots, i_n) \mid w_1^{i_1} \cdots w_n^{i_n} \in L\}$ is a semilinear set.

Basic knowledge of nondeterministic finite automata (NFA) is assumed (see [5] for them). A *(one-way) multitape NFA* is, as the term indicates, an NFA equipped with multiple input tapes each of which has its own (one-way) read-only head. We assume that input tapes of a multitape NFA have right end markers, though they are not indispensable (see Sect. 6). For $k \geq 0$, a multitape machine M (with a right end marker on each tape) is *k-synchronized* if, for any word it accepts, there exists an accepting computation during which the distance between any pair of heads that have not reached the end marker is at most k.

3   Bounded Generators

A standard representation of a semilinear set Q is by a generator $G = \{(c_1, V_1), \ldots, (c_m, V_m)\}$, where $(c_1, V_1), \ldots, (c_m, V_m)$ are linear generators. We call this a *generator without states* or *stateless generator* in contrast to another automata-like representation we will propose shortly.

Definition 1. *A generator G of a semilinear set Q is k-bounded if for every n-tuple $(x_1, \ldots, x_n)$ in Q, there exists a linear generator $(c_i, V_i) \in G$ and periodic vectors $v_{i_1}, \ldots, v_{i_r} \in V_i$ such that the following holds:*

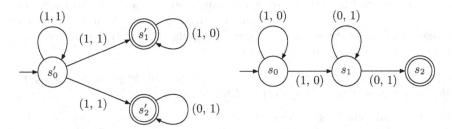

Fig. 2. Examples of generators with states for the semilinear set $Q = \{(i,j) \mid i, j \geq 1\}$, where double circles indicate accepting states. The left generator is 0-bounded whereas the right one is not k-bounded for any $k \geq 0$.

1. $(x_1, \ldots, x_n) = c_i + v_{i_1} + \cdots + v_{i_r}$.
2. For $1 \leq j < r$, if $c_i + v_{i_1} + \cdots + v_{i_j} = (y_1, \ldots, y_n)$, then every $1 \leq p < q \leq n$ such that $y_p \neq x_p$ and $y_q \neq x_q$ satisfies $|y_p - y_q| \leq k$.

Thus, every n-tuple in Q can be obtained by adding to c_i ($1 \leq i \leq m$) periodic vectors in V_i one after another in such a way that after each vector addition, the resulting n-tuple has the property that the difference of any of its two components neither of which has reached its final value is at most k. This property is not trivial as seen in the following example.

Example 1. The linear generator $((0,0), \{(0,1),(1,0),(1,1)\})$ generates a linear set $Q_1 = \{(i,j) \mid i, j \geq 1\}$ and actually it is 0-bounded; a tuple (i,j) with $1 \leq i \leq j$ can be generated as $(i,j) = (0,0)+(1,1)+\cdots+(1,1)+(0,1)+\cdots+(0,1)$, where $(1,1)$ occurs i times and $(0,1)$ occurs $j - i$ times; and an analogous way to sum periodic vectors works for the other case when $0 \leq j \leq i$. Similarly, $Q_2 = \{(i,i) \mid i \geq 0\}$ can be generated by a 0-bounded linear generator. In contrast, as for $Q_3 = \{(i,2i) \mid i \geq 0\}$, even k-bounded generator (not-necessarily linear) does not exist for any k. This will be rigorously shown in Example 2.

We will show the relationships between the boundedness of generators of n-dimensional semilinear sets and the head-synchronization of n-tape NFAs over unary inputs $a_1^* \times \cdots \times a_n^*$ ($a_1, \ldots, a_n$ are letters which do not have to be pairwise distinct). For this purpose, we generalize the specification of generators by adding the notion of state transition. A *generator with states* is specified by a 5-tuple $G_s = (S, T, \bar{0}, s_0, F)$, where S is a finite set of states, the zero vector $\bar{0} = (0, \ldots, 0)$ is the starting vector, $s_0 \in S$ is the initial state, $F \subseteq S$ is the set of final or accepting states, and T is a finite set of transitions of the form: $s \rightarrow (s', v)$, where $s, s' \in S$ are states and v is a vector in $\mathbb{N}^n$. The set generated by G_s consists of the vectors in $\mathbb{N}^n$ that can be obtained from $\bar{0}$ by adding the assigned vector every time a transition occurs until a final state is reached. We denote the set by $Q(G_s)$. Two generators with states are illustrated in Fig. 2, which are for the same semilinear set $Q = \{(i,j) \mid i, j \geq 1\}$. The generator with states is a variant of vector addition system with states (VASS) [6].

As illustrated in Fig. 1, the vector generation process by a (stateless) generator $G = \{(c_1, V_1), \ldots, (c_m, V_m)\}$ can be simulated by a generator with $m+1$ states. The converse is also true; in fact, introducing the notion of states does not expand the class of generable vector sets.

Proposition 1. *The class of the sets of vectors that can be generated by a generator with states is equal to the class of semilinear sets.*

Proof. It is sufficient to show that the set of vectors that can be generated by a generator G_s with states is semilinear. We construct an n-tape NFA which, when given an n-tuple $(a_1^{i_1}, \ldots, a_n^{i_n})$, simulates G_s and checks if the input can be generated by G_s. (Note that M can only move a head at most one cell to the right at each step, but by using more states, M can simulate a vector addition by G_s in a finite number of steps.) When G_s accepts, M moves all its heads to the right and accepts if they are all on the end marker. It follows from [4] that the set $Q = \{(i_1, \ldots, i_n) \mid (a_1^{i_1}, \ldots, a_n^{i_n}) \in L(M)\}$ is semilinear. □

Given a generator G_s with states, interpreting the vector v assigned to a state transition as moving the i-th head to the right by $v[i]$ enables us to regard G_s as an n-tape NFA that accepts $\{(a_1^{i_1}, \ldots, a_n^{i_n}) \mid (i_1, \ldots, i_n) \in Q(G_s)\}$, and the converse interpretation is also valid (note that the guessing power of a multitape NFA eliminates the need for end markers, see Sect. 6). If the NFA thus interpreted is k-synchronized, then any vector v in $Q(G_s)$ admits a generation process by G_s in such a manner that for any intermediate vector u and $1 \le i, j \le n$ with $u[i] < v[i]$ and $u[j] < v[j]$, the inequality $|u[i] - u[j]| \le k$ holds. We call this property the k-*boundedness* of generator with states. For instance, the left generator in Fig. 2 is 0-bounded, while the right one is not k-bounded for any $k \ge 0$. This definition of bounded generators is consistent with the one given in Definition 1 (as shown in Fig. 1, a stateless generator can be regarded as a generator with states).

It is known that any k-synchronized n-tape NFA admits an equivalent 0-synchronized one [11]. Hence, we have:

Proposition 2. *For any k-bounded generator with states, there exists a 0-bounded generator with states that generates the same semilinear set.*

In Example 1, we claimed that there does not exist $k \ge 0$ such that the set $\{(i, 2i) \mid i \ge 0\}$ could be generated by a k-bounded (stateless) generator. Here, we prove this claim, even for generator with states.

Example 2. For the sake of contradiction, suppose that the semilinear set $Q = \{(i, 2i) \mid i \ge 0\}$ has a k-bounded generator G_s with p states for some $k \ge 0$. Proposition 2 enables us to assume that G_s is 0-bounded. Consider the tuple $(p, 2p)$ in Q. Then G_s would generate a tuple $(p + k, 2p + k)$ for some $k \ge 1$ but this is not in Q.

In contrast to Proposition 2, we will see later (Theorem 3) that there exists an infinite hierarchy of semilinear sets with respect to the degree k of boundedness for stateless generators.

Lemma 1. *Given a generator G_s with states,*

1. it is decidable whether G_s is k-bounded for a given $k \geq 0$;
2. it is decidable whether G_s is k-bounded for some $k \geq 0$.

Proof. As mentioned above, G_s can be considered as an n-tape NFA M and G_s is k-bounded if and only if M is k-synchronized.

We construct another n-tape NFA M' that simulates M and makes sure that during the simulation, the separation between any pair of heads that have not reached the end marker is at most k; otherwise M' rejects. It follows that G_s is k-bounded if and only if $L(M) = L(M')$. The first decidability holds because equivalence of n-tape NFAs over bounded inputs is decidable [8].

As for the second decidability, it suffices to decide whether M' is k-synchronized for some $k \geq 0$, and this is known to be decidable [2]. (The result in [2] was for $n = 2$, but can be generalized for an arbitrary n.)   □

In this proof, we can see that an exhaustive search brings us an integer $k \geq 0$ such that G_s is k-bounded, if such k exists, and a k-synchronized n-tape NFA M' that accepts the language $\{(a_1^{i_1}, \ldots, a_n^{i_n}) \mid (i_1, \ldots, i_n) \in Q(G_s)\}$. Recall that M' can be effectively converted into an equivalent 0-synchronized n-tape NFA. Thus, the next result holds.

Theorem 1. *For $k \geq 0$, a k-bounded generator G_s with states can be effectively converted into a 0-synchronized n-tape NFA that accepts $L = \{(a_1^{i_1}, \ldots, a_n^{i_n}) \mid (i_1, \cdots, i_n) \in Q(G_s)\}$.*

4   Boundedness of Stateless Generators and Head-Synchronization of Multitape NFAs

The conversion of Theorem 1 trivially works for any k-bounded *stateless* generator. Interestingly, the following converse is also true.

Theorem 2. *If a language $L \subseteq a_1^* \times \cdots \times a_n^*$ is accepted by a 0-synchronized n-tape NFA M, we can effectively compute $k \geq 0$ and a k-bounded (stateless) generator for the semilinear set $Q(L) = \{(i_1, \ldots, i_n) \mid (a_1^{i_1}, \ldots, a_n^{i_n}) \in L\}$.*

The proof of this theorem requires some preliminary notions and lemmas. First of all, as pointed out in [12], we can regard a 0-synchronized n-tape NFA as an NFA that works on one tape over an extended alphabet Π of n-*track symbols*, which is defined as:

$$\Pi = \left\{ \begin{bmatrix} a_1 \\ \vdots \\ a_n \end{bmatrix} \middle| a_1, \ldots, a_n \in \Sigma \cup \{\square\} \right\},$$

where $\square \notin \Sigma$ is the special letter for the blank symbol. Track symbols are distinguished from tuples of letters by square brackets, and for the space sake, written as $t = [a_1, \ldots, a_n]^T$; the superscript T will be omitted unless confusion arises. For an index $1 \le i \le n$, $t[i]$ denotes the letter on the i-th track of t, that is, a_i.

We endow Π with the partial order $\preceq$, which is defined as: for track symbols $t_1, t_2 \in \Pi$, $t_1 \preceq t_2$ if $t_1[i] = \square$ implies $t_2[i] = \square$ for all $1 \le i \le n$. For example, $[\square, b, c]$ is smaller than $[\square, \square, c]$ according to this order but incomparable with $[a, \square, c]$. An n-track word $t_1 t_2 \cdots t_m$ is *left-aligned* if $t_1 \preceq t_2 \preceq \cdots \preceq t_m$ holds. Informally speaking, on a left-aligned track word, once we find $\square$ on a track, then to its right we will find nothing but $\square$'s. The left-aligned n-track word $t_1 t_2 \cdots t_m$ can be converted into the n-tuple $(h(t_1[1]t_2[1] \cdots t_m[1]), \ldots, h(t_1[n]t_2[n] \cdots t_m[n]))$ of words over Σ using the homomorphism $h : \Sigma \cup \{\square\} \to \Sigma$ to erase $\square$. For instance, $[a, b, c][\square, b, c][\square, \square, c]$ is thus converted into $(h(a \square \square), h(bb\square), h(ccc))$, which is (a, bb, ccc). By reversing this process, we can retrieve the original left-aligned n-track word from the resulting n-tuple of words. This one-to-one correspondence makes possible to assume that 0-synchronized n-tape NFAs, being regarded as a 1-tape NFA over track symbols, accept only left-aligned inputs.

Powers of t are left-aligned. This is because $t \preceq t$ (reflexiveness) holds for any track symbol $t \in \Pi$ as $\preceq$ is a partial order. The m-th power of $t \in \Pi$ corresponds to the n-tuple $(t[1]^m, t[2]^m, \ldots, t[n]^m)$. The next lemma should be straightforward.

Lemma 2. *Let t be an n-track symbol and $m \ge 0$. Then for any $1 \le j \le n$,*

$$|t^m[j]| = \begin{cases} 0 & \text{if } t[j] = \square, \\ m & \text{otherwise.} \end{cases}$$

First, we show a property of a linear set that corresponds to a given unary n-track language L, that is, $L \subseteq t^*$ for some track symbol $t \in \Pi$.

Lemma 3. *Let L be a language over one n-track symbol $t = [a_1, \ldots, a_n] \in \Pi$. If a linear generator (c, V) generates $\{(i_1, \ldots, i_n) \mid [a_1^{i_1}, \ldots, a_n^{i_n}] \in L\}$, then for any vectors $v_1, v_2 \in \{c\} \cup V$ and $1 \le i, j \le n$,*

1. *$v_1[i] \ne 0$ if and only if $v_2[i] \ne 0$.*
2. *if $v_1[i], v_1[j] \ne 0$, then $v_1[i] = v_1[j]$;*

Proof. Let us begin with the proof for 1. For the sake of contradiction, suppose that there were a periodic vector v and $1 \le i \le n$ such that either

a). $c[i] = 0$ but $v[i] \ne 0$, or
b). $c[i] \ne 0$ but $v[i] = 0$.

The word $[a_1^{c[1]}, \ldots, a_1^{c[n]}]$ belongs to L, and in the case a), its i-th component is the empty word. Then there are two subcases to be examined depending on whether c is the zero vector or not. We consider only the subcase when it is not, and see the subcase lead us to a contradiction; the other case can be proved also contradictory by comparing two periodic vectors instead. So, if $c[i] = 0$ but c is not zero, then the word $(a_1^{c[1]}, \ldots, a_1^{c[n]})$ in L is not the 0-th power of

t; nevertheless its i-th component is the empty word. This means that the i-th component of the track symbol t is the empty word, and as a result, so is the i-th component of all track words in L. Since $[a_1^{c[1]+v[1]}, \ldots, a_n^{c[n]+v[n]}] \in L$, $c[i] + v[i] = 0$, but this contradicts $v[i] \neq 0$. Thus, if $c[i] = 0$, then $v[i] = 0$.

In the case b), all the words in L that correspond to $c + \{v\}^*$ have $a_1^{c[i]}$ as their i-th component. Lemma 2 implies that nonzero components of these words are all of length $|a_1^{c[i]}|$. This means that this linear set is finite, and hence, v would be the zero vector.

As for the property 2, since L contains the word $[a_1^{c[1]}, \ldots, a_n^{c[n]}]$ and L is unary, if neither $a_i^{c[i]}$ nor $a_j^{c[j]}$ is the empty word, then $c[i] = c[j]$ (Lemma 2). For a periodic vector v, the property 1 gives $c[i], c[j] \neq 0$, and hence, $c[i] = c[j]$ as just proved. We can use Lemma 2 to derive $c[i] + v[i] = c[j] + v[j]$. Combining these two equations together results in $v[i] = v[j]$. □

Corollary 1. *For a language L over one track symbol $t = [a_1, \ldots, a_n] \in \Pi$, any linear generator that generates $\{(i_1, \ldots, i_n) \mid [a_1^{i_1}, \ldots, a_n^{i_n}] \in L\}$ is 0-bounded.*

The closure property of the set of 0-bounded semilinear sets under finite union strengthens this corollary further as follows.

Corollary 2. *For a language L over one track symbol $t = [a_1, \ldots, a_n] \in \Pi$, any generator $G = \{(c_1, V_1), \ldots, (c_m, V_m)\}$ that generates $\{(i_1, \ldots, i_n) \mid [a_1^{i_1}, \ldots, a_n^{i_n}] \in L\}$ is 0-bounded.*

Lemma 4. *Let L be a language accepted by an n-track 1-tape NFA over one symbol. Then the set $Q(L) = \{(i_1, \ldots, i_n) \mid [a_1^{i_1}, \ldots, a_n^{i_n}] \in L\}$ is a 0-bounded semilinear set.*

Proof. The semilinearity of $Q(L)$ follows from a result in [4], and then the above argument gives its 0-boundedness. □

Now we are ready to prove Theorem 2.

Proof of Theorem 2. Since M is 0-synchronized, we regard it rather as an n-track 1-tape NFA. We also assume that M is free from any state from which no accepting state is reachable. Let $M = (\Pi, S, s_0, \delta, F)$, where S is a set of states, $s_0 \in S$ is an initial state, δ is the transition function, and F is the set of final states.

For some $m \geq 0$, let $t_1, t_2, \ldots, t_m \in \Pi$ be distinct track symbols with $t_1 \preceq t_2 \preceq \cdots \preceq t_m$. That is to say, $t_1 = [a_1, \ldots, a_n]$, and t_{i+1} contains at least one more tracks with $\square$ than t_i. Thus, $m \leq n$. Needless to say, there are only finite number of such choices of $t_1, \ldots, t_m$ from Π since Σ is finite. Due to the closure property of the set of 0-bounded generators under finite union, it suffices to show a way to compute $k \geq 0$ and a k-bounded generator for $Q(L(M) \cap t_1^+ \cdots t_m^+)$.

Let $L_1 = L \cap t_1^+ \cdots t_m^+$. In the acceptance computation for L_1, the finite-state control of M traverses from its sub-NFA over t_1 to its another sub-NFA over t_2, and so forth. Formally, for a track symbol $t \in \Pi$, the sub-NFA of M over t consists of all t-transitions of M and all vertices of M associated with them, and we denote it by $M(t)$. Moreover, for states $p, q \in S$, by $M(t)[p, q]$, we denote the sub-NFA that can be obtained from $M(t)$ by appointing p, q as initial state and final state, respectively. For any word $w \in L_1$, there exist states $s_1, s_2, \ldots, s_{m-1} \in S$ and a final state $s_f \in F$ such that an acceptance computation of M for w goes through $M(t_1)[s_0, s_1], M(t_2)[s_1, s_2], \ldots, M(t_m)[s_{m-1}, s_f]$. Each of these m sub-NFAs accepts a unary track language so that any of their generators are 0-bounded due to Lemma 4 and they can be effectively computed [8].

Now we have generators (collections of linear generators) for each of the m unary languages. For each $1 \leq i \leq m$, from the corresponding collection we choose a linear generator (c_i, V_i). Then from them we construct a linear generator (c, V), where $c = \sum_{1 \leq i \leq m} c_i$ and $V = \bigcup_{1 \leq i \leq m} V_i$. We claim that this linear generator is k-bounded for $k = \max\{|c[i] - c[j]| \mid 1 \leq i, j \leq n, c[i], c[j] \neq 0\}$. By definition, any vector u in the set $c + V^*$ admits vectors $v_{i_1}, \ldots, v_{i_r}$ some of whose are firstly taken from V_1, then taken from V_2, and so on such that $c + v_{i_1} + \cdots + v_{i_r} = u$. The offset created by the constant vector c, which is at most k, is not enlarged as long as these vectors are added in the order because of the properties mentioned in Lemma 3 and $t_1 \preceq t_2 \preceq \cdots \preceq t_m$. As such, a bounded linear generator is constructed for each choice of linear generators from the m generators. Taking their union yields a generator for $Q(L(M) \cap t_1^+ \cdots t_m^+)$, and the resulting generator is bounded by the maximum bound degree of the summands (linear generators). $\square$

Having proved Theorem 2, now we combine it with Theorems 1 as:

Corollary 3. *A language $L \subseteq a_1^* \times \cdots \times a_n^*$ is accepted by a 0-synchronized n-tape NFA M if and only if the semilinear set $Q(L) = \{(i_1, \ldots, i_n) \mid (a_1^{i_1}, \ldots, a_n^{i_n}) \in L\}$ can be generated by a k-bounded (stateless) generator for some $k \geq 0$.*

It can be shown that it is decidable whether the language accepted by an n-tape NFA over $a_1^* \times \cdots \times a_n^*$ is accepted by a 0-synchronized NFA or not. Thus, the next corollary holds.

Corollary 4. *It is decidable whether, for a given semilinear set Q, there exists a k-bounded (stateless) generator for some $k \geq 0$.*

Corollary 3 also strengthens Proposition 2 as follows.

Corollary 5. *For a semilinear set Q, the following statements are equivalent:*

1. *Q can be generated by a k-bounded generator with states for some $k \geq 0$;*
2. *Q can be generated by a 0-bounded generator with states;*
3. *Q can be generated by a k'-bounded (stateless) generator for some $k' \geq 0$;*

As we will be convinced of by Theorem 3, the integer k' in the third statement of this corollary cannot be replaced by 0. That is, there exists a hierarchy of (stateless) generators with respect to the degree of boundedness, as compared with the collapse of corresponding hierarchy among generators with states (Proposition 2). This signifies structures more complex than the one shown in Fig. 1. Let us denote by $\mathcal{Q}_k$ the class of k-bounded semilinear sets.

Theorem 3. *For $k \geq 0$, $\mathcal{Q}_{k+1}$ properly contains $\mathcal{Q}_k$.*

Proof. We prove that the semilinear set $\{(k+i, i) \mid i \geq 0\}$ can be generated by a k-bounded (stateless) generator but for any $r < k$, there is no r-bounded (stateless) generator for that. Clearly, a linear generator $((k, 0), \{(1, 1)\})$ generates the set and k-bounded.

Suppose the set could be generated by an r-bounded generator G for some $r < k$. Consider the tuple $x = (k + n, n)$, where n is greater than the largest of the two components in all constant vectors in the linear generators of G. By assumption, x can be generated by an r-bounded generation, where $r < s$. Let c be the constant vector used in such a generation. Since $c[1], c[2] < n$, at least one periodic vector v is used in the generation. In fact, since $|c[1] - c[2]| \leq r < k$, at least one periodic vector v with $v[1] > v[2]$ is used in such a generation. It follows that $(c[1], c[2]) + j(v[1], v[2]) = (c[1] + jv[1], c[2] + jv[2])$ would be in the semilinear set for all j. But then, except for a finite number of j's, $(c[1] + jv[1], c[2] + jv[2])$ are not of the form $(k + i, i)$, which is not possible. $\qquad\square$

5   Boundedness of a Semilinear Set

In Fig. 2, we illustrated two generators with states that generate the same semilinear set, but one of which is 0-bounded and the other is not k-bounded for any $k \geq 0$. We conjecture that such a phenomenon cannot happen for (stateless) generators.

Conjecture 1. If a semilinear set Q is generated by a (stateless) generator that is not k-bounded for any k, then every (stateless) generator of Q would not be k-bounded for any k, either.

If this conjecture is true, then the boundedness becomes a property of semilinear set as long as stateless generators are concerned. We conclude this paper by proving that Conjecture 1 is true for linear sets by providing one sufficient condition for a semilinear set to satisfy the conjectured property (Lemma 5) and observing that any linear set satisfies the condition.

Lemma 5. *Let Q be a semilinear set that admits two generators $G_1 = \{(c_1, V_1), \ldots, (c_m, V_m)\}$ and $G_2 = \{(d_1, U_1), \ldots, (d_n, U_n)\}$ such that, for any $1 \leq i \leq m$, there exists $1 \leq j_i \leq n$ satisfying:*

1. *$c_i \in d_{j_i} + U_{j_i}^*$;*
2. *$V_i \subseteq U_{j_i}^*$.*

If G_1 is k-bounded, then G_2 is k'-bounded for some $k' \geq 0$.

Proof. Since G_1 is k-bounded, any element in Q admits a k-bounded derivation $c_i + v_1 + v_2 + \cdots + v_\ell$, where $\ell \geq 0$ and $v_1, \ldots, v_\ell \in V_i$. Due to the first property, c_i can be written as $c_i = d_{j_i} + x_1 + x_2 + \cdots + x_s$, where $x_1, x_2, \ldots, x_s \in U_{j_i}$. We replace the constant vector c_i in the above derivation with this sum, and obtain the derivation $(d_{j_i} + x_1 + x_2 + \cdots + x_s) + v_1 + v_2 + \cdots + v_\ell$. Some of the newly-introduced intermediate products $d_{j_i}, d_{j_i} + x_1, \ldots, d_{j_i} + x_1 + x_2 + \cdots + x_{s-1}$ may not be k-bounded, but it should be clear the existence of an integer $k' \geq k$ such that they are k'-bounded. We can replace $v_1, v_2, \ldots, v_\ell$ likewise, but using the second property instead, one by one while preserving the bounded property. □

When Q is linear, we show in Lemma 6 below that any of its generators satisfy the two properties stated in Lemma 5.

Lemma 6. *If a linear set has two linear generators (c_1, V_1) and (c_2, V_2), then $c_1 = c_2$, $V_1 \subseteq V_2^*$, and $V_2 \subseteq V_1^*$ hold.*

Proof. Suppose $c_1 \neq c_2$. Then either $c_1 \subseteq c_2 + V_2^+$ or $c_2 \subseteq c_1 + V_1^+$ must hold. It suffices to consider the former. Then c_2 is smaller than c_1 with respect to the components comparison. Thus, $c_2 \subsetneq c_1 + V_1^*$, but this contradicts that the two generators specify the same linear set.

Let $c = c_1 = c_2$. We have $c + V_1^* = c + V_2^*$. If $V_1 \subseteq V_2^*$ did not hold, then there would exist a vector $v_1 \in V_1$ such that $v_1 \notin V_2^*$, and hence, $c + v_1 \notin c + V_2^*$, a contradiction. In the same manner, we can prove $V_2 \subseteq V_1^*$. □

Corollary 6. *If a linear set Q is generated by a k-bounded generator $G = (c, V)$ for some $k \geq 0$, then for any other linear generator G' for Q, there is an integer $k' \geq 0$ such that $G' = (c', V')$ is k'-bounded.*

In other words, if a linear set admits one unbounded linear generator, then all of its linear generators are also unbounded.

6 Appendix

An n-tape NFA without end markers accepts if it enters an accepting state after it has scanned all the tapes.

Proposition 3. *The following two statements hold:*

1. *If M is an n-tape NFA without end markers, we can construct an n-tape NFA M' with end markers such that $L(M') = L(M)$. Moreover, M' is k-synchronized if and only if M is k synchronized ($k \geq 0$).*
2. *If M is an n-tape NFA with end markers, we can construct an n-tape NFA M' without end markers such that $L(M') = L(M)$. Moreover, M' is k-synchronized if and only if M is k-synchronized ($k \geq 0$).*

Proof. The first statement is obvious. Given M, we construct M' that faithfully simulates M. When M enters an accepting state, M' moves all the tape heads to the right and accepts if they are all on the end marker.

For the second statement, let M be an n-tape NFA with end markers. We describe the construction of an n-tape NFA M' without end markers that simulates M. Let H be a set of head indices to specify which heads have reached the end marker according to M's guesses, being initialized empty. H will be stored in the finite control of M' and will be updated during the computation.

(**) M' nondeterministically guesses to execute (1) or (2) below.

1. M' guesses that after the next move, some but not all heads of M will reach the end marker. In this case M' simulates the move of M and also updates H to include the indices of the heads that would reach the end marker after the move according to the guess. In subsequent simulations of the moves of M, M' assumes that the heads in H are on the end marker. M' then proceeds to (**).
2. M' guesses that after the next move, all the remaining heads of M not in H will reach the end marker. In this case M' simulates in *one move* the next move of M and the moves that follow when all heads are on the endmarker, and accepts if M accepts.

Clearly, $L(M') = L(M)$, and M' is k-synchronized if and only if so is M.   □

References

1. Doty, D., Patitz, M.J., Summers, S.M.: Limitations of self-assembly at temperature 1. Theoretical Computer Science 412(1-2), 145–158 (2011)
2. Eğecioğlu, Ö., Ibarra, O.H., Tran, N.Q.: Multitape NFA: Weak synchronization of the input heads. In: Bieliková, M., Friedrich, G., Gottlob, G., Katzenbeisser, S., Turán, G. (eds.) SOFSEM 2012. LNCS, vol. 7147, pp. 238–250. Springer, Heidelberg (2012)
3. Ginsburg, S.: The Mathematical Theory of Context-Free Languages. McGraw-Hill, New York (1966)
4. Harju, T., Ibarra, O.H., Karhumäki, J., Salomaa, A.: Some decision problems concerning semilinearity and commutation. Journal of Computer and System Sciences 65(2), 278–294 (2002)
5. Hopcroft, J., Ullman, J.: Introduction to Automata Theory, Languages, and Computation. Addison-Wesley (1979)
6. Hopcroft, J., Pansiot, J.-J.: On the reachability problem for 5-dimensional vector addition systems. Theoretical Computer Science 8, 135–159 (1979)
7. Huynh, D.T.: The complexity of semilinear sets. In: de Bakker, J.W., van Leeuwen, J. (eds.) ICALP 1980. LNCS, vol. 85, pp. 324–337. Springer, Heidelberg (1980)
8. Ibarra, O.H.: Reversal-bounded multicounter machines and their decision problems. J. Assoc. Comput. Mach. 25, 116–133 (1978)
9. Ibarra, O.H., Seki, S.: Characterizations of bounded semilinear languages by one-way and two-way deterministic machines. International Journal of Foundations of Computer Science 23(6), 1291–1305 (2012)
10. Ibarra, O.H., Tran, N.Q.: Weak synchronization and synchronizability of multitape pushdown automata and Turing machines. In: Dediu, A.-H., Martín-Vide, C. (eds.) LATA 2012. LNCS, vol. 7183, pp. 337–350. Springer, Heidelberg (2012)
11. Ibarra, O.H., Tran, N.Q.: How to synchronize the heads of a multitape automaton. In: Moreira, N., Reis, R. (eds.) CIAA 2012. LNCS, vol. 7381, pp. 192–204. Springer, Heidelberg (2012)
12. Ibarra, O.H., Tran, N.Q.: On synchronized multitape and multihead automata. In: Holzer, M. (ed.) DCFS 2011. LNCS, vol. 6808, pp. 184–197. Springer, Heidelberg (2011)
13. Lavado, G.J., Pighizzini, G., Seki, S.: Converting nondeterministic automata and context-free grammars into Parikh equivalent deterministic automata. In: Yen, H.-C., Ibarra, O.H. (eds.) DLT 2012. LNCS, vol. 7410, pp. 284–295. Springer, Heidelberg (2012)

14. Parikh, R.J.: On context-free languages. J. Assoc. Comput. Mach. 13, 570–581 (1966)
15. To, A.W.: Model Checking Infinite-State Systems: Generic and Specific Approaches, Ph.D. thesis, School of Informatics, University of Edinburgh (2010)
16. Yu, F., Bultan, T., Ibarra, O.H.: Relational string verification using multi-track automata. Int. J. Found. Comput. S. 22, 1909–1924 (2011)

Turing Machines Can Be Efficiently Simulated by the General Purpose Analog Computer

Olivier Bournez[1], Daniel S. Graça[2,3], and Amaury Pouly[1,2]

[1] Ecole Polytechnique, LIX, 91128 Palaiseau Cedex, France
`Olivier.Bournez@lix.polytechnique.fr`
[2] CEDMES/FCT, Universidade do Algarve, C. Gambelas, 8005-139 Faro, Portugal
`dgraca@ualg.pt`
[3] SQIG /Instituto de Telecomunicações, Lisbon, Portugal

Abstract. The Church-Turing thesis states that any sufficiently powerful computational model which captures the notion of algorithm is computationally equivalent to the Turing machine. This equivalence usually holds both at a computability level and at a computational complexity level modulo polynomial reductions. However, the situation is less clear in what concerns models of computation using real numbers, and no analog of the Church-Turing thesis exists for this case. Recently it was shown that some models of computation with real numbers were equivalent from a computability perspective. In particular it was shown that Shannon's General Purpose Analog Computer (GPAC) is equivalent to Computable Analysis. However, little is known about what happens at a computational complexity level. In this paper we shed some light on the connections between this two models, from a computational complexity level, by showing that, modulo polynomial reductions, computations of Turing machines can be simulated by GPACs, without the need of using more (space) resources than those used in the original Turing computation, as long as we are talking about bounded computations. In other words, computations done by the GPAC are as space-efficient as computations done in the context of Computable Analysis.

1 Introduction

The Church-Turing thesis is a cornerstone statement in theoretical computer science, stating that any (discrete time, digital) sufficiently powerful computational model which captures the notion of algorithm is computationally equivalent to the Turing machine (see e.g. [19], [23]). It also relates various aspects of models in a very surprising and strong way.

The Church-Turing thesis, although not formally a theorem, follows from many equivalence results for discrete models and is considered to be valid by the scientific community [19]. When considering non-discrete time or non-digital models, the situation is far from being so clear. In particular, when considering models working over real numbers, several models are clearly not equivalent [9].

However, a question of interest is whether physically *realistic* models of computation over the real numbers are equivalent, or can be related. Some of the

T-H.H. Chan, L.C. Lau, and L. Trevisan (Eds.): TAMC 2013, LNCS 7876, pp. 169–180, 2013.

results of non-equivalence involve models, like the BSS model [5], [4], which are claimed not to be physically realistic [9] (although they certainly are interesting from an algebraic perspective), or models that depend critically of computations which use exact numbers to obtain super-Turing power, e.g. [1], [3].

Realistic models of computation over the reals clearly include the *General Purpose Analog Computer (GPAC)* [21], an analog continuous-time model of computation and *Computable Analysis* (see e.g. [24]). The GPAC is a mathematical model introduced by Claude Shannon of an earlier analog computer, the Differential Analyzer. The first general-purpose Differential Analyzer is generally attributed to Vannevar Bush [10]. Differential Analyzers have been used intensively up to the 1950's as computational machines to solve various problems from ballistic to aircraft design, before the era of the digital computer [18].

Computable analysis, based on Turing machines, can be considered as today's most used model for talking about computability and complexity over reals. In this approach, real numbers are encoded as sequences of discrete quantities and a discrete model is used to compute over these sequences. More details can be found in the books [20], [17], [24]. As this model is based on classical (digital and discrete time) models like Turing machines, which are considered to be realistic models of today's computers, one can consider that Computable Analysis is a realistic model (or, more correctly, a theory) of computation.

Understanding whether there could exist something similar to a Church-Turing thesis models of computation involving real numbers, or whether analog models of computation could be more powerful than today's classical models of computation motivated us to try to relate GPAC computable functions to functions computable in the sense of computable analysis.

The paper [6] was a first step towards the objective of obtaining a version of the Church-Turing thesis for physically feasible models over the real numbers. This paper proves that, from a computability perspective, Computable Analysis and the GPAC are equivalent: GPAC computable functions are computable and, conversely, functions computable by Turing machines or in the computable analysis sense can be computed by GPACs. However this is about *computability*, and not *computational complexity*. This proves that one cannot solve more problems using the GPAC than those we can solve using discrete-based approaches such as Computable Analysis. But this leaves open the question whether one could solve some problems *faster* using analog models of computations (see e.g. what happens for quantum models of computations...). In other words, the question of whether the above models are equivalent at a computational complexity level remained open. Part of the difficulty stems from finding an appropriate notion of complexity (see e.g. [22], [2]) for analog models of computations.

In the present paper we study both the GPAC and Computable Analysis at a complexity level. In particular, we introduce measures for space complexity and show that, using these measures, both models are equivalent, even at a computational complexity level, as long as we consider time-bounded simulations. Since we already have shown in our previous paper [7] that Turing machines can simulate efficiently GPACs, this paper is a big step towards showing the converse direction: GPACs can simulate Turing machines in an efficient manner.

More concretely we show that computations of Turing machines can be simulated in polynomial space by GPACs as long as we use bounded (but arbitrary) time. We firmly believe that this construction can be used as a building brick to show the more general result that the computations of Turing machines can be simulated in polynomial space by GPACs, removing the hypothesis of arbitrary but fixed time. This latter construction would probably be much more involved, and we intend to focus on it in the near future since this result would show that computations done by the GPAC and in the context of Computable Analysis are equivalent modulo polynomial space reductions.

We believe that these results open the way for some sort of more general Church-Turing thesis, which applies not only to discrete-based models of computation but also to physically realistic models of computation, and which holds both at a computability and computational complexity (modulo polynomial reductions) level.

Incidently, these kind of results can also be the first step towards a well-founded complexity theory for analog models of computations and for continuous dynamical systems.

Notice that it has been observed in several papers that, since continuous time systems might undergo arbitrary space and time contractions, Turing machines, as well as even accelerating Turing machines[1] [14], [13], [12] or even oracle Turing machines, can actually be simulated in an arbitrary short time by ordinary differential equations in an arbitrary short time or space. This is sometimes also called *Zeno's phenomenon*: an infinite number of discrete transitions may happen in a finite time: see e.g. [8]. Such constructions or facts have been deep obstacles to various attempts to build a well founded complexity theory for analog models of computations: see [8] for discussions. One way to interpret our results is then the following: all these time and space phenomena, or Zeno's phenomena do not hold (or, at least, they do not hold in a problematic manner) for ordinary differential equations corresponding to GPACs, that is to say for *realistic* models, for carefully chosen measures of complexity. Moreover, these measures of complexity relate naturally to standard computational complexity measures involving discrete models of computation

2   Preliminaries

2.1   Notation

Throughout the paper we will use the following notation:

$$\|(x_1,\ldots,x_n)\| = \max_{1\leqslant i\leqslant n} |x_i| \qquad \|(x_1,\ldots,x_n)\|_2 = \sqrt{|x_1|^2 + \cdots + |x_n|^2}$$

$$\pi_i(x_1,\ldots,x_k) = x_i \qquad \text{int}(x) = \lfloor x \rfloor \qquad \text{frac}(x) = x - \lfloor x \rfloor$$

---

[1] Similar possibilities of simulating accelerating Turing machines through quantum mechanics are discussed in [11].

$$\text{int}_n(x) = \min(n, \text{int}(x)) \qquad \text{frac}_n(x) = x - \text{int}_n(x)$$

$$f^{[n]} = \begin{cases} \text{id} & \text{if } n = 0 \\ f^{[n-1]} & \text{otherwise} \end{cases}$$

$$\text{sgn}(x) = \begin{cases} -1 & \text{if } x < 0 \\ 0 & \text{if } x = 0 \\ 1 & \text{if } x > 0 \end{cases} \qquad \mathbb{R}^* = \mathbb{R} \setminus \{0\}$$

2.2   Computational Complexity Measures for the GPAC

It is known [16] that a function is generable by a GPAC iff it is a component of the solution a polynomial initial-value problem. In other words, a function $f : I \to \mathbb{R}$ is GPAC-generable iff it belongs to following class.

Definition 1. *Let $I \subseteq \mathbb{R}$ be an open interval and $f : I \to \mathbb{R}$. We say that $f \in \text{GPAC}(I)$ if there exists $d \in \mathbb{N}$, a vector of polynomials p, $t_0 \in I$ and $y_0 \in \mathbb{R}^d$ such that for all $t \in I$ one has $f(t) = y_1(t)$, where $y : I \to \mathbb{R}$ is the unique solution over I of*

$$\begin{cases} \dot{y} = p(y) \\ y(t_0) = y_0 \end{cases} \tag{1}$$

Next we introduce a subclass of GPAC generable functions which allow us to talk about space complexity. The idea is that a function f generated by a GPAC belongs to the class $\text{GSPACE}(I, g)$ if f can be generated by a GPAC in I and does not grow faster that g. Since the value of f in physical implementations of the GPAC correspond to some physical quantity (e.g. electric tension), limiting the growth of f corresponds to effectively limiting the size of resources (i.e. magnitude of signals) needed to compute f by a GPAC.

Definition 2. *Let $I \subseteq \mathbb{R}$ be an open interval and $f, g : I \to \mathbb{R}$ be functions. The function f belongs to the class $\text{GSPACE}(I, g)$ if there exist $d \in \mathbb{N}$, a vector of polynomials p, $t_0 \in I$ and $y_0 \in \mathbb{R}^d$ such that for all $t \in I$ one has $f(t) = y_1(t)$ and $\|y(t)\| \leqslant g(t)$, where $y : I \to \mathbb{R}$ is the unique solution over I of (1). More generally, a function $f : I \to \mathbb{R}^d$ belongs to $f \in \text{GSPACE}(I, g)$ if all its components are also in the same class.*

We can generalize the complexity class GSPACE to multidimensional open sets I defined over $\mathbb{R}^d$. The idea is to reduce it to the one-dimensional case defined above through the introduction of a subset $J \subseteq \mathbb{R}$ and of a map $g : J \to I$.

Definition 3. *Let $I \subseteq \mathbb{R}^d$ be an open set and $f, s_f : I \to \mathbb{R}$ be functions. Then $f \in \text{GSPACE}(I, s_f)$ if for any open interval $J \subseteq \mathbb{R}$ and any function $(g : J \to \mathbb{R}^d \in \text{GSPACE}(J, s_g)$ such that $g(J) \subseteq I$, one has $f \circ g \in \text{GSPACE}(J, \max(s_g, s_f \circ s_g))$.*

The following closure results can be proved (proofs are omitted for reasons of space).

Lemma 1. *Let $I, J \subseteq \mathbb{R}^d$ be open sets, and $(f : I \to \mathbb{R}^n)$ and $(g : J \to \mathbb{R}^m)$ be functions which belong to GSPACE (I, s_f) and GSPACE (J, s_g), respectively. Then:*

- $f + g, f - g \in$ GSPACE $(I \cap J, s_f + s_g)$ *if $n = m$.*
- $fg \in$ GSPACE $(I \cap J, \max(s_f, s_g, s_f s_g))$ *if $n = m$.*
- $f \circ g \in$ GSPACE $(J, \max(s_g, s_f \circ s_g))$ *if $m = d$ and $g(J) \subseteq I$.*

2.3   Main Result

Our main result states that any Turing machine can be simulated by a GPAC using a space bounded by a polynomial, where T and S are respectively the time and the space used by the Turing machine.

If one prefers, (formal statement in Theorem 3):

Theorem 1. *Let $\mathcal{M}$ be a Turing Machine. Then there is a GPAC-generable function $f_\mathcal{M}$ and a polynomial p with the following properties:*

1. *Let S, T be arbitrary positive integers. Then $f_\mathcal{M}(S, T, [e], n)$ gives the configuration of $\mathcal{M}$ on input e at step n, as long as $n \leq T$ and $\mathcal{M}$ uses space bounded by S.*
2. *$f_\mathcal{M}(S, T, [e], t)$ is bounded by $p(T + S)$ as long as $0 \leq t \leq n$.*

The first condition of the theorem states that the GPAC simulates TMs on bounded space and time, while the second condition states that amount of resources used by the GPAC computation is polynomial on the amount of resources used by original Turing computation.

3   The Construction

3.1   Helper Functions

Our simulation will be performed on a real domain and may be subject to (small) errors. Thus, to simulate a Turing machine over a large number of steps, we need tools which allow us to keep errors under control. In this section we present functions which are specially designed to fulfill this objective. We call these functions *helper functions*. Notice that since functions generated by GPACs are analytic, all helper functions are required to be analytic. As a building block for creating more complex functions, it will be useful to obtain analytic approximations of the functions $\text{int}(x)$ and $\text{frac}(x)$. Notice that we are only concerned about non-negative numbers so there is no need to discuss the definition of these functions on negative numbers.

Definition 4. *For any $x, y, \lambda \in \mathbb{R}$ define $\xi(x, y, \lambda) = \tanh(xy\lambda)$.*

Lemma 2. *For any $x \in \mathbb{R}$ and $\lambda > 0, y \geqslant 1$,*

$$| \operatorname{sgn}(x) - \xi(x, y, \lambda)| < 1$$

Furthermore if $|x| \geqslant \lambda^{-1}$ then

$$| \operatorname{sgn}(x) - \xi(x, y, \lambda)| < e^{-y}$$

and $\xi \in \mathrm{GSPACE}\left(\mathbb{R}^3, 1\right)$.

Definition 5. *For any $x, y, \lambda \in \mathbb{R}$, define*

$$\sigma_1(x, y, \lambda) = \frac{1 + \xi(x - 1, y, \lambda)}{2}$$

Corollary 1. *For any $x \in \mathbb{R}$ and $y > 0, \lambda > 2$,*

$$| \operatorname{int}_1(x) - \sigma_1(x, y, \lambda)| \leqslant 1/2$$

Furthermore if $|1 - x| \geqslant \lambda^{-1}$ then

$$| \operatorname{int}_1(x) - \sigma_1(x, y, \lambda)| < e^{-y}$$

and $\sigma_1 \in \mathrm{GSPACE}\left(\mathbb{R}^3, 1\right)$.

Definition 6. *For any $p \in \mathbb{N}$, $x, y, \lambda \in \mathbb{R}$, define*

$$\sigma_p(x, y, \lambda) = \sum_{i=0}^{k-1} \sigma_1(x - i, y + \ln p, \lambda)$$

Lemma 3. *For any $p \in \mathbb{N}$, $x \in \mathbb{R}$ and $y > 0, \lambda > 2$,*

$$| \operatorname{int}_p(x) - \sigma_p(x, y, \lambda)| \leqslant 1/2 + e^{-y}$$

Furthermore if $x < 1 - \lambda^{-1}$ or $x > p + \lambda^{-1}$ or $d(x, \mathbb{N}) > \lambda^{-1}$ then

$$| \operatorname{int}_p(x) - \sigma_p(x, y, \lambda)| < e^{-y}$$

and $\sigma_p \in \mathrm{GSPACE}\left(\mathbb{R}^3, p\right)$.

Finally, we build a square wave like function which we be useful later on.

Definition 7. *For any $t \in \mathbb{R}$, and $\lambda > 0$, define $\theta(t, \lambda) = e^{-\lambda(1 - \sin(2\pi t))^2}$*

Lemma 4. *For any $\lambda > 0$, $\theta(\cdot, \lambda)$ is a positive and 1-periodic function bounded by 1, furthermore*

$$\forall t \in [1/2, 1], |\theta(t, \lambda)| \leqslant \frac{e^{-\lambda}}{2}$$

$$\int_0^{\frac{1}{2}} \theta(t, \lambda) dt \geqslant \frac{(e\lambda)^{-\frac{1}{4}}}{\pi}$$

and $\theta \in \mathrm{GSPACE}\left(\mathbb{R} \times \mathbb{R}_+^*, (t, \lambda) \mapsto \max(1, \lambda)\right)$.

3.2   Polynomial Interpolation

In order to implement the transition function of the Turing Machine, we will use polynomial interpolation techniques (Lagrange interpolation). But since our simulation may have to deal with some amount of error in inputs, we have to investigate how this error propagates through the interpolating polynomial.

Definition 8 (Lagrange polynomial). *Let* $d \in \mathbb{N}$ *and* $f : G \to \mathbb{R}$ *where* G *is a finite subset of* $\mathbb{R}^d$, *we define*

$$L_f(x) = \sum_{\bar{x} \in G} f(\bar{x}) \prod_{i=1}^{d} \prod_{\substack{y \in G \\ y \neq \bar{x}}} \frac{x_i - y_i}{\bar{x}_i - y_i}$$

Lemma 5. *Let* $n \in \mathbb{N}$, $x, y \in \mathbb{R}^n$, $K > 0$ *be such that* $\|x\|, \|y\| \leqslant K$, *then*

$$\left| \prod_{i=1}^{n} x_i - \prod_{i=1}^{n} y_i \right| \leqslant K^{n-1} \sum_{i=1}^{n} |x_i - y_i|$$

3.3   Turing Machines — Assumptions

Let $\mathcal{M} = (Q, \Sigma, b, \delta, q_0, F)$ be a Turing Machine which will be fixed for the whole simulation. Without loss of generality we assume that:

- When the machine reaches a final state, it stays in this state
- $Q = \{0, \dots, m-1\}$ are the states of the machines; $q_0 \in Q$ is the initial state; $F \subseteq Q$ are the accepting states
- $\Sigma = \{0, \dots, k-2\}$ is the alphabet and $b = 0$ is the blank symbol.
- $\delta : Q \times \Sigma \to Q \times \Sigma \times \{L, R\}$ is the transition function, and we identify $\{L, R\}$ with $\{0, 1\}$ ($L = 0$ and $R = 1$). The components of δ are denoted by $\delta_1, \delta_2, \delta_3$. That is $\delta(q, \sigma) = (\delta_1(q, \sigma), \delta_2(q, \sigma), \delta_3(q, \sigma))$ where δ_1 is the new state, δ_2 the new symbol and δ_3 the head move direction.

Notice that the alphabet of the Turing machine has $k - 1$ symbols. This will be important for lemma 6. Consider a configuration $c = (x, \sigma, y, q)$ of the machine. We can encode it as a triple of integers as done in [15] (e.g. if $x_0, x_1, \dots$ are the digits of x in base k, encode x as the number $x_0 + x_1 k + x_2 k^2 + \dots + x_n k^n$), but this encoding is not suitable for our needs. We define the *rational encoding* $[c]$ of c as follows.

Definition 9. *Let* $c = (x, s, y, q)$ *be a configuration of* $\mathcal{M}$, *we define the* rational encoding $[c]$ *of* c *as* $[c] = (0.x, s, 0.y, q)$ *where:*

$$0.x = x_0 k^{-1} + x_1 k^{-2} + \dots + x_n k^{-n-1} \in \mathbb{Q} \quad if \quad x = x_0 + x_1 k + \dots + x_n k^n \in \mathbb{N}$$

The following lemma explains the consequences on the rational encoding of configurations of the assumptions we made for $\mathcal{M}$.

Lemma 6. *Let* c *be a reachable configuration of* $\mathcal{M}$ *and* $[c] = (0.x, \sigma, 0.y, q)$, *then* $0.x \in [0, \frac{k-1}{k}]$ *and similarly for* $0.y$.

3.4 Simulation of Turing Machines — Step 1: Capturing the Transition Function

The first step towards a simulation of a Turing Machine $\mathcal{M}$ using a GPAC is to simulate the transition function of $\mathcal{M}$ with a GPAC-computable function $\text{step}_{\mathcal{M}}$. The next step is to iterate the function $\text{step}_{\mathcal{M}}$ with a GPAC. Instead of considering configurations c of the machine, we will consider its rational configurations $[c]$ and use the helper functions defined previously. Theoretically, because $[c]$ is rational, we just need that the simulation works over rationals. But, in practice, because errors are allowed on inputs, the function $\text{step}_{\mathcal{M}}$ has to simulate the transition function of $\mathcal{M}$ in a manner which tolerates small errors on the input. We recall that δ is the transition function of the $\mathcal{M}$ and we write δ_i the ith component of δ.

Definition 10. *We define:*

$$
\text{step}_{\mathcal{M}} : \left\{
\begin{array}{ccc}
\mathbb{R}^4 & \longrightarrow & \mathbb{R}^4 \\
\begin{pmatrix} x \\ s \\ y \\ q \end{pmatrix} & \longmapsto &
\begin{pmatrix}
\text{choose}\left[\text{frac}(kx), \frac{x+L_{\delta_2}(q,s)}{k}\right] \\
\text{choose}\left[\text{int}(kx), \text{int}(ky)\right] \\
\text{choose}\left[\frac{y+L_{\delta_2}(q,s)}{k}, \text{frac}(ky)\right] \\
L_{\delta_1}(q,s)
\end{pmatrix}
\end{array}
\right.
$$

where $\text{choose}[a,b] = (1-L_{\delta_3}(q,s))a + L_{\delta_3}(q,s)b$ *and* L_{δ_i} *is given by definition 8.*

The function $\text{step}_{\mathcal{M}}$ simulates the transition function of the Turing Machine $\mathcal{M}$, as shown in the following result.

Lemma 7. *Let* $c_0, c_1, \ldots$ *be the sequence of configurations of* $\mathcal{M}$ *starting from* c_0. *Then*

$$
\forall n \in \mathbb{N}, [c_n] = \text{step}_{\mathcal{M}}^{[n]}([c_0])
$$

Now we want to extend the function $\text{step}_{\mathcal{M}}$ to work not only on rationals encodings of configurations but also on reals close to configurations, in a way which tolerates small errors on the input. That is we want to build a robust approximation of $\text{step}_{\mathcal{M}}$. We also have some results on $\text{int}(\cdot)$ and $\text{frac}(\cdot)$. However, we need to pay attention to the case of nearly empty tapes. This can be done by a shifting x by a small amount $(1/(2k))$ before computing the interger/fractional part. Then lemma 6 and lemma 2 ensure that the result is correct.

Definition 11. *Define:*

$$
\overline{\text{step}}_{\mathcal{M}}(\tau, \lambda) : \left\{
\begin{array}{ccc}
\mathbb{R}^4 & \longrightarrow & \mathbb{R}^4 \\
\begin{pmatrix} x \\ s \\ y \\ q \end{pmatrix} & \longmapsto &
\begin{pmatrix}
\text{choose}\left[\overline{\text{frac}}(kx), \frac{x+L_{\delta_2}(q,s)}{k}, q, s\right] \\
\text{choose}\left[\overline{\text{int}}(kx), \overline{\text{int}}(ky), q, s\right] \\
\text{choose}\left[\frac{y+L_{\delta_2}(q,s)}{k}, \overline{\text{frac}}(ky), q, s\right] \\
L_{\delta_1}(q,s)
\end{pmatrix}
\end{array}
\right.
$$

where

$$\text{choose}[a, b, q, s] = (1 - L_{\delta_3}(q, s))a + L_{\delta_3}(q, s)b$$

$$\overline{\text{int}}(x) = \sigma_k\left(x + \frac{1}{2k}, \tau, \lambda\right)$$

$$\overline{\text{frac}}(x) = x - \overline{\text{int}}(x)$$

We now show that $\overline{\text{step}}_{\mathcal{M}}$ is a robust version of $\text{step}_{\mathcal{M}}$. We first begin with a lemma about function *choose*.

Lemma 8. *There exists $A_3 > 0$ and $B_3 > 0$ such that $\forall q, \bar{q}, s, \bar{s}, a, b, \bar{a}, \bar{b} \in \mathbb{R}$, if*

$$\|(\bar{a}, \bar{b})\| \leqslant M \qquad \text{and} \qquad q \in Q, s \in \Sigma \qquad \text{and} \qquad \|(q, s) - (\bar{q}, \bar{s})\| \leqslant 1$$

then

$$\left|\text{choose}[a, b, q, s] - \text{choose}[\bar{a}, \bar{b}, \bar{q}, \bar{s}]\right| \leqslant \|(a, b) - (\bar{a}, \bar{b})\| + 2MA_3\|(q, s) - (\bar{q}, \bar{s})\|$$

Furthermore, choose is computable in polynomial space by a GPAC.

Lemma 9. *There exists $a, b, c, d, e > 0$ such that for any $\tau, \lambda > 0$, any valid rational configuration $c = (x, s, y, q) \in \mathbb{R}^4$ and any $\bar{c} = (\bar{x}, \bar{s}, \bar{y}, \bar{q}) \in \mathbb{R}^4$, if*

$$\|(x, y) - (\bar{x}, \bar{y})\| \leqslant \frac{1}{2k^2} - \frac{1}{k\lambda} \qquad \text{and} \qquad \|(q, s) - (\bar{q}, \bar{s})\| \leqslant 1$$

then, for $p \in \{1, 3\}$

$$\begin{aligned}
&\left|\text{step}_{\mathcal{M}}(c)_p - \overline{\text{step}}_{\mathcal{M}}(\tau, \lambda)(\bar{c})_p\right| \leqslant k\|(x, y) - (\bar{x}, \bar{y})\| + a\|(q, s) - (\bar{q}, \bar{s})\| + b \\
&\left|\text{step}_{\mathcal{M}}(c)_2 - \overline{\text{step}}_{\mathcal{M}}(\tau, \lambda)(\bar{c})_2\right| \leqslant c\|(q, s) - (\bar{q}, \bar{s})\| + d \\
&\left|\text{step}_{\mathcal{M}}(c)_4 - \overline{\text{step}}_{\mathcal{M}}(\tau, \lambda)(\bar{c})_4\right| \leqslant e\|(q, s) - (\bar{q}, \bar{s})\|
\end{aligned}$$

Furthermore, $\overline{\text{step}}_{\mathcal{M}}$ is computable in polynomial space by a GPAC.

We summarize the previous lemma into the following simpler form.

Corollary 2. *For any $\tau, \lambda > 0$, any valid rational configuration $c = (x, s, y, q) \in \mathbb{R}^4$ and any $\bar{c} = (\bar{x}, \bar{s}, \bar{y}, \bar{q}) \in \mathbb{R}^4$, if*

$$\|(x, y) - (\bar{x}, \bar{y})\| \leqslant \frac{1}{2k^2} - \frac{1}{k\lambda} \qquad \text{and} \qquad \|(q, s) - (\bar{q}, \bar{s})\| \leqslant 1$$

then

$$\left\|\text{step}_{\mathcal{M}}(c) - \overline{\text{step}}_{\mathcal{M}}(\tau, \lambda)(\bar{c})\right\| \leqslant O(1)(e^{-\tau} + \|c - \bar{c}\|)$$

Furthermore,

$$\overline{\text{step}}_{\mathcal{M}} \in \text{GSPACE}\left((\mathbb{R}_+^*)^2 \times [-1, 1] \times [-m, m] \times [-1, 1] \times [-k, k], O(1)\right)$$

3.5   Simulation of Turing Machines — Step 2: Iterating Functions with Differential Equations

We will use a special kind of differential equations to perform the iteration of a map with differential equations. In essence, it relies on the following core differential equation

$$\dot{x}(t) = A\phi(t)(g - x(t)) \qquad\qquad \text{(Reach)}$$

We will see that with proper assumptions, the solution converges very quickly to the *goal* g. However, (Reach) is a simplistic idealization of the system so we need to consider a perturbed equation where the goal is not a constant anymore and the derivative is subject to small errors

$$\dot{x}(t) = A\phi(t)(\bar{g}(t) - x(t)) + E(t) \qquad\qquad \text{(ReachPerturbed)}$$

We will again see that, with proper assumptions, the solution converges quickly to the *goal* within a small error. Finally we will see how to build a differential equation which iterates a map within a small error.

We first focus on (Reach) and then (ReachPerturbed) to show that they behave as expected. In this section we assume ϕ is a positive C^1 function.

Lemma 10. *Let x be a solution of* (Reach), *let $T, \lambda > 0$ and assume $A \geqslant \frac{\lambda}{\int_0^T \phi(u)du}$ then $|x(T) - g| \leqslant |g - x(0)|e^{-\lambda}$.*

Lemma 11. *Let $T, \lambda > 0$ and let x be the solution of* (ReachPerturbed) *with initial condition $x(0) = x_0$. Assume $|\bar{g}(t) - g| \leqslant \eta$, $A \geqslant \frac{\lambda}{\int_0^T \phi(u)du}$ and $E(t) = 0$ for $t \in [0, T]$. Then*

$$|x(T) - g| \leqslant \eta(1 + e^{-\lambda}) + |x_0 - g|e^{-\lambda}$$

We can now define a system that simulates the iteration of a function using a system based on (ReachPerturbed). It work as described in [15]. There are two variables for simulating each component f_i, $i = 1, \ldots, n$, of the function f to be iterated. There will be periods in which the function is iterated one time. In half of the period, half (n) of the variables will stay (nearly) constant and close to values $\alpha_1, \ldots, \alpha_n$, while the other remaining n variables update their value to $f_i(\alpha_1, \ldots, \alpha_n)$, for $i = 1, \ldots, n$. In the other half period, the second subset of variables is then kept constant, and now it is the first subset of variables which is updated to $f_i(\alpha_1, \ldots, \alpha_n)$, for $i = 1, \ldots, n$.

Definition 12. *Let $d \in \mathbb{N}$, $F : \mathbb{R}^d \to \mathbb{R}^d$, $\lambda \geqslant 1, \mu \geqslant 0$ and $u_0 \in \mathbb{R}^d$, we define*

$$\begin{cases} z(0) = u_0 \\ u(0) = u_0 \end{cases} \quad \begin{cases} \dot{z}_i(t) = A\theta(t, B)(F_i(u(t)) - z_i(t)) \\ \dot{u}_i(t) = A\theta(t - 1/2, B)(z_i(t) - u_i(t)) \end{cases} \qquad \text{(Iterate)}$$

where $A = 10(\lambda + \mu)^2$ and $B = 4(\lambda + \mu)$.

Theorem 2. *Let $d \in \mathbb{N}$, $F : \mathbb{R}^d \to \mathbb{R}^d$, $\lambda \geqslant 1$, $\mu \geqslant 0$, $u_0, c_0 \in \mathbb{R}^d$. Assume z, u are solutions to* (Iterate) *and let ΔF and $M \geqslant 1$ be such that*

$$\forall k \in \mathbb{N}, \forall \varepsilon > 0, \forall x \in]-\varepsilon, \varepsilon[^d, \left\| F^{[k+1]}(c_0) - F\left(F^{[k]}(c_0) + x\right) \right\| \leqslant \Delta F(\varepsilon)$$

$$\forall t \geqslant 0, \|u(t)\|, \|z(t)\|, \|F(u(t))\| \leqslant M = e^\mu$$

and consider

$$\begin{cases} \varepsilon_0 = \|u_0 - c_0\| \\ \varepsilon_{k+1} = (1 + 3e^{-\lambda})\Delta F(\varepsilon_k + 2e^{-\lambda}) + 5e^{-\lambda} \end{cases}.$$

Then

$$\forall k \in \mathbb{N}, \left\| u(k) - F^{[k]}(c_0) \right\| \leqslant \varepsilon_k$$

Furthermore, if $F \in \text{GSPACE}\left([-M, M]^d, s_F\right)$ for $s_F : [-M, M] \to \mathbb{R}$ then $((\lambda, \mu, t, u_0) \mapsto u(t))$ is computable in polynomial space by a GPAC.

3.6 Simulation of Turing Machines — Step 3: Putting All Pieces Together

In this section, we will use results of both section 3.3 and section 3.5 to simulate Turing Machines with differential equations. Indeed, in section 3.3 we showed that we could simulate a Turing Machine by iterating a robust real map, and in section 3.5 we showed how to efficiently iterate a robust map with differential equations. Now we just have to put these results together.

Theorem 3. *Let $\mathcal{M}$ be a Turing Machine as in section 3.3, then there are functions $s_f : I \to \mathbb{R}^4$ and $f_\mathcal{M} \in \text{GSPACE}\left(\mathbb{R}^4, s_f\right)$ such that for any sequence $c_0, c_1, \ldots,$ of configurations of $\mathcal{M}$ starting with input e:*

$$\forall S, T \in \mathbb{R}_+^*, \forall n \leqslant T, \|[c_n] - f_\mathcal{M}(S, T, n, e)\| \leqslant e^{-S}$$

and

$$\forall S, T \in \mathbb{R}_+^*, \forall n \leqslant T, s_f(S, T, n, e) = O(poly(S, T))$$

Acknowledgments. D.S. Graça was partially supported by *Fundação para a Ciência e a Tecnologia* and EU FEDER POCTI/POCI via SQIG - Instituto de Telecomunicações through the FCT project PEst-OE/EEI/LA0008/2011.

Olivier Bournez and Amaury Pouly were partially supported by ANR project SHAMAN, by DGA, and by DIM LSC DISCOVER project.

References

1. Asarin, E., Maler, O.: Achilles and the tortoise climbing up the arithmetical hierarchy. J. Comput. System Sci. 57(3), 389–398 (1998)
2. Ben-Hur, A., Siegelmann, H.T., Fishman, S.: A theory of complexity for continuous time systems. J. Complexity 18(1), 51–86 (2002)

3. Blondel, V.D., Bournez, O., Koiran, P., Tsitsiklis, J.N.: The stability of saturated linear dynamical systems is undecidable. J. Comput. System Sci. 62, 442–462 (2001)
4. Blum, L., Cucker, F., Shub, M., Smale, S.: Complexity and Real Computation. Springer (1998)
5. Blum, L., Shub, M., Smale, S.: On a theory of computation and complexity over the real numbers: NP-completeness, recursive functions and universal machines. Bull. Amer. Math. Soc. 21(1), 1–46 (1989)
6. Bournez, O., Campagnolo, M.L., Graça, D.S., Hainry, E.: Polynomial differential equations compute all real computable functions on computable compact intervals. J. Complexity 23(3), 317–335 (2007)
7. Bournez, O., Graça, D.S., Pouly, A.: On the complexity of solving initial value problems. Submitted to the Conference ISSAC 2012: International Symposium on Symbolic and Algebraic Computation (2012)
8. Bournez, O., Campagnolo, M.L.: A Survey on Continuous Time Computations. In: New Computational Paradigms. Changing Conceptions of What is Computable, pp. 383–423. Springer, New York (2008)
9. Brattka, V.: The emperor's new recursiveness: the epigraph of the exponential function in two models of computability. In: Ito, M., Imaoka, T. (eds.) Words, Languages & Combinatorics III, ICWLC 2000, Kyoto, Japan (2000)
10. Bush, V.: The differential analyzer. A new machine for solving differential equations. J. Franklin Inst. 212, 447–488 (1931)
11. Calude, C.S., Pavlov, B.: Coins, quantum measurements, and Turing's barrier. Quantum Information Processing 1(1-2), 107–127 (2002)
12. Copeland, B.J.: Accelerating Turing machines. Minds and Machines 12, 281–301 (2002)
13. Copeland, J.: Even Turing machines can compute uncomputable functions. In: Casti, J., Calude, C., Dinneen, M. (eds.) Unconventional Models of Computation (UMC 1998), pp. 150–164. Springer (1998)
14. Davies, E.B.: Building infinite machines. The British Journal for the Philosophy of Science 52, 671–682 (2001)
15. Graça, D.S., Campagnolo, M.L., Buescu, J.: Computability with polynomial differential equations. Adv. Appl. Math. 40(3), 330–349 (2008)
16. Graça, D.S., Costa, J.F.: Analog computers and recursive functions over the reals. J. Complexity 19(5), 644–664 (2003)
17. Ko, K.-I.: Computational Complexity of Real Functions. Birkhäuser (1991)
18. Nyce, J.M.: Guest editor's introduction. IEEE Ann. Hist. Comput. 18, 3–4 (1996)
19. Odifreddi, P.: Classical Recursion Theory, vol. 1. Elsevier (1989)
20. Pour-El, M.B., Richards, J.I.: Computability in Analysis and Physics. Springer (1989)
21. Shannon, C.E.: Mathematical theory of the differential analyzer. J. Math. Phys. MIT 20, 337–354 (1941)
22. Siegelmann, H.T., Ben-Hur, A., Fishman, S.: Computational complexity for continuous time dynamics. Phys. Rev. Lett. 83(7), 1463–1466 (1999)
23. Sipser, M.: Introduction to the Theory of Computation, 2nd edn. Course Technology (2005)
24. Weihrauch, K.: Computable Analysis: an Introduction. Springer (2000)

Computing with and without Arbitrary Large Numbers

Michael Brand

Faculty of IT, Monash University
Clayton, VIC 3800, Australia
michael.brand@alumni.weizmann.ac.il

Abstract. In the study of random access machines (RAMs) it has been shown that the availability of an extra input integer, having no special properties other than being sufficiently large, is enough to reduce the computational complexity of some problems. However, this has only been shown so far for specific problems. We provide a characterization of the power of such extra inputs for general problems.

To do so, we first correct a classical result by Simon and Szegedy (1992) as well as one by Simon (1981). In the former we show mistakes in the proof and correct these by an entirely new construction, with no great change to the results. In the latter, the original proof direction stands with only minor modifications, but the new results are far stronger than those of Simon (1981).

In both cases, the new constructions provide the theoretical tools required to characterize the power of arbitrary large numbers.

Keywords: integer RAM, complexity, arbitrary large number.

1   Introduction

The Turing machine (TM), first introduced in [1], is undoubtedly the most familiar computational model. However, for algorithm analysis it often fails to adequately represent real-life complexities, for which reason the random access machine (RAM), closely resembling the intuitive notion of an idealized computer, has become the common choice in algorithm design. Ben-Amram and Galil [2] write "The RAM is intended to model what we are used to in conventional programming, idealized in order to be better accessible for theoretical study."

Here, "what we are used to in conventional programming" refers, among other things, to the ability to manipulate high-level objects by basic commands. However, this ability comes with some unexpected side effects. For example, one can consider a RAM that takes as an extra input an integer that has no special property other than being "large enough". Contrary to intuition, it has been shown that such arbitrary large numbers (ALNs) can lower problem time complexities. For example, [3] shows that the availability of ALNs lowers the arithmetic time complexity[1] of calculating 2^{2^x} from $\Theta(x)$ to $\Theta(\sqrt{x})$. However, all previous

---

[1] Arithmetic complexity is the computational complexity of a problem under the RAM$[+, \overset{.}{-}, \times, \div]$ model, which is defined later on in this section.

T-H.H. Chan, L.C. Lau, and L. Trevisan (Eds.): TAMC 2013, LNCS 7876, pp. 181–192, 2013.
© Springer-Verlag Berlin Heidelberg 2013

attempts to characterize the contribution of ALNs dealt with problem-specific methods of exploiting such inputs, whereas the present work gives, for the first time, a broad characterization of the scenarios in which arbitrary numbers do and those in which they do not increase computational power.

In order to present our results, we first redefine, briefly, the RAM model. (See [4] for a more formal introduction.)

Computations on RAMs are described by *programs*. RAM programs are sets of *commands*, each given a *label*. Without loss of generality, labels are taken to be consecutive integers. The bulk of RAM commands belong to one of two types. One type is an *assignment*. It is described by a triplet containing a k-ary operator, k operands and a target. The other type is a *comparison*. It is given two operands and a comparison operator, and is equipped with labels to proceed to if the comparison is evaluated as either true or false. Other command-types include unconditional jumps and execution halt commands.

The execution model for RAM programs is as follows. The RAM is considered to have access to an infinite set of *registers*, each marked by a non-negative integer. The input to the program is given as the initial state of the first registers. The rest of the registers are initialized to 0. Program execution begins with the command labeled 1 and proceeds sequentially, except in comparisons (where execution proceeds according to the result of the comparison) and in jumps. When executing assignments, the k-ary operator is evaluated based on the values of the k operands and the result is placed in the target register. The output of the program is the state of the first registers at program termination.

In order to discuss the computational power of RAMs, we consider only RAMs that are comparable in their input and output types to TMs. Namely, these will be the RAMs whose inputs and outputs both lie entirely in their first register. We compare these to TMs working on one-sided-infinite tapes over a binary alphabet, where "0" doubles as the blank. A RAM will be considered equivalent to a TM if, given as an input an integer whose binary encoding is the initial state of the TM's tape, the RAM halts with a non-zero output value if and only if the TM accepts on the input.

Furthermore, we assume, following e.g. [5], that all explicit constants used as operands in RAM programs belong to the set $\{0, 1\}$. This assumption does not make a material difference to the results, but it simplifies the presentation.

In this paper we deal with RAMs that use non-negative integers as their register contents. This is by far the most common choice. A RAM will be indicated by RAM[*op*], where *op* is the set of basic operations supported by the RAM. These basic operations are assumed to execute in a single unit of time. We use the syntax $f(n)$-RAM[*op*] to denote the set of problems solvable in $f(n)$ time by a RAM[*op*], where n is the bit-length of the input. Replacing "RAM[*op*]" by "TM" indicates that the computational model used is a Turing machine.

Note that because registers only store non-negative integers, such operations as subtraction cannot be supported without tweaking. The customary solution is to replace subtraction by "natural subtraction", denoted "$\dotdiv$" and defined by $a \dotdiv b \stackrel{\text{def}}{=} \max(a - b, 0)$. We note that if the comparison operator "$\leq$"

(testing whether the first operand is less than or equal to the second operand) is not supported by the RAM directly, the comparison "$a \leq b$" can be simulated by the equivalent equality test "$a \dot{-} b = 0$". Testing for equality is always assumed to be supported.

By the same token, regular bitwise negation is not allowed, and $\neg a$ is tweaked to mean that the bits of a are negated only up to and including its most significant "1" bit.

Operands to each operation can be explicit integer constants, the contents of explicitly named registers or the contents of registers whose numbers are specified by other registers. This last mode, which can also be used to define the target register, is known as "indirect addressing". In [6] it is proved that for the RAMs considered here indirect addressing has no effect. We therefore assume throughout that it is unavailable to the RAMs.

The following are two classical results regarding RAMs. Operations appearing in brackets within the operation list are optional, in the sense that the theorem holds both when the operation is part of *op* and when it is not.

Theorem 1 ([7]). *PTIME-RAM*$[+, [\dot{-}], [\times], \leftarrow, [\rightarrow], Bool] = PSPACE$

and

Theorem 2 ([8]). *PTIME-RAM*$[+, \dot{-}, /, \leftarrow, Bool; \leq] = ER$, *where ER is the set of problems solvable by Turing machines in*

$$2^{2^{\cdot^{\cdot^{2}}}} \Big\} n \tag{1}$$

time, where n is the length of the input.

Here, "/" indicates exact division, which is the same as integer division (denoted "$\div$") but is only defined when the two operands divide exactly. The operations "$\leftarrow$" and "$\rightarrow$" indicate left shift ($a \leftarrow b = a \times 2^b$) and right shift ($a \rightarrow b = a \div 2^b$), respectively, and *Bool* is shorthand for the set of all bitwise Boolean functions.

In this paper, we show that while Theorem 1 is correct, its original proof is not. Theorem 2, on the other hand, despite being a classic result and one sometimes quoted verbatim (see, e.g., [9]), is, in fact, erroneous.

We re-prove the former here, and replace the latter by a stronger result, for the introduction of which we first require several definitions.

Definition 1 (Expansion Limit). *Let $M = M_{op}(t, inp)$ be the largest number that can appear in any register of a RAM[op] working on inp as its input, during the course of its first t execution steps.*

We define $EL_{op}(f(n))$ to be the maximum of $M_{op}(f(n), inp)$ over all values of inp for which $len(inp) \leq n$. This is the maximum number that can appear in any register of a RAM[op] that was initialized by an input of length at most n, after $f(n)$ execution steps.

The subscript 'op' may be omitted if understood from the context.

As a slight abuse of notation, we use EL(t) to be the maximum of $M_{op}(t, inp)$ over all inp of length at most n, when n is understood from the context and t is independent of n. (The following definition exemplifies this.)

Definition 2 (RAM-Constructability). *A set of operations op is RAM-constructable if the following two conditions are satisfied: (1) there exists a RAM program that, given inp and t as its inputs, with n being the length of inp, returns in $O(t)$ time a value no smaller than $EL_{op}(t)$, and (2) each operation in op is computable in $EL(O(l))$ space on a Turing machine, where l is the total length of all operands and of the result.*

Our results are as follows.

Theorem 3. *For a RAM-constructable op $\supseteq \{+, /, \leftarrow, Bool\}$ and any function $f(n)$,*

$$
\begin{aligned}
O(f(n))\text{-}RAM[op] &= EL_{op}(O(f(n)))\text{-}TM \\
&= N\text{-}EL_{op}(O(f(n)))\text{-}TM \qquad (2) \\
&= EL_{op}(O(f(n)))\text{-}SPACE\text{-}TM \\
&= N\text{-}EL_{op}(O(f(n)))\text{-}SPACE\text{-}TM \,,
\end{aligned}
$$

where the new notations refer to nondeterministic Turing machines, to space-bounded Turing machines and to nondeterministic space-bounded Turing machines, respectively.

Among other things, this result implies for polynomial-time RAMs that their computational power is far greater than ER, as was previously believed.

The theoretical tools built for proving Theorem 3 and re-proving Theorem 1 then allow us to present the following new results regarding the power of arbitrary large numbers.

Theorem 4. *PTIME-ARAM$[+, [\overset{\cdot}{-}], [\times], \leftarrow, [\rightarrow], Bool] = PSPACE$.*

Theorem 5. *Any recursively enumerable (r.e.) set can be recognized in $O(1)$ time by an ARAM$[+, /, \leftarrow, Bool]$.*

Here, "ARAM" is the RAM model assisted by an arbitrary large number. Formally, we say that a set S is computable by an ARAM$[op]$ in $f(n)$ time if there exists a Boolean function $g(inp, x)$, computable in $f(n)$ time on a RAM$[op]$, such that $inp \in S$ implies $g(inp, x) \neq 0$ for almost all x (all but a finite number of x) whereas $inp \notin S$ implies $g(inp, x) = 0$ for almost all x. Here, n conventionally denotes the bit length of the input, but other metrics are also applicable.

We see, therefore, that the availability of arbitrary numbers has no effect on the computational power of a RAM without division. However, for a RAM equipped with integer division, the boost in power is considerable, to the extent that any problem solvable by a Turing machine in any amount of time or space can be solved by an ARAM in $O(1)$ time.

2   Models without Division

2.1   Errata on [7]

We begin with a definition.

Definition 3 (Straight Line Program). *A Straight Line Program (SLP), denoted SLP[op], is a list of tuples, $s_2, \ldots, s_n$, where each s_i is composed of an operator, $s_i^{op} \in op$, and k integers, $s_i^1, \ldots, s_i^k$, all in the range $0 \leq s_i^j < i$, where k is the number of operands taken by s_i^{op}. This list is to be interpreted as a set of computations, whose targets are $v_0, \ldots, v_n$, which are calculated as follows: $v_0 = 0$, $v_1 = 1$, and for each $i > 1$, v_i is the result of evaluating the operator s_i^{op} on the inputs $v_{s_i^1}, \ldots, v_{s_i^k}$. The output of an SLP is the value of v_n.*

A technique first formulated in a general form in [10] allows results on SLPs to be generalized to RAMs. Schönhage's theorem, as worded for the special case that interests us, is that if there exists a Turing machine, running on a polynomial-sized tape and in finite time, that takes an SLP[op] as input and halts in an accepting state if and only if v_n is nonzero, then there also exists a TM running on a polynomial-sized tape that simulates a RAM[op]. This technique is used both in [7] and in our new proof.

The proof of [7] follows this scheme, and attempts to create such a Turing machine. In doing so, this TM stores monomial-based representations of certain powers of two. These are referred to by the paper as "monomials" but are, for our purposes, integers.

The main error in [7] begins with the definition of a relation, called "vicinity", between monomials, which is formulated as follows.

> We define an equivalence relation called *vicinity* between monomials. Let M_1 and M_2 be two monomials. Let B be a given parameter. If $M_1/M_2 < 2^{2^B}$ [...], then M_1 is in the vicinity of M_2. The symmetric and transitive closure of this relation gives us the full vicinity relation. As it is an equivalence relation, we can talk about two monomials being in the same vicinity (in the same equivalence class).

It is unclear from the text whether the authors' original intention was to define this relation in a universal sense, as it applies to the set of all monomials (essentially, the set of all powers of two), or whether it is only defined over the set of monomials actually used by any given program. If the former is correct, any two monomials are necessarily in the same vicinity, because one can bridge the gap between them by monomials that are only a single order of magnitude apart. If the latter is correct, it is less clear what the final result is. The paper does not argue any claim that would characterize the symmetric and transitive closure in this case.

However, the paper does implicitly assume throughout that the vicinity relation, as originally defined (in the $M_1/M_2 < 2^{2^B}$ sense) is *its own* symmetric and transitive closure. This is used in the analysis by assuming for any M_i and

M_j which are in the same vicinity (in the equivalence relation sense) that they also satisfy $2^{-(2^B)} < M_i/M_j < 2^{2^B}$, i.e. they are in the same vicinity also in the restrictive sense.

Unfortunately, this claim is untrue. It is quite possible to construct an SLP that violates this assumption, and because the assumption is central to the entire algorithm, the proof does not hold.

We therefore provide here an alternate algorithm, significantly different from the original, that bypasses the entire "vicinity" issue.

2.2   Our New Construction

Our proof adapts techniques from two previous papers: [11] (which uses lazy evaluation to perform computations on operands that are too long to fit into a polynomial-sized tape) and [12] (which stores operands in a hierarchical format that notes only the positions of "interesting bits", these being bit positions whose values are different than those of the less significant bit directly preceding them). The former method is able to handle multiplication but not bit shifting and the latter the reverse. We prove Theorem 1 using a sequence of lemmas.

Lemma 1. *In an SLP$[+, \overset{\centerdot}{-}, \times, \leftarrow, \rightarrow, Bool]$, the number of interesting bits in the output v_n grows at most exponentially with n. There exists a Turing machine working in polynomial space that takes such an SLP as its input, and that outputs an exponential-sized set of descriptions of bit positions, where bit positions are described as functions of $v_0, \ldots, v_{n-1}$, such that the set is a superset of the interesting bit positions of v_n.*

The fact that the number of interesting bits grows only exponentially given this operation set was noted in [7]. Our proof follows the reasoning of the original paper.

Proof. Consider, for simplicity, the instruction set $op = \{+, \times, \leftarrow\}$. Suppose that we were to change the meaning of the operator "$\leftarrow$", so that, instead of calculating $a \leftarrow b = a \times 2^b$, its result would be $a \leftarrow b = aX$, where X is a formal parameter, and a new formal parameter is generated every time the "$\leftarrow$" operator is used. The end result of the calculation will now no longer be an integer but rather a polynomial in the formal parameters. The following are some observations regarding this polynomial.

1. The number of formal parameters is at most n, the length of the SLP.
2. The power of each formal parameter is at most 2^{n-k}, where k is the step number in which the parameter was defined. (This exponent is at most doubled at each step in the SLP. Doubling may happen, for example, if the parameter is multiplied by itself.)
3. The sum of all multiplicative coefficients in the polynomial is at most $2^{2^{n-2}}$. (During multiplication, the sum of the product polynomial's coefficients is the product of the sums of the operands' coefficients. As such, this value can at most square itself at each operation. The maximal value it can attain at step 2 is 2.)

If we were to take each formal variable, X, that was created at an "$a \leftarrow b$" operation, and substitute in it the value 2^b (a substitution that [7] refers to as the "standard evaluation"), then the value of the polynomial will equal the value of the SLP's output. We claim that if p is an interesting bit position, then there is some product of formal variables appearing as a monomial in the result polynomial such that its standard evaluation is 2^x, and $p \geq x \geq p - 2^n$.

The claim is clearly true for $n = 0$ and $n = 1$. For $n > 1$, we will make the stronger claim $p \geq x \geq p - 2^{n-2} - 2$. To prove this, note that any monomial whose standard evaluation is greater than 2^p cannot influence the value of bit p and cannot make it "interesting". On the other hand, if all remaining monomials are smaller than $p - 2^{n-2} - 2$, the total value that they carry within the polynomial is smaller than $2^{p-2^{n-2}-2}$ times the sum of their coefficients, hence smaller than 2^{p-2}. Bits $p - 1$ and p, however, are both zero. Therefore, p is not an interesting bit.

We proved the claim for the restricted operation set $\{+, \times, \leftarrow\}$. Adding logical AND ("$\wedge$") and logical OR ("$\vee$") can clearly not change the fact that bits $p - 1$ and p are both zero, nor can it make the polynomial coefficients larger than $2^{2^{n-2}}$.

Incorporating "$\overset{\bullet}{-}$" and "$\neg$" into the operation set has a more interesting effect: the values of bit $p - 1$ and p can both become "1". This will still not make bit p interesting, but it does require a small change in the argument. Instead of considering polynomials whose coefficients are between 0 and $2^{2^{n-2}}$, we can now consider polynomials whose coefficients are between $-2^{2^{n-2}}$ and $2^{2^{n-2}}$. This changes the original argument only slightly, in that we now need to argue that in taking the product over two polynomials the sum of the absolute values of the coefficients of the product is no greater than the product of the sums of the absolute values of the coefficients of the operands.

Similarly, adding "$\rightarrow$" into consideration, we no longer consider only formal variables of the form $a \leftarrow b = aX$ but also $a \rightarrow b = \lfloor aY \rfloor$, where the standard evaluation of Y is 2^{-b} and $\lfloor \cdot \rfloor$ is treated as a bitwise Boolean operation (in the sense that, conceptually, it zeroes all bit positions that are "to the right of the decimal point" in the product).

We can therefore index the set of interesting bits by use of a tuple, as follows. If $i_1, \ldots, i_k$ are the set of steps for which $s_{i_j}^{op} \in \{\leftarrow, \rightarrow\}$, the tuple will contain one number between -2^{n-i_j} and 2^{n-i_j} for each $1 \leq j \leq k$, to indicate the exponent of the formal parameter added at step i_j, and an additional $k + 1$'th element, between 0 and 2^n to indicate a bit offset from this bit position.

Though this tuple may contain many non-interesting bits, or may describe a single bit position by many names, it is a description of a super-set of the interesting bits in polynomial space.   □

We refer to the set of bit positions thus described as the *potentially-interesting* bits, or *po-bits*, of the SLP.

Lemma 2. *Let $\mathcal{O}$ be an Oracle that takes an $\mathcal{S} \in SLP[+, \overset{\bullet}{-}, \times, \leftarrow, \rightarrow, Bool]$ as input and outputs the descriptions of all its po-bits in order, from least-significant* to most-significant, without repetitions. There exists a TM working in polynomial space but with access to $\mathcal{O}$ that takes as inputs an $S' \in SLP[+, \div, \times, \leftarrow, \rightarrow, Bool]$ and the description of a po-bit position, i, of S', and that outputs the i'th bit of the output of S'.

Proof. Given a way to iterate over the po-bits in order, the standard algorithms for most operations required work as expected. For example, addition can be performed bit-by-bit if the bits of the operands are not stored, but are, rather, calculated recursively whenever they are needed. The depth of the recursion required in this case is at most n.

The fact that iterating only over the po-bits, instead of over all bit positions, makes no difference to the results is exemplified in Fig. 1.

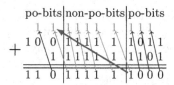

Fig. 1. An example of summing two numbers

As can be seen, not only are the non-po-bits all equal to the last po-bit preceding them, in addition, the carry bit going over from the last po-bit to the first non-po-bit is the same as the carry bit carried over from the last non-po-bit to the first po-bit. Because of this, the sequential carry bits across non-po-bits (depicted in light blue in Fig. 1) can be replaced by a single non-contiguous carry operation (the thick red arrow).

This logic works just as well for subtraction and Boolean operations. The only operation acting differently is multiplication. Implementing multiplication directly leads to incorrect results. Instead, we re-encode the operand bits in a way that reflects our original observation, that the operands can be taken to be polynomials with small coefficients in absolute value, though these coefficients may not necessarily be nonnegative.

The new encoding is as follows: going from least significant bit to most significant bit, a "0" bit is encoded as a 1 if preceded by a "1" and as 0, otherwise. A "1" bit is encoded as a 0 if preceded by a "1" and as -1, otherwise. It is easy to see that a number, A, encoded in regular binary notation but including a leading zero by a $\{0,1\}$ sequence, $a_0, \ldots, a_k$, denoting coefficients of a power series $A = \sum_{i=0}^{k} a_i 2^i$, does not change its value if the a_i are switched for the b_i that are the result of the re-encoding procedure described. The main difference is that now the value of all non-po-bits is 0.

Proving that multiplication works correctly after re-encoding is done by observing its base cases and bilinear properties. The carry in the calculation is exponential in size, so can be stored using a polynomial number of bits.   □

Lemma 3. *Let Q be an Oracle that takes an $S \in SLP[+, \stackrel{.}{-}, \times, \leftarrow, \rightarrow, Bool]$ and two po-bit positions of S and determines which position is the more significant. Given access to Q, Oracle O, described in Lemma 2, can be implemented as a polynomial space Turing machine.*

Proof. Given an Oracle able to compare between indices, the ability to enumerate over the indices in an arbitrary order allows creation of an ordered enumeration. Essentially, we begin by choosing the smallest value, then continue sequentially by choosing, at each iteration, the smallest value that is still greater than the current value. This value is found by iterating over all index values in an arbitrary order and trying each in turn.   □

Lemma 4. *Oracle Q, described in Lemma 3, can be implemented as a polynomial space Turing machine.*

Proof. Recall that an index position is an affine function of the coefficients of the formal variables introduced, in their standard evaluations. To determine which of two indices is larger, we subtract these, again reaching an affine function of the same form. The coefficients themselves are small, and can be stored directly. Determining whether the subtraction result is negative or not is a problem of the same kind as was solved earlier: subtraction, multiplication and addition need to be calculated over variables; in this case the variables are the coefficients, instead of the original formal variables.

However, there is a distinct difference in working with coefficients, in that they, themselves, are calculable as polynomials over formal variables. The calculation can, therefore, be transformed into addition, multiplication and subtraction, once again over the original formal variables.

Although it may seem as though this conclusion returns us to the original problem, it does not. Consider, among all formal variables, the one defined last. This variable cannot appear in the exponentiation coefficients of any of the new polynomials. Therefore, the new equation is of the same type as the old equation but with at least one formal parameter less. Repeating the process over at most n recursion steps (a polynomial number) allows us to compare any two indices for their sizes.   □

Proof (of Theorem 1). The equality P-RAM$[+, \leftarrow, Bool]$ = PSPACE was already shown in [12]. Hence, we only need to prove P-RAM$[+, \stackrel{.}{-}, \times, \leftarrow, \rightarrow, Bool] \subseteq$ PSPACE. This is done, as per Schönhage's method [10], by simulating a polynomial time SLP$[+, \stackrel{.}{-}, \times, \leftarrow, \rightarrow, Bool]$ on a polynomial space Turing machine.

Lemmas 1–4, jointly, demonstrate that this can be done.   □

We remark that Theorem 1 is a striking result, in that right shifting is part of the SLP being simulated, and right shifting is a special case of integer division. Compare this with the power of exact division, described in Theorem 3, which is also a special case of integer division.

2.3   Incorporating Arbitrary Numbers

The framework described in Section 2.1 can readily incorporate simulation of arbitrary large number computation. We use it now, to prove Theorem 4.

Proof (of Theorem 4). Having proved Theorem 1, what remains to be shown is

$$\text{PTIME-ARAM}[+, \div, \times, \leftarrow, \rightarrow, Bool] \subseteq \text{PSPACE} . \tag{3}$$

As in the proof of Theorem 1, it is enough to show that an SLP that is able to handle all operations can be simulated in PSPACE.

We begin by noting that because the PTIME-ARAM must work properly for all but a finite range of numbers as its ALN input, it is enough to show one infinite family of numbers that can be simulated properly. We choose $X = 2^{\omega}$, for any sufficiently large ω. In the simulation, we treat this X as a new formal variable, as was done with outputs of "$a \leftarrow b$" operations.

Lemmas 1–3 continue to hold in this new model. They rely on the ability to compare between two indices, which, in the previous model, was guaranteed by Lemma 4. The technique by which Lemma 4 was previously proved was to show that comparison of two indices is tantamount to evaluating the sign of an affine combination of the exponents associated with a list of formal variables, when using their standard evaluation. This was performed recursively. The recursion was guaranteed to terminate, because at each step the new affine combination must omit at least one formal variable, namely the last one to be defined. Ultimately, the sign to be evaluated is of a scalar, and this can be performed directly.

When adding the new formal variable $X = 2^{\omega}$, the same recursion continues to hold, but the terminating condition must be changed. Instead of evaluating the sign of a scalar, we must evaluate the sign of a formal expression of the form $a\omega + b$. For a sufficiently large ω (which we assume ω to be), the sign is the result of lexicographic evaluation. $\qquad\square$

3   Models with Division

Our proof of Theorem 3 resembles that of [8] in that it uses Simon's ingenious argument that, for any given n, the value $\sum_{i=0}^{2^n-1} i \times 2^{ni}$ can be calculated in $O(1)$-time by considering geometric series summation techniques. The result is an integer that includes, in windows of length n bits, every possible bit-string of length n. The simulating RAM acts by verifying whether any of these bit-strings is a valid tableau for an accepting computation by the simulated TM. This verification is performed using bitwise Boolean operations, in parallel over all options.

Instead of reiterating the entire proof, we give here the most salient differences between the two arguments, these being the places where our argument corrects errors in Simon's original proof. These are as follows.

1. Simon does not show how a TM can simulate an arbitrary RAM in ER-time, making his result a lower-bound only. Indeed, this is, in general, impossible to do. On the other hand, given $\mathrm{EL}_{op}(\mathrm{O}(f(n)))$ tape, a TM can simulate a RAM by storing uncompressed address-value pairs for every non-zero register. The equivalence between the various TMs in Theorem 3 is given by Savitch's Theorem [13], as well as by the well-known relation $\mathrm{TIME}(n) \subseteq \mathrm{SPACE}(n) \subseteq \mathrm{TIME}(\mathrm{EXP}(n))$, so the RAM can be simulated equally well by a time-bounded TM.

2. Simon uses what he calls "oblivious Turing machines" (which are different than those of [14]) in a way that simultaneously limits the TM's tape size and maximum execution time (only the latter condition being considered in the proof), and, moreover, are defined in a way that is non-uniform, in the sense that adding more tape may require a different TM, with potentially more states, a fact not accounted for in the proof. This is corrected by working with non-oblivious, deterministic Turing machines, bounded by a tape of size s. Let c be the number of bits required to store the state of the TM's finite control, then

$$tape\text{-}contents + state \times 2^{s+c+head\text{-}pos-1} + 2^{2(s+c-1)+head\text{-}pos} \qquad (4)$$

is a $3(s+c-1)$-bit number encoding the complete instantaneous description of the TM in a way that allows advancing the TM solely by Boolean operations and bit shifting by offsets dependent only on s and c. This allows verification of an entire tableau, and, indeed, the entire set of all possible tableaus, simultaneously in $\mathrm{O}(1)$ time, when given s, or any number at least as large as s, as input. The complexity of the RAM's execution time is due to the $\mathrm{EL}_{op}(\mathrm{O}(f(n)))$ steps required to reach any number that is as large as s.

3. Most importantly, Simon underestimates the length needed for the tableau, taking it to be the value of the input. TMs are notorious for using up far more tape than the value of their inputs (see [15]). By contrast, our proof uses the fact that a tape bounded TM has only a finite number of possible instantaneous descriptions, so can only progress a bounded number of steps before either halting or entering an infinite loop. By simulating $2^{3(s+c-1)}$ steps of the TM's execution, we are guaranteed to determine its halting state.

Ultimately, Theorem 3 proves that the power of a RAM[op], where op is RAM-constructable and includes $\{+, /, \leftarrow, Bool\}$, is limited only by the maximal size of values that it can produce (relating to the maximal tableau size that it can generate and check). Considering this, the proof of Theorem 5 becomes a trivial corollary: instead of generating a number as large as s by $\mathrm{EL}_{op}(\mathrm{O}(f(n)))$ RAM operations, it is possible to assign to s the value of the ALN. Following this single instruction, simulating the TM's entire execution is done as before, in $\mathrm{O}(1)$ time.

We have shown, therefore, that while arbitrary numbers have no effect on computational power without division, with division they provide Turing completeness in $\mathrm{O}(1)$ computational resources.

References

1. Turing, A.M.: On computable numbers, with an application to the Entscheidungsproblem. Proc. London Math. Soc. 42, 230–265 (1936)
2. Ben-Amram, A.M., Galil, Z.: On the power of the shift instruction. Inf. Comput. 117, 19–36 (1995)
3. Bshouty, N.H., Mansour, Y., Schieber, B., Tiwari, P.: Fast exponentiation using the truncation operation. Comput. Complexity 2(3), 244–255 (1992)
4. Aho, A.V., Hopcroft, J.E., Ullman, J.D.: The Design and Analysis of Computer Algorithms. Addison-Wesley Publishing Co., Reading (1975); Second printing, Addison-Wesley Series in Computer Science and Information Processing
5. Mansour, Y., Schieber, B., Tiwari, P.: Lower bounds for computations with the floor operation. SIAM J. Comput. 20(2), 315–327 (1991)
6. Brand, M.: Does indirect addressing matter? Acta Inform. 49(7-8), 485–491 (2012)
7. Simon, J., Szegedy, M.: On the complexity of RAM with various operation sets. In: Proceedings of the Twenty-Fourth Annual ACM Symposium on Theory of Computing, STOC 1992, pp. 624–631. ACM, New York (1992)
8. Simon, J.: Division in idealized unit cost RAMs. J. Comput. System Sci. 22(3), 421–441 (1981); Special issue dedicated to Michael Machtey
9. Trahan, J.L., Loui, M.C., Ramachandran, V.: Multiplication, division, and shift instructions in parallel random-access machines. Theor. Comp. Sci. 100(1), 1–44 (1992)
10. Schönhage, A.: On the power of random access machines. In: Maurer, H.A. (ed.) ICALP 1979. LNCS, vol. 71, pp. 520–529. Springer, Heidelberg (1979)
11. Hartmanis, J., Simon, J.: On the power of multiplication in random access machines. In: 15th Annual Symposium on Switching and Automata Theory, pp. 13–23. IEEE Comput. Soc., Long Beach (1974)
12. Simon, J.: On feasible numbers (preliminary version). In: Conference Record of the Ninth Annual ACM Symposium on Theory of Computing (Boulder, Colo., 1977), pp. 195–207. Assoc. Comput. Mach., New York (1977)
13. Savitch, W.J.: Relationships between nondeterministic and deterministic tape complexities. J. Comput. System Sci. 4, 177–192 (1970)
14. Pippenger, N., Fischer, M.J.: Relations among complexity measures. J. Assoc. Comput. Mach. 26(2), 361–381 (1979)
15. Radó, T.: On non-computable functions. Bell System Tech. J. 41, 877–884 (1962)

On the Sublinear Processor Gap
for Parallel Architectures

Alejandro López-Ortiz and Alejandro Salinger

David R. Cheriton School of Computer Science, University of Waterloo
{alopez-o,ajsalinger}@uwaterloo.ca

Abstract. In the past, parallel algorithms were developed, for the most part, under the assumption that the number of processors is $\Theta(n)$ (where n is the size of the input) and that if in practice the actual number was smaller, this could be resolved using Brent's Lemma to simulate the highly parallel solution on a lower-degree parallel architecture. In this paper, however, we argue that design and implementation issues of algorithms and architectures are significantly different—both in theory and in practice—between computational models with high and low degrees of parallelism. We report an observed gap in the behavior of a parallel architecture depending on the number of processors. This gap appears repeatedly in both empirical cases, when studying practical aspects of architecture design and program implementation as well as in theoretical instances when studying the behaviour of various parallel algorithms. It separates the performance, design and analysis of systems with a sublinear number of processors and systems with linearly many processors. More specifically we observe that systems with either logarithmically many cores or with $O(n^\alpha)$ cores (with $\alpha < 1$) exhibit a qualitatively different behavior than a system with a linear number of cores on the size of the input, i.e., $\Theta(n)$. The evidence we present suggests the existence of a sharp theoretical gap between the classes of problems that can be efficiently parallelized with $o(n)$ processors and with $\Theta(n)$ processors unless $P = NC$.

1 Introduction

There is a vast experience in the study and development of algorithms for the PRAM architecture. In this case, the standard assumption (though often unstated) was that the number of processors p was linear on the size of the input, i.e., $p = \Theta(n)$ (see, e.g., [18] for a thorough discussion). Indeed, the definition of the class NC, which is often equated with the class of problems that can be efficiently parallelized on a PRAM, allows for up to polynomially many processors. Hence, algorithms were designed to handle the case when $p = \Theta(n)$ or $p = \Theta(n^k)$ for $k \geq 1$ and if the actual number of processors available was lower, this could readily be handled by Brent's Lemma using a suitable scheduler [11,6]. A fruitful theory was developed under these assumptions, and papers in which $p = o(n)$ were relatively rare.

T-H.H. Chan, L.C. Lau, and L. Trevisan (Eds.): TAMC 2013, LNCS 7876, pp. 193–204, 2013.

Table 1. Optimal performance for each case according to processor count $(0 < \alpha < 1)$

Processor count	$\Theta(n)$	$\Theta(n^\alpha)$	$\Theta(\log n)$
Merge sort	X	X	✓
Master theorem			
-Case 1	X	✓	✓
-Case 2	X	✓	✓
-Case 3	X	X	X
Amdahl's law	X	✓ (if $\alpha \leq 1/2$)	✓
Collision	X	✓ (if $\alpha \leq 1/2$)	✓
Buffering	X	✓	✓
Network size	X	✓ (if $\alpha \ll 1/2$)	✓
TM simulation	X	X	✓

In this paper we analyze and report on the influence of the assumed number of processors on several aspects of the performance of various types of parallel architectures. Because of its current prevalence, we focus especially on multi-core architectures, which actually feature a relatively small number of processors, and hence advantages that can be identified for parallel systems with a small number of processor count can lead to benefits in parallel computation in these architectures. However, we also report on aspects of parallel computation that are relevant in general in other architectures, such as memory collisions, communication in distributed architectures, and network sizes, as well as in more theoretical aspects like complexity classes and simulations of other models. Our observations suggest the existence of fundamental differences in the qualities of parallel systems with sublinear and linear number of processors, and that exploiting the advantages of the former can lead to more practical and conceptually simpler designs of both parallel architectures and algorithms, ultimately increasing their adoption and reducing development costs.

2   Overview of Arguments

In this section we briefly list the arguments in favour of considering a limited degree of parallelism. We emphasize that we did not start from the outset with this goal, but rather we sought to develop algorithms and tools (both practical and theoretical) for current multi-core architectures. The observations within are derived from both theoretical investigations and practical experiences in which time and time again we found that there seems to be a qualitative difference between a model with $O(\log(n))$ processors and one with $\Theta(n)$ processors, with, surprisingly, the advantage being for the weaker, i.e., $O(\log(n))$ model. Table 1 shows a summary of our observations for the considered processor counts. There is strong evidence of a sublinear cliff, beyond which development and implementation of efficient PRAM algorithms for many problems is substantially harder if not completely impossible, unless P = NC. In several instances among the evidence observed, the phenomenon had been observed earlier by others [18,20,14].

We now list our arguments briefly, before we expand on each of them individually in the next section.

1. The number of cores in current multi-core processors is nearly a constant, but first, if it is truly a constant, there is not much we can say about parallel speedups, and second, it seems to be steadily though slowly growing.
2. In analogous fashion to the word-RAM, the number of bits in a word could be an arbitrary w but really it is most likely $\Theta(\log n)$, since it is also an index into memory, and memory is usually polynomial on n.
3. The probability of collision on a memory access is only acceptably low for up to $O(\sqrt{n})$ processors.
4. The number of interconnects on a CPU network is prohibitively large for a large number of processors.
5. Serialization at the network end is too costly, i.e., if more than two processors want to talk to a single processor at the same time, this processor has to listen to them serially.
6. There are natural $\log n$ and n^ϵ barriers in the complexity of designing algorithms.
7. Efficient cache performance requires bounded number of processors in terms of cache sizes, which are always assumed to be below n, and often as well in terms of the ratio of shared and private cache sizes, which is well below 100.
8. We define the class of problems which can be sped up using a logarithmic number of processors and show that it contains ENC and EP [20] and, furthermore, this containment is strict.
9. For Turing machines we can automatically increase performance when simulating with a parallel computer using random access memories, with natural constraints limiting the speedup to a $\Theta(\log n)$ factor.
10. Amdahl's law suggests that programs can only noticeably benefit from parallelism if the number of processors is proportional to the relative difference between the execution time of the serial and parallel portions of a program.

3   Exposition

In this section we briefly expand on each of the points above. We aim to keep each argument as short as possible, since the entirety of the case is more important than any individual point.

3.1   Limited Parallelism

In principle, it is possible to build a computer with an arbitrary degree of parallelism. In practice, PRAMs algorithms and architectures focused on $\Theta(n)$-processor architectures, while relying on Brent's Lemma for cases when the number of processors was below that. In contrast, multi-core processors have aimed for a much smaller number of cores. In principle, this number could be modeled as a constant. However, this is unrealistic as the number of cores continues to grow—albeit slowly—with desktop computers having transitioned over the last decade from single core to dual core to quad core and presently eight cores and sixteen cores already shipping at the higher end of the spectrum. Additionally, it has been observed that generally speaking larger inputs justify larger investments in RAM and CPU capacity, so a function of n is much more reflective of real life constraints. This suggests that the number of cores is a function which grows slowly on the input size n, since there is a high processor cost. Let $\mathcal{P}(n)$ denote this function. Natural candidates for $\mathcal{P}(n)$ are $\Theta(\log n)$ and $\Theta(n^\alpha)$ for $\alpha < 1$, though there are other possibilities. Over the next subsections we shall consider various candidates for $\mathcal{P}(n)$.

3.2   Natural Constraints

The ability to index memory using a computer word as an address in a program's virtual memory suggests that the size of the word is $w = \Omega(\log M)$, where M is the memory size, though this does not necessarily need to be the case[1]. Memory itself is usually a polynomial function of the input size, i.e., $M = \Theta(n^k)$ for some $k \geq 1$, with $k = 1$ being a common value. Substituting $M = \Theta(n^k)$ in $w = \Omega(\log M)$ gives $w = \Omega(\log n)$. This is assumed in the word-RAM model, in order for algorithms to be able to refer to any input element. A common assumption in word-RAM papers is actually $w \approx \log n$, which enables constant-time lookup-table implementations of some functions on words while keeping table sizes sublinear (see, e.g., [21]), and restricts the size of pointers in succinct data structures that could otherwise increase their space usage (see, e.g., [10]).

Hence, the word size, which in the early days of computing was treated as a constant, namely 4 or 8 bits, became better understood as in fact proportional to the logarithm of the input size, that is $\Theta(\log n)$. Similarly, in modern multi-core computers, the number of processors has remained relatively bounded (in contrast to commercial PRAMs or GPUs which support anywhere from hundreds to thousands of processors). This relatively slow growth (at least as compared to most other usually exponential growing performance hardware indices) on the number of processors can thus be best modeled as $\log n$ in similar fashion to the word size.

3.3   Write Conflicts

We now analyze memory contention between threads as a function of the number of processors, when write memory accesses are assumed to be distributed uniformly at random among memory cells.

Consider a multi-threaded server application receiving requests from several clients simultaneously. Assume that these requests are served by parallel threads

---

[1] In practice, there have been architectures in which the memory size was strictly greater than 2^w. Currently, in the Intel architecture the size w places a limit on the largest addressable space, but this has not always been the case (e.g., the 8088 processor).

running on p processors that share the system's memory. Such an application is likely to have several portions of the computation accessing shared data such as database tables, buffers, and other shared data structures. Write access to shared data involves synchronization to avoid race conditions, usually implemented by synchronization primitives such as barriers and locks. In general, regardless of how synchronization is implemented, a simultaneous memory access to the same memory cell involves an overhead, either due to serialization or data invalidation. Let us call a simultaneous access by a pair of threads a *collision*. We define a collision in terms of pairs of threads. Thus, a simultaneous access to the same memory cell by t threads is counted as $\binom{t}{2}$ collisions.

We are interested in analyzing the influence of the number of processors on the number of collisions during a period of computation. The uncertainty added by the timing of client requests suggests that write access to shared memory can be modeled as a random process with a certain probability of collision. A crude but reasonable approximation is to model the memory accesses of each process as uniformly distributed over memory cells at each step.

We investigate the expected number of collisions for p threads accessing m memory cells, uniformly at random at each timestep of a period of service time. Clearly, the smaller the number of processors the lower the probability of collision. The question is for what value of p as a function of m does this probability become negligible. Note that in general the size of the memory is usually modeled as a growing function of a program's input size, with $m = O(n^k)$ being a common assumption. Thus, it is reasonable to analyze the number of collisions as m grows.

This reduces to a balls-and-bins scenario (see, e.g., [16]). Let us first consider the total number of overall collisions in one step. Let C be a random variable denoting this number. The probability that two memory accesses are to the same cell is $1/m$. Since there are $\binom{p}{2}$ pairs of memory accesses, the expected number of collisions in one step is $E[C] = \frac{p(p-1)}{2m}$. As m grows, this expression tends to 0 if $p = o(\sqrt{m})$, tends to infinity if $p = \omega(\sqrt{m})$, and it converges to a positive constant for $p = \Theta(\sqrt{m})$.

Now we consider an alternative expression for memory access conflicts, namely the number of cells involved in collisions at each step. Thus, if three or more accesses are to the same cell, the event counts as one conflict. Let X be a random variable denoting the number of memory cells which suffer a collision when there are p simultaneous memory accesses. The probability of a memory cell not being accessed is $(1 - 1/m)^p$, and thus the expected number of accessed cells is $m - m(1 - 1/m)^p$. Then, for p accesses the expected number of cells for which there is more than one access is $E[X] = p - m + m(1 - 1/m)^p$. Assume that $p = m^\alpha$ with $\alpha \leq 1$. The expression above is then $E[X] \approx m^\alpha - m + me^{-m^{\alpha-1}}$. Using the Taylor expansion of $e^{-m^{\alpha-1}}$ we obtain $E[X] \approx \frac{m^{2(\alpha-1)}}{2}$.

Again, when m tends to infinity, the above tends to 0, 1/2, or diverges if α is less, equal, or greater than 1/2, and thus the threshold again is for $p = \Theta(\sqrt{m})$.

Suppose that every instruction takes unit time if there is no collision and $s \geq 1$ units of time otherwise. The expected number of collisions per processor per step is $\frac{(p-1)}{2m}$, and thus the expected slowdown in performance due to collisions is $\frac{s(p-1)}{2m}$, which is negligible for $p = o(m/s)$.

3.4   Processor Communication Network

Traditionally, parallel computers use either shared memory or a processor communication network (or both) to exchange information between the various processing units. The advantage of shared memory is that no additional hardware is required for it; the disadvantages are issues of synchronization and memory contention. Hence, a widely explored alternative is the use of an ad-hoc processor communication network connecting the processors. In general, from the perspective of performance, a full communication network is the preferable network architecture. However, when the number of processors is assumed to be very large this is unfeasible. For example, for the case of $\Theta(n)$-processors of many commercial PRAM implementations, the number of interconnects required would have been $\Theta(n^2)$ which is prohibitive. Thus there was extensive study of alternative network topologies which reduced the complexity of the network while attempting to minimize the penalty in performance derived from the smaller network.

We observe now that full processor communication network becomes a realistic possibility if the number of processors is $O(\log n)$ or even possibly $O(n^\alpha)$ for some $\alpha \ll 1/2$. For example, for a modest (by present standards) input size of $n = 2^{27} = 134,217,728$, even $n^{1/2}$ processors would require an impossible number of interconnects on the full graph ($\binom{\sqrt{n}}{2} \approx 6.7 \times 10^7$). A complete network of $\log n = 27$ processors, on the other hand, would require 351 interconnects, which are well within the realm of current architectures.

3.5   Buffer Overflow

Aside from issues of network topology, in practice it is natural to assume that each processor in a communication network can handle at most a small constant number of messages at once. If more than a constant number of processors send messages to a single processor, said messages would queue at the receiving end for further processing. In this section we consider a natural communication model in which in each instruction cycle a processor may send a message to at most one other processor. In practice, depending on the specific application the probability of collision may range anywhere from zero for the execution of independent threads to one for, say, a master processor serializing requests to some shared lock. As a compromise, we model again this process as if the processors chose their destination uniformly at random. Let p be the number of processors; then the maximum number of collisions observed at the most loaded buffer is $(\ln p / \ln \ln p)(1 + o(1))$ with high probability [22]. For input sizes $n > 2^{22}$, buffer handling with $p = n$ can introduce delays of about twice as many instruction cycles than with $p = \log n$, with the difference growing unboundedly (albeit slowly) for larger input sizes.

3.6 Divide-and-Conquer Algorithms

Consider a divide-and-conquer algorithm whose time complexity can be written as $T(n) = aT(n/b) + f(n)$. The master theorem yields the time bounds for a sequential execution of such an algorithm. A parallel version of this theorem can be obtained by analyzing the parallel time $T_p(n)$ of an execution in which recursive calls are executed in parallel and scheduled with the scheduler in [14] or work stealing [12] with a bounded number of processors [14]:

$$
T_p(n) = \begin{cases} O(T(n)/p), & \text{if } f(n) = O(n^{\log_b(a)-\epsilon}) \text{ and } p = O(n^\epsilon) \text{ (Case 1)} \\ O(T(n)/p), & \text{if } f(n) = \Theta(n^{\log_b a}) \text{ and } p = O(\log n) \text{ (Case 2)} \\ \Theta(f(n)), & \text{if } f(n) = \Omega(n^{\log_b(a)+\epsilon}) \text{ and} \\ & af(n/b) \le cf(n), \text{ for some } c < 1 \qquad \text{(Case 3)} \end{cases} \tag{1}
$$

Optimal speedups are achieved in Cases 1 and 2 only for $p = O(n^\epsilon)$ for $\epsilon > 0$, and $p = O(\log n)$, respectively. In Case 3, the time is dominated by the sequential divide and conquer time $f(n)$ at the top of the recursion [14].

We note that it is possible to obtain optimal speedups with larger numbers of processors for many divide-and-conquer algorithms. However, this invariably requires parallelizing the divide and combine phases of the algorithm, as otherwise the sequential time $f(n)$ of the divide and combine phases dominates the parallel time. In fact, if an optimal parallel algorithm for the divide and combination phases is known, then all cases above yield optimal speedup, and the bounds of the processors can be relaxed. Then the parallel time in Case 3 becomes $T_p(n) = \Theta(f(n)/p)$ [14]. Now Case 1 requires $p = O\left(\frac{n^{\log_b a}}{\log n}\right)$, Case 2 requires $p = O\left(n^{\log_b a}\right)$, while Case 3 requires $p = O(f(n)/\log n)$.

The result for a small number of processors in (1) shows that for a system with a small number of processors the implementation of parallel divide-and-conquer algorithms that achieve the full speedup offered by the architecture is simple and can be implemented without the unnecessary complexity of implementing specific parallel algorithms for the divide and combine phases of the algorithms.

When considering cache performance of divide-and-conquer algorithms, a bounded number of processors can also be advantageous. Blelloch et al. [7] show that the class of *hierarchical* divide-and-conquer algorithms —algorithms in which the divide and combine phases can also be implemented as divide-and-conquer algorithms— can be parallelized to obtain optimal speedups and good cache performance when scheduled with a Controlled-PDF scheduler. While a Brent's Lemma type of implementation of some of the algorithms in [7] can achieve optimal speedups for a large number of processors (e.g., matrix addition and cache oblivious matrix multiplication algorithms can both be sped up optimally up to n^2 processors) [7], the optimal speedup and cache performance bounds under the Controlled-PDF scheduler is only achieved for a much smaller number of processors, bounded by the ratio between shared and private cache sizes, and even smaller in some cases, as we shall see in the next section.

3.7   Cache Imposed Bounds

Cache contention is a key factor in the efficiency of multi-core systems. Various multi-core cache models have been studied which focus on algorithms and schedulers with provable cache performance. Many of the results involving shared and private caches performance require bounds on the number of processors related to the size of the input and/or to the relative sizes of private and shared caches.

The Parallel External Memory (PEM) model [3] models p processors, each with a private cache of size M, partitioned in blocks of size B. A sorting algorithm given in this model is asymptotically optimal for the I/O bounds for at most $p \leq n/B^2$ processors, and it is actually proven that $p \leq n/(B \log B)$ is an upper bound for optimal processor utilization for any sorting algorithm in the PEM model [3]. This algorithm is used in further results in the model for graph and geometry problems [4,1,2]. Thus the assumption that $p \leq n/B^2$ is carried on to these results as well, some of which actually require $p \leq n/(B \log n)$ and even $p \leq n/(B^2 \log B \log^{(t)} n)$, where $\log^{(t)} n$ denotes the composition of t log functions, and t is a constant.

Shared cache performance is studied in [8], which compares the number of cache misses of a multi-threaded computation running on a system with p processors and shared cache of size C_2 to those of a sequential computation with a private cache of size C_1. It is shown that under the PDF-scheduler [9], the parallel number of misses is at most the sequential one if $C_p \geq C_1 + pd$, where d is the critical path of the computation. This implies that $p \leq (C_p - C_1)/d$, which is less than n (as otherwise all the input would fit in the cache) and is usually sublinear, as d is rarely constant and is $\Omega(\log n)$ for many algorithms. Thus, for many algorithms the bound on the parallel misses holds for $p = O(n/\log n)$.

As mentioned in Sect. 3.6, Blelloch et al. [7] study hierarchical divide-and-conquer algorithms in a multi-core cache model of p processors with private L_1 caches of size C_1 and a shared L_2 cache of size C_2. An assumption of the model is that $p \leq \frac{C_2}{C_1} \ll n$, since the input size is assumed not to fit in L_2. It is shown that under a Controlled-PDF scheduler, parallel implementations achieve optimal speedup and cache complexity within constant factors of the sequential cache complexity for a class of hierarchical divide-and-conquer algorithms. Optimality for some algorithms, such as Strassen's matrix multiplication and associative matrix inversion even require $p \leq (C_2/C_1)^{\frac{1}{1+\epsilon}}$ [7]. This multi-core model with the same $p \leq \frac{C_2}{C_1}$ assumption has been used to design cache efficient dynamic programming algorithms [13]. Although the time complexity of the obtained algorithms allows a large number of processors for optimal speedups, the efficiency in cache performance restricts the level of parallelism.

Observe that presently the ratio between L_2 shared cache and private L_1 cache is in the order of 4 to 100 depending on the specific processor architecture.

3.8   The Class $E(p(n))$

The class NC can be defined as the class of problems which can be solved in polylogarithmic time using polynomially many processors. It is believed that

NC $\neq$ P and hence that there are known problems which do not admit a solution in time $O(\log^k n)$, for some $k \geq 1$. In our case we are interested in the study of problems which can be sped up using $O(\log n)$ or $O(n^\alpha)$ processors for $\alpha < 1$. Kruskal et al. [20] introduced the classes ENC and EP which encode the classes of problems that allow optimal speed up (up to constant factors) using polynomially many processors on a CRCW PRAM. The class ENC has polylogarithmic running time, while the class EP has polynomial running time. They also define the related classes SNC, ANC, SP, and AP, which are analogous to ENC and EP in terms of the required running times but allow for some inefficiency. In general, one could introduce the class $\mathcal{C}(p(n), S(n))$ as the class of problems that allow a speedup of $S(n)$ with $p(n)$ processors. Thus, following the notation in [20] we define the class $E(p(n)) = \mathcal{C}(p(n), p(n))$, which is the class of problems that can be solved using $O(p(n))$ processors in time $O(T(n)/p(n))$ where $T(n)$ is the running time of the best sequential solution to the problem. In this work we are particularly interested in the classes $E(\log n)$ and $E(n^\alpha)$ for $\alpha < 1$. For consistency in the class comparisons, we assume a CRCW PRAM as in [20], though these classes can be defined for other PRAM types (EREW,CREW), as well as for asynchronous models (such as multi-cores).

The class ENC is a sharpening of the well known class NC. Recall that the class NC requires maximal speedup down to polylogarithmic time even at the cost of a polynomial amount of inefficiency (i.e., the ratio between parallel and sequential work). In contrast, ENC requires the same speedup but bounds the inefficiency to a constant factor. The class $E(\log n)$ bounds the inefficiency to a constant which implies a speed up of $\Theta(\log n)$ on the sequential solution to the problem. By Brent's Lemma we can show that $E(\log n)$ includes the problems in classes ENC and EP. Since we investigate problems that are most worth parallelizing, we restrict this inclusion to problems with at least sequential linear time[2].

Theorem 1. *Let Π be a problem with sequential time $t(n) = \Omega(n)$. Then, (1) $\Pi \in ENC \Rightarrow \Pi \in E(\log n)$ and (2) $\Pi \in EP \Rightarrow \Pi \in E(\log n)$.*

The reverse is not the case, i.e., not all problems that are in $E(\log n)$ are in ENC, unless P $=$ NC: there are known P-complete problems which allow optimal speedup using a polynomial number of processors [17], and thus they are in EP (and hence in $E(\log n)$). If any such problem is in ENC, this would imply P $=$ NC. We conjecture that the same is the case for $E(\log n)$ and EP. This gives a theoretical separation between the problems that can be sped up optimally using polynomially many processors and those that can be sped up using a logarithmic number of processors.

Similarly, $E(n^\alpha)$ bounds the inefficiency to a constant which implies a speed up of $O(n^\alpha)$ on the sequential solution to the problem. We show that $E(n^\alpha)$ includes most problems (with at least linear time sequential complexity) in ENC. For the same reasons described above, not all problems in $E(n^\alpha)$ are in ENC, for any $\alpha < 1$.

---

[2] Proofs are omitted due to space constraints.

Theorem 2. *Let Π be a problem with sequential time $t(n) = \Omega(n)$. Then, $\Pi \in ENC \Rightarrow \Pi \in E(n^{\alpha})$.*

3.9   Parallelism in Turing Machine Simulations

In Sect. 3.2 we argued that there are natural constraints in the amount of inherent parallelism of computing models. In this section we extend these arguments to show the limitations of the speedup that can be obtained from the Four Russians technique [5] when used for Turing machine (TM) simulations [19,15]. Here we briefly outline a simulation of a TM by a multi-core computer that is similar to those in [19,15] and argue about its limitations based on realistic assumptions about the number of processors as well as word and memory sizes.

Let M be a single-tape deterministic TM that performs $T(n)$ steps on an input of length n (and hence it always halts). Assume that M's alphabet is binary. The idea is to treat contiguous blocks of b bits of M's tape as a word in RAM. By precomputing M's resulting configuration after b steps when starting with each possible block, we can then simulate b steps of M at a time by successively looking up the next configuration of M. Since in b steps M can only alter the contents of b cells, for a given position within the tape we need only to consider the content of $2b + 1$ cells around the position. A block configuration consists of this $(2b + 1)$-bit string representing the contents of M's tape around some position plus d bits to specify the state. Thus, each configuration uses $2b + 1 + d \leq kb$ bits, where k is a constant. For each possible configuration c, we store in $A[c]$ the resulting configuration when running M starting from c for b steps, plus information about how many positions the head moved, and in which direction. There are at most 2^{kb} starting configurations. Since all entries in A can be computed independently in parallel, preprocessing takes $g(n) = 2^{kb}b/p$ steps using p processors. The simulation proceeds by successively looking up configurations and updating M's tape (with one processor) until an accepting or rejecting configuration is reached. The total time is then $T_p(n) = T(n)/b + 2^{kb}b/p$.

There are natural restrictions that limit the speedup that can be achieved with the above technique: the word size, the size of table A, and the efficiency in terms of processor use. The number of configurations is 2^{kb} and hence A requires that many words of memory. This implies that, for a memory of size n^r, for some r, $b \leq (r/k)\log n = O(\log n)$. Moreover, in order to be able to access entries of A in constant time using block configurations as addresses, we require $bk \leq w$, where w is the word size, which is consistent with the common assumption $w = \Theta(\log n)$ (see Sect. 3.2). Furthermore, assume that in order to enable larger speedups we allow $b = \omega(\log n)$ and allow a table of superpolynomial size. Then, in order for the simulation time to dominate over preprocessing we require $2^{kb}b/p = O(T(n)/b)$, and thus $p \geq n^{\omega(1)}/T(n)$, which would be prohibitive for any polynomial time $T(n)$.

The parallelism exploited by this approach is both in terms of the parallel computation of the table A and in terms of the ability to manipulate various bits simultaneously to perform a constant time table lookup (which, as we argue above, can only be exploited up to the manipulation of $\Theta(\log n)$ bits). The use of various processors is only for the precomputation phase, which is embarrassingly parallel. Thus, in principle, we could benefit from the use of a polynomial number of processors. However, the maximum speedup factor that the approach can lead to is the size of the block b. Hence, for optimal processor utilization, the maximum number of processors that we can use is $p = b$. In this case we have that total time is $T_p(n) = O(T(n)/p + 2^{kp}p/p)$. For the simulation time to dominate, we then require $2^{kp} = O(T(n)/p)$, and thus $p = O(\log T(n))$. For any polynomial time TM, this implies $p = O(\log n)$. We note that these arguments do not preclude the existence of other approaches that could result in optimal simulation times without the restrictions described above.

3.10 Amdahl's Law

Consider a program whose execution has a serial part that cannot be parallelized (unless $\mathsf{P} = \mathsf{NC}$) represented by $S(n)$ and a fully parallelizable part denoted by $P(n)$ then the parallel time with p processors is $T_p(n) = S(n) + P(n)/p$ and the speedup is represented by $T_1(n)/T_p(n) = (S(n) + P(n))(S(n) + P(n)/p)$.

Observe now that for $p = \Theta(n)$ we get that the parallel program is noticeably faster only if $S(n) = O(P(n)/n)$. For $p = \Theta(n^\alpha)$ we get that the parallel program is noticeably faster only if $S(n) = O(P(n)/n^\alpha)$. Lastly, for $p = \Theta(\log n)$ we get that the parallel program is noticeable faster if $S(n) = O(P(n)/\log n)$. Observe that most practical algorithms on large data sets run in time $O(n \log n)$ or less, with the sequential part often corresponding to I/O operations, i.e., reading the input. This means that the likeliest value for which one can obtain optimal speedup corresponds to $p = P(n)/S(n)$ which is often (though not always) $\log n$.

4 Conclusions

We presented a list of theoretical arguments and practical evidence as to the existence of a qualitative difference between the classes of problems that can be sped up with a sublinear number of processors and those that can be sped up with polynomially many processors. We also showed that in various specific instances even though there are optimal algorithms for either case, it is conceptually and practically much simpler to design an algorithm for a sublinear number of processors. The benefits of a low processor count extend to issues of processor communication, buffering, memory access, and cache bounds. We introduced classes that describe the problems that allow for optimal speed up, up to a constant factors, for logarithmic and sublinear number of processors and show that they contain a strictly larger class of problems that the PRAM equivalents introduced by Kruskal et al. in 1990 [20], unless $\mathsf{NC} = \mathsf{P}$.

The discontinuities identified in behaviour and performance of parallel systems for logarithmic and sublinear number of processors make these particular processor count functions theoretically interesting, practically relevant, and worth of further exploration.

Acknowledgments. We would like to thank Daniel Remenik for helpful discussions.

References

1. Ajwani, D., Sitchinava, N., Zeh, N.: Geometric algorithms for private-cache chip multiprocessors. In: de Berg, M., Meyer, U. (eds.) ESA 2010, Part II. LNCS, vol. 6347, pp. 75–86. Springer, Heidelberg (2010)
2. Ajwani, D., Sitchinava, N., Zeh, N.: I/O-optimal distribution sweeping on private-cache chip multiprocessors. In: IPDPS, pp. 1114–1123. IEEE (2011)
3. Arge, L., Goodrich, M.T., Nelson, M.J., Sitchinava, N.: Fundamental parallel algorithms for private-cache chip multiprocessors. In: SPAA, pp. 197–206 (2008)
4. Arge, L., Goodrich, M.T., Sitchinava, N.: Parallel external memory graph algorithms. In: IPDPS, pp. 1–11. IEEE (2010)
5. Arlazarov, V., Dinic, E., Kronrod, M., Faradzev, I.: On economic construction of the transitive closure of a directed graph. Dokl. Akad. Nauk SSSR 194, 487–488 (1970) (in Russian); English translation in Soviet Math. Dokl. 11, 1209–1210 (1975)
6. Bender, M.A., Phillips, C.A.: Scheduling DAGs on asynchronous processors. In: SPAA, pp. 35–45. ACM (2007)
7. Blelloch, G.E., Chowdhury, R.A., Gibbons, P.B., Ramachandran, V., Chen, S., Kozuch, M.: Provably good multicore cache performance for divide-and-conquer algorithms. In: SODA. ACM (2008)
8. Blelloch, G.E., Gibbons, P.B.: Effectively sharing a cache among threads. In: SPAA, pp. 235–244. ACM (2004)
9. Blelloch, G.E., Gibbons, P.B., Matias, Y.: Provably efficient scheduling for languages with fine-grained parallelism. J. ACM 46, 281–321 (1999)
10. Bose, P., Chen, E.Y., He, M., Maheshwari, A., Morin, P.: Succinct geometric indexes supporting point location queries. In: SODA, pp. 635–644. SIAM (2009)
11. Brent, R.P.: The parallel evaluation of general arithmetic expressions. J. ACM 21(2), 201–206 (1974)
12. Burton, F.W., Sleep, M.R.: Executing functional programs on a virtual tree of processors. In: FPCA, pp. 187–194. ACM (1981)
13. Chowdhury, R.A., Ramachandran, V.: Cache-efficient dynamic programming algorithms for multicores. In: SPAA, pp. 207–216. ACM (2008)
14. Dorrigiv, R., López-Ortiz, A., Salinger, A.: Optimal speedup on a low-degree multicore parallel architecture (LoPRAM). In: SPAA, pp. 185–187. ACM (2008)
15. Dymond, P.W., Tompa, M.: Speedups of deterministic machines by synchronous parallel machines. In: STOC, pp. 336–343. ACM (1983)
16. Feller, W.: An Introduction to Probability Theory and Its Applications, vol. 1. Wiley (1968)
17. Fujiwara, A., Inoue, M., Masuzawa, T.: Parallelizability of some P-complete problems. In: Rolim, J.D.P. (ed.) IPDPS 2000 Workshops. LNCS, vol. 1800, pp. 116–122. Springer, Heidelberg (2000)
18. Greenlaw, R., Hoover, H.J., Ruzzo, W.L.: Limits to parallel computation: P-completeness theory. Oxford University Press, Inc., New York (1995)
19. Hopcroft, J.E., Paul, W.J., Valiant, L.G.: On time versus space and related problems. In: FOCS, pp. 57–64. IEEE (1975)
20. Kruskal, C.P., Rudolph, L., Snir, M.: A complexity theory of efficient parallel algorithms. Theor. Comput. Sci. 71(1), 95–132 (1990)
21. Munro, J.I.: Tables. In: Chandru, V., Vinay, V. (eds.) FSTTCS 1996. LNCS, vol. 1180, pp. 37–42. Springer, Heidelberg (1996)
22. Raab, M., Steger, A.: "Balls into Bins" - A Simple and Tight Analysis. In: Rolim, J.D.P., Serna, M., Luby, M. (eds.) RANDOM 1998. LNCS, vol. 1518, pp. 159–170. Springer, Heidelberg (1998)

On Efficient Constructions of Short Lists Containing Mostly Ramsey Graphs

Marius Zimand*

Department of Computer and Information Sciences, Towson University, Baltimore, MD, USA

Abstract. One of the earliest and best-known application of the probabilistic method is the proof of existence of a $2 \log n$-Ramsey graph, i.e., a graph with n nodes that contains no clique or independent set of size $2 \log n$. The explicit construction of such a graph is a major open problem. We show that a reasonable hardness assumption implies that in polynomial time one can construct a list containing polylog(n) graphs such that most of them are $2 \log n$-Ramsey.

1 Introduction

A k-Ramsey graph is a graph G that has no clique of size k and no independent set of size k. It is known that for all sufficiently large n, there exists a $2 \log n$-Ramsey graph with n vertices. The proof is nonconstructive, but of course such a graph can be built in exponential time by exhaustive search. A major line of research is dedicated to constructing a k-Ramsey graph having n vertices with k as small as possible and in time that is bounded by a small function in n, for example in polynomial time, or in quasi-polynomial time, DTIME$[2^{\text{polylog}(n)}]$. Till recently, the best polynomial-time construction of a k-Ramsey graph with n vertices has been the one by Frankl and Wilson [FW81], for $k = 2^{\tilde{O}(\sqrt{\log n})}$. Using deep results from additive combinatorics and the theory of randomness extractors and dispersers, Barak, Rao, Shaltiel and Wigderson [BRSW06] improved this to $k = 2^{(\log n)^{o(1)}}$. Notice that this is still far off from $k = 2 \log n$.

As usual when dealing with very difficult problems, it is natural to consider easier versions. In this case, one would like to see if it is possible to efficiently construct a small list of n-vertices graphs with the guarantee that one of them is $2 \log n$-Ramsey. The following positive results hold.

Theorem 1. *There exists a quasipolynomial-time algorithm that on input 1^n returns a list with $2^{O(\log^3 n)}$ graphs with n vertices, and most of them are $2 \log n$-Ramsey. In fact, since in quasipolynomial time one can check whether a graph is $2 \log n$-Ramsey, the algorithm can be modified to return one graph that is $2 \log n$-Ramsey.*

---

* The author is supported in part by NSF grant CCF 1016158. URL: `http://triton.towson.edu/~mzimand`

Theorem 2. *Under a reasonable hardness assumption H, there exists a constant c and a polynomial-time algorithm that on input 1^n returns a list with $\log^c n$ graphs with n vertices, and most of them (say, 90%) are $2\log n$-Ramsey.*

The proofs of these two results use basic off-the-shelf derandomization techniques. The proof (one of them) of Theorem 1 notices that the probabilistic argument that shows the existence of $2\log n$-Ramsey graphs only needs a distribution on the set of n-vertices graphs that is $2\log^2 n$-wise independent. There exist such distributions whose support have the following properties: (a) the size is $2^{O(\log^3 n)}$ and (b) it can be indexed by strings of size $O(\log^3 n)$. Therefore if we make an exhaustive search among these indeces, we obtain the result.

Theorem 2 uses a pseudo-random generator g that can fool NP-predicates. The assumption H, which states that there exists a function in E that, for some $\epsilon > 0$, requires circuits with SAT gates of size $2^{\epsilon n}$, implies the existence of such pseudo-random generators. Then going back to the previous proof, it can be observed that the property that an index corresponds to a graph that is not $2\log n$-Ramsey is an NP predicate. Since most indeces correspond to graphs that are $2\log n$-Ramsey, it follows that for most seeds s, $g(s)$ is also $2\log n$-Ramsey. Therefore, it suffices to make an exhaustive search among all possible seeds. Since a seed has length $O(\log|\text{index}|) = O(\log\log^3 n)$, the result follows. We state and prove Theorem 2 with 90% of the graphs in the list being $2\log n$-Ramsey, but the fraction of $2\log n$-Ramsey graphs in the list can be shown to be at least $1 - O(1/\log n)$.

Theorem 2 can be strengthened to produce a list of concise representations of graphs. A string t is a *concise representation* of a graph $G = (V, E)$ with $V = \{1, \ldots, n\}$ if there is an algorithm A running in time $\text{poly}(\log n)$ such that for every $u \in V, v \in V$, $A(t, u, v) = 1$ if $(u, v) \in E$ and $A(t, u, v) = 0$ if $(u, v) \notin E$. With basically the same proof as that of Theorem 2 one can show the following result.

Theorem 3. *Under a reasonable hardness assumption H, there exists a constant c and an algorithm running in time $\text{poly}(\log n)$ that on input n (written in binary notation) returns a list $t_1, \ldots, t_{\log^c n}$, and most elements of the list are concise representations of $2\log n$-Ramsey graphs.*

Theorem 1 is folklore. It appears implicitly in the paper of M. Naor [Nao92]. Theorem 2 may also be known, but we are not aware of any published statement of it. Fortnow in the *Computational Complexity* blog [For06] and Santhanam [San12] mention a weaker version of Theorem 2, in which the same hardness assumption is used but the size of the list is polynomial instead of polylogarithmic. This motivated us to write this note.

Theorem 2 is also related to a question of Moore and Russell [MR12]. From an improved version of Theorem 1 (see our note after the proof of Theorem 1), they note that $2\log n$-Ramsey graphs can be build using $O(\log^2 n)$ random bits, and they ask if such a construction can be done with $o(\log^2 n)$ random bits. Theorem 2 shows that under a plausible assumption, one needs only $O(\log\log n)$ random bits for the constuction.

Section 4 contains some additional remarks. First we analyze the implication of Theorem 2 when plugged in a construction of M. Naor [Nao92] that builds a k-Ramsey graph from a list of graphs, most of which are k'-Ramsey graphs, which is exactly what Theorem 2 delivers. We notice that the parameters obtained in this way are inferior to the result of Barak et al. [BRSW06]. Secondly, we consider the problem of explicit lower bounds for the van der Waerden Theorem, a problem which is related to the explicit construction of Ramsey graphs. We notice that the hardness assumption which derandomizes BPP implies lower bounds for the van der Waerden Theorem that match the non-constructive lower bounds obtained via the Lovasz Local Lemma. The original proof of the Lovasz Local Lemma does not seem to yield this result. Instead we use a proof of Gasarch and Haeupler [GH11], based on the methods of Moser [Mos09] and Moser and Tardos [MT10].

2 The Hardness Assumption

The hardness assumption needed in theorem 2 is that there exists a function f computable in E (where $E = \bigcup_c \text{DTIME}[2^{cn}]$) that, for some $\epsilon > 0$, cannot be computed by circuits of size $2^{\epsilon n}$ that also have SAT gates (in addition to the standard logical gates). More formally let us denote by $C_f^{\text{SAT}}(n)$ the size of the smallest circuit with SAT gates that computes the function f for inputs of length n.

Assumption H: There exists a function f in E such that, for some $\epsilon > 0$, for every n, $C_f^{\text{SAT}}(n) > 2^{\epsilon n}$.

Klivans and van Melkebeek [KvM02], generalizing the work of Nisan and Wigderson [NW94] and Impagliazzo and Wigderson [IW97], have shown that, under assumption H, for every k, there is a constant c and a pseudo-random generator $g : \{0,1\}^{c \log n} \to \{0,1\}^n$, computable in time polynomial in n, that fools all n^k-size circuits with SAT gates. Formally, for every circuit C with SAT gates of size n^k,

$$|\text{Prob}_{s \in \{0,1\}^{c \log n}}[C(g(s)) = 1] - \text{Prob}_{z \in \{0,1\}^n}[C(z) = 1]| < 1/n^k.$$

We note that assumption H is realistic. Miltersen [Mil01] has shown that it is implied by the following natural assumption, involving uniform complexity classes: for every $\epsilon > 0$, there is a function $f \in E$ that cannot be computed in space $2^{\epsilon n}$ for infinitely many lengths n.

3 Proofs

Proof of Theorem 1.

Let us first review the probabilistic argument showing the existence of $2 \log n$-Ramsey graphs. A graph G with n vertices can be represented by a string of length $\binom{n}{2}$. If we take at random such a graph and fix a subset of k vertices, the probability that the set forms a clique or an independent set is $2^{-\binom{k}{2}+1}$. The probability that this holds for some k-subset is bounded by

$$\binom{n}{k} \cdot 2^{-\binom{k}{2}+1} \leq (\tfrac{en}{k})^k \cdot 2^{-\binom{k}{2}+1}$$
$$= 2^{k \log \frac{en}{k} - \binom{k}{2}+1}.$$

For $k = 2 \log n$, the above expression goes to 0. Thus, for n large enough, the probability that a graph G is $2 \log n$-Ramsey is ≥ 0.99.

The key observation is that this argument remains valid if we take a distribution that is $2 \log^2 n$-wise independent. Thus, we can take a polynomial $p(X)$ of degree $2 \log^2 n$ over the field $\mathrm{GF}[2^q]$, where $q = \log \binom{n}{2}$. To the polynomial p we associate the string $\tilde{p} = p(a_1)_1 \ldots p(a_{\binom{n}{2}})_1$, where $a_1, \ldots, a_{\binom{n}{2}}$ are the elements of the field and $(p(a))_1$ is the first bit of $p(a)$. When p is random, this yields a distribution over strings of length $\binom{n}{2}$ that is $2 \log^2 n$-wise independent. Observe that a polynomial p is given by a string of length $\bar{n} = (2 \log^2 n + 1) \log \binom{n}{2} = O(\log^3 n)$. It follows that

$$\mathrm{Prob}_{p \in \{0,1\}^{\bar{n}}}[\tilde{p} \text{ is } 2 \log n\text{-Ramsey}] \geq 0.99.$$

In quasipolynomial time we can enumerate the graphs $\tilde{p}$, and 99% of them are $2 \log n$-Ramsey. ∎

Note. By using an almost k-wise independent distribution (see [NN93], [AGHR92]), one can reduce the size of the list to $2^{O(\log^2 n)}$.

Proof of Theorem 2 and of Theorem 3.

Let $p, \tilde{p}, \bar{n}$ be as in the proof of Theorem 1. Thus:

- $p \in \{0,1\}^{\bar{n}}$ represents a polynomial,

- $\tilde{p}$ is built from the values taken by p at all the elements of the underlying field, and represents a graph with n vertices,

- $\bar{n} = O(\log^3 n)$.

Let us call a string p *good* if $\tilde{p}$ is a $2 \log n$-Ramsey graph.

Checking that a string p is not *good* is an NP predicate. Indeed, p is not *good* iff $\exists (i_1, \ldots, i_{2 \log n}) \in [n]^{2 \log n}$ [vertices $i_1, \ldots, i_{2 \log n}$ in $\tilde{p}$ form a clique or an independent set]. The $\exists$ is over a string of length polynomial in $|p|$ and the property in the right parentheses can be checked by computing $O(\log^2 n)$ values of the polynomial p, which can be done in time polynomial in $|p|$.

Assumption H implies that there exists a pseudo-random generator $g : \{0,1\}^{c \log \bar{n}} \to \{0,1\}^{\bar{n}}$, computable in time polynomial in $\bar{n}$, that fools all NP predicates, and, in particular, also the one above. Since 99% of the p are *good*, it follows that for 90% of the seeds $s \in \{0,1\}^{c \log \bar{n}}$, $g(s)$ is *good*, i.e., for 90% of s, $\widetilde{g(s)}$ is $2 \log n$-Ramsey. Note that from a seed s we can compute $g(s)$ and next $\widetilde{g(s)}$ in time polynomial in n. If we do this for every seed $s \in \{0,1\}^{c \log \bar{n}}$, we obtain a list with $\bar{n}^c = O(\log^{3c} n)$ graphs of which at least 90% are $2 \log n$-Ramsey graphs.

Theorem 3 is obtained by observing that $\{g(s) \mid s \in \{0,1\}^{c\log\overline{n}}\}$ is a list that can be computed in poly($\log n$) time, and most of its elements are concise representations of $2\log n$-Ramsey graphs. ∎

4   Additional Remarks

4.1   Constructing a Single Ramsey Graph from a List of Graphs of Which the Majority Are Ramsey Graphs

M. Naor [Nao92] has shown how to construct a Ramsey graph from a list of m graphs such that all the graphs in the list, except at most αm of them, are k-Ramsey. We analyze what parameters are obtained, if we apply Naor's construction to the list of graphs in Theorem 2.

The main idea of Naor's construction is to use the product of two graphs $G_1 = (V_1, E_1)$ and $G_2 = (V_2, E_2)$, which is the graph whose set of vertices is $V_1 \times V_2$ and edges defined as follows: there is an edge between (u_1, u_2) and (v_1, v_2) if and only if $(u_1, v_1) \in E_1$ or $(u_1 = v_1)$ and $(u_2, v_2) \in E_2$. Then, one can observe that if G_1 is k_1-Ramsey and G_2 is k_2-Ramsey, the product graph, $G_1 \times G_2$ is $k_1 k_2$-Ramsey. Extending to the product of multiple graphs $G_1, G_2, \ldots, G_m$ where each G_i is k_i-Ramsey, we obtain that the product graph is $k_1 k_2 \ldots k_m$-Ramsey.

If we apply this construction to a list of m graphs $G_1, G_2, \ldots, G_m$, each having n vertices and such that $\mathrm{Prob}_i[G_i$ is not k-Ramsey$] \leq \alpha$, we obtain that the product of $G_1, G_2, \ldots, G_m$ is a graph G with $N = n^m$ vertices that is t-Ramsey for $t = n^{\alpha m} k^{(1-\alpha)m}$. For $\alpha \leq 1/\log n$, we have $t \leq (2k)^m$. The list produced in Theorem 2 has $m = \log^c n$, $k = 2\log n$, and one can show that $\alpha \leq 1/\log n$. The product graph G has $N = 2^{\log^c n \log n}$ vertices and is t-Ramsey for $t \leq 2^{\log^c n \cdot \log\log n + O(1)} < 2^{(\log N)^{1-\beta}}$, for some positive constant β.

Thus, under assumption H, there is a positive constant β and a polynomial time algorithm that on input 1^N constructs a graph with N vertices that is $2^{(\log N)^{1-\beta}}$-Ramsey. Note that this is inferior to the parameters achieved by the unconditional construction of Barak, Rao, Shaltiel and Wigderson [BRSW06].

4.2   Constructive Lower Bounds for the van der Waerden Theorem

Van der Waerden Theorem is another classical result in Ramsey theory. It states that for every c and k there exists a number n such that for any coloring of $\{1, \ldots, n\}$ with c colors, there exists k elements in arithmetic progression (k-AP) that have the same color. Let $W(c, k)$ be the smallest such n. One question is to find a constructive lower bound for $W(c, k)$. To simplify the discussion, let us focus on $W(2, k)$.

In other words, the problem that we want to solve is the following:

For any k, we want to find a value of $n = n(k)$ as large as possible and a 2-coloring of $\{1, \ldots, n\}$ such that no k-AP is monochromatic. Furthermore, we want the 2-coloring to be computable in time polynomial in n.

Gasarch and Haeupler [GH11] have studied this problem. They present a probabilistic polynomial time construction for $n = \frac{2^{k-1}}{ek} - 1$ (i.e., the 2-coloring is obtained by a probabilistic algorithm running in $2^{O(k)}$ time) and a (deterministic) polynomial time construction for $n = \frac{2^{(k-1)(1-\epsilon)}}{4k}$ (i.e., the 2-coloring is obtained in deterministic $2^{O(k/\epsilon)}$ time). Their constructions are based on the constructive version of the Lovasz Local Lemma due to Moser [Mos09] and Moser and Tardos [MT10]. The probabilistic algorithm of Gasarch and Haeupler is "BPP-like", in the sense that it succeeds with probability 2/3 and the correctness of the 2-coloring produced by it can be checked in polynomial time. It follows that it can be derandomized under the hardness assumption that derandomizes BPP, using the Impagliazzo-Wigderson pseudo-random generator [IW97]. It is interesting to remark that the new proof by Moser and Tardos of the Local Lovasz Lemma is essential here, because the success probability guaranteed by the classical proof is too small to be used in combination with the Impagliazzo-Wigderson pseudo-random generator.

We proceed with the details.

We use the following hardness assumption H' (weaker than assumption H), which is the one used to derandomize BPP [IW97].

Assumption H': There exists a function f in E such that, for some $\epsilon > 0$, for every n, $C_f(n) > 2^{\epsilon n}$.

Impagliazzo and Wigderson [IW97] have shown that, under assumption H', for every k, there is a constant c and a pseudo-random generator $g : \{0,1\}^{c \log n} \to \{0,1\}^n$ that fools all n^k-size circuits and that is computable in time polynomial in n.

Proposition 1. *Assume assumption H'. For every k, let $n = n(k) = \frac{2^{k-1}}{ek} - 1$. There exists a polynomial-time algorithm that on input 1^n 2-colors the set $\{1, \ldots, n\}$ such that no k-AP is monochromatic.*

Proof. The algorithm of Gasarch and Haeupler [GH11], on input 1^n, uses a random string z of size $|z| = n^c$, for some constant c, and, with probability at least 2/3, succeeds to 2-color the set $\{1, \ldots, n\}$ such that no k-AP is monochromatic. Let us call a string z to be *good* for n if the Gasarch-Haeupler algorithm on input 1^n and randomness z, produces a 2-coloring with no monochromatic k-APs. Note that there exists a polynomial-time algorithm A that checks if a string z is *good* or not, because the Gasarch-Haeupler algorithm runs in polynomial time and the number of k-APs inside $\{1, \ldots, n\}$ is bounded by n^2/k. Using assumption H' and invoking the result of Impagliazzo and Wigderson [IW97], we derive that there exists a constant d and a pseudo-random generator $g : \{0,1\}^{d \log n} \to \{0,1\}^{n^c}$ such that

$$\text{Prob}_{s \in \{0,1\}^{d \log n}}[A(g(s)) = \text{ good for } n] \geq 2/3 - 1/10 > 0.$$

Therefore if we try all possible seeds s of length $d \log n$, we will find one s such that $g(s)$ induces the Gasarch-Haeupler algorithm to 2-color the set $\{1, \ldots, n\}$ such that no k-AP is monochromatic.

References

[AGHR92] Alon, N., Goldreich, O., Håstad, J., Peralta, R.: Simple constructions of almost k-wise independent random variables. Random Structures and Algorithms 3(3), 289–304 (1992)

[BRSW06] Barak, B., Rao, A., Shaltiel, R., Wigderson, A.: 2-source dispersers for sub-polynomial entropy and Ramsey graphs beating the Frankl-Wilson construction. In: Kleinberg, J.M. (ed.) STOC, pp. 671–680. ACM (2006)

[For06]   Fortnow, L.: Full derandomization. Computational Complexity blog (July 31, 2006)

[FW81]   Frankl, P., Wilson, R.M.: Intersection theorems with geometric consequences. Combinatorica 1(4), 357–368 (1981)

[GH11]   Gasarch, W.I., Haeupler, B.: Lower bounds on van der Waerden numbers: Randomized- and deterministic-constructive. Electr. J. Comb. 18(1) (2011)

[IW97]   Impagliazzo, R., Wigderson, A.: P = BPP if E requires exponential circuits: Derandomizing the XOR lemma. In: Proceedings of the 29th Annual ACM Symposium on the Theory of Computing (STOC 1997), pp. 220–229. Association for Computing Machinery, New York (1997)

[KvM02]   Klivans, A., van Melkebeek, D.: Graph nonisomorphism has subexponential size proofs unless the polynomial-time hierarchy collapses. SIAM J. Comput. 31(5), 1501–1526 (2002)

[Mil01]   Miltersen, P.B.: Derandomizing complexity classes. In: Pardalos, P., Reif, J., Rolim, J. (eds.) Handbook on Randomized Computing, Volume II. Kluwer Academic Publishers (2001)

[Mos09]   Moser, R.A.: A constructive proof of the Lovász local lemma. In: Mitzenmacher, M. (ed.) STOC, pp. 343–350. ACM (2009)

[MR12]   Moore, C., Russell, A.: Optimal epsilon-biased sets with just a little randomness. CoRR, abs/1205.6218 (2012)

[MT10]   Moser, R.A., Tardos, G.: A constructive proof of the general Lovász local lemma. J. ACM 57(2) (2010)

[Nao92]   Naor, M.: Constructing Ramsey graphs from small probability spaces. Technical report, IBM Research Report RJ 8810 (70940) (1992)

[NN93]   Naor, J., Naor, M.: Small-bias probability spaces: Efficient constructions and applications. SIAM Journal on Computing 22(4), 838–856 (1993)

[NW94]   Nisan, N., Wigderson, A.: Hardness vs. randomness. Journal of Computer and System Sciences 49, 149–167 (1994)

[San12]   Santhanam, R.: The complexity of explicit constructions. Theory Comput. Syst. 51(3), 297–312 (2012)

On Martin-Löf Convergence of Solomonoff's Mixture

Tor Lattimore and Marcus Hutter

Australian National University
{tor.lattimore,marcus.hutter}@anu.edu.au

Abstract. We study the convergence of Solomonoff's universal mixture on individual Martin-Löf random sequences. A new result is presented extending the work of Hutter and Muchnik (2004) by showing that there does not exist a universal mixture that converges on all Martin-Löf random sequences.

Keywords: Solomonoff induction, Kolmogorov complexity, theory of computation.

1 Introduction

Sequence prediction is the task of predicting symbol α_n having seen $\alpha_{1:n-1} = \alpha_1 \cdots \alpha_{n-1}$. Solomonoff approached this problem by taking a Bayesian mixture over all lower semicomputable semimeasures where complex semimeasures were assigned lower prior probability than simple ones.[1] He then showed that, with probability one, the predictive mixture converges (fast) to the truth for any computable measure [9]. Solomonoff induction arguably solves the sequence prediction problem and has numerous attractive properties, both technical [9, 2, 5] and philosophical [8]. There is, however, some hidden unpleasantness, which we explore in this paper.

Martin-Löf randomness is the usual characterisation of the randomness of individual sequences [6]. A sequence is Martin-Löf random if it passes all effective tests, such as the laws of large numbers and the iterated logarithm. Intuitively, a sequence is Martin-Löf random with respect to measure μ if it satisfies all the properties one would expect of an infinite sequence sampled from μ. It has previously been conjectured that the set of Martin-Löf random sequences is precisely, or contained within, the set on which the Bayesian mixture converges.

This question has seen a number of attempts with a partial negative solution and a more detailed history of the problem by Hutter and Muchnik [3]. They showed that there exists a universal lower semicomputable semimeasure M and Martin-Löf random sequence α (with respect to the Lebesgue measure λ) for which $M(\alpha_n|\alpha_{<n}) \not\to \lambda(\alpha_n|\alpha_{<n})$. The α used in their proof is computable from

---

[1] Actually, Solomonoff mixed over proper measures. The use of semimeasures was introduced later by Levin to ensure that the mixture itself was lower semicomputable [14].

T-H.H. Chan, L.C. Lau, and L. Trevisan (Eds.): TAMC 2013, LNCS 7876, pp. 212–223, 2013.

the halting problem, which presumably inspired the work in [7] where it is shown that if α is 2-random, then every universal lower semicomputable semimeasure converges on α. It is worth remarking that there are known semimeasures that do converge on all Martin-Löf random sequences, some of which are even lower semicomputable. Unfortunately, however, convergence rates for these semimeasures are unknown. For a detailed discussion see [3].

While Hutter and Muchnik showed that there exists a universal lower semicomputable semimeasure and Martin-Löf random sequence on which it fails to converge, the question of whether or not this failure occurs for all such semimeasures has remained open. We prove that for every universal lower semicomputable Bayesian mixture there exists a Martin-Lof random sequence on which it fails to converge. This result is interesting for a few reasons. The choice of universal mixture is akin to choosing an optimal universal Turing machine when computing Kolmogorov complexity. In both cases, asymptotic results are rarely dependent on this choice and so it is useful to confirm this trend here. On the other hand, if the result had been positive then the existence of a universal mixture that did converge on all Martin-Löf random strings would be a nice property that might justify the choice of one universal mixture over another.

2   Notation

Overviews of algorithmic information theory can be found in [5, 1].

General. The natural, rational and real numbers are denoted by $\mathbb{N}$, $\mathbb{Q}$ and $\mathbb{R}$. Logarithms are taken with base 2. A real $\theta \in (0, 1)$ has entropy $H(\theta) := -\theta \log \theta - (1 - \theta) \log(1 - \theta)$. The indicator function is $[\![expr]\!]$, which takes value 1 if $expr$ is true and 0 otherwise. For sets A and B we write $A - B$ for their difference and $|A|$ for the size of A. The natural density of $A \subseteq \mathbb{N}$ is $d(A) := \lim_{n \to \infty} |\{a \in A : a \leq n\}| / n$. and $\bar{d}(A) := \limsup_{n \to \infty} |\{a \in A : a \leq n\}| / n$. We use $\vee$ and $\wedge$ for logical or and and respectively.

Strings. A finite binary string x is a finite sequence $x_1 x_2 x_3 \cdots x_n$ with $x_i \in \mathcal{B} := \{0, 1\}$. Its length is $\ell(x)$. An infinite binary string ω is an infinite sequence $\omega_1 \omega_2 \omega_3 \cdots$. The empty string of length zero is denoted by ε. The sets $\mathcal{B}^n$, $\mathcal{B}^*$ and $\mathcal{B}^\infty$ are the sets of all strings of length n, all finite strings and all infinite strings respectively. Substrings of $x \in \mathcal{B}^* \cup \mathcal{B}^\infty$ are denoted by $x_{s:t} := x_s x_{s+1} \cdots x_{t-1} x_t$ where $s, t \in \mathbb{N}$ and $s \leq t$. If $s > t$, then $x_{s:t} := \varepsilon$. A useful shorthand is $x_{<t} := x_{1:t-1}$. Let $x, y \in \mathcal{B}^*$, then $\#x(y)$ is the number of (possibly overlapping and wrapping around) occurrences of x in y and xy is their concatenation. For example, $\#010(1010) = 2$. If $\ell(y) \geq \ell(x)$ and $x_{1:\ell(x)} = y_{1:\ell(x)}$, then we write $x \sqsubseteq y$ and say x is a prefix of y. Otherwise we write $x \not\sqsubseteq y$. A string $\omega \in \mathcal{B}^\infty$ is normal if $\forall x \in \mathcal{B}^*$, $\lim_{n \to \infty} \#x(\omega_{1:n})/n = 2^{-\ell(x)}$.

Measures and Semimeasures. A semimeasure is a function $\mu : \mathcal{B}^* \to [0, 1]$ satisfying $\mu(\varepsilon) \leq 1$ and $\mu(x) \geq \mu(x0) + \mu(x1)$ for all $x \in \mathcal{B}^*$. It is a measure if both inequalities are replaced by equalities. A function $\mu : \mathcal{B}^* \to \mathbb{R}$ is lower semicomputable if the set $\{(x, r) : r < \mu(x), r \in \mathbb{Q}, x \in \mathcal{B}^*\}$ is recursively enumerable. In this case there exists a recursively enumerable sequence $\mu_1, \mu_2, \cdots$

of computable functions approximating μ from below. For $b \in \mathcal{B}$ and $x \in \mathcal{B}^*$, $\mu(b|x) := \mu(xb)/\mu(x)$ is the μ-probability that x is followed by b. The Lebesgue measure is $\lambda(x) := 2^{-\ell(x)}$.

Complexity. A Turing machine T is a recursively enumerable set of pairs of binary strings $T := \{(p^1, x^1), (p^2, x^2), \cdots\}$ where p^k is the program for x^k. It is a prefix machine if the set of programs is prefix free, $p^k \not\sqsubseteq p^j$ for all $j \neq k$. T is a monotone machine if $p^k \sqsubseteq p^j \implies x^k \sqsubseteq x^j \vee x^j \sqsubseteq x^k$. For prefix machine T the prefix complexity with respect to T is a function $K_T : \mathcal{B}^* \to \mathbb{N}$ defined by

$$K_T(x) := \min_p \{\ell(p) : (p, x) \in T\}$$

If T is a monotone machine, then the monotone complexity with respect to T is defined by

$$Km_T(x) := \min_p \{\ell(p) : (p, y) \in T \wedge x \sqsubseteq y\}$$

There exists an additively optimal prefix machine U such that for all prefix machines T there exists a constant c_T with $K_U(x) < K_T(x) + c_T$. In identical fashion there exists an additively optimal monotone machine. As is usual in algorithmic information theory, we fix a pair of additively optimal prefix and monotone machines and write $K(x) := K_U(x)$ and $Km(x) := Km_U(x)$. The choice of reference machine is irrelevant for this work.

A lower semicomputable semimeasure M is universal if for every lower semicomputable semimeasure μ there exists a constant $c_\mu > 0$ such that $\forall x, M(x) > c_\mu \mu(x)$. Zvonkin and Levin [14] showed that the set of all lower semicomputable semimeasures is recursively enumerable (possibly with repetition). Let $\nu_1, \nu_2, \cdots$ be such an enumeration and $w : \mathbb{N} \to [0, 1]$ be a lower semicomputable sequence satisfying $\sum_{i \in \mathbb{N}} w_i \leq 1$, which we view as a prior on the lower semicomputable semimeasures. Then the universal mixture is defined by

$$M(x) := \sum_{i \in \mathbb{N}} w_i \nu_i(x). \tag{1}$$

There are, of course, many possible enumerations and priors, and hence there are many universal mixtures. This paper aims to prove certain inconsistency results about all universal mixtures, regardless of the choice of prior. Defining $w_i(x) := w_i \nu_i(x)/M(x)$ and substituting into Eq. 1 leads to

$$M(b|x) = \sum_{i \in \mathbb{N}} w_i(x) \nu_i(b|x). \tag{2}$$

There exist universal lower semicomputable semimeasures that are not representable as universal mixtures, but we do not consider these here [13].

Martin-Löf Randomness. Let μ be a computable measure and M a universal lower semicomputable semimeasure. An infinite binary string ω is μ-Martin-Löf random (μ-random) if and only if there exists a $c > 0$ such that

$$\mu(\omega_{<n})/M(\omega_{<n}) > c, \quad \forall n \in \mathbb{N}. \tag{3}$$

Observe that the definition does not depend on the choice of universal lower semicomputable semimeasure since for any two universal lower semicomputable semimeasures M and M' there exists a constant $c > 0$ such that $cM'(x) > M(x) > M'(x)/c$, $\forall x$ [5]. We write $\mathcal{R}_\mu \subset \mathcal{B}^\infty$ for the set of μ-random strings.

Lemma 1. *The following hold:*

1. *If $\omega \in \mathcal{B}^\infty$ is λ-random, then it is normal.*
2. *If $x \in \mathcal{B}^*$ with $\ell(x) = n$ and $\theta := \#1(x)/n$, then $Km(x) < nH(\theta) + \frac{1}{2}\log n + c$ for some $c > 0$ independent of x and n.*
3. *Let $A, B \subseteq \mathbb{N}$ and $\phi_n := [\![n \in A]\!]$. If $d(A) = 0$ and $\bar{d}(B) > 0$, then*
   (a) $\bar{d}(B - A) > 0$.
   (b) $\lim_{n \to \infty} Km(\phi_{1:n})/n = 0$.

Proof. Part 1 is well known [5, §2.6]. For part 2 we use the KT-estimator, which is defined by

$$\mu(x) := \int_0^1 \frac{1}{\pi\sqrt{(1-\theta)\theta}} \theta^{\#1(x)}(1-\theta)^{\#0(x)} d\theta.$$

Because μ is a measure and is finitely computable using a recursive formula [12], we can apply Theorem 4.5.4 in [5] to show that there exists a constant $c_\mu > 0$ such that

$$Km(x) < -\log \mu(x) + c_\mu \leq \frac{1}{2}\log n + 1 + \log \theta^{\#1(x)}(1-\theta)^{\#0(x)} + c_\mu$$

$$= \frac{1}{2}\log n + 1 + nH(\theta) + c_\mu,$$

where we used the redundancy bound for the KT-estimator [12] and the definition of $H(\theta)$. Part 3a is immediate from the definition of the natural density. For 3b, let $\theta_n := \#1(\phi_{1:n})/n$ and note that $d(A) = 0$ implies that $\lim_{n \to \infty} \theta_n = 0$ and so $\lim_{n \to \infty} H(\theta_n) = 0$. Finally apply part 2 to complete the proof. ∎

3   Almost Sure Convergence

Before Martin-Löf convergence is considered we present a version of the celebrated theorem of Solomonoff with which we will contrast our results [10].

Theorem 2 (Solomonoff, 1978). *If M is a universal lower semicomputable semimeasure and α is sampled from computable measure μ, then*

$$\lim_{n \to \infty} \sum_{b \in \mathcal{B}} (M(b|\alpha_{<n}) - \mu(b|\alpha_{<n}))^2 = 0, \quad w.\mu.p.1.$$

A subtle point is that convergence in Theorem 2 holds both off-sequence and on-sequence. A weaker (on-sequence only) statement would be that $\lim_{n \to \infty} (M(\alpha_n|\alpha_{<n}) - \mu(\alpha_n|\alpha_{<n}))^2 = 0$, $w.\mu.p.1$. Unfortunately, both results only hold with probability 1 while we are primarily interested in convergence on individual sequences.

4   Martin-Löf Convergence

We now ask whether there exists a universal mixture such that $M(\alpha_n|\alpha_{<n}) \to \mu(\alpha_n|\alpha_{<n})$ for all μ-random α. Two new theorems are presented, the first is subsumed by the second, but admits an easy proof and serves as a nice warm-up.

Theorem 3. *Let M be a universal mixture. Then there exists a λ-random α such that $\lim_{n\to\infty} \sum_{b \in \mathcal{B}} (M(b|\alpha_{<n}) - 1/2)^2 \neq 0$.*

Proof. We use the same λ-random string α as Hutter and Muchnik [3], which is defined inductively by $\alpha_n := [\![M(\alpha_{<n}0) > 2^{-n}]\!]$. Define $\nu : \mathcal{B}^* \to [0,1]$ by

$$\nu(x) := M(x)[\![\forall n \leq \ell(x) : x_n = 0 \vee M(x_{<n}0) > 2^{-n}]\!].$$

It is straightforward to check that ν is both lower semicomputable and a semimeasure. Therefore there exists a $j \in \mathbb{N}$ such that $\nu = \nu_j$ in the enumeration of all lower semicomputable semimeasures used by M. By the definition of ν we have that $\nu(\alpha_{1:n}) = M(\alpha_{1:n})$ for all n. Furthermore,

$$\alpha_n = 0 \implies M(\alpha_{<n}0) \leq 2^{-n} \implies \nu(\alpha_{<n}1) = 0 \implies \nu(1|\alpha_{<n}) = 0,$$

where we used the definitions of α, ν and the conditional probability respectively. Therefore if $\alpha_n = 0$, then

$$
\begin{aligned}
M(0|\alpha_{<n}) + M(1|\alpha_{<n}) &\overset{(a)}{=} \sum_{i \in \mathbb{N}} w_i(\alpha_{<n}) \left(\nu_i(0|\alpha_{<n}) + \nu_i(1|\alpha_{<n}) \right) \\
&\overset{(b)}{\leq} \sum_{i \in \mathbb{N}} w_i(\alpha_{<n}) - w_j(1 - M(0|\alpha_{<n})) \\
&\overset{(c)}{=} 1 - w_j(1 - M(0|\alpha_{<n})) \overset{(d)}{\leq} 1 - w_j M(1|\alpha_{<n}), \quad (4)
\end{aligned}
$$

where (a) follows directly from Eq. 2. (b) follows by extracting $w_j(\alpha_{<n})$ from the sum and using the facts that $\nu_j(0|\alpha_{<n}) + \nu_j(1|\alpha_{<n}) = M(0|\alpha_{<n})$ and $\nu_i(0|\alpha_{<n}) + \nu_i(1|\alpha_{<n}) \leq 1$ for all i. (c) follows from the fact that $\sum_{i \in \mathbb{N}} w_i(x) = 1$. For (d) we note that M is a semimeasure, which implies that $1 - M(0|\alpha_{<n}) \geq M(1|\alpha_{<n})$. Because α is λ-random, it must contain infinitely many zeros by part 1 of Lemma 1 and the definition of a normal string. Let n_i be the position of the ith 0 in α and $k \in \mathbb{N}$ be such that $\nu_k = \lambda$. Therefore there exists a $c > 0$ such that

$$M(1|\alpha_{<n_i}) \overset{(a)}{=} \sum_{i \in \mathbb{N}} w_i(\alpha_{<n})\nu(1|\alpha_{<n_i}) \overset{(b)}{\geq} w_k(\alpha_{<n})\lambda(1|\alpha_{<n_i}) \overset{(c)}{>} c,$$

where (a) is the same as Eq. 2 and (b) follows by extracting the contribution of the Lebesgue measure λ. (c) follows by recalling that $\lambda(1|\alpha_{<n_i}) = 1/2$ and the fact that α is λ-random combined with Eq. 3. Then by Eq. 4,

$$\liminf_{i\to\infty} M(0|\alpha_{<n_i}) + M(1|\alpha_{<n_i}) \leq 1 - w_j c < 1.$$

Therefore $\lim_{n\to\infty} M(0|\alpha_{<n}) + M(1|\alpha_{<n}) \neq 1$ and so $\lim_{n\to\infty} \sum_{b\in\mathcal{B}} (M(b|\alpha_{<n}) - 1/2)^2 \neq 0$, as required.   ∎

Coincidentally, the proof of Theorem 3 demonstrates the existence random sequences on which M fails to converge to a proper measure. This is interesting as it is a straightforward corollary of Theorem 2 that M converges to a measure with μ-probability one with respect to any computable measure μ.

We now present the on-sequence version of Theorem 3, which uses the same α for a counter-example, but turns out to be significantly harder to prove.

Theorem 4. *Let M be a universal mixture. Then there exists a λ-random α such that $\lim_{n\to\infty} M(\alpha_n|\alpha_{<n}) \neq 1/2$.*

Initially we follow the proof in [3] by constructing a lower semicomputable semimeasure ν that dominates M on α infinitely often, but where $\nu(0|\alpha_{<n}) = 1$ if $\alpha_n = 0$.

Definition 5. *Let M_t be a sequence of computable functions approximating M from below and define $\alpha^t \in \mathcal{B}^\infty$ similarly to α by $\alpha_n^t := [\![M_t(\alpha_{<n}^t 0) > 2^{-n}]\!]$. Now define $\nu_t : \mathcal{B}^* \to [0,1]$ by*

$$\nu_t(x) := \begin{cases} 2^{-t} & \text{if } \ell(x) = t \wedge x < \alpha_{1:t}^t \\ \nu_t(x0) + \nu_t(x1) & \text{if } \ell(x) < t \\ 0 & \text{otherwise,} \end{cases}$$

where $x < \alpha_{1:t}^t$ is decided by lexicographical order.

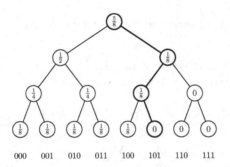

Fig. 1. ν_3 if $\alpha_{1:3}^3 = 101$

It is shown in [3] that $\lim_{t\to\infty} \alpha^t = \alpha$ and $2^{-n} \equiv \lambda(\alpha_{1:n}) > M(\alpha_{1:n})$. Additionally, $\nu := \lim_{t\to\infty} \nu_t$ exists and is a lower semicomputable semimeasure with $\nu(x) = \nu(x0) + \nu(x1)$ and $\nu(\alpha_{1:n}) < 2^{-n}$. Hutter and Muchnik then argued that if $\alpha_{n:n+1} = 01$, then $\nu(\alpha_n|\alpha_{<n}) = 1$ and $\nu(\alpha_{<n}) \geq M(\alpha_{<n})/2$. They then set $M' := \gamma M + (1-\gamma)\nu$ for suitable γ and so poisoned convergence of either M or M'. Here we diverge from their work and consider ν when predicting ones. For the remainder of this article α and ν refer to those defined above.

Lemma 6. *The following hold:*

1. *If $\alpha_{n:n+1} = 10$ then $\nu(1|\alpha_{<n}) \in (0, 1/3)$.*
2. *If $\alpha_n = 1$ then $\nu(1|\alpha_{<n}) \in (0, 1/2)$.*

Proof. For part one,

$$\nu(1|\alpha_{<n}) \stackrel{(a)}{=} \frac{\nu(\alpha_{<n}1)}{\nu(\alpha_{<n})} \stackrel{(b)}{=} \frac{\nu(\alpha_{<n}10) + \nu(\alpha_{<n}11)}{\nu(\alpha_{<n}0) + \nu(\alpha_{<n}10) + \nu(\alpha_{<n}11)}$$
$$\stackrel{(c)}{=} \frac{\nu(\alpha_{<n}10)}{\nu(\alpha_{<n}0) + \nu(\alpha_{<n}10)} \stackrel{(d)}{=} \frac{\nu(\alpha_{<n}10)}{2^{-n} + \nu(\alpha_{<n}10)} \stackrel{(e)}{<} \frac{2^{-n-1}}{2^{-n} + 2^{-n-1}} = \frac{1}{3}.$$

(a) is the definition of the conditional measure. (b) follows because $\nu(x) = \nu(x0) + \nu(x1)$. (c) and (d) are true, since $\alpha_{n:n+1} = 10$ and so $\alpha_{<n}11 > \alpha_{1:n+1}$ and $\alpha_{<n}0 < \alpha_{1:n}$, which imply that $\nu(\alpha_{<n}11) = 0$ and $\nu(\alpha_{<n}0) = 2^{-n}$. (e) follows from algebra and because $\nu(\alpha_{1:n}) < 2^{-n}$ for all n. For the second part we use the same reasoning to obtain

$$\nu(1|\alpha_{<n}) = \frac{\nu(\alpha_{<n}1)}{\nu(\alpha_{<n}0) + \nu(\alpha_{<n}1)} = \frac{\nu(\alpha_{<n}1)}{2^{-n} + \nu(\alpha_{<n}1)} < \frac{1}{2},$$

as required.   ∎

Lemma 7. *Let n_i be the position of the ith 1 in α and $j \in \mathbb{N}$ be such that $\nu = \nu_j$ in the enumeration of all lower semicomputable semimeasures used by M. If $\lim_{n\to\infty} M(\alpha_n|\alpha_{<n}) = \frac{1}{2}$, then the function $\bar{M}(x) := M(x) - w_j \nu(x)/2$ satisfies the following properties:*

1. *$\bar{M}$ is a universal mixture.*
2. *$\liminf_{i\to\infty} \bar{M}(1|\alpha_{<n_i}) \geq \frac{1}{2}$.*
3. *There exists $\gamma \in (0, 1)$ such that for all sufficiently large n with $\alpha_{n:n+1} = 10$, $\bar{M}(1|\alpha_{<n}) > \frac{1}{2\gamma^2}$.*

Proof. The first part is trivial. For the third part, let w_k be the prior weight that M assigns to itself, n be such that $\alpha_{n:n+1} = 10$ and $\epsilon_n := \frac{1}{2} - \frac{M(\alpha_{<n}1)}{M(\alpha_{<n})}$. Then

$$\bar{M}(1|\alpha_{<n}) \stackrel{(a)}{=} \frac{M(\alpha_{<n}1) - w_j\nu(\alpha_{<n}1)/2}{M(\alpha_{<n}) - w_j\nu(\alpha_{<n})/2} \stackrel{(b)}{>} \frac{M(\alpha_{<n}1) - w_j\nu(\alpha_{<n})/6}{M(\alpha_{<n}) - w_j\nu(\alpha_{<n})/2}$$
$$\stackrel{(c)}{=} \frac{1}{2} + \frac{w_j\nu(\alpha_{<n}) - 12M(\alpha_{<n})\epsilon_n}{12\bar{M}(\alpha_{<n})} \stackrel{(d)}{>} \frac{1}{2} + \frac{w_j}{24} - w_k\epsilon_n,$$

(a) is the definition of $\bar{M}$ and conditional probability. (b) follows from part 1 of Lemma 6. In (c) we substituted ϵ_n. (d) by substituting inequalities $\nu(\alpha_{<n}) \geq \nu(\alpha_{<n}0) = 2^{-n} > M(\alpha_{<n})/2 > \bar{M}(\alpha_{<n})/2$ and $\bar{M}(\alpha_{<n}) > w_k M(\alpha_{<n})$. Since $\epsilon_n \to 0$ for sufficiently large n with $\alpha_{n:n+1} = 10$, we have $\bar{M}(1|\alpha_{<n}) > 1/2 + w_j/48 = \frac{1}{2\gamma^2}$ where $\gamma^2 := \frac{1}{1+w_j/24} \in (0, 1)$. For the second part

$$\bar{M}(1|\alpha_{<n_i}) = \frac{M(\alpha_{<n_i}1) - w_j\nu(\alpha_{<n_i}1)/2}{M(\alpha_{<n_i}) - w_j\nu(\alpha_{<n_i})/2} \stackrel{(a)}{>} \frac{M(\alpha_{<n_i}1) - w_j\nu(\alpha_{<n_i})/4}{M(\alpha_{<n_i}) - w_j\nu(\alpha_{<n_i})/2}$$
$$= \frac{1}{2} - \frac{M(\alpha_{<n_i})\epsilon_{n_i}}{\bar{M}(\alpha_{<n_i})} \geq \frac{1}{2} - w_k\epsilon_{n_i},$$

where (a) follows from part 2 of Lemma 6. Taking the limit as $i \to \infty$ completes the result. ∎

To prove the main theorem we construct a pair of infinite binary sequences χ and ψ such that $\alpha_{1:n}$ is computable from $\chi_{1:n}$ and $\psi_{1:n}$. This implies that $Km(\alpha_{1:n}) < Km(\chi_{1:n}) + K(\psi_{1:n}) + O(1)$, which holds because you can construct a program for $\alpha_{1:n}$ using the concatenation of a prefix program for $\psi_{1:n}$ and a monotone program for $\chi_{1:n}$. Finally we assume that M converges on-sequence to λ on α and show that this implies $\liminf_{n\to\infty} Km(\chi_{1:n})/n < 1$ and $\lim_{n\to\infty} K(\psi_{1:n})/n = 0$. But α is λ-random, so $\lim_{n\to\infty} Km(\alpha_{1:n})/n = 1$, which leads to a contradiction.

Proof of Theorem 4. Let α be as in the proof of Theorem 3. Define $\{m_i\}$ and $\{n_i\}$ inductively by

$$m_1 := \min\{m : \alpha_m = 1\}$$
$$n_i := \min\{n \geq m_i : \alpha_{n+1} = 0\}$$
$$m_i := \min\{m > n_{i-1} : \alpha_m = 1\},$$

which are chosen so that $\alpha_{m_i-1:n_i+1} = 01^{n_i-m_i+1}0$. Since α is λ-random, by part 1 of Lemma 1, $d(\{n_i : i \in \mathbb{N}\}) > 0$. Furthermore, $\bar{M}$ is universal so by Eq. 3 there exists an $\epsilon > 0$ such that $1 \geq 2^{n_i}\bar{M}(\alpha_{1:n_i}) > \epsilon$ for all i. Let γ be as in the proof of Lemma 7. Therefore we can choose a $c \in \mathbb{Q}$ such that:

1. $\bar{d}\left(A := \{i : c < 2^{n_i}\bar{M}(\alpha_{1:n_i}) \leq c/\gamma\}\right) > 0$.
2. $d\left(B := \{i : 2^{n_i}\bar{M}(\alpha_{1:n_i}) > c/\gamma\}\right) = 0$.

Define $F \subset \mathbb{N}$ by

$$F := \{i : \exists j \in \{m_i, \cdots, n_i - 1\} \text{ such that } 2^j \bar{M}(\alpha_{1:j}) > c\} - B.$$

Now define indicators χ and ψ by

$$\chi_n := [\![\alpha_n = 1 \vee \exists i : (n = n_i + 1 \wedge i \in A - F)]\!]$$
$$\psi_n := [\![\exists i : n = m_i \wedge i \in F \cup B]\!].$$

Let M_t and $\bar{M}_t$ be computable approximations of M and $\bar{M}$ from below respectively and $m(x) := \max\{m \leq \ell(x) : x_{m-1} = 0 \vee m = 1\}$. Then

$$\alpha_n = \begin{cases} 0 & \text{if } \chi_n = 0 \\ 1 & \text{if } \chi_n = 1 \wedge \psi_{m(\alpha_{<n}1)} = 1 \\ 1 & \text{if } \chi_n = 1 \wedge \exists t : M_t(\alpha_{<n}0) > 2^{-n} \\ 0 & \text{if } \chi_n = 1 \wedge \exists t : 2^{n-1}\bar{M}_t(\alpha_{<n}) > c. \end{cases}$$

The equation above is computable given $\chi_{1:n}$, $\psi_{1:n}$ and $\alpha_{<n}$ by the following argument.

1. The first two cases are straightforward since $m(\alpha_{<n}1)$ is computable.
2. If neither the first nor second case match, then by the definitions of χ and ψ exactly one of the 3rd or 4th cases must hold. Therefore the conditions can be computed in parallel for increasing t until one completes.

Since $\alpha_{1:n}$ is λ-random and can be computed from $\psi_{1:n}$ and $\chi_{1:n}$ using the equation above, there exist constants $c_1, c_2 > 0$ such that $Km(\chi_{1:n}) + K(\psi_{1:n}) + c_2 > Km(\alpha_{1:n}) > n - c_1$, where the second inequality follows from [5, Example 4.5.3]. We now work by contradiction and show that if $\lim_{n\to\infty} M(\alpha_n|\alpha_{<n}) = \frac{1}{2}$ then $Km(\chi_{1:n}) + K(\psi_{1:n})$ is smaller than $n - c_1 - c_2$ for sufficiently large n.

We start by showing that $d(F) = 0$. By Lemma 7, for each $k \in \mathbb{N}$ there exists an N_k such that if $i > N_k$, then $\bar{M}(\alpha_{n_i}|\alpha_{<n_i}) > 1/(2\gamma^2)$ and $\bar{M}(1|\alpha_{<n}) > \gamma^{1/k}/2$ whenever $\alpha_n = 1$ and $n \geq m_i$. Suppose $N_k < i \notin B$ and $j \in \{m_i, \cdots, n_i - 1\}$ with $\ell_i := n_i - m_i + 1 \leq k$, then

$$2^j \bar{M}(\alpha_{1:j}) \overset{(a)}{=} 2^j \frac{\bar{M}(\alpha_{1:n_i})}{\bar{M}(1|\alpha_{<n_i})} \prod_{n=j+1}^{n_i-1} \frac{1}{\bar{M}(1|\alpha_{<n})}$$

$$\overset{(b)}{<} \gamma^2 2^{n_i} \bar{M}(\alpha_{1:n_i}) \gamma^{-(n_i-j-1)/k} \overset{(c)}{\leq} c\gamma^{1-(n_i-j-1)/k} \overset{(d)}{\leq} c,$$

where (a) follows from the definition of the conditional measure. (b) follows from the inequalities $\bar{M}(1|\alpha_{<n_i}) > 1/(2\gamma^2)$ and $\bar{M}(1|\alpha_{<n}) > \gamma^{1/k}/2$. (c) is true by the assumption that $i \notin B$, which implies that $2^{n_i} \bar{M}(\alpha_{1:n_i}) \leq c/\gamma$. Finally (d) follows because $n_i - j - 1 \leq n_i - m_i + 1 \leq k$. Therefore $i \notin F$ and

$$\frac{1}{\mathcal{I}} \sum_{i=1}^{\mathcal{I}} [\![i \in F]\!] \overset{(a)}{\leq} \frac{1}{\mathcal{I}} \sum_{i=1}^{\mathcal{I}} \left([\![\ell_i \leq k]\!][\![i \in F]\!] + [\![\ell_i > k]\!] \right)$$

$$\overset{(b)}{\leq} \frac{N_k}{\mathcal{I}} + \frac{1}{\mathcal{I}} \sum_{i=1}^{\mathcal{I}} [\![\ell_i > k]\!] \overset{(c)}{=} \frac{N_k}{\mathcal{I}} + 1 - \sum_{\kappa=1}^{k} \frac{1}{\mathcal{I}} \sum_{i=1}^{\mathcal{I}} [\![\ell_i = \kappa]\!].$$

(a) and (c) follow by algebra. (b) because if $i > N_k$ and $\ell_i \leq k$, then $i \notin F$. Now ℓ_i is the length of a contiguous block of 1's surrounded by zeros. Since α is λ-random, by Lemma 1 the asymptotic proportion of such contiguous blocks of length κ is $2^{-\kappa}$ by the following argument.

$$\lim_{\mathcal{I}\to\infty} \frac{1}{\mathcal{I}} \sum_{i=1}^{\mathcal{I}} [\![\ell_i = \kappa]\!] \overset{(a)}{=} \lim_{\mathcal{I}\to\infty} \frac{1}{\mathcal{I}} \#01^\kappa 0(\alpha_{1:n_\mathcal{I}+1})$$

$$\overset{(b)}{=} \lim_{\mathcal{I}\to\infty} \frac{(n_\mathcal{I}+1)}{\#10(\alpha_{1:n_\mathcal{I}+1})} \cdot \frac{\#01^\kappa 0(\alpha_{1:n_\mathcal{I}+1})}{(n_\mathcal{I}+1)} \overset{(c)}{=} 2^{-\kappa},$$

where (a) and (b) follow from the definitions of the intervals and (c) follows the definition of normal numbers and from part 1 of Lemma 1. Therefore $\frac{1}{\mathcal{I}} \sum_{i=1}^{\mathcal{I}} [\![\ell_i > k]\!] < 2^{1-k}$ for sufficiently large $\mathcal{I}$. Sending $k \to \infty$ gives $d(F) := \lim_{\mathcal{I}\to\infty} \sum_{i=1}^{\mathcal{I}} [\![i \in F]\!]/\mathcal{I} = 0$. It follows from $d(B) = d(F) = 0$ and Lemma 1 that $d(B \cup F) = 0$ and $\lim_{n\to\infty} Km(\psi_{1:n})/n = 0$. Since

$|Km(x) - K(x)| < O(\log \ell(x))$ for all x [5, §4.5.5], $\lim_{n\to\infty} K(\psi_{1:n})/n = 0$ as well. Let $\theta_n := \#1(\chi_{1:n})/n$. By Lemma 1 we have that $\bar{d}(A - F) > 0$. Therefore there exists a $0 < c_3 \in \mathbb{Q}$ such that $\limsup_{n\to\infty} \theta_n > \frac{1}{2} + c_3$, where we also used the fact that $\alpha_n = 1 \implies \chi_n = 1$ and $\bar{d}(\{n_i : i \in \mathbb{N}\}) > 0$. If $\theta_n > \frac{1}{2} + c_3$ then by Lemma 1 there exists a $c_4 > 0$ such that $Km(\chi_{1:n}) < nH\left(\frac{1}{2} + c_3\right) + \frac{1}{2}\log n + c_4$. Therefore for all $\epsilon > 0$ there exists an arbitrarily large n such that

$$n - c_1 < Km(\alpha_{1:n}) < Km(\chi_{1:n}) + K(\psi_{1:n}) + c_2$$
$$< \epsilon n + nH\left(\frac{1}{2} + c_3\right) + \frac{1}{2}\log n + c_2 + c_4.$$

This is a contradiction since $H\left(\frac{1}{2} + c_3\right) < 1$. Therefore $\lim_{n\to\infty} M(\alpha_n|\alpha_{<n}) \neq \frac{1}{2}$ as required. ∎

5 Summary

We have shown that for every universal mixture there exists an infinite λ-random sequence on which it fails to converge.

Open Problems. There are a number of natural questions remaining. Suppose M is a universal lower semi-computable semimeasure and define $\mathcal{C}_M$ and $\mathcal{C}$ by

$$\mathcal{C}_M := \left\{\omega : \lim_{t\to\infty} M(\omega_n|\omega_{<n}) = \frac{1}{2}\right\} \quad \text{and} \quad \mathcal{C} := \bigcap_M \mathcal{C}_M$$

where the intersection is taken over all universal lower semi-computable semimeasures. What is the nature of $\mathcal{C}_M$ and $\mathcal{C}$? It follows from [3] that there exists an M such that $\mathcal{R}_\lambda \not\subseteq \mathcal{C}_M$, which implies that $\mathcal{R}_\lambda \not\subseteq \mathcal{C}$. In [7] it is shown that the 2-random reals are a subset of $\mathcal{C}$. In this work we showed that for all universal mixtures $\mathcal{R}_\lambda \not\subseteq \mathcal{C}_M$. Obvious open questions are:

1. Does there exists a universal lower semi-computable semimeasure (not a mixture) such that $\mathcal{R}_\lambda \subseteq \mathcal{C}_M$? An example of a non-trivial universal enumerable semimeasure that is not (essentially) a mixture may also be of interest.
2. As above, but where $\mathcal{R}_\lambda$ is replaced with a different class of random reals somewhere on the hierachy between Martin-Löf random and 2-random reals, such as the weak 2-random reals.

Unfortunately, an elegant characterisation of $\mathcal{C}_M$ and $\mathcal{C}$ seems unlikely because there exists an $\alpha \in \mathcal{C}$ that is not λ-random. See Proposition 8 in the appendix, which is adapted from Theorem 7 in [3]. Note that it is known that there exists a lower semicomputable semimeasure W that converges on all λ-random sequences, but W is not universal [4].

References

[1] Calude, C.: Information and Randomness: An Algorithmic Perspective, 2nd edn. Springer-Verlag New York, Inc., Secaucus (2002)

[2] Hutter, M.: On universal prediction and Bayesian confirmation. Theoretical Computer Science 384(1), 33–48 (2007)

[3] Hutter, M., Muchnik, A.: Universal convergence of semimeasures on individual random sequences. In: Ben-David, S., Case, J., Maruoka, A. (eds.) ALT 2004. LNCS (LNAI), vol. 3244, pp. 234–248. Springer, Heidelberg (2004)

[4] Hutter, M., Muchnik, A.: On semimeasures predicting Martin-Löf random sequences. Theoretical Computer Science 382(3), 247–261 (2007)

[5] Li, M., Vitanyi, P.: An Introduction to Kolmogorov Complexity and Its Applications, 3rd edn. Springer (2008)

[6] Martin-Löf, P.: The definition of random sequences. Information and Control 9(6), 602–619 (1966)

[7] Miyabe, K.: An optimal superfarthingale and its convergence over a computable topological space. In: Solomonoff Memorial. LNCS. Springer, Heidelberg (2011)

[8] Rathmanner, S., Hutter, M.: A philosophical treatise of universal induction. Entropy 13(6), 1076–1136 (2011)

[9] Solomonoff, R.: A formal theory of inductive inference, Part I. Information and Control 7(1), 1–22 (1964)

[10] Solomonoff, R.: Complexity-based induction systems: Comparisons and convergence theorems. IEEE Transactions on Information Theory 24(4), 422–432 (1978)

[11] Vovk, V.: On a randomness criterion. Soviet Mathematics Doklady 35, 656–660 (1987)

[12] Willems, F., Shtarkov, Y., Tjalkens, T.: The context tree weighting method: Basic properties. IEEE Transactions on Information Theory 41, 653–664 (1995)

[13] Wood, I., Sunehag, P., Hutter, M. (Non-)equivalence of universal priors. In: Solomonoff Memorial. LNCS. Springer, Heidelberg (2011)

[14] Zvonkin, A., Levin, L.: The complexity of finite objects and the development of the concepts of information and randomness by means of the theory of algorithms. Russian Mathematical Surveys 25(6), 83 (1970)

A Convergence on Non-random Sequences

Proposition 8. *There exists an $\alpha \in \mathcal{B}^\infty$ such that*

1. α is not λ-random.
2. For all universal lower semi-computable semimeasures M

$$\lim_{t \to \infty} \sum_{b \in \mathcal{B}} \left(M(b|\alpha_{<t}) - \frac{1}{2} \right)^2 = 0.$$

Proof. Define computable measure ν inductively by

$$\mu(1|x) := \frac{1}{2} + \frac{1}{2\sqrt{1 + \ell(x)}}$$

For universal lower semi-computable semimeasure M define the set of μ-random sequences on which M converges to μ by

$$A_M := \left\{ \omega : \lim_{t \to \infty} \sum_{b \in \mathcal{B}} (M(b|\omega_{<t}) - \mu(1|\omega_{<t}))^2 = 0 \wedge \omega \text{ is } \mu\text{-random} \right\}.$$

Now $\mu(A_M) = 1$ by Theorem 2 and the well-known fact that $\mu(\mathcal{R}_\mu) = 1$ for all computable measures μ. Therefore since there are only countably many universal lower semi-computable semimeasures, we have $\mu \left(A := \bigcap_M A_M \right) = 1$. Let $\alpha \in A$, which is μ-random. Then

$$\sum_{t=1}^{\infty} \sum_{b \in \mathcal{B}} \left(\sqrt{\mu(b|\alpha_{<t})} - \sqrt{\lambda(b|\alpha_{<t})} \right)^2 \geq \sum_{t=1}^{\infty} \left(\sqrt{\frac{1}{2} + \frac{1}{2\sqrt{t}}} - \sqrt{\frac{1}{2}} \right)^2 = \infty.$$

Therefore α is not λ-random by Theorem 3 of [11]. Finally by the definition of $\alpha \in A$ and μ we have that for all universal lower semi-computable semimeasures M

$$\lim_{t \to \infty} \sum_{b \in \mathcal{B}} \left(M(b|\alpha_{<t}) - \frac{1}{2} \right)^2 = \lim_{t \to \infty} \sum_{b \in \mathcal{B}} (M(b|\alpha_{<t}) - \mu(b|\alpha_{<t}))^2 = 0$$

as required.   ∎

Any Monotone Property of 3-Uniform Hypergraphs Is Weakly Evasive

Raghav Kulkarni[1,*], Youming Qiao[1], and Xiaoming Sun[2,**]

[1] Centre for Quantum Technologies, the National University of Singapore
{kulraghav,jimmyqiao86}@gmail.com
[2] Institute of Computing Technology, Chinese Academy of Sciences
sunxiaoming@ict.ac.cn

Abstract. For a Boolean function f, let $D(f)$ denote its deterministic decision tree complexity, i.e., minimum number of (adaptive) queries required in worst case in order to determine f. In a classic paper, Rivest and Vuillemin [18] show that any non-constant monotone property $\mathcal{P} : \{0,1\}^{\binom{n}{2}} \to \{0,1\}$ of n-vertex graphs has $D(\mathcal{P}) = \Omega(n^2)$.

We extend their result to 3-uniform hypergraphs. In particular, we show that any non-constant monotone property $\mathcal{P} : \{0,1\}^{\binom{n}{3}} \to \{0,1\}$ of n-vertex 3-uniform hypergraphs has $D(\mathcal{P}) = \Omega(n^3)$.

Our proof combines the combinatorial approach of Rivest and Vuillemin with the topological approach of Kahn, Saks, and Sturtevant. Interestingly, our proof makes use of Vinogradov's Theorem (weak Goldbach Conjecture), inspired by its recent use by Babai et. al. [1] in the context of the topological approach. Our work leaves the generalization to k-uniform hypergraphs as an intriguing open question.

1 Introduction

The *decision tree model* aka *query model* [3], perhaps due to its simplicity and fundamental nature, has been extensively studied over decades; yet there remain some outstanding open questions about it.

Fix a Boolean function $f : \{0,1\}^n \to \{0,1\}$. A deterministic decision tree D_f for f takes $x = (x_1, \ldots, x_n)$ as an input and determines the value of $f(x_1, \ldots, x_n)$ using queries of the form " is $x_i = 1$? ". Let $C(D_f, x)$ denote the cost of the computation, that is the number of queries made by D_f on input x. The *deterministic decision tree complexity* of f is defined as $D(f) = \min_{D_f} \max_x C(D_f, x)$.

The function f is called *evasive* if $D(f) = n$, i.e., one must query all the variables in worst case in order to determine the value of the function.

---

* Research at the Centre for Quantum Technologies is funded by the Singapore Ministry of Education and the National Research Foundation.
** Part of this work was done while the author was visiting the Centre for Quantum Techologies, National University of Singapore. This work was supported in part by the National Natural Science Foundation of China Grant 61170062, 61222202.

T-H.H. Chan, L.C. Lau, and L. Trevisan (Eds.): TAMC 2013, LNCS 7876, pp. 224–235, 2013.
© Springer-Verlag Berlin Heidelberg 2013

1.1   The Anderaa-Rosenberg-Karp Conjecture

A Boolean fuction f is said to be monotone (increasing) if for any $x \leq y$ we have $f(x) \leq f(y)$, where $x \leq y$ iff for all $i : x_i \leq y_i$. A property of n-vertex graphs is a Boolean function $\mathcal{P} : \{0,1\}^{\binom{n}{2}} \to \{0,1\}$ whose variables are identified with the $\binom{n}{2}$ potential edges of n-vertex graphs and the function $\mathcal{P}$ is invariant under relabeling of the vertices. $\mathcal{P}(G) = 0$ means that the graph G satisfies the property. A natural theme in the study of decision tree complexity is to exploit the structure within f to prove strong lower bounds on its query complexity. A classic example is the following conjecture attributed to Anderaa, Rosenberg, and Karp, asserting the *evasiveness* of monotone graph properties:

Conjecture 1 (ARK Conjecture). (cf. [8]) Every non-trivial monotone graph property is evasive.

Some natural examples of monotone graph properties are: connectedness, planarity, 3-colorability, containment of a fixed subgraph etc.

Since its origin around 1975, the ARK Conjecture has caught the imagination of generations of researchers resulting in beautiful mathematical ideas; yet - to this date - remains unsolved. A major breakthrough on ARK Conjecture was obtained by Kahn, Saks, and Sturvevant [8] via their novel *topological approach*. They settled the conjecture when the number of vertices of the graphs is a power of prime number. The topological approach subsequently turned out useful for solving some other variants and special cases of the conjecture. For example: Yao confirms the variant of the conjecture for monotone properties of bipartite graphs [24]. More recently, building on Chakraborty, Khot, and Shi's work [4], Babai et. al. [1] show that under some well-known conjectures in number theory, *forbidden subgraph* property - containment of a fixed subgraph in the graph - is evasive.

1.2   The Evasiveness Conjecture

The key feature of monotone graph properties is that they are sufficiently *symmetric*. In particular, they are *transitive* Boolean functions, i.e., there is a group acting transitively on the set of variables under which the function remains invariant. A natural question was raised: how much *symmetry* is necessary in order to guarantee the evasiveness? The following generalization (cf. [12]) of ARK Conjecture asserts that only transitivity suffices.

Conjecture 2 (Evasiveness Conjecture (EC)). If f is a non-trivial monotone transitive Boolean function, then f is evasive.

Rivest and Vuillemin [18] confirm the above conjecture when the number of variables is a power of prime number. The general case remains widely open.

1.3   The Weak Evasiveness Conjecture

Recently Kulkarni [7] proposes to investigate the following:

Conjecture 3 (Weak Evasiveness Conjecture). If $\{f_n\}$ is a sequence of non-trivial monotone transitive Boolean functions then for every $\epsilon > 0$

$$D(f_n) \geq n^{1-\epsilon}.$$

The best known lower bound in this context is $D(f) \geq R(f) \geq n^{2/3}$, which follows from the work of O'Donnell et. al. [17]. It turns out that [7] the above conjecture is equivalent to the EC! Furthermore: the Rivest and Vuillemin [18] result, which settles the ARK conjecture up to a constant factor, in fact confirms the Weak-EC for graph properties.

Theorem 1 (Rivest and Vuillemin). *If $\mathcal{P} : \{0,1\}^{\binom{n}{2}} \to \{0,1\}$ is a non-trivial monotone property of graphs on n vertices then $D(\mathcal{P}) = \Omega(n^2)$.*

It is interesting to note that the proof of equivalence in Kulkarni [7] does not hold between ARK and Weak-ARK. Hence: even though Weak-ARK is settled, the ARK is still wide open.

1.4   Our Results on the Weak EC

In this paper we prove an analogue of Rivest and Vuillemin's result (Theorem 1) for 3-uniform hypergraphs. A property of 3-uniform hypergraphs on n vertices is a Boolean function $\mathcal{P} : \{0,1\}^{\binom{n}{3}} \to \{0,1\}$ whose variables are labeled by the $\binom{n}{3}$ potential edges of n-vertex 3-uniform hypergraphs and $\mathcal{P}$ is invariant under relabeling of the vertices.

Theorem 2. *If $\mathcal{P} : \{0,1\}^{\binom{n}{3}} \to \{0,1\}$ is a non-trivial monotone property of 3-uniform hypergraphs on n vertices, then*

$$D(\mathcal{P}) = \Omega(n^3).$$

Our proof technique can be briefly described as follows: First we combine the combinatorial approach of Rivest and Vuillemin with the topological approach of Kahn, Saks, and Sturtevant to prove the result when $n = 3^k$. Then we use the 3^k case to prove the result for arbitrary n via an interesting application of the famous Vinogradov's Theorem that asserts that every odd integer can be expressed as sum of three prime numbers.

Interestingly, we do not yet know how to generalize our proof technique to k-uniform hypergraphs. But in this context we are able to prove a partial result on 4-uniform hypergraphs.

Theorem 3. *Let $\mathcal{P} : \{0,1\}^{n \times n \times n \times n} \to \{0,1\}$ be a 4-uniform 4-partite hypergraph property of $4n$-vertex hypergraphs. If $\mathcal{P}$ is non-trivial and monotone, then*

$$D(\mathcal{P}) = \Omega(n^4).$$

The organisation of this paper is as follows. Section 2 contains the preliminaries. Section 3 contains the proof of $n = 3^k$ case. Section 4 uses 3^k case to prove the general case, in particular it contains the proof of Theorem 2. Section 5 contains some partial results for 4-uniform hypergraphs. Section 6 contains conclusion and open ends.

2 Preliminaries

In this paper $[n]$ denotes the set $\{1, \ldots, n\}$.

2.1 Rivest-Vuillemin: Combinatorial Approach

In a beautiful paper, Rivest and Vuillemin show that the ARK Conjecture holds up to a constant factor, i.e., any non-trivial monotone graph property is weakly evasive. As an intermediate step [18] show the following:

Theorem 4 (Rivest-Vuillemin). *If n is a power of a prime number and $f : \{0,1\}^n \to \{0,1\}$ is any function invariant under a transitive permutation group such that $f(0, \ldots, 0) \neq f(1, \ldots, 1)$, then $D(f) = n$.*

In this paper we prove the weak-evasiveness of monotone properties of 3-uniform hyper-graphs, which extends the result of Rivest and Vuillemin for graph properties. Our proof is inspired by the one by Rivest and Vuillemin. Interestingly we use, in addition to the combinatorial approach of Rivest and Vuillemin, the powerful topological approach of Kahn, Saks, and Sturtevant combined with a deep theorem in number theory.

2.2 Kahn-Saks-Sturtevant: Topological Approach

In a seminal paper, Kahn, Saks, and Sturtevant [8] introduce a novel topological approach to settle the ARK Conjecture when the number of vertices of graphs is a power of a prime number. Their crucial observation was that non-evasiveness of monotone properties has a strong topological consequence, namely the corresponding simplicial complex is contractible to a point. Further they exploit this topological consequence via Oliver's Fixed Point Theorem [16] under the actions of certain special type of groups.

We say that a group Γ satisfies **Olivers Condition** if there exist (not necessarily distinct) primes p, q such that Γ has a (not necessarily proper) chain of subgroups $\Gamma_2 \lhd \Gamma_1 \lhd \Gamma$ such that Γ_2 is a p-group, Γ_1/Γ_2 is cyclic, and Γ/Γ_1 is a q-group, where p-group means a group whose order is a power of a prime p.

Theorem 5 (Kahn-Saks-Sturtevant). *If Γ satisfies Oliver's Condition and acts transitively on the set S of variables, then for any non-trivial monotone Γ-invariant function $f : \{0,1\}^S \to \{0,1\}$, we have: $D(f) = |S|$.*

Kahn, Saks, and Sturtevant made the assumption that the number of vertices of the graph - n is a prime power and used the following group that satisfies Oliver's Condition:

$$AFF(n) := AGL(1, n),$$

the group of affine transformations $x \mapsto ax + b$ over the field $\mathbb{F}_n$ of order n; $a \in \mathbb{F}_n - \{0\}, b \in \mathbb{F}_n$. The two key properties of this group are that it is a cyclic extension of a p-group, i.e., it satisfies Oliver's Condition; moreover it acts doubly transitively on $[n]$, i.e., any (i, j) can be mapped to any (i', j') for $i \neq j$ and $i' \neq j'$.

In this paper, we make use of the $AFF(n)$, as well as another group-theoretic construction called wreath product. We recall the definition, and refer the readers to [19, Section 1.6] detailed discussion. For a finite set S, let $\mathrm{Sym}(S)$ be the symmetric group on S. Let $G \leq \mathrm{Sym}(S)$ and $H \leq \mathrm{Sym}(T)$. The *wreath product* $G \wr H$ is a permutation group acting on $S \times T$, defined as follows. The *base group* of the wreath product is the direct product G^T, that is $|T|$ copies of G. For $t \in T$, the G_t independently acts on the corresponding copy $S \times \{t\}$. Specifically, for $(\omega, \delta) \in S \times T$, and $f \in G^T$, $(s, t)^f = (s^{f(t)}, t)$. $G \wr H$ also contains a subgroup H^* isomorphic to H, acting only on the second component of $S \times T$. That is for $h \in H$, $(s, t)^h = (s, t^h)$. $G \wr H$ is the group generated by G^S and H^*.

2.3   Prime-Partition via Vinogradov's Theorem

The Goldbach Conjecture asserts that every even integer can be written as the sum of two primes. Vinogradov's Theorem [23] says that every sufficiently large odd integer m is the sum of three primes $m = p_1 + p_2 + p_3$. We use here Haselgroves version [5] of Vinogradov's theorem which states that we can require the primes to be roughly equal: $p_i \sim m/3$. This can be combined with the Prime Number Theorem to conclude that every sufficiently large even integer m is a sum of four roughly equal primes.

This fact was first used by Babai et. al. [1] to construct the group actions satisfying Oliver's Condition in order to show that any monotone property of sparse graphs is evasive.

3   3-Uniform Hypergraphs: $n = 3^k$

We prove the following theorem in this section.

Theorem 6. *Let $n = 3^k$, and $\mathcal{P}$ be 3-uniform hypergraph property of n-vertex hypergraphs. If $\mathcal{P}$ is non-trivial and monotone, then $D(\mathcal{P}) = \Omega(n^3)$.*

Proof. Our proof strategy is inspired by the one by Rivest and Vuillemin's proof that non-trivial and monotone graph properties of graphs with $n = 2^k$ vertices are weakly evasive. The basic strategy is to set up a family of graphs $G_0 \subset G_1 \subset \ldots \subset G_k$, among which there are two *adjacent* graphs G_ℓ and $G_{\ell+1}$ such that G_ℓ satisfies $\mathcal{P}$ whereas $G_{\ell+1}$ does not. Now we start with the smaller graph G_ℓ and gradually *add* edges to it under the assumption that $D(\mathcal{P})$ is not $\Omega(n^3)$ and conclude that even after adding these edges the property $\mathcal{P}$ is satisfied. Eventually, after adding sufficiently many edges this would lead to a contradiction as we would be able to conclude that $G_{\ell+1}$ satisfies the property.

Rivest and Vuillemin choose G_ℓ to be the disjoint union of $2^{n-\ell}$ cliques on 2^ℓ vertices. Further they use Theorem 4 to *add* the edges to finally lead to a contradiction. Similar to Rivest and Vuillemin, we start our proof by using Theorem 4 to add certain type of edges. However, while handling the 3-uniform hypergraph properties, we face more complications. The natural choice of G_i to be disjoint union of hyper-cliques seems to fail and Theorem 4 seems inadequate in dealing with all types of edges. We overcome this obstacle by suitably changing the family of graphs and by making use of the topological approach of Kahn, Saks, and Sturtevant (Theorem 5) to deal with the other type of edges.

3.1 Our Choice of the Graph Family: Cliques with Spikes

To prove the theorem we consider the following family of hypergraphs on n vertices. For $j \in \{0, 1, \ldots, k\}$, let G_j be the hypergraph defined as follows: firstly G_j contains a disjoint union of 3^{k-j} copies of cliques on 3^j vertices. Then if an edge $\{u, v, w\}$ satisfies that u, v are in the one clique while w is in another one, it is also included in G_j. We call such edges *spikes*.

As G_0 is the empty hypergraph, and G_k is the complete hypergraph, we see that G_0 satisfies $\mathcal{P}$ while G_k does not as $\mathcal{P}$ is non-trivial. This suggests that there exists $\ell \in \{0, 1, \ldots, k-1\}$ such that G_ℓ satisfies the property while $G_{\ell+1}$ does not as $\mathcal{P}$ is monotone.

Now collect the cliques in G_ℓ into three groups $V_1 \cup V_2 \cup V_3$, each group containing $3^{k-\ell-1}$ cliques. We then consider the property $\mathcal{P}_1$ induced by $\mathcal{P}$ after fixing the values at the edges $\{\{u, v, w\} \mid u, v, w \in V_i, \text{or } \{u, v, w\} \in G_\ell\}$ as in G_ℓ. Note that $\mathcal{P}_1$ is a non-trivial property, because $\mathcal{P}$ is monotone and the graph $G_{\ell+1}$ is contained in the graph $G_\ell \bigcup E$ where E denotes the edges corresponding to the domain of $\mathcal{P}_1$.

3.2 Two Types of Edges

The edges not fixed in $\mathcal{P}_1$ are of two types:

Type 1 $T_1 = \{\{v_1, v_2, v_3\} \mid v_i \in V_i, i \in [3]\}$.
Type 2 $T_2 = \{\{u, v, w\} \mid u, v \in V_i, w \in V_j, i \neq j\}$. Note that v, w cannot come from the same clique otherwise it would have been fixed.

Before going on we define two group actions on V_1. Firstly, $H_1 = \mathbb{Z}_{3^\ell} \wr \mathbb{Z}_{3^{k-1-\ell}}$ acts on V_1, where $3^{k-1-\ell}$ copies of $\mathbb{Z}_{3^\ell}$ act independently on the $3^{k-1-\ell}$ cliques, and $\mathbb{Z}_{3^{k-1-\ell}}$ permutes among the cliques. Secondly, we define the group action of $H_2 = \mathbb{Z}_{3^\ell} \wr AFF(3^{k-1-\ell})$ on V_1 similarly to H_1. That is, $3^{k-1-\ell}$ copies of $\mathbb{Z}_{3^\ell}$ act independently on the $3^{k-1-\ell}$ cliques, and $AFF(3^{k-1-\ell})$ acts on the cliques in doubly-transitive way. Recall that for a vector space V, $AFF(V)$ is

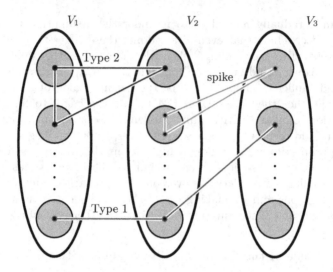

Fig. 1. An illustration of the spike (green), and Type 1 (red) and Type 2 (blue) edges

the affine group on V. H_1 and H_2 are both subgroups of the automorphism group of induced subgraph of G_ℓ on V_1. Then note that H_1 is a 3-group, and H_2 is transitive on $\{\{u, v\} \mid u$ and v are from different cliques$\}$. Finally, it can be verified that H_2 belongs to the group class as in Theorem 5.

Table 1. Groups used to handle Type 1 and Type 2 edges

Type 1	$(\mathbb{Z}_{3^\ell} \wr \mathbb{Z}_{3^{k-1-\ell}}) \times (\mathbb{Z}_{3^\ell} \wr \mathbb{Z}_{3^{k-1-\ell}}) \times (\mathbb{Z}_{3^\ell} \wr \mathbb{Z}_{3^{k-1-\ell}})$
Type 2	$(\mathbb{Z}_{3^\ell} \wr AFF(3^{k-1-\ell})) \times (\mathbb{Z}_{3^\ell} \wr \mathbb{Z}_{3^{k-1-\ell}})$

3.3 Adding Type 1 Edges

Now we consider the property $\mathcal{P}_2$ induced by $\mathcal{P}_1$ by setting Type 2 edges to be absent. Note that the number of Type 1 edges is $3^{3(k-1)}$, thus a prime power. Let $H_1 \times H_1 \times H_1$ act on $V_1 \times V_2 \times V_3$ in a natural way: each copy of H_1 acts on vertices of V_i independently. It is seen that this action preserves the fixed subgraph, and $\mathcal{P}_2$ is invariant under this action. If after adding all Type 1 edges $\mathcal{P}_2$ would *not* be satisfied, then by the Rivest-Vuillemin theorem, $\mathcal{P}_2$ is evasive. That is $D(\mathcal{P}) \geq D(\mathcal{P}_2) = 3^{3(k-1)} = \Omega(n^3)$ and we would be done.

3.4 Adding Type 2 Edges

Let $\mathcal{P}_3$ be the property induced by $\mathcal{P}_1$ by setting Type 1 edges to be present. The discussion from last paragraph suggests that $\mathcal{P}_3$ is a non-trivial property, and note that $\mathcal{P}_3$ only has Type 2 edges left unfixed. For $i, j \in [3], i \neq j$, let $T_2(i, j) = \{\{u, v, w\} \mid u, v \in V_i, w \in V_j\}$. Let $\mathcal{P}_4$ be the property induced by $\mathcal{P}_3$ by setting edges in $T_2 \setminus T_2(1, 2)$ to be absent. Note that $|T_2(1, 2)| = \Omega(n^3)$.

(Oliver's Condition Holds) Consider the group $H = H_2 \times H_1$ acting on $V_1 \times V_2$ in a natural way. It is verified that H preserves the structure of the fixed graph, and $\mathcal{P}_3$ is invariant under H. It is easy to check that H belongs to the group class described in Theorem 5. This allows us to apply Theorem 5 (the Kahn, Saks and Sturtevant Theorem) to conclude that either $\mathcal{P}_4$ is trivial; if not then its query complexity is $\Omega(n^3)$ and we would be done.

(Orbits are large) Here we use a key property of the action of H, namely that the orbit of any edge is of large size: $\Omega(n^3)$. Thus we can add edges in $T_2(1,2)$ to get another restriction $\mathcal{P}_5$. Then we use the same group as above for $V_2 \times V_3$ to add edges in $T_2(2,3)$.

3.5 Deriving a Contradiction

Continuing this way we can keep adding $T_2(i,j)$ edges while maintaining that the hyper-graph still satisfies the property. But then we would get $G_{\ell+1}$ as a subgraph which by our choice of ℓ, does not satisfy the property. Contradiction!

4 3-Uniform Hypergraphs: General n

In this section we prove the main theorem. **Theorem 1.5, restated.** If $\mathcal{P}$: $\{0,1\}^{\binom{n}{3}} \to \{0,1\}$ is a non-trivial monotone property of 3-uniform hypergraphs on n vertices, then $D(\mathcal{P}) = \Omega(n^3)$.

Proof. The natural way of extending Rivest and Vuillemin's argument for 3-uniform hyper graphs for arbitrary n leads to analysis of several types of edges. We do not know an easy way to handle this via combinatorial approach. We can use the topological approach together with an interesting theorem about partitioning an integer into prime numbers to patch up the 3^k case to arbitrary n.

4.1 Prime-Partition of n via Vinogradov's Theorem

We distinguish two cases: (Case 1) n is even and (Case 2) n is odd. Let us consider Case 1: n is even. The other case can be handled in a similar fashion. Let k be the largest power of 3 that does not exceed n. Since n is odd, we can write (using the above mentioned Hasegrove's Version of Vinogradov's Theorem) $n = p_1 + p_2 + p_3 + 3^{k-1}$, where p_is are prime numbers and $p_i \sim p_j$. Moreoever: note that by our choice of k we can assume: $p_i \leq 3^k$.

4.2 Patching Up 3^k Case to General n

We partition $[n]$ into parts of size p_1, p_2, p_3 and 3^{k-1} as described in the previous section. Let $\mathcal{P}$ be a non-trivial monotone property of 3-uniform hyper-graphs on n vertices. Theorem 6 allows us to conclude that either (a) $D(\mathcal{P}) = \Omega(n^3)$ or

(b) *any* 3^k vertex (hyper) clique satisfies $\mathcal{P}$. In Case (a) we are done. So let us assume that we are in Case (b). Since $p_1 \leq 3^k$ and since $\mathcal{P}$ is monotone, we may assume that the clique on p_1 vertices satisfies the property. Now we assume that the clique on p_1 vertices is present and restrict our attention to the induced property $\mathcal{P}_2$ of 3-uniform hypergraphs on $p_2 + p_3 + 3^{k-1}$ vertices. Again using the fact that $p_2 \leq 3^k$ we can assume that the clique on p_2 vertices is also present in addition to the clique on p_1 vertices. Now we move our attention to the induced property $\mathcal{P}_2$ on $p_3 + 3^{k-1}$ vertex graphs. In one more step, we can move our attention to the induced property $\mathcal{P}_3$ on 3^{k-1} graphs which assumes that the cliques on the p_1, p_2, and p_3 vertices are present. Finally, with the use of Theorem 6, we can conclude that the clique on the 3^{k-1} vertices is also present; if not then we could already conclude $D(\mathcal{P}) = \Omega(n^3)$.

4.3 Two Types of Edges

Now we have a restriction $\mathcal{P}'$ of our original property $\mathcal{P}$ in which the cliques on p_1, p_2, p_3 and 3^{k-1} vertices are present. We partition the absent edges into two types:

Type A the three endpoints of the edges belong to different cliques;
Type B two of the three endpoints belong to one clique and the remaining endpoint belongs to a different clique.

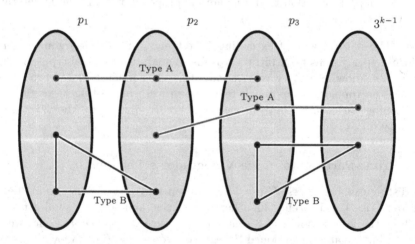

Fig. 2. An illustration of Type A (red) and Type B (blue) edges

4.4 Adding Type A Edges

Firstly: we conclude that all Type A edges must also be present; if not then $D(\mathcal{P}) = \Omega(n^3)$. For this we use the following two types of groups: $\mathbb{Z}_{p_1} \times \mathbb{Z}_{p_2} \times \mathbb{Z}_{p_3}$ and $\mathbb{Z}_{p_i} \times \mathbb{Z}_{p_j} \times \mathbb{Z}_{3^{k-1}}$.

Table 2. Groups used to handle Type A and Type B edges

Type A	$\mathbb{Z}_{p_1} \times \mathbb{Z}_{p_2} \times \mathbb{Z}_{p_3}$, and $\mathbb{Z}_{p_i} \times \mathbb{Z}_{p_j} \times \mathbb{Z}_{3^{k-1}}$
Type B	$AFF(p_i) \times \mathbb{Z}_{p_j}$, and $AFF(p_i) \times \mathbb{Z}_{3^{k-1}}$, and $AFF(3^{k-1}) \times \mathbb{Z}_{p_i}$

4.5 Adding Type B Edges

Secondly: after adding all Type A edges we can conclude that all Type B edges must also be present; if not then $D(\mathcal{P}) = \Omega(n^3)$. For this we use the following three types of groups: $AFF(p_i) \times \mathbb{Z}_{p_j}$ and $AFF(p_i) \times \mathbb{Z}_{3^{k-1}}$ and $AFF(3^{k-1}) \times \mathbb{Z}_{p_i}$.

(Oliver's Condition Holds) It is easy to check that all the groups that we use are the right ones for using the topological approach, i.e., they are "q-group extension of cyclic extension of p-groups," i.e., they satisfy Oliver's Condition.

(Orbits are large) A crucial property that we used in our proof is that the orbit of any edge under any of our group actions is large: $\Omega(n^3)$.

4.6 Deriving a Contradiction

After adding both Type A and Type B edges to the cliques on the p_1, p_2, p_3 and 3^{k-1} vertices, we can conclude that the clique on n vertices must satisfy $\mathcal{P}$; this contradicts with our initial assumption that $\mathcal{P}$ is non-trivial.

This completes the proof of Theorem 2.

5 4-Uniform 4-Partite Hypergraphs

In this section we prove the weak evasiveness for properties of 4-uniform 4-partite hypergraphs. **Theorem 1.6, restated.** Let $\mathcal{P} : \{0,1\}^{n \times n \times n \times n} \to \{0,1\}$ be a 4-uniform 4-partite hypergraph property of $4n$-vertex hypergraphs. If $\mathcal{P}$ is non-trivial and monotone, then $D(\mathcal{P}) = \Omega(n^4)$.

Proof: If n is prime, the result directly follows from Theorem 4.

In the case when n is not prime, let p be a prime number such that $p < n < 2p$. Let $V = V_1 \cup V_2 \cup V_3 \cup V_4$, $|V_i| = n$ be the vertex set. The strategy is again by contrapositive: assume $D(\mathcal{P})$ is not of $\Omega(n^4)$. Then we shall start from the empty graph, and then add the edges with different types while keeping the value of the property not change. Finally we will get that the complete graph satisfies the property, which contradicts to the condition of being a non-trivial property. Let G_0 be the empty graph; thus $f(G_0) = 0$. Let $V_i = A_i \cup B_i$, where A_i is a vertex set of size p $(i = 1, 2, 3, 4)$, and $B_i = V_i \setminus A_i$.

5.1 Adding Edges in $A_1 \times A_2 \times A_3 \times A_4$

Consider a restriction $\mathcal{P}_1$ of $\mathcal{P}$ where all the variables outside $A_1 \times A_2 \times A_3 \times A_4$ are set to be 0. $\mathcal{P}_1$ is a monotone transitive invariant function with p^4 variables, by Theorem 4 $\mathcal{P}_1$ is trivial, otherwise $D(\mathcal{P}_1) = p^4 = \Omega(n^4)$. Let G_1 be the graph with edges $A_1 \times A_2 \times A_3 \times A_4$, thus $f(G_1) = 0$.

5.2 Adding Edges in $B_1 \times A_2 \times A_3 \times A_4$

We will add all the edges in $B_1 \times A_2 \times A_3 \times A_4$ to G_1, resulting in a graph G_2 with edges $V_1 \times A_2 \times A_3 \times A_4$. Before doing that we consider a graph G_1' with edges $B_1 \times A_2 \times A_3 \times A_4$. Since $p > n - p$, from the monotone and symmetry condition, we have $f(G_1') \leq f(G_1) = 0$. Consider a restriction $\mathcal{P}_2$ of $\mathcal{P}$ where all the edges in G_1' are set to 1 and all the edges outside $V_1 \times A_2 \times A_3 \times A_4$ are set to 0. It is clear that $\mathcal{P}_2$ is a monotone transitive invariant function with p^4 variables, thus from Theorem 4 $\mathcal{P}_2$ is a constant, otherwise $D(\mathcal{P}_2) = p^4 = \Omega(n^4)$. Hence we get $f(G_2) = 0$.

5.3 Adding Edges in $V_1 \times B_2 \times A_3 \times A_4$

Similar to the previous step, we first use the monotone and symmetry condition to "delete" some edges from G_2. Let G_2' be the graph with edges $V_1 \times B_2 \times A_3 \times A_4$. From the monotone and symmetry condition, $f(G_2') \leq f(G_2) = 0$. Consider the restriction $\mathcal{P}_3$ of $\mathcal{P}$ where all the edges in G_2' are set to 1 and all the edges outside $V_1 \times V_2 \times A_3 \times A_4$ are set to 0. It is easy to see that $\mathcal{P}_3$ can be further partitioned into two properties isomorphism to $\mathcal{P}_1$ and $\mathcal{P}_2$, respectively. By repeating the steps in Section 5.1 and 5.2 we conclude that $\mathcal{P}_3$ is trivial, otherwise $D(\mathcal{P}_3) = \Omega(n^4)$. Hence $f(G_3) = 0$, where G_3 is the graph with edges $V_1 \times V_2 \times A_3 \times A_4$.

Adding Edges in $V_1 \times V_2 \times B_3 \times A_4$ and $V_1 \times V_2 \times V_3 \times B_4$

These two steps are similar to the previous step, and we omit them here. After doing these steps, we get that the value of the complete graph is also 0, which contradicts the non-trivial condition.   □

Remark 1. We note that the proof strategy for Theorem 1.6 can be extended to show that any k-uniform k-partite hypergraph property is weakly evasive, when k is a constant. On the other hand, we do not know how to prove the weak evasiveness for 4-uniform hypergraph properties.

6 Conclusion

In this paper we are able to confirm a special case of the Weak-EC. In particular, we have shown that any non-trivial monotone property of 3-uniform hypergraphs is weakly evasive. It is interesting to see how far can one generalize our results.

Question 1. Is any non-trivial monotone property of k-uniform hypergraphs weakly evasive?

References

1. Babai, L., Banerjee, A., Kulkarni, R., Naik, V.: Evasiveness and the Distribution of Prime Numbers. In: STACS 2010, pp. 71–82 (2010)
2. Benjamini, I., Kalai, G., Schramm, O.: Noise sensitivity of Boolean functions and its application to percolation. Inst. Hautes tudes Sci. Publ. (MATH), 90 (1999)
3. Buhrman, H., de Wolf, R.: Complexity measures and decision tree complexity: a survey. Theor. Comput. Sci. 288(1), 21–43 (2002)
4. Chakrabarti, A., Khot, S., Shi, Y.: Evasiveness of Subgraph Containment and Related Properties. SIAM J. Comput. 31(3), 866–875 (2001)
5. Haselgrove, C.B.: Some theorems on the analytic theory of numbers. J. London Math. Soc. 36, 273–277 (1951)
6. Hayes, T.P., Kutin, S., van Melkebeek, D.: The Quantum Black-Box Complexity of Majority. Algorithmica 34(4), 480–501 (2002)
7. Kulkarni, R.: Evasiveness Through A Circuit Lens. To appear in ITCS 2013 (2013)
8. Kahn, J., Saks, M.E., Sturtevant, D.: A topological approach to evasiveness. Combinatorica 4(4), 297–306 (1984)
9. Kushilevitz, E., Mansour, Y.: Learning Decision Trees Using the Fourier Spectrum. SIAM J. Comput. 22(6), 1331–1348 (1993)
10. Kulkarni, R., Santha, M.: Query complexity of matroids. Electronic Colloquium on Computational Complexity (ECCC) 19, 63 (2012)
11. Linial, N., Mansour, Y., Nisan, N.: Constant Depth Circuits, Fourier Transform, and Learnability. J. ACM 40(3), 607–620 (1993)
12. Lutz, F.H.: Some Results Related to the Evasiveness Conjecture. Comb. Theory, Ser. B 81(1), 110–124 (2001)
13. Montanaro, A., Osborne, T.: On the communication complexity of XOR functions. CoRR abs/0909.3392 (2009)
14. Nisan, N., Szegedy, M.: On the Degree of Boolean Functions as Real Polynomials. Computational Complexity 4, 301–313 (1994)
15. Nisan, N., Wigderson, A.: On Rank vs. Communication Complexity. Combinatorica 15(4), 557–565 (1995)
16. Oliver, R.: Fixed-point sets of group actions on finite acyclic complexes. Comment. Math. Helv. 50, 155–177 (1975)
17. O'Donnell, R., Saks, M.E., Schramm, O., Servedio, R.A.: Every decision tree has an influential variable. In: FOCS 2005, pp. 31–39 (2005)
18. Rivest, R.L., Vuillemin, J.: On Recognizing Graph Properties from Adjacency Matrices. Theor. Comput. Sci. 3(3), 371–384 (1976)
19. Robinson, D.J.S.: A Course in the Theory of Groups, 2nd edn. Springer (1996)
20. Saks, M.E., Wigderson, A.: Probabilistic Boolean Decision Trees and the Complexity of Evaluating Game Trees. In: FOCS 1986, pp. 29–38 (1986)
21. Shi, Y., Zhang, Z.: Communication Complexities of XOR functions. CoRR abs/0808.1762 (2008)
22. Zhang, Z., Shi, Y.: On the parity complexity measures of Boolean functions. Theor. Comput. Sci. 411(26-28), 2612–2618 (2010)
23. Vinogradov, I.M.: The Method of Trigonometrical Sums in the Theory of Numbers. Trav. Inst. Math. Stekloff 10 (1937) (Russian)
24. Yao, A.C.-C.: Monotone Bipartite Graph Properties are Evasive. SIAM J. Comput. 17(3), 517–520 (1988)

The Algorithm for the Two-Sided Scaffold Filling Problem

Nan Liu[1,2] and Daming Zhu[1]

[1] School of Computer Science and Technology, Shandong University, Jinan, China
liunansdu@gmail.com, dmzhu@sdu.edu.cn
[2] School of Computer Science and Technology, Shandong Jianzhu University, Jinan, China

Abstract. Scaffold filling is a new combinatorial optimization problem in genome sequencing and can improve the accuracy of the sequencing results. The two-sided Scaffold Filling to Maximize the Number of String Adjacencies(SF-MNSA) problem can be described as: given two incomplete gene sequences A and B, respectively fill the missing genes into A and B such that the number of adjacencies between the resulting sequences A' and B' is maximized. The two-sided scaffold filling problem is NP-complete for genomes with duplicated genes and there is no effective approximation algorithm. In this paper, we propose a new version problem that symbol # is added to each end of each input sequence for any instance of two-sided SF-MNSA problem and design a polynomial algorithm for one special case of this new version problem. For any instance, we present a better lower bound of the optimal solution and devise a factor-1.5 approximation algorithm by exploiting greedy strategy.

1 Introduction

In the process of biological sequencing, a gene fragment or genome is generally sequenced and assembled many times. Finally, we often obtain more than one incomplete sequence (scaffolds or contigs). The accuracy of final sequence will affect the results of biological analysis. Therefore, it is important that using scaffold filling technology improves the accuracy of final gene sequence.

Muñoz *et al.* first investigate the one-sided permutation scaffold filling problem, and propose an exact algorithm to minimize the genome rearrangement (DCJ) distance [10]. Subsequently, Jiang *et al.* solve the two-sided permutation scaffolding filling problem under the DCJ distance in polynomial time [8]. When genomes contain some duplicated genes, the scenario is completely different. There are three general criteria (or distance) to measure the similarity of genomes: the exemplar genomic distance [11], the minimum common string partition (MCSP) distance [3] and the maximum number of string adjacencies [1,9]. Unfortunately, unless P=NP, there does not exist any polynomial time approximation (regardless of the factor) algorithm for computing the exemplar genomic distance even when each gene is allowed to repeat three times [5,4] or even two times [2,7]. The MCSP problem is NP-complete even if each gene repeats at most two times [6]. Jiang *et al.* prove the SF-MNSA problem for genomes with

T-H.H. Chan, L.C. Lau, and L. Trevisan (Eds.): TAMC 2013, LNCS 7876, pp. 236–247, 2013.
© Springer-Verlag Berlin Heidelberg 2013

gene repetitions is also NP-complete and design a 1.33-approximation algorithm for one-sided problem [9]. But we find the approximation algorithm should be for the one-sided SF-MNSA with symbol # problem. So far, there is no effective algorithm for two-sided sequence scaffold filling problem.

In this paper, we propose a new version of the two-sided SF-MNSA problem and design a polynomial algorithm for the special instance of this new version problem. For any instance, an approximation algorithm with 1.5-factor is devised, by analyzing the two-sided characteristic and using greedy method.

2 Preliminaries

At first, we review some necessary definitions [9]. Throughout this paper, all genes and genomes are unsigned. Given a gene set Σ, a string P is called *permutation* if each element in Σ appears exactly once in P. We use $c(P)$ to denote the set of elements in permutation P. A string S is called *sequence* if some genes appear more than once in S, and $c(S)$ denotes genes of S, which is a multiset of elements in Σ. For example, $\Sigma = \{a, b, c, d\}$, $S = abcdacd$, $c(S) = \{a, a, b, c, c, d, d\}$. A *scaffold* (with gene repetitions) is an incomplete sequence, typically obtained by some sequencing and assembling process. A substring with m genes is called an *m-substring*, and a 2-substring is also called a *pair*. As the genes are unsigned, the relative order of the two genes of a pair does not matter, i.e., the pair xy is equal to the pair yx. Given a scaffold $A=a_1a_2a_3\cdots a_n$, let $P_A = \{a_1a_2, a_2a_3, \ldots, a_{n-1}a_n\}$ be the set of pairs in A.

Definition 1. *Given two scaffolds $A=a_1a_2\cdots a_n$ and $B=b_1b_2\cdots b_m$, if $a_ia_{i+1} = b_jb_{j+1}$ (or $a_ia_{i+1}=b_{j+1}b_j$), where $a_ia_{i+1} \in P_A$ and $b_jb_{j+1} \in P_B$, a_ia_{i+1} and b_jb_{j+1} are matched to each other. In a maximum matching of pairs in P_A and P_B, a matched pair is called an **adjacency**, and an unmatched pair is called a **breakpoint** in A and B respectively.*

It follows from the definition that scaffolds A and B contain the same set of adjacencies but distinct breakpoints. The maximum matched pairs in B (or equally, in A) form the *adjacency set* between A and B, denoted as $a(A, B)$. We use $b_A(A, B)$ and $b_B(A, B)$ to denote the set of breakpoints in A and B respectively. A gene is called a *bp-gene*, if it appears in a breakpoint. Each maximal substring W of A (or B) is called a *bp-string*, if each pair in it is a breakpoint. The leftmost and rightmost genes of a bp-string W are called the *end-genes* of W, the other genes in W are called the *mid-genes* of W. For example, we have scaffold $A = abcedaba$, $B = cbabda$, $P_A = \{ab, bc, ce, ed, da, ab, ba\}$, $P_B = \{cb, ba, ab, bd, da\}$, then matched pairs are $(ab, ba), (bc, cb), (da, da), (ab, ab)$. For the scaffold A, ab, bc, da, ab are adjacency and the set of A's breakpoints $b_A(A, B) = \{ce, ed, ba\}$, string ced and ba are bp-string. For the scaffold B, cb, ba, ab, da are adjacency and $b_B(A, B) = \{bd\}$, string bd is a bp-string.

For a scaffold A and a missing genes multiset X, let A' be a resulting scaffold after filling all the genes in X into A, then $A' = A + X$. The process is called *Scaffold Filling*. We use "+" to denote the scaffold filling operation.

Definition 2. *Scaffold Filling to Maximize the Number of (String) Adjacencies (SF-MNSA).*
Input: *two scaffolds A and B over a gene set Σ and two multi-sets of elements X and Y, where $X = c(B) - c(A) \neq \emptyset$ and $Y = c(A) - c(B) \neq \emptyset$.*
Question: *Find $A' \in A + X$ and $B' \in B + Y$ such that $|a(A', B')|$ is maximized.*

Given two scaffolds $A = a_1 a_2 \cdots a_n$ and $B = b_1 b_2 \cdots b_m$, as we can see, each gene except the four ending ones is involved in two adjacencies or two breakpoints or one adjacency and one breakpoint. To get rid of this imbalance, we add '#' to both ends of A and B, The problem of this version is called *two-sided SF-MNSA with symbol #*. For the new version problem, it is ensured that one gene insertion can generate at least one new adjacency (see the proof of Lemma 1). The fact is needed by the approximation algorithm.

We list a few basic properties of this problem. Note that A and B are sequences that have been added symbol # and all propositions, lemmas, theorems are for any instance of SF-MNSA with symbol # problem in the following paper.

Proposition 1. *If a gene appears the same times in both input sequences, then the gene constitutes no breakpoint in one input sequence when it constitutes no breakpoint in another input sequence.*

Proof. let A, B be input sequences and a appear n times in A and B respectively. Let a constitute no breakpoint in A, then a's form in A be $\# \cdots u_1 a v_1 \cdots u_2 a v_2 \cdots u_i a v_i \cdots w_1 A_1 z_1 \cdots w_2 A_2 z_2 \cdots w_j A_j z_j \cdots \#$, where $A_1, A_2, \ldots, A_j$ are gene string composed of more than one gene a and the numbers of gene a are $|A_1| = k_1, |A_2| = k_2, \cdots, |A_j| = k_j$. Then the number of all gene a is $n = i + k_1 + k_2 + \cdots + k_j$. Because gene a constitutes no breakpoint in A, a maximum matching about gene a between A and B is shown in the Figure1. Gene a's number in

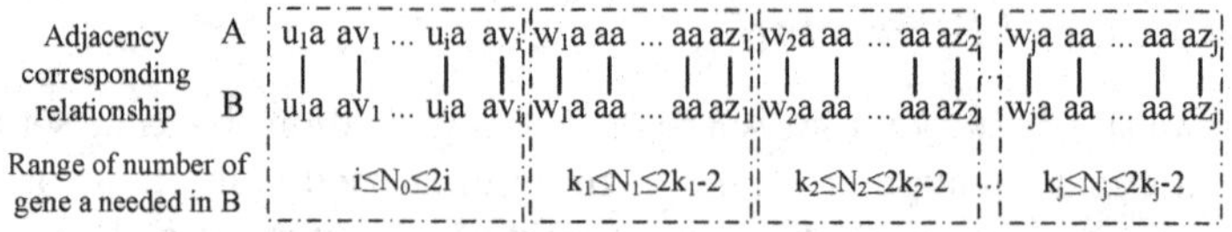

Fig. 1. Adjacency matching and range of a's number in B

B that constitutes all adjacencies must satisfy: $i + k_1 + k_2 + \cdots + k_j \leq N_0 + N_1 + N_2 + \cdots + N_j \leq 2i + 2k_1 - 2 + 2k_2 - 2 + \cdots + 2k_j - 2$, Then, $n \leq N_0 + N_1 + N_2 + \cdots + N_j \leq 2(n - j)$, i.e., a's number in B that constitutes adjacency is at least n. However, there exist only n gene a's in B. So, all n gene a's constitute adjacencies. Therefore, gene a constitutes no breakpoint in B when it constitutes no breakpoint in A. The proof is similar that gene a constitutes no breakpoint in A when it constitutes no breakpoint in B.   □

Proposition 2. *Each bp-gene in A appears in either Y or some breakpoint in B. And each bp-gene in B appears in either X or some breakpoint in A.*

Proof. Let gene a be a bp-gene in A. Assume that the proposition is not true, *i.e.* gene a appears in neither Y nor any breakpoint in B. So, gene a's number in A is same with in B and all gene a's constitute adjacencies in B. According to Proposition 1, all gene a's constitute adjacencies in A. This contradicts with that gene a is a bp-gene in A. The case that each bp-gene in B appears in either X or some breakpoints in A can be similarly proved. $\square$

We know that each breakpoint contains two genes. All breakpoints in each input sequence can be divided into three sets according to Proposition 2. For two bp-genes of the breakpoint in A, we have

$BP_1(A)$: one appears in Y, the other appears in some breakpoint in B.

$BP_2(A)$: both genes appear in Y.

$BP_3(A)$: both genes appear in breakpoints in B.

For two bp-genes of the breakpoint in B, we have

$BP_1(B)$: one appears in X, the other appears in some breakpoint in A.

$BP_2(B)$: both genes appear in X.

$BP_3(B)$: both genes appear in breakpoints in A.

3   A Polynomial Time Algorithm for the Special Case

In this section, we present a polynomial time algorithm for a special case of the two-sided SF-MNSA problem with symbol $\#$. The proof about the correctness of the algorithm is in the Appendix.

The special case is that there are no breakpoint in $BP_1(A)$ and $BP_1(B)$. For two strings s_1 and s_2, if the right end-gene $r(s_1)$ of s_1 is the same as the left end-gene $\ell(s_2)$ of s_2, we use $s_1 \bowtie s_2$ to represent the string obtained by first concatenating s_1 with s_2 and then delete one copy of $r(s_1)$ and $\ell(s_2)$. For example, $s_1 = acbd, s_2 = decb$, then $s_1 \bowtie s_2 = abcdecb$. If $BP_1(A) = \emptyset$ and $BP_1(B) = \emptyset$, we have the algorithm:

Algorithm 1
Input: sequence A, B, with X, Y and set of breakpoints $BP_2(A)$, $BP_2(B)$
Output: A', B'
1. Compute sets of bp-strings according to $BP_2(A)$, $BP_2(B)$: BS_A, BS_B.
2. WHILE($BS_A \neq \emptyset$){
    2.1 Call the function $BPS(BP, S)$ to compute a string S that is composed of bp-strings in BS_A, where $BP = BS_A$. Then, let $AS = S$.
    2.2 Replace some gene identical to $\ell(AS)$ in B by string AS to obtain B'. Update BS_A.}
3. WHILE($BS_B \neq \emptyset$){
    3.1 Call the function $BPS(BP, S)$ to compute a string S that is composed of bp-strings in BS_B, where $BP = BS_B$. Then, let $BS = S$.
    3.2 Replace some gene identical to $\ell(BS)$ in A by string BS to obtain A'. Update BS_B.}
4. Return A', B'.
Function $BPS(BP, S)$

1 Choose any bp-string of BP, say s_j. Let $S = s_j = x_{j,1} \cdots x_{j,u_j}$.
2 WHILE($\ell(S) \neq r(S)$)
    2.1 Find a bp-string s_i or $\overline{s_i} = x_{i,1} \cdots x_{i,u_i}$ in BP that $\overline{s_i}$ is s_i"s reversal. Update $S \leftarrow S \bowtie s_i$ or $S \leftarrow S \bowtie \overline{s_i}$.
    2.2 WLOG, let $S = s_1 \bowtie s_2 \bowtie \cdots \bowtie s_i$. Compare $r(S)$ with $\ell(s_j)$ for $j = i, i-1, \cdots$. If $r(S) = \ell(s_k)$, then update $S \leftarrow s_k \bowtie s_{k+1} \cdots \bowtie s_i$.
    2.3 Update the set BP.
3 return S.

When the instance is not satisfied with the special condition, *i.e.*, $BP_1(A) \neq \emptyset$ or $BP_1(B) \neq \emptyset$, we design an approximation algorithm for this problem.

4   A 1.5-Approximation Algorithm

In this section, we firstly prove some premises for our algorithm. Next, we present an approximation algorithm and prove the approximation factor is 1.5.

4.1   Premises

When symbol # is added to the two-sided SF-MNSA problem, we have

Lemma 1. *There exists a polynomial time algorithm that all missing genes can be inserted to obtain at least $|X|+|Y|$ new adjacencies, the number of breakpoints doesn't increase, and each insertion of an m-substring can generate at least m new adjacencies.*

Proof. For all breakpoints in A or B, we process all breakpoints of $BP_1(A)$ or $BP_1(B)$ at first. Let $a_i a_j$ be any breakpoint of $BP_1(A)$, where a_i belongs to Y and a_j belongs to some breakpoint $(a_k a_j)$ in B. So, a_i can be inserted into breakpoint $a_k a_j$ to generate at least one adjacency $a_i a_j$ and keep the number of breakpoints in B. After all breakpoints of $BP_1(A)$ and $BP_1(B)$ are eliminated, left breakpoints belonging to $BP_2(\bullet)$ or $BP_3(\bullet)$ and left missing genes can be inserted in A or B to obtain new adjacencies with generating no breakpoint according to Algorithm 1. Each insertion of an m-substring can generate at least m new adjacencies and keep the number of breakpoints not to increase.   □

Obviously, inserting a 1-substring will generate at most two adjacencies, and inserting an m-substring will generate at most $m+1$ adjacencies. Therefore, we will have two types of inserted strings.

1. Type-1: a string of k missing genes $x_1 x_2 \cdots x_k$ are inserted between $y_i y_{i+1}$ in A or B to obtain $k+1$ adjacencies(i.e., $y_i x_1, x_1 x_2, \ldots, x_{k-1} x_k, x_k y_{i+1}$), where $y_i y_{i+1}$ is a breakpoint. In this case, $x_1 x_2 \cdots x_k$ is called a k-Type-1 string.
2. Type-2: a string of l missing genes $z_1 z_2 \cdots z_l$ are inserted between $y_j y_{j+1}$ in A or B to obtain l adjacencies(i.e., $y_j z_1$ or $z_l y_{j+1}, z_1 z_2, \ldots, z_{l-1} z_l$), where $y_j y_{j+1}$ is a breakpoint; or a string of l missing genes $z_1 z_2 \cdots z_l$ are inserted between $y_j y_{j+1}$ in A or B to obtain $l+1$ adjacencies(i.e., $y_j z_1, z_1 z_2, \ldots, z_{l-1} z_l, z_l y_{j+1}$), where $y_j y_{j+1}$ is an adjacency.

It is easy to know that there may exist some original adjacencies in two input sequences. For the original adjacencies and new adjacencies, we have

Theorem 1. *Let the optimal solution value be OPT, let k_0 be the number of original adjacencies, and let k be the number of new adjacencies generated by inserting missing genes to the input sequences, then $OPT = k_0 + k$.*

Proof. Let $A' \in A + X$ and $B' \in B + Y$ be the final scaffolds in the optimal solution after inserting all missing genes. Compared to A, all genes belonging to X appear as substrings in A'. Let $x_1 x_2 \ldots x_l$ be a string inserted between $y_i y_{i+1}$ in A', then either $y_i x_1$ or $x_l y_{i+1}$ or both are adjacencies. Otherwise, we could delete this string from A' (number of adjacencies decreases by at most l-1), re-insert it following the algorithm in Lemma 1(number of adjacencies increases by at least l), and obtain one more adjacency. Thus, each substring in A' composed of genes of X is either Type-1 or Type-2. Similarly, each substring in B' composed of genes of Y is either Type-1 or Type-2. According to the definitions of Type-1 and Type-2, $OPT = k_0 + k$. □

Therefore, in our algorithm we only need to consider how to insert genes into the input sequences to produce as many as possible new adjacencies.

4.2 A Better Lower Bound

At first, we analyze the characteristic of two-sided SF-MNSA with symbol $\#$ problem. We find that final sequence $A' = A + X = (A - Y) + (X + Y)$ and $B' = B + Y = (B - X) + (X + Y)$, *i.e.* respectively inserting genes of X and Y into sequence A and B can be seen as restrictively inserting genes of $X + Y$ into $A - Y$ and $B - X$. Restrictive insertion means that we need to keep the original position of every gene of Y and X in sequence A and B when we insert $X + Y$ into $A - Y$ and $B - X$. For example: $A = \#14z2\#$, $B = \#1xy234\#$, $X = \{x, y, 3\}$, $Y = \{z\}$, $X + Y = \{x, y, z, 3\}$, $A - Y = \#142\#$, $B - X = \#124\#$. We restrictively insert genes of $X + Y$ to $A - Y$ to obtain $A' = \#1xy43z2\#$ and keep the original position of z. We restrictively insert genes of $X + Y$ to $B - X$ to obtain $B' = \#1xy2z34\#$ and the positions of $x, y, 3$ are changeless. The number of new adjacency between A'' and B'' is 5. Next, we'll consider the feasibility of the transformation.

Lemma 2. *Let $a_{new}(A + X, B + Y)$ be the number of new adjacencies generated by respectively inserting genes of X and Y into sequence A and B, and let $a_{new}((A - Y) + (X + Y), (B - X) + (X + Y))$ be the number of new adjacencies generated by restrictively inserting genes of $X + Y$ into $A - Y$ and $B - X$, then*

$$a_{new}(A + X, B + Y) = a_{new}((A - Y) + (X + Y), (B - X) + (X + Y)) \quad (1)$$

Proof. To prove $a_{new}(A + X, B + Y) = a_{new}((A - Y) + (X + Y), (B - X) + (X + Y))$, which is equivalent to prove genes of $Y(or X)$ in sequence $A(or B)$ do not generate any original adjacency. Suppose that genes of $Y(or X)$ in sequence

$A(or B)$ generate original adjacencies, these genes should not be in $Y(or X)$. It is contradict with the definition of Y(or X). Because restrictive insertion keeps the original locations of these genes, Lemma 2 is proven.   □

After inserting $X + Y$ into $A - Y$ and $B - X$, we find genes of $X + Y$ can constitute two kinds of gene substrings in final sequences.

Class-1: gene substrings composed of genes only belonging to X or only belonging to Y, e.g. gene substring xy in the example above.

Class-2: gene substrings composed of genes belonging to X and genes belonging to Y, e.g. gene substring $z3$.

Furthermore, every kind of gene substring can also be classified into two kinds of type (Type-1 and Type-2). In order to describe the relationship between the number of every kind of gene substring and the optimal solution value, we need parameters a little more:

k_0: the number of original adjacencies between A and B.

k_1: the number of genes of X. k_2: the number of genes of Y.

b_i: the number of i-Type-1, *Class-1* substrings in X , and let p be the maximum length of this kind of substrings.

E_1: the number of genes belonging to Type-2, *Class-1* substrings in X.

d_i: the number of i-Type-1, *Class-1* substrings in Y , and let q be the maximum length of this kind of substrings.

E_2: the number of genes belonging to Type-2, *Class-1* substrings in Y.

C_i: the number of i-Type-1, *Class-2* substrings in $X + Y$, and let r be the maximum length of this kind of substrings. Let S_1 be the number of genes constituting this kind of substrings in X and S_2 be the number of genes constituting this kind of substrings in Y.

F: the total number of genes belonging to Type-2, *Class-2* substrings in $X+Y$. Let T_1 be the number of genes constituting this kind of substrings in X and T_2 be the number of genes constituting this kind of substrings in Y.

We need to compute the value of $a_{new}((A-Y)+(X+Y),(B-X)+(X+Y))$. Obviously, it is obtained by computing the number of adjacencies generated by genes of all kinds of substrings in $X + Y$.

Lemma 3. *Let OPT be the optimal solution value of any instance, then*

$$OPT - k_0 = k_1 + b_1 + b_2 + \cdots + b_p + k_2 + d_1 + d_2 + \cdots + d_q$$
$$+C_2 + C_3 + \cdots + C_r \leq \frac{3}{2}(k_1 + k_2 + \frac{1}{3}b_1 + \frac{1}{3}d_1). \tag{2}$$

Proof. According to above definitions, we have the total number of genes in X
$k_1 = \sum_{i=1}^{p}(i \times b_i) + E_1 + S_1 + T_1 \Longrightarrow \sum_{i=2}^{p} b_i \leq \frac{1}{2}(k_1 - S_1 - b_1)$.
Obviously, $k_1 = b_1 + 2b_2 + \cdots + pb_p + E_1 + S_1 + T_1 \geq b_1 + 2(b_2 + \cdots + b_p) + E_1 + S_1 + T_1 \Longrightarrow \sum_{i=2}^{p} b_i \leq \frac{1}{2}(k_1 - S_1 - b_1 - E_1 - T_1) \leq \frac{1}{2}(k_1 - S_1 - b_1)$.
Similarly, the total number of genes in Y is k_2.
$k_2 = \sum_{i=1}^{q}(i \times d_i) + E_2 + S_2 + T_2 \Longrightarrow \sum_{i=2}^{q} d_i \leq \frac{1}{2}(k_2 - S_2 - d_1)$.
Moreover, the number of genes of *Class-2*, i-Type-1 substrings is
$2C_2 + 3C_3 + \cdots + rC_r = S_1 + S_2 \Longrightarrow \sum_{i=2}^{r} C_i \leq \frac{1}{2}(S_1 + S_2)$.

Obviously, $F = T_1 + T_2$. So, by the Theorem 1, we have

$OPT = k_0 + a_{new}(A + X, B + Y) = k_0 + a_{new}((A - Y) + (X + Y), (B - X) + (X+Y)) = k_0 + \sum_{i=1}^{p}(i+1) \times b_i + E_1 + \sum_{i=1}^{q}(i+1) \times d_i + E_2 + \sum_{i=1}^{r}(i+1) \times C_i + F.$

$\Longrightarrow OPT - k_0 = \sum_{i=1}^{p}(i \times b_i) + \sum_{i=1}^{p} b_i + E_1 + \sum_{i=1}^{q}(i \times d_i) + \sum_{i=1}^{q} d_i + E_2$
$\qquad + \sum_{i=1}^{r}(i \times C_i) + \sum_{i=1}^{r} C_i + T_1 + T_2$
$\qquad = k_1 + b_1 + b_2 + \cdots + b_p + k_2 + d_1 + d_2 + \cdots + d_q$
$\qquad + C_2 + C_3 + C_4 + \cdots + C_r \leq \frac{3}{2}(k_1 + k_2 + \frac{1}{3}b_1 + \frac{1}{3}d_1).$ $\qquad\qquad\square$

Lemma 3 shows that if the number of 1-Type-1 substrings from the approximation algorithm isn't less than $\frac{1}{3}(b_1 + d_1)$, the approximation factor is $\frac{3}{2}$.

4.3   Description of the Algorithm

In this section, we present the main idea of our algorithm, which uses the greedy strategy. According to previous analysis, we should insert at least $\frac{1}{3}(b_1 + d_1)$ 1-Type-1 substrings. Because we will only analyze the number of 1-Type-1 substring and each 1-Type-1 substring contains only one gene which belongs to either X or Y, we can still insert genes of X into sequence A and insert genes of Y into sequence B. The main steps of the algorithm are as follows.

Algorithm 2:
Input: sequence A,B, with X, Y and set of breakpoints $b_A(A, B), b_B(A, B)$
Output: A+X, B+Y
1. For each gene of X ,we scan sequence A from left to right to find a breakpoint and we can insert the gene into the breakpoint to generate 2 adjacencies. we insert the remaining missing genes of X into A in arbitrary fashion, provided that each inserted missing gene generates one adjacency.
2. For each gene of Y, we scan sequence B from left to right to find a breakpoint and we can insert the gene into the breakpoint to generate 2 adjacencies. we insert the remaining missing genes of Y into B in arbitrary fashion, provided that each inserted missing gene generates one adjacency.

For example, $A = \#2x43z5\#$, $B = \#24y3w5\#$, $X = \{y, w\}$, $Y = \{x, z\}$. At first step, we insert gene y of X into breakpoint 43 in A. Gene y is a 1-Type-1 substring. The remaining missing gene w is inserted into breakpoint $3z$. Gene w is a Type-2 substring. At second step, we insert gene x of Y into breakpoint 24 in B. Gene x is a 1-Type-1 substring. The remaining missing gene z is inserted into breakpoint $w5$. Gene z is a Type-2 substring.

4.4   Proof of the Approximation Factor

Here we prove that the algorithm must be able to find at least $\frac{1}{3}(b_1 + d_1)$ 1-Type-1 substrings and prove the approximation factor of our algorithm is 3/2.

Lemma 4. *Let b_1, d_1 denote the number of Class-1,1-Type-1 substrings inserted in sequence A,B of some optimal solution, b'_1, d'_1 denote the number of Class-1,1-Type-1 substrings inserted in sequence A,B obtained by our algorithm. Then $b'_1 + d'_1 \geq \frac{1}{2}(b_1 + d_1)$.*

Proof. Adjacencies at two locations are affected by insertion of a gene x at most, one is at x's location to have been inserted in and the other is at x's location should be in the OPT solution. So, every 1-Type-1 substring obtained by our algorithm can destroy at most two Type-1 substrings of some optimal solution. Cases of substrings which are destroyed by a 1-Type-1 substring are described in the following Table 1, where $b'_{1j}s$ and $d'_{1j}s$ denote the number of 1-Type-1 substrings of each case j for sequence A and B.

Table 1. Cases of substrings destroyed by a 1-Type-1 substring in A or B

number of 1-Type-1 substrings	one substring destroyed	another substring destroyed
b'_{11} or d'_{11}	1-Type-1	1-Type-1
b'_{12} or d'_{12}	1-Type-1	i-Type-1, $i > 1$ or none
b'_{13} or d'_{13}	i-Type-1, $i > 1$	k-Type-1, $k > 1$

Let b'_0, d'_0 be the number of 1-Type-1 substrings consistent with optimal solution obtained by our algorithm in final sequence A', B'. Because other cases except in the Table 1. may exist when 1-Type-1 substrings are inserted according to our algorithm, we have $b'_1 \geq b'_0 + b'_{11} + b'_{12} + b'_{13}$, $d'_1 \geq d'_0 + d'_{11} + d'_{12} + d'_{13}$. It is not difficult to understand that the number of 1-Type-1 substrings of the optimal solution should not larger than the sum of the number of 1-Type-1 substrings consistent with optimal solution obtained by our algorithm and the number of 1-type-1 substrings destroyed by our algorithm, *i.e.* $b_1 \leq 2b'_{11} + b'_{12} + b'_0$, $d_1 \leq 2d'_{11} + d'_{12} + d'_0$. Then $b_1 + d_1 \leq 2(b'_{11} + d'_{11}) + b'_{12} + d'_{12} + b'_0 + d'_0 \leq 2(b'_1 + d'_1)$. The Lemma 4 is proven. □

Theorem 2. *The two-sided SF-MNSA with symbol # problem admits a polynomial time factor-1.5 approximation.*

Proof. Following the approximation algorithm and Lemmas $2 - 4$, we have the approximation solution value APP, which satisfies the following inequalities.
$$APP - k_0 = k_1 + k_2 + b'_1 + d'_1 \geq k_1 + k_2 + \tfrac{1}{2}(b_1 + d_1) \geq k_1 + k_2 + \tfrac{1}{3}(b_1 + d_1)$$
$$\geq \tfrac{2}{3}(OPT - k_0).$$
Hence $\frac{OPT}{APP} \leq 1.5$, and the theorem is proven. □

5   Concluding Remarks

In this paper, we conduct a further research on the two-sided SF-MNSA problem and we propose a new version of this problem. Some propositions of this version problem are presented at first. We design a polynomial algorithm for the special instance of this new version problem. For any instance, we apply greedy method to design an approximation algorithm with factor 1.5. One interesting open problem is whether one can improve the 1.5 factor further, another problem is to find new approximation algorithm for the problem without symbol #.

Acknowledgents. This research is partially supported by the Doctoral Found of Ministry of Education of China under grant 20090131110009, by China Post-doctoral Science Foundation under grant $2011M501133$ and 2012T50614, and by NSF grant DMS-0918034, NSF of China under grant 61070019 and 61202014.

References

1. Angibaud, S., Fertin, G., Rusu, I., Thevenin, A., Vialette, S.: On the approxima-bility of comparing genomes with duplicates. J. Graph Algorithms and Applica-tions 13(1), 19–53 (2009)
2. Blin, G., Fertin, G., Sikora, F., Vialette, S.: The EXEMPLAR BREAKPOINT DISTANCE for non-trivial genomes cannot be approximated. In: Das, S., Uehara, R. (eds.) WALCOM 2009. LNCS, vol. 5431, pp. 357–368. Springer, Heidelberg (2009)
3. Cormode, G., Muthukrishnan, S.: The string edit distance matching problem with moves. In: Proc. 13th ACM-SIAM Symp. on Discrete Algorithms (SODA 2002), pp. 667–676 (2002)
4. Chen, Z., Fowler, R., Fu, B., Zhu, B.: On the inapproximability of the exemplar con-served interval distance problem of genomes. J. Combinatorial Optimization 15(2), 201–221 (2008)
5. Chen, Z., Fu, B., Zhu, B.: The approximability of the exemplar breakpoint distance problem. In: Cheng, S.-W., Poon, C.K. (eds.) AAIM 2006. LNCS, vol. 4041, pp. 291–302. Springer, Heidelberg (2006)
6. Goldstein, A., Kolman, P., Zheng, J.: Minimum common string partition problem: Hardness and approximations. In: Fleischer, R., Trippen, G. (eds.) ISAAC 2004. LNCS, vol. 3341, pp. 484–495. Springer, Heidelberg (2004); also in: The Electronic Journal of Combinatorics 12 (2005), paper R50
7. Jiang, M.: The zero exemplar distance problem. In: Tannier, E. (ed.) RECOMB-CG 2010. LNCS, vol. 6398, pp. 74–82. Springer, Heidelberg (2010)
8. Jiang, H., Zheng, C., Sankoff, D., Zhu, B.: Scaffold Filling under the Break-point and Related Distances. IEEE/ACM Trans. Comput. Biology Bioinform. 9(4), 1220–1229 (2012)
9. Jiang, H., Zhong, F., Zhu, B.: Filling scaffolds with gene repetitions: maximizing the number of adjacencies. In: Giancarlo, R., Manzini, G. (eds.) CPM 2011. LNCS, vol. 6661, pp. 55–64. Springer, Heidelberg (2011)
10. Muñoz, A., Zheng, C., Zhu, Q., Albert, V., Rounsley, S., Sankoff, D.: Scaffold filling, contig fusion and gene order comparison. BMC Bioinformatics 11, 304 (2010)
11. Sankoff, D.: Genome rearrangement with gene families. Bioinformatics 15(11), 909–917 (1999)

Appendix

Proof of Algorithm 1:
Now, we prove the algorithm is feasible and the solution is optimal if $BP_1(A) = \emptyset$ and $BP_1(B) = \emptyset$. At first, we prove Algorithm 1 is feasible. Because the symbol # must not be missing gene, it appears in neither X nor Y. According to the definition of the set $BP_2(A)$ and $BP_2(B)$, we have

Lemma 5. *If $BP_1(A) = \emptyset$ and $BP_1(B) = \emptyset$, then any gene which appears in set $BP_2(A)$(resp. $BP_2(B)$) must not be symbol # and appears at least once in Y(resp. X).*

Lemma 6. *If $BP_1(A) = \emptyset$ and $BP_1(B) = \emptyset$, the number of occurrence that end-gene of bp-string in BS_A (resp. BS_B) appears at endpoints of bp-strings in BS_A (resp. BS_B) is even.*

Proof. Let gene a be an end-gene of any bp-string in BS_A and gene b be an end-gene of any bp-string in BS_B. Assume that the lemma is not true, *i.e.*, the number of occurrence that a (*resp. b*) appears at the endpoints of bp-strings in BS_A (*resp. BS_B*) is odd. At first, gene a and b are not symbol # according to Lemma 5. Let the form of a in A be $\# \cdots e_1 a X_1 a f_1 \cdots e_2 a X_2 a f_2 \cdots e_i a X_i a f_i \cdots k_1 a Y_1 b_1 \cdots k_2 a Y_2 b_2 \cdots k_j a Y_j b_j \cdots u_1 a v_1 \cdots u_2 a v_2 \cdots u_c a v_c \cdots w_1 A_1 z_1 \cdots w_2 A_2 z_2 \cdots w_d A_d z_d \cdots \#$, where $A_1, A_2, \ldots, A_d$ are strings that are composed of gene a and $|A_1| = k_1, |A_2| = k_2, \ldots, |A_d| = k_d$, strings like aXa, aYb are bp-strings composed of breakpoints in $BP_2(A)$ and strings like uav, wAz are adjacency strings. According to above assumption, the number of gene a that is end-gene of breakpoint string in BS_A is $2i+j$ and j is odd. The maximal matching about gene a in A and B are $(e_1 a, e_1 a), (af_1, af_1), \ldots, (e_i a, e_i a), (af_i, af_i), (k_1 a, k_1 a), \ldots, (k_j a, k_j a), (u_1 a, u_1 a), (av_1, av_1), \ldots, (u_c a, u_c a), (av_c, av_c), (w_1 a, w_1 a), (aa, aa), \ldots, (aa, aa), (a z_1, a z_1), \ldots, (w_d a, w_d a), (aa, aa), \ldots, (aa, aa), (a z_d, a z_d)$. We can prove a must not appear in $BP_2(B)$ and $BP_3(B)$. If a appears in $BP_2(B)$ or $BP_3(B)$, the breakpoint including a in $BP_2(A)$ would have a common gene a with the breakpoint including a in $BP_2(B)$ or $BP_3(B)$ and the breakpoint including a in $BP_2(A)$ should belongs to $BP_1(A)$. But $BP_1(A) = \emptyset$. So, a must not appear in $BP_2(B)$ or $BP_3(B)$. Moreover, $BP_1(B) = \emptyset$. Then, a constitutes no breakpoint in B. The form of gene a in B is like uav or wAz. The number of adjacencies about gene a whose form is one element is gene a and the other element is not gene a must be even in B. But the number of adjacencies about gene a in the maximal matching like that form is odd because j is odd. This is a contradiction. The fact about gene b is similarly proved. Therefore, lemma 6 is proven.   □

Lemma 7. *If $BP_1(A) = \emptyset$ and $BP_1(B) = \emptyset$, then mid-genes $x_1, x_2, \ldots, x_n$ of any bp-string $a x_1 x_2 \cdots x_n b$ in BS_B satisfy $x_1, x_2, \ldots, x_n \in X$ and mid-genes $y_1, y_2, \ldots, y_m$ of any bp-string $c y_1 y_2 \cdots y_m d$ in BS_A satisfy $y_1, y_2, \ldots, y_m \in Y$.*

Proof. Assume that the lemma is not true, *i.e.*, at least one of $x_1, x_2, \ldots, x_n$ does not belong to X, let it be x_i and at least one of $y_1, y_2, \ldots, y_m$ dose not belong to Y, let it be y_j. Because x_i does not belong to X and x_i only constitutes adjacency in A and its number of occurrence in B is same as in A. According to Proposition 1, x_i constitutes no breakpoint in B. This is a contradiction with that x_i is a bp-gene in BS_B. The proof about y_j is similar.   □

Lemma 8. *If $BP_1(A) = \emptyset$ and $BP_1(B) = \emptyset$, then any end-gene of bp-string in BS_A(resp. BS_B) appears at least once in B (resp. A).*

Proof. Let gene a (*resp.* b) be any end-gene of bp-string in BS_A (*resp.* BS_B). Assume that the lemma is not true, *i.e.*, gene a does not appear in B and gene b does not appear in A. Let the form of gene a in A be the one as in the proof of Lemma 6. Let the number of adjacencies about a in A be N_1, $N_1 = 2i + j + 2c + 2d + N(a) = 2(i + c + d) + N(a) + j$, where $N(a)$ is the number of adjacencies about all string $aa \cdots a$. Because a is an end-gene of bp-string, i and j are not zero at same time. So, $N_1 \neq 0$. According to the assumption, a constitutes no adjacency in B because it doesn't appear in B. So, we have $N_1 = 0$ contradicting with $N_1 \neq 0$. The proof about b is same as a's.   □

In the Algorithm 1, main task is computing string AS and BS. According to Lemma 6, end-gene of bp-string in BS_A(*resp.* BS_B) appears at the endpoints of bp-strings in BS_A(*resp.* BS_B) for even times. So, string AS and BS must can be found by the Algorithm 1. Lemma 7 ensures that the mid-genes of bp-strings merged to string $AS(BS)$ can be inserted in $B(A)$ and be deleted from $Y(X)$. Because Lemma 5 only ensures that the end-gens of bp-strings appears once in the set of missing genes, step 2.1-2 in function $BPS(\bullet)$ are needed to delete repeated end-genes. Lemma 8 ensures that step 2.2 and 3.2 can run. Because string AS or BS is composed of breakpoints, it is inserted in some adjacency in input sequences not to generate any breakpoint but new adjacencies. The number of new adjacencies is $|AS| + |BS| - 2$. Finally, all missing genes can be inserted in input sequence not to generate any breakpoint. The total number of new adjacencies is $|X| + |Y|$. The Algorithm 1 is feasible.

Next, we will prove that the solution computed by Algorithm 1 is optimal. Obviously, we have the number of breakpoints $|b_{A+X}(A + X, B + Y)| = |b_{B+Y}(A + X, B+Y)|$. And, $|a(A+X, B+Y)|$ is maximum value when $|b_{A+X}(A+X, B+Y)|$ and $|b_{B+Y}(A + X, B + Y)|$ are minimum value. If $min|b_{A+X}(A + X, B + X)| = |BP_3(A)|$ or $min|b_{B+Y}(A + X, B + X)| = |BP_3(B)|$, then the solution is optimal because the breakpoints in set $BP_3(A)$ or $BP_3(B)$ are unchanging in the algorithm. Assume that $min|b_{A+X}(A+X, B+Y)| < |BP_3(A)|$, *i.e.*, there is another algorithm that at least one breakpoint in set $BP_3(A)$ can been eliminated. This is impossible because genes of set $BP_3(A)$ do not appear in Y but appear in breakpoints of B. Any breakpoint in $BP_3(A)$ can not be eliminated because the breakpoint will not belong to $BP_3(A)$ if the breakpoint is eliminated by inserting some gene. Similarly, any breakpoint in $BP_3(B)$ can not be eliminated too. So the solution obtained by Algorithm 1 is optimal solution and the solution value $|a(A + X, B + Y)| = k_0 + |X| + |Y|$ is optimal, where k_0 is the number of the original adjacencies between A and B.

Energy-Efficient Threshold Circuits Detecting Global Pattern in 1-Dimentional Arrays*

Akira Suzuki[1], Kei Uchizawa[2], and Xiao Zhou[1]

[1] Graduate School of Information Sciences, Tohoku University
Aramaki-aza Aoba 6-6-05,
Aoba-ku, Sendai, 980-8579, Japan
{a.suzuki,zhou}@ecei.tohoku.ac.jp
[2] Graduate School of Science and Engineering, Yamagata University
Johnan 4-3-16, Yonezawa-shi, Yamagata, 992-8510, Japan
uchizawa@yz.yamagata-u.ac.jp

Abstract. In this paper, we investigate a relationship between energy and size of a threshold circuit processing a simple task, called P_{LR}^n, that was introduced in a context of pattern recognition. Formally, $P_{LR}^n : \{0,1\}^n \times \{0,1\}^n \to \{0,1\}$ is defined as follows: For every $\boldsymbol{x} = (x_1, x_2, \ldots, x_n) \in \{0,1\}^n$ and $\boldsymbol{y} = (y_1, y_2, \ldots, y_n) \in \{0,1\}^n$, $P_{LR}^n(\boldsymbol{x}, \boldsymbol{y}) = 1$ if there exists a pair of indices i and j such that $i < j$ and $x_i = y_j = 1$; and $P_{LR}^n(\boldsymbol{x}, \boldsymbol{y}) = 0$ otherwise. We prove that P_{LR}^n can be computed by a threshold circuit of energy e and size $s = O\left(e \cdot n^{2/(e-1)}\right)$ for any integer e, $3 \le e \le 2\log_2 n + 1$. Our result implies that one can construct an energy-efficient circuit computing P_{LR}^n if it is allowable to use large size. Moreover, we focus on an extreme case where a threshold circuit has energy $e = 1$, and show that P_{LR}^n can be computed by a threshold circuit of energy $e = 1$ and size $s = \lceil n/2 \rceil$, while P_{LR}^n cannot be computed by any threshold circuit of energy $e = 1$ and size $s \le \lceil n/2 \rceil - 1$.

1 Introduction

Neurons communicate with each other by "firing" (i.e., emitting an electrical signal) for information processing, and a circuit consisting of neurons is often modelled by a combinatorial logic circuit, called a *threshold circuit*. Motivated by a biological fact that a neuron consumes substantially more energy to fire than not to fire [1–3], Uchizawa, Douglas and Maass proposed a complexity measure, called *energy complexity*, for threshold circuits, and initiate a study for the following question: what computational tasks can or cannot be computed by reasonably small threshold circuits with small energy complexity? Formally, the *energy* e of a threshold circuit C is defined as the maximum number of gates outputting "1" in C, where the maximum is taken over all inputs to C [4]. In previous research, it is shown that there exists a tradeoff between the energy

* This work is partially supported by JSPS Grant-in-Aid for Scientific Research, Grant Numbers 24.3660 (A.Suzuki), 23700003 (K.Uchizawa) and 23500001 (X.Zhou).

T-H.H. Chan, L.C. Lau, and L. Trevisan (Eds.): TAMC 2013, LNCS 7876, pp. 248–259, 2013.
© Springer-Verlag Berlin Heidelberg 2013

and size (i.e., the number of gates) of a threshold circuit computing the Parity function; more formally, it is proved that the Parity function of n variables is computable by a threshold circuit of energy e and size

$$s = O\left(e \cdot n^{\frac{1}{(e-1)}}\right) \tag{1}$$

for every integer $e \geq 2$ [5], while any threshold circuit C of energy $e \geq 2$ computing the Parity function of n variables has size $s = \Omega\left(e \cdot n^{1/e}\right)$ [6]. The result implies that the energy complexity has an interesting relationship with the major complexity measure, the size, of threshold circuits computing the Parity function. However, the Parity function is a typical arithmetic function, and hence it was not clear if such a tradeoff holds for other computational tasks, especially, that arise in a context of biological information processing.

In this paper, we consider a Boolean function, called P_{LR}^n, which Legenstrin and Maass introduced to model a simple task for a pattern recognition on 1-dimensional array [7]. Suppose there are two types of local feature detectors $\boldsymbol{x} = (x_1, x_2, \ldots, x_n) \in \{0,1\}^n$ and $\boldsymbol{y} = (y_1, y_2, \ldots, y_n) \in \{0,1\}^n$, where each of $x_1, x_2, \ldots, x_n$ represents a detector for one feature, while each of $y_1, y_2, \ldots, y_n$ does a detector for the other feature: We have $x_i = 1$ ($y_j = 1$, respectively) if a detector on the i-th (j-th) position is activated. Then the function $P_{LR}^n : \{0,1\}^n \times \{0,1\}^n \to \{0,1\}$ is defined as follows: For every pair of $\boldsymbol{x} = (x_1, x_2, \ldots, x_n) \in \{0,1\}^n$ and $\boldsymbol{y} = (y_1, y_2, \ldots, y_n) \in \{0,1\}^n$, $P_{LR}^n(\boldsymbol{x}, \boldsymbol{y}) = 1$ if there exists a pair of indices i and j such that $1 \leq i < j \leq n$ and $x_i = y_j = 1$; and $P_{LR}^n(\boldsymbol{x}, \boldsymbol{y}) = 0$ otherwise. Intuitively, P_{LR}^n models a task for determining a relative position between the two features. Legenstrin and Maass study threshold circuits computing P_{LR}^n, and show that

(a) P_{LR}^n is computable by a threshold circuit of size $O(\log n)$, and
(b) the size of the circuit given in (a) is asymptotically optimal, that is, any threshold circuit computing P_{LR}^n has size $\Omega(\log n)$.

(In fact, they also show that their circuit design has advantage for total wire length in certain VLSI models.)

Following the tradeoff result for the Parity function described above, we investigate a relationship between the energy and size of a threshold circuit computing P_{LR}^n. We then show that, as in the case for the Parity function, one can construct an energy-efficient circuit computing P_{LR}^n if it is allowable to use large size: we prove that P_{LR}^n can be computed by a threshold circuit of energy e and size

$$s = O\left(e \cdot n^{\frac{2}{(e-1)}}\right) \tag{2}$$

for any integer e, $3 \leq e \leq 2\log_2 n + 1$. Our result clearly implies that there exists a threshold circuit C of small energy e if C is allowed to have a large size s (e.g., $e = 3$ and $s = O(n)$), while there exists a threshold circuit C of small size s if C is allowed to use large energy e (e.g., $s = O(\log_2 n)$ and $e = O(\log_2 n)$). It worth mentioning that Eq. (2) has a quite similar form to the one (i.e., Eq. (1)) for the Parity function.

Moreover, we consider an extreme case where a threshold circuit has energy $e = 1$. In this case, we provide an exact value of size of an optimal threshold circuit: we prove that P_{LR}^n can be computed by a threshold circuit of energy $e = 1$ and size

$$s = \left\lceil \frac{n}{2} \right\rceil,$$

while P_{LR}^n cannot be computed by any threshold circuit of energy $e = 1$ and size

$$s \leq \left\lceil \frac{n}{2} \right\rceil - 1.$$

The rest of this paper is organized as follows. In Section 2, we define some terms on threshold circuits and P_{LR}^n functions. In Section 3, we give the construction of energy-efficient threshold circuits for arbitrary energy $e \geq 3$. In Section 4, we give the upper and lower bounds for threshold circuits of energy $e = 1$. In Section 5, we conclude with some remarks.

2   Preliminaries

A *threshold circuit* is a combinatorial circuit of threshold gates. A threshold circuit C is expressed by a directed acyclic graph; let n be the number of input variables to C, then each node of in-degree 0 in C corresponds to one of the n input variables $x_1, x_2, \cdots, x_n$, and the other nodes correspond to threshold gates. We define *size* $s(C)$, simply denoted by s, of a threshold circuit C as the number of threshold gates in C. Let $g_1, g_2, \ldots, g_s$ be the gates in C. One may assume without loss of generality that $g_1, g_2, \ldots, g_s$ are topologically ordered with respects to the underlying graph of C. Let i be an integer such that $1 \leq i \leq s$. For each gate g_i, we denote by $w_{i,1}, w_{i,2}, \ldots, w_{i,l_i}$ the weights and by t_i the threshold of the gate g_i, respectively, where the weights and the threshold are real numbers and l_i is the fan-in of the gate g_i. Let $z_i(x) = (z_{i,1}(x), z_{i,2}(x), \cdots, z_{i,l_i}(x)) \in \{0,1\}^{l_i}$ be an input to g_i for a circuit input x, where each $z_{i,j}(x)$, $1 \leq j \leq l_i$, is either a value of an input variable or an output of a gate $g_{i'}$, $i' < i$. While the output $g_i(z_i(x))$ of g_i is determined by $z_i(x)$, we simply denote $g_i(z_i(x))$ by $g_i[x]$. When an input $z_i(x)$ is given to the threshold gate g_i for a circuit input x, the output $g_i[x]$ of the gate is defined as

$$g_i[x] = \text{sign} \left(\sum_{j=1}^{l_i} w_{i,j} z_{i,j}(x) - t_i \right),$$

where $\text{sign}(z) = 1$ if $z \geq 0$ and $\text{sign}(z) = 0$ if $z < 0$. For every input $x \in \{0,1\}^n$, the *output* $C(x)$ of C is denoted by $g_s[x]$. The gates g_s is called *top gate* of C. Let $f : \{0,1\}^n \to \{0,1\}$ be a Boolean function of n inputs. A threshold circuit C *computes* a Boolean function f if $C(x) = f(x)$ for every input $x \in \{0,1\}^n$. We define the *energy* $e(C)$ of C as

$$e(C) = \max_{\boldsymbol{x} \in \{0,1\}^n} \sum_{i=1}^{s(C)} g_i[\boldsymbol{x}].$$

For any positive integer n, we define P_{LR}^n as follows: Let $N = \{1, 2, \ldots, n\}$. For a pair of $\boldsymbol{x} = (x_1, x_2, \ldots, x_n) \in \{0,1\}^n$ and $\boldsymbol{y} = (y_1, y_2, \ldots, y_n) \in \{0,1\}^n$, $P_{LR}^n(\boldsymbol{x}, \boldsymbol{y}) = 1$ if there exists a pair of indices i and j such that $1 \le i < j \le n$ and $x_i = y_j = 1$; and $P_{LR}^n(\boldsymbol{x}, \boldsymbol{y}) = 0$ otherwise. More formally,

$$P_{LR}^n(\boldsymbol{x}, \boldsymbol{y}) = \begin{cases} 1 & \text{if } l(\boldsymbol{x}) < r(\boldsymbol{y}); \\ 0 & \text{otherwise.} \end{cases}$$

where

$$l(\boldsymbol{x}) = \begin{cases} n & \text{if } \boldsymbol{x} = (0,0,\ldots,0); \\ \min\{i \in N \mid x_i = 1\} & \text{otherwise} \end{cases}$$

and

$$r(\boldsymbol{y}) = \begin{cases} 0 & \text{if } \boldsymbol{y} = (0,0,\ldots,0); \\ \max\{i \in N \mid y_i = 1\} & \text{otherwise.} \end{cases}$$

3   Energy-Efficient Circuits of Bounded Size

In this section, we give a construction of energy-efficient threshold circuits computing P_{LR}^n. The following theorem gives an upper bound on the size of threshold circuits computing P_{LR}^n with energy e for any $e \ge 3$.

Theorem 1. *Let n be a positive integer. Then, there is a threshold circuit C computing P_{LR}^n such that C has energy $e \ge 3$ and size*

$$s = \left\lfloor \frac{e-1}{2} \right\rfloor \left\lceil (n+1)^{\frac{1}{\lfloor (e-1)/2 \rfloor}} \right\rceil + \left\lceil \frac{e-1}{2} \right\rceil \left\lceil (n+1)^{\frac{1}{\lceil (e-1)/2 \rceil}} \right\rceil + 1 = O\left(e \cdot n^{\frac{2}{e-1}}\right).$$

Proof. Let n and e be integers where $n \ge 1$ and $e \ge 3$. We prove Theorem 1 by constructing the desired circuit C. To simplify our proof, we only consider the case where e is odd; the proof for the other case is similar. Since e is odd, we have

$$\left\lfloor \frac{e-1}{2} \right\rfloor = \left\lceil \frac{e-1}{2} \right\rceil = \frac{e-1}{2}.$$

Thus, it suffices to construct a threshold circuit C of energy

$$e = 2\alpha + 1 \tag{3}$$

and size

$$s = 2\alpha\beta + 1 \tag{4}$$

where

$$\alpha = \frac{e-1}{2} \quad \text{and} \quad \beta = \left\lceil (n+1)^{\frac{2}{e-1}} \right\rceil.$$

Note that

$$\beta^\alpha - 1 = \left\lceil (n+1)^{\frac{2}{e-1}} \right\rceil^{\frac{e-1}{2}} - 1$$
$$\geq (n+1)^{\frac{2}{e-1} \cdot \frac{e-1}{2}} - 1$$
$$= n.$$

For each j, $0 \leq j \leq \alpha - 1$, we define $h_j : \{0, 1, \ldots, \beta^\alpha - 1\} \to \{0, 1, \ldots, \beta - 1\}$ as a function mapping an integer p to the j-th figure of p in notation system of base β, that is,

$$h_j(p) \equiv \lfloor p/\beta^j \rfloor \pmod{\beta}$$

for every $p \in \{0, 1, \ldots, \beta^\alpha - 1\}$. Note that

$$\sum_{j=0}^{\alpha-1} h_j(p) \cdot \beta^j = p.$$

Below we construct $\alpha\beta$ threshold gates to represent $h_1(l(\boldsymbol{x}))$, $h_2(l(\boldsymbol{x}))$, ..., $h_{\alpha-1}(l(\boldsymbol{x}))$. For each pair of j and k, $0 \leq j \leq \alpha - 1$ and $0 \leq k \leq \beta - 1$, a threshold gate $g_{j,k}^x$ has a threshold

$$t_{j,k}^x = \begin{cases} 0 & \text{if } k = h_j(n); \\ 1 & \text{otherwise} \end{cases} \tag{5}$$

and receives $x_1, x_2, \ldots, x_n$ as its input, where for each i, $1 \leq i \leq n$, the weight $w_{i,j,k}^x$ for x_i is given as

$$w_{i,j,k}^x = \begin{cases} 2^{n-i} & \text{if } k = h_j(i); \\ -2^{n-i} & \text{otherwise.} \end{cases}$$

Then the following claim holds.

Claim 1. *For each pair of j and k, $0 \leq j \leq \alpha - 1$ and $0 \leq k \leq \beta - 1$,*

$$g_{j,k}^x[\boldsymbol{x}] = \begin{cases} 1 & \text{if } k = h_j(l(\boldsymbol{x})); \\ 0 & \text{otherwise.} \end{cases} \tag{6}$$

We omit the proof of claim due to the page limitation. Thus, $g_{j,k}[\boldsymbol{x}] = 1$ if and only if the j-th figure of $l(\boldsymbol{x})$ is k in notation system of base β.

Similarly, we construct another set of $\alpha\beta$ gates to represent $h_1(r(\boldsymbol{y}))$, $h_2(r(\boldsymbol{y})), \ldots, h_{\alpha-1}(r(\boldsymbol{y}))$. For each pair of j and k, $0 \leq j \leq \alpha - 1$ and $0 \leq k \leq \beta - 1$, a threshold gate $g_{j,k}^y$ has a threshold

$$t_{j,k}^y = \begin{cases} 0 & \text{if } k = h_j(0) \text{ (i.e., } k = 0\text{)}; \\ 1 & \text{otherwise} \end{cases}$$

and receives $y_1, y_2, \ldots, y_n$ as its input, where for each i, $1 \leq i \leq n$, the weight for y_i is given as

$$w_{i,j,k}^y = \begin{cases} 2^i & \text{if } k = h_j(i); \\ -2^i & \text{otherwise.} \end{cases}$$

Similarly to Claim 1 above, the following claim holds

Claim 2. *For each pair of j and k, $0 \leq j \leq \alpha - 1$ and $0 \leq k \leq \beta - 1$,*

$$g_{j,k}^y[\boldsymbol{y}] = \begin{cases} 1 & \text{if } h_j(r(\boldsymbol{y})) = k; \\ 0 & \text{otherwise.} \end{cases}$$

We omit the proof of the claim; the proof is similar to the one for Claim 1. Thus, $g_{j,k}^y[\boldsymbol{y}] = 1$ if and only if the j-th figure of $r(\boldsymbol{y})$ is k in notation system of base β.

Finally, we construct the top gate g to compare $l(\boldsymbol{x})$ and $r(\boldsymbol{y})$ (see Fig. 1): For each pair of j and k, $0 \leq j \leq \alpha - 1$ and $0 \leq k \leq \beta - 1$, g receives an output of $g_{j,k}^x$ with weight

$$w_{j,k}^x = -k \cdot \beta^j$$

and receives an output of $g_{j,k}^y$ with weight

$$w_{j,k}^y = k \cdot \beta^j;$$

and g has threshold one. Consequently, the output of g is given as follows:

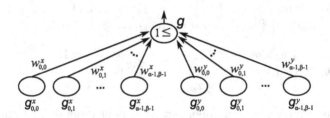

Fig. 1. The top gate g of C

$$g[\boldsymbol{x}, \boldsymbol{y}] = \operatorname{sign}\left(\sum_{j=0}^{\alpha-1}\sum_{k=0}^{\beta-1} w_{j,k}^x \cdot g_{j,k}^x[\boldsymbol{x}] + \sum_{j=0}^{\alpha-1}\sum_{k=0}^{\beta-1} w_{j,k}^y \cdot g_{j,k}^y[\boldsymbol{y}] - 1\right). \tag{7}$$

Then C clearly computes P_{LR}^n: Consider the value in the sign function of Eq. (7), then the two claims imply that

$$\sum_{j=0}^{\alpha-1}\sum_{k=0}^{\beta-1} w_{j,k}^x \cdot g_{j,k}^x[\boldsymbol{x}] = -l(\boldsymbol{x}) \quad \text{and} \quad \sum_{j=0}^{\alpha-1}\sum_{k=0}^{\beta-1} w_{j,k}^y \cdot g_{j,k}^y[\boldsymbol{y}] = r(\boldsymbol{y});$$

hence we have

$$g[\boldsymbol{x}, \boldsymbol{y}] = \mathrm{sign}\,(-l(\boldsymbol{x}) + r(\boldsymbol{y}) - 1) = \begin{cases} 1 & \text{if } l(\boldsymbol{x}) < r(\boldsymbol{y}); \\ 0 & \text{otherwise.} \end{cases}$$

We now verify Eqs. (3) and (4). For each j, $0 \le j \le \alpha - 1$, $g^x_{j,k}[\boldsymbol{x}] = 1$ if $k = l(\boldsymbol{x})$ and $g^x_{j,k}[\boldsymbol{x}] = 0$ otherwise; and hence only one of $g^x_{j,1}, g^x_{j,2}, \ldots, g^x_{j,\beta-1}$ outputs one. Similarly, for each j, $0 \le j \le \alpha - 1$, only one of $g^y_{j,0}, g^y_{j,1}, \ldots, g^y_{j,\beta-1}$ outputs one. Since the top gate g may output one, the energy of C is $2\alpha+1$, and hence Eq. (3) holds. Clearly, C consists of $2\alpha\beta + 1$ gates; $g^x_{j,k}$ and $g^y_{j,k}$ and for $0 \le j \le \alpha - 1$ and $0 \le k \le \beta - 1$ together with the top gate g. Thus, Eq. (4) holds.   □

4   Circuits of Energy One

In this section, we consider the extreme case where a threshold circuit has energy $e = 1$. In the following theorem, we prove by construction that a linear number of gates are sufficient to compute P^n_{LR} for a threshold circuit of energy $e = 1$.

Theorem 2. *Let n be a positive integer. Then, there is a threshold circuit C computing P^n_{LR} such that C has energy $e = 1$ and size*

$$s = \left\lceil \frac{n}{2} \right\rceil.$$

Before proving Theorem 2, we introduce some terms. For a Boolean variable a, we denote by $\neg a$ the negation of a. For each i, $1 \le i \le n$, we define a Boolean function $f_i : \{0,1\}^n \times \{0,1\}^n \to \{0,1\}$ as follows: For every $\boldsymbol{x} = (x_1, x_2, \ldots, x_n) \in \{0,1\}^n$ and $\boldsymbol{y} = (y_1, y_2, \ldots, y_n) \in \{0,1\}^n$,

$$f_i(\boldsymbol{x}, \boldsymbol{y}) = \begin{cases} y_2 \vee y_3 \vee \ldots \vee y_n & \text{if } i = 1; \\ x_1 \vee \ldots \vee x_{i-1} \vee y_{i+1} \vee \ldots \vee y_n & \text{if } i = 2, 3, \ldots, n-1; \\ x_1 \vee x_2 \vee \ldots \vee x_{n-1} & \text{if } i = n. \end{cases} \quad (8)$$

The following lemma plays an important role in our proof of Theorem 2.

Lemma 1. *Let n be a positive integer. Then, for every pair of $\boldsymbol{x} \in \{0,1\}^n$ and $\boldsymbol{y} \in \{0,1\}^n$,*

$$P^n_{LR}(\boldsymbol{x}, \boldsymbol{y}) = \bigwedge_{i \in N} f_i(\boldsymbol{x}, \boldsymbol{y}). \quad (9)$$

We omit the proof of Lemma 1 due to the page limitation.
    Using Lemma 1, we prove Theorem 2 as follows.

Proof of Theorem 2. We prove Theorem 2 by constructing a threshold circuit C of energy $e = 1$ and size $s = \lceil n/2 \rceil$. In this proof, we consider only the case where n is even, since the proof is similar for the other case. Lemma 1 implies that it suffices to construct C that computes $\bigwedge_{i \in N} f_i(\boldsymbol{x}, \boldsymbol{y})$.

For each k, $1 \leq k \leq n/2$, we define α_k as

$$\alpha_k(\boldsymbol{x}, \boldsymbol{y}) = f_{2k-1}(\boldsymbol{x}, \boldsymbol{y}) \wedge f_{2k}(\boldsymbol{x}, \boldsymbol{y}). \tag{10}$$

We also recursively define $\beta_1, \beta_2, \ldots, \beta_{n/2}$ as:

$$\beta_k(\boldsymbol{x}, \boldsymbol{y}) = \begin{cases} \neg\alpha_1(\boldsymbol{x}, \boldsymbol{y}) & \text{if } k = 1; \\ \neg\alpha_k(\boldsymbol{x}, \boldsymbol{y}) \wedge \neg\left(\bigvee_{j=1}^{k-1} \beta_j(\boldsymbol{x}, \boldsymbol{y})\right) & \text{if } 2 \leq k \leq n/2 - 1; \\ \alpha_{n/2}(\boldsymbol{x}, \boldsymbol{y}) \wedge \neg\left(\bigvee_{j=1}^{(n/2)-1} \beta_j(\boldsymbol{x}, \boldsymbol{y})\right) & \text{if } k = n/2. \end{cases} \tag{11}$$

Clearly, at most one of $\beta_1, \beta_2, \ldots, \beta_{n/2}$, has the value one as follows: Let k^*, $1 \leq k^* \leq n/2$, be the minimum index satisfying $\beta_{k^*}(\boldsymbol{x}, \boldsymbol{y}) = 1$, then Eq. (11) implies that $\beta_k(\boldsymbol{x}, \boldsymbol{y}) = 0$ for every k, $k^* + 1 \leq k \leq n/2$.

Below we will show that, for every $\boldsymbol{x} \in \{0,1\}^n$ and $\boldsymbol{y} \in \{0,1\}^n$,

$$\beta_{n/2}(\boldsymbol{x}, \boldsymbol{y}) = P_{LR}^n(\boldsymbol{x}, \boldsymbol{y}). \tag{12}$$

After the proof for Eq. (12), we use $n/2$ gates $g_1, g_2, \ldots, g_{n/2}$ to obtain the threshold circuit C so that for each gate g_k, $1 \leq k \leq n/2$, computes β_k, which clearly complete the proof.

[Proof of Eq. (12)]
    First, we consider the case where $\beta_{n/2}(\boldsymbol{x}, \boldsymbol{y}) = 1$. In this case, Eq. (11) implies that

$$\alpha_{n/2}(\boldsymbol{x}, \boldsymbol{y}) = 1 \tag{13}$$

and

$$\beta_j(\boldsymbol{x}, \boldsymbol{y}) = 0 \tag{14}$$

for each j, $1 \leq j \leq n/2 - 1$. By Eq. (14), for each k, $1 \leq k \leq n/2 - 1$,

$$\alpha_k(\boldsymbol{x}, \boldsymbol{y}) = 1. \tag{15}$$

Equations (10), (13) and (15) imply that, for each k, $1 \leq k \leq n$, $f_k(\boldsymbol{x}, \boldsymbol{y}) = 1$. Thus, by Eq. (9), $P_{LR}^n(\boldsymbol{x}, \boldsymbol{y}) = 1$.

    Next, we consider the case where $\beta_{n/2}(\boldsymbol{x}, \boldsymbol{y}) = 0$. In this case, Eq. (11) implies that $\alpha_{n/2}(\boldsymbol{x}, \boldsymbol{y}) = 0$ or $\beta_{j^*}(\boldsymbol{x}, \boldsymbol{y}) = 1$ for some j^*, $1 \leq j^* \leq n/2 - 1$. If $\alpha_{n/2}(\boldsymbol{x}, \boldsymbol{y}) = 0$, then $f_{n-1}(\boldsymbol{x}, \boldsymbol{y})$ or $f_n(\boldsymbol{x}, \boldsymbol{y})$ is 0, and if $\beta_{j^*}(\boldsymbol{x}, \boldsymbol{y}) = 1$ then $f_{2j^*-1}(\boldsymbol{x}, \boldsymbol{y}) = 0$ or $f_{2j^*}(\boldsymbol{x}, \boldsymbol{y}) = 0$. Thus by Eq. (9), $P_{LR}^n(\boldsymbol{x}, \boldsymbol{y}) = 0$.

[Construction of C] We construct the gates $g_1, g_2, \ldots, g_{n/2}$ so that g_k computes β_k for each k, $1 \leq k \leq n/2$, that is, $g_k[\boldsymbol{x}, \boldsymbol{y}] = \beta_k(\boldsymbol{x}, \boldsymbol{y})$.
    We first prove

$$g_1[\boldsymbol{x}, \boldsymbol{y}] = \beta_1(\boldsymbol{x}, \boldsymbol{y}). \tag{16}$$

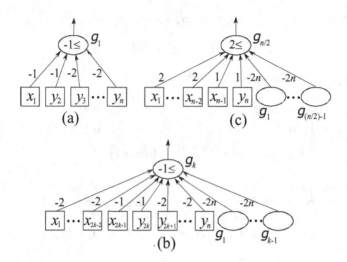

Fig. 2. (a) The gate g_1, (b) the gate g_k, $2 \leq k \leq (n/2) - 1$, and (c) the gate $g_{n/2}$

The gate g_1 has a threshold -1 and receives input from x_1 and y_2 with weights -1 and from $y_3, y_4, \ldots, y_n$ with weights -2, that is,

$$g_1[\boldsymbol{x}, \boldsymbol{y}] = \text{sign}\left(-(x_1 + y_2) - 2\left(\sum_{i=3}^{n} y_i\right) + 1\right).$$

(See Fig. 2 (a).) By the construction, g_1 outputs 0 if either $x_1 = y_2 = 1$ or at least one of $y_3, y_4, \ldots y_n$ is 1, and otherwise, g_1 outputs 1, that is,

$$g_1[\boldsymbol{x}, \boldsymbol{y}] = \neg\left((x_1 \wedge y_2) \vee y_3 \vee y_4 \vee \ldots \vee y_n\right). \tag{17}$$

On the other hand, Eq. (11) implies that

$$\beta_1(\boldsymbol{x}, \boldsymbol{y}) = \neg\alpha_1(\boldsymbol{x}, \boldsymbol{y}) = \neg\left((y_2 \vee y_3 \vee \ldots \vee y_n) \wedge (x_1 \vee y_3 \vee y_4 \vee \ldots \vee y_n)\right).$$

Since $a \vee (a \wedge b) = a$ for any pair of Boolean variables a and b, we have

$$\beta_1(\boldsymbol{x}, \boldsymbol{y}) = \neg\left((x_1 \wedge y_2) \vee y_3 \vee y_4 \vee \ldots \vee y_n\right). \tag{18}$$

Thus Eqs. (17) and (18) imply that g_1 compute β_1 and hence Eq. (16) holds.

Let k, $2 \leq k \leq n/2 - 1$ be an arbitrary integer. We construct the gate g_k as follows. The gate g_k has a threshold -1 and receive inputs from x_{2k-1} and y_{2k} with weights -1, from $x_1, x_2, \ldots, x_{2k-2}$ and $y_{2k+1}, y_{2k+2}, \ldots, y_n$ with weights -2 and from $g_1, g_2, \ldots, g_{k-1}$ with weights $-2n$, that is,

$$g_k[\boldsymbol{x}, \boldsymbol{y}]$$
$$= \text{sign}\left(-(x_{2k-1} + y_{2k}) - 2\left(\sum_{i=1}^{2k-2} x_i + \sum_{i=2k+1}^{n} y_i\right) - 2n\left(\sum_{i=1}^{k-1} g_i[\boldsymbol{x}, \boldsymbol{y}]\right) + 1\right).$$

(See Fig. 2 (b).) Then, g_k outputs 0 if and only if at least one of the following three conditions holds:

(1) $x_{2k-1} = y_{2k} = 1$;
(2) at least one of $x_1, x_2, \ldots, x_{2k-2}, y_{2k+1}, y_{2k+2}, \ldots y_n$ is 1;
(3) at least one of $g_1[\boldsymbol{x}, \boldsymbol{y}], g_2[\boldsymbol{x}, \boldsymbol{y}], \ldots, g_{k-1}[\boldsymbol{x}, \boldsymbol{y}]$ is 1.

In other words, we have

$$g_k[\boldsymbol{x}, \boldsymbol{y}] = \neg \left((x_{2k-1} \wedge y_{2k}) \vee \bigvee_{i=1}^{2k-2} x_i \vee \bigvee_{i=2k+1}^{n} y_i \vee \bigvee_{i=1}^{k-1} g_i[\boldsymbol{x}, \boldsymbol{y}] \right). \quad (19)$$

We now prove $g_k[\boldsymbol{x}, \boldsymbol{y}] = \beta_k(\boldsymbol{x}, \boldsymbol{y})$ and hence by Eq. (19) it suffices to prove

$$\beta_k(\boldsymbol{x}, \boldsymbol{y}) = \neg \left((x_{2k-1} \wedge y_{2k}) \vee \bigvee_{i=1}^{2k-2} x_i \vee \bigvee_{i=2k+1}^{n} y_i \vee \bigvee_{i=1}^{k-1} g_i[\boldsymbol{x}, \boldsymbol{y}] \right). \quad (20)$$

by an induction on k. We start from the case of $k = 2$ as the basis.
[Basis: $k = 2$]
Equations (11) and (16) implies that

$$\beta_2(\boldsymbol{x}, \boldsymbol{y}) = \neg \alpha_2(\boldsymbol{x}, \boldsymbol{y}) \wedge \neg \beta_1(\boldsymbol{x}, \boldsymbol{y})$$
$$= \neg \left(\left(\left(\bigvee_{i=1}^{2} x_i \vee \bigvee_{i=4}^{n} y_i \right) \wedge \left(\bigvee_{i=1}^{3} x_i \vee \bigvee_{i=5}^{n} y_i \right) \right) \vee \beta_1[\boldsymbol{x}, \boldsymbol{y}] \right)$$
$$= \neg \left((x_3 \wedge y_4) \vee \bigvee_{i=1}^{2} x_i \vee \bigvee_{i=5}^{n} y_i \vee g_1[\boldsymbol{x}, \boldsymbol{y}] \right).$$

[Inductive Step: $k \geq 3$]
Equation (11) implies that

$$\beta_k(\boldsymbol{x}, \boldsymbol{y}) = \neg \left(\alpha_k(\boldsymbol{x}, \boldsymbol{y}) \vee \bigvee_{i=1}^{k-1} \beta_i(\boldsymbol{x}, \boldsymbol{y}) \right)$$
$$= \neg \left(\left(\left(\bigvee_{i=1}^{2k-2} x_i \vee \bigvee_{i=2k}^{n} y_i \right) \wedge \left(\bigvee_{i=1}^{2k-1} x_i \vee \bigvee_{i=2k+1}^{n} y_i \right) \right) \vee \bigvee_{i=1}^{k-1} \beta_i(\boldsymbol{x}, \boldsymbol{y}) \right). \quad (21)$$

By the induction hypothesis, for each $i \leq k - 1$, we have

$$g_i[\boldsymbol{x}, \boldsymbol{y}] = \beta_i(\boldsymbol{x}, \boldsymbol{y}).$$

Thus, from Eq. (21) we have

$$\beta_k(\boldsymbol{x}, \boldsymbol{y}) = \neg \left((x_{2k-1} \wedge y_{2k}) \vee \bigvee_{i=1}^{2k-2} x_i \vee \bigvee_{i=2k+1}^{n} y_i \vee \bigvee_{i=1}^{k-1} g_i[\boldsymbol{x}, \boldsymbol{y}] \right). \quad (22)$$

Thus by Eq. (22), Eq. (20) holds true.

Finally, we construct $g_{n/2}$. The gate $g_{n/2}$ has a threshold 2 and receives inputs from x_{n-1} and y_n with weights 1 from $x_1, x_2, \ldots, x_{n-2}$ with weights 2, and from $g_1, g_2, \ldots, g_{n/2-1}$ with weights $-2n$, that is,

$$g_{n/2}[\boldsymbol{x}, \boldsymbol{y}] = \text{sign}\left((x_{n-1} + y_n) + 2\left(\sum_{i=1}^{n-2} x_i\right) - 2n\left(\sum_{i=1}^{(n/2)-1} g_i[\boldsymbol{x}, \boldsymbol{y}]\right) - 2 \right).$$

(See Fig. 2 (c).) By the construction, $g_{n/2}$ outputs 1 if and only if all of $g_1[\boldsymbol{x}, \boldsymbol{y}], g_2[\boldsymbol{x}, \boldsymbol{y}], \ldots, g_{(n/2)-1}[\boldsymbol{x}, \boldsymbol{y}]$ are 0s. and at least one of the following two conditions holds:

(1) $x_{n-1} = y_n = 1$;
(2) at least one of $x_1, x_2, \ldots, x_{(n/2)-2}$ is 1.

In other words, we have

$$g_{n/2}[\boldsymbol{x}, \boldsymbol{y}] = \left((x_{n-1} \wedge y_n) \vee \bigvee_{i=1}^{n-2} x_i \right) \wedge \neg \left(\bigvee_{i=1}^{(n/2)-1} g_i[\boldsymbol{x}, \boldsymbol{y}] \right). \tag{23}$$

On the other hand, Eq. (11) implies that

$$\beta_{n/2}(\boldsymbol{x}, \boldsymbol{y}) = \alpha_{n/2}(\boldsymbol{x}, \boldsymbol{y}) \wedge \neg \left(\bigvee_{i=1}^{(n/2)-1} \beta_i(\boldsymbol{x}, \boldsymbol{y}) \right)$$

$$= \left(\left(\bigvee_{i=1}^{n-2} x_i \vee y_n \right) \wedge \left(\bigvee_{i=1}^{n-1} x_i \right) \right) \wedge \neg \left(\bigvee_{i=1}^{(n/2)-1} \beta_i(\boldsymbol{x}, \boldsymbol{y}) \right). \tag{24}$$

Since $\beta_i(\boldsymbol{x}, \boldsymbol{y}) = g_i[\boldsymbol{x}, \boldsymbol{y}]$ for each $i \leq (n/2) - 1$, we have from Eq. (24)

$$\beta_{n/2}(\boldsymbol{x}, \boldsymbol{y}) = \left((x_{n-1} \wedge y_n) \vee \bigvee_{i=1}^{n-2} x_i \right) \wedge \neg \left(\bigvee_{i=1}^{(n/2)-1} g_i[\boldsymbol{x}, \boldsymbol{y}] \right). \tag{25}$$

Equations (23) and (25) imply that g_k compute β_k. $\qquad \square$

The following theorem implies that the size of C given in Theorem 2 is optimal, that is, P_{LR}^n cannot be computed by any threshold circuit of energy $e = 1$ and size $s \leq \lceil n/2 \rceil - 1$.

Theorem 3. *Let n be a positive integer. Let C be any threshold circuit computing P_{LR}^n with energy $e = 1$. Then, C has size*

$$s \geq \left\lceil \frac{n}{2} \right\rceil.$$

We omit the proof of Theorem 3 due to the page limitation.

5   Conclusions

In this paper, we design energy-efficient threshold circuits computing P_{LR}^n functions. We also give an optimal circuit for energy $e = 1$. Our upper bound

$$s = O\left(e \cdot n^{\frac{2}{(e-1)}}\right).$$

for $e \geq 3$ has a quite similar form to the one for the Parity function given in [5], that is,

$$s = O\left(e \cdot n^{\frac{1}{(e-1)}}\right).$$

However, unlike the case for the Parity function, the tightness of our bound for $e \geq 3$ remains open.

References

1. Laughlin, S.B., Sejnowski, T.J.: Communication in neuronal networks. Science 301(5641), 1870–1874 (2003)
2. Lennie, P.: The cost of cortical computation. Current Biology 13, 493–497 (2003)
3. Margrie, T.W., Brecht, M., Sakmann, B.: In vivo, low-resistance, whole-cell recordings from neurons in the anaesthetized and awake mammalian brain. European Journal of Physiology 444(4), 491–498 (2002)
4. Uchizawa, K., Douglas, R., Maass, W.: On the computational power of threshold circuits with sparse activity. Neural Computation 18(12), 2994–3008 (2006)
5. Suzuki, A., Uchizawa, K., Zhou, X.: Energy-efficient threshold circuits computing mod functions. In: Proceedings of the 17th Computing: the Australasian Theory Symposium. CRPIT, vol. 119, pp. 105–110. Australian Computer Society (2011)
6. Uchizawa, K., Takimoto, E., Nishizeki, T.: Size-energy tradeoffs of unate circuits computing symmetric Boolean functions. Theoretical Computer Science 412, 773–782 (2011)
7. Legenstein, R.A., Maass, W.: Neural circuits for pattern recognition with small total wire length. Theoretical Computer Science 287, 239–249 (2002)

Resolving Rooted Triplet Inconsistency
by Dissolving Multigraphs*

Andrew Chester[1], Riccardo Dondi[2], and Anthony Wirth[1]

[1] Department of Computing and Information Systems, The University of Melbourne
a.chester@student.unimelb.edu.au, awirth@unimelb.edu.au
[2] Dipartimento di Scienze Umane e Sociali. Università degli Studi di Bergamo
riccardo.dondi@unibg.it

Abstract. The MINIMUM ROOTED TRIPLET INCONSISTENCY (MINRTI) problem represents a key computational task in the construction of phylogenetic trees. Inspired by Aho et al's seminal paper and Bryant's thesis, we describe an edge-labelled multigraph problem, MINIMUM DISSOLVING GRAPH (MINDG) and show that it is equivalent to MINRTI. We prove that on an n-vertex graph, for every $\varepsilon > 0$, MINDG is hard to approximate within a factor in $O(2^{\log^{1-\varepsilon} n})$, even on trees formed by multi-edges. Via a further reduction, this result applies to MINRTI, resolving the open question of whether there is a sub-linear approximation factor for MINRTI. In addition, we provide polynomial-time algorithms that return optimal solutions when the input multigraph is restricted to a multi-edge path or a simple tree.

1   Introduction

One of the central tasks of computational evolutionary biology is to construct phylogenetic trees. These represent the evolutionary history of a given set of species. Often the goal is to construct trees for a huge set of species, such as in *The Tree of Life* web project [1]. Direct construction of the trees from DNA evidence is possible using sequence based methods, but is prohibitively expensive for large numbers of species [2]. An alternative approach is to construct a set of smaller phylogenetic trees from sequence data, and then apply a *supertree method* to infer a larger tree from that set. The smallest possible informative trees are either *triplets*, rooted binary trees on three labels, or *quartets*, unrooted ternary trees on four labels.

In many cases, due to experimental errors, or because the trees represent the evolution of different genes, a collection of input trees might not lead to a consistent structure. Hence, the supertree approach aims to merge the information represented by the maximum possible set of (consistent) input trees into a large phylogenetic tree. Over the last decade, quartet methods have received prominent attention [3–7]. Triplet methods have certain interesting properties which

---

* Supported in part by an Australian Research Council Future Fellowship and The Melbourne School of Engineering.

T.-H.H. Chan, L.C. Lau, and L. Trevisan (Eds.): TAMC 2013, LNCS 7876, pp. 260–271, 2013.

make them typically computationally cheaper. A given set of triplets is *compatible* if they have no contradictory structures: there is a single phylogenetic tree that extends them all (defined formally in Section 2). Aho et al. presented a polynomial-time algorithm to find such a tree [8]. For quartets, however, the compatibility problem is NP-complete [9]. In addition there is often no information to identify the root of the input trees, which can make determination of the root particularly problematic [7].

In this paper, we focus on the MINIMUM ROOTED TRIPLET INCONSISTENCY problem (MINRTI). Given a set of triplets $\mathcal{T}$ over n items, MINRTI seeks the smallest subset of triplets to delete from $\mathcal{T}$ to leave a compatible remaining set. The problem is known to be NP-hard [10–12] and not approximable within a factor in $O(\log n)$ [13]. Moreover, for dense instances of MINRTI, in which there is at least one rooted triplet for every subset of $\mathcal{T}$ of cardinality three, there are fixed-parameter algorithms [14]. The MINRTI problem has a dual, with the same optimum solutions, called MAXIMUM ROOTED TRIPLET CONSISTENCY (MAXRTC). Table 1 summarizes the complexity and approximability results.

Based on techniques similar to semidefinite programming for the MAX CUT problem, Snir and Rao introduced heuristics for MAXRTC and the quartet version [6, 7]. Byrka et al.'s survey paper has an extensive list of references [13]. In addition, the authors propose the following open problems (we answer the second one in the negative): (1) Does MAXRTC have an approximation algorithm whose ratio is significantly less than 3? (2) Is there a polylogarithmic approximation algorithm for MINRTI? (3) Is there a constant-ratio approximation for dense inputs?

1.1 New Results

In this paper, inspired by Aho et al. [8] and Bryant [10], we describe an edge-labelled multigraph problem, MINIMUM DISSOLVING GRAPH (MINDG). It asks for the smallest set of edges whose deletion leaves a multigraph that has a self-dissolving property. In Section 3, we show that MINDG is equivalent to MINRTI. Then, in Section 4, we consider the approximability of the MINDG problem, and hence of MINRTI. When the input consists of an arbitrary simple graph, we provide an L-reduction from the TARGET SET SELECTION problem. This result implies that MINDG (and hence MINRTI) cannot be approximated within a ratio $O(2^{\log^{1-\varepsilon} n})$, for every fixed $\varepsilon > 0$, unless NP is in DTIME($n^{\text{polylog}(n)}$). On the positive side, in Section 5 we give a polynomial time algorithm for MINDG when the input graph is restricted to a simple tree, and likewise in Section 6 for paths in multigraphs. A reduction from MINDG to a directed version of TARGET SET SELECTION, alluded to in Section 5, will appear in the full version of this paper.

1.2 Related Topics

The idea of deleting a subset from a given family of relations, to obtain a consistent subfamily is not new. In the area of constrained clustering, we are given a

Table 1. Summary of previous approximation results for MAXRTC and MINRTI. *We show here that, assuming NP is not in DTIME($n^{\text{polylog}(n)}$), MINRTI admits no polynomial-time algorithm whose approximation factor is in $O(2^{\log^{1-\varepsilon} n})$, for fixed $\varepsilon > 0$.

		Negative results	Positive results
MAXRTC:	General	APX-hard [15]	$(3 - \frac{2}{n-2})$-approx [13]
	Dense	NP-hard [16]	PTAS [17]
	Minimally Dense	NP-hard [13]	PTAS [17]
	Consistent	P [8]	Exact Solution [8]
MINRTI:	General	Inapprox. $c \cdot \ln n$ [13] *	$(n-2)$-approx [13, 18]
	Dense	NP-hard [16]	$(n-2)$-approx [13, 18]
	Minimally Dense	NP-hard [13]	$(n-2)$-approx [13, 18]
	Consistent	P [8]	Exact Solution [8]

set of pairwise recommendations of the form: "Items x and y should be clustered together," or "Items x and y should be in distinct clusters" [19]. In particular the MIN DISAGREEMENTS variant of Correlation Clustering [20] seeks a clustering that is inconsistent with the smallest such set of such constraints. The task of deleting some constraints to leave a consistent family, one that admits a clean clustering, is known as the CLUSTER EDITING problem.

Related to constrained clustering are ranking problems. Specifically, we are given a set of pairwise relations of the form "x should be ranked higher than y" and asked to produce an ordering of the items that is consistent with as many of the relations as possible. It is convenient to view this as a graph problem, with the ranking relation represented as directed edge from x to y. If the relations are indeed consistent, this problem is merely topological sorting. In general, it is the same as FEEDBACK ARC SET problem, which has been studied in particular on tournament graphs [21].

In some sense, MINRTI is from the same family of problems, but the relations are on triples of items and impose constraints on a hierarchical structure.

2   Definitions

A phylogenetic tree is a rooted, unordered tree, which in this paper we assume is a binary tree. Each leaf has a single, unique label. In MINRTI, we are given a set of triplets $\mathcal{T}$ over a size-n universe of possible leaf labels $\mathcal{X}$.

Definition 1. *A triplet $t = (x, y \mid z)$ and a binary tree T, whose leaves are labelled with items from $\mathcal{X}$, are consistent if the lowest common ancestor of x and y in T is a proper descendant of x and z (and that of y and z). A family of triplets $\mathcal{T}$ is compatible if there exists a leaf-labelled binary tree consistent with every triplet in $\mathcal{T}$.*

Given an input set of triplets $\mathcal{T}$, MINRTI asks for the smallest subset of $\mathcal{T}$, whose deletion leaves it compatible.

Consider a multigraph $G(\mathcal{T})$ on $\mathcal{X}$ whose edges are undirected and labelled with elements from $\mathcal{X}$. An edge e between x and y with label z is written as $(x, y \mid z)$. A sequence of vertices $x_1, \ldots, x_k$ is a *path* if each pair x_i, x_{i+1}, $i < k$, is adjacent (shares some edge). In such a multigraph, a tree is therefore a set of vertices between every pair of which there is at most one path.

A multigraph G *dissolves* if there is a sequence of steps of the following form that leads to a graph with no edges remaining:

> If there is no path in the graph between x and z, or equivalently between y and z, then *remove* the edge $(x, y \mid z)$.

Definition 2. *The* MINDG *problem asks for the smallest set of edges whose deletion leaves a graph that dissolves.*

Note the distinction between *removing* an edge, for no cost, during the dissolving process, and *deleting* an edge, at cost, in order to leave a graph that dissolves. In Corollary 1 below, we show that MINRTI and MINDG are essentially the same problem. In Section 2.5.1 of his PhD thesis, Bryant explores a similar characterization of compatible triplet sets [10].

3   Multigraph Representation

We now prove the equivalence between MINRTI and MINDG.

Definition 3. *Instances* $G(\mathcal{T})$ *of* MINDG *and* $\mathcal{T}$ *of* MINRTI, *on the same space* $\mathcal{X}$, *correspond whenever there is an edge* $(x, y \mid z)$ *in* $G(\mathcal{T})$ *if and only if there is a triplet* $(x, y \mid z)$ *in* $\mathcal{T}$.

Lemma 1 (Multigraph equivalence). *A set of triplets* $\mathcal{T}$ *is compatible if and only if the corresponding multigraph* $G(\mathcal{T})$ *dissolves.*

Proof. **Only if**   This is a proof by induction, over the size of the set $\mathcal{X}$.

Base case. If $\mathcal{X}$ has just three elements, then $\mathcal{T}$ is compatible if and only if it has at most one triplet. Likewise, the graph G on three vertices dissolves if and only if there is at most one edge.

Induction step. If $\mathcal{T}$ is compatible, then there is a binary tree T in which the left subtree has leaves $\mathcal{X}_L$, the right subtree $\mathcal{X}_R$, and for every triplet $(x, y \mid z)$, either (1) x, y, z are in the same subtree, and $(x, y \mid z)$ is consistent with that subtree, or (2) x, y are in the same subtree, but different from z (the triplet $(x, y \mid z)$ is thus *consistent* with the tree T, and is ignored when considering the subtrees).

If x and y were to be split, the triplet would not be consistent with T.

Translating to $G(\mathcal{T})$: Consider partitioning the vertices into subsets $\mathcal{X}_L$ and $\mathcal{X}_R$. Since there is no triplet with x and y in different subtrees, there is no edge across the partition, so it is a cut in $G(\mathcal{T})$. Therefore, we can focus on the induced subgraphs on each of $\mathcal{X}_L$ and $\mathcal{X}_R$: in particular, each edge $(x, y \mid z)$ with z on a different side of the partition from x, y can be *removed*, and ignored in each subgraph. The resulting two induced subgraphs correspond triplets belonging to each of the subtrees of T and subsequent edge removals in each subgraph are independent, so we have completed the inductive step.

If Now suppose $G(\mathcal{T})$ dissolves. As the algorithm proceeds, each edge $(x, y \mid z)$ falls into one of three categories: *removed*, as previously described; *doomed*, not yet removed, but with z disconnected from x and from y, the edge is eligible for removal; *unmarked*, not yet doomed, nor removed.

All edges are initially unmarked, and edges may only may be removed once they are doomed. At the beginning of the process, and as subsequent edges are removed, the graph splits into two components. If there are more than two components at the beginning, we consider repeatedly splitting the graph into two components in some arbitrary order before the edge removal begins. As the graph splits, this will cause some unmarked edges to become doomed. We construct two subtrees corresponding to this split: each doomed edge represents a triplet that is immediately satisfied by the split in the tree. Since eventually all edges are removed, each must have been doomed, and therefore at some stage each triplet must have been satisfied. The triplets $\mathcal{T}$ are therefore compatible.

Corollary 1. *The optimal solutions to* MINRTI *and* MINDG *are equivalent and have equal optimal values.*

3.1 Weighted Edges

Though by default we assume $G(\mathcal{T})$ has unit-weight edges, in some MINDG instances the edges may have weights. Nevertheless, there is a reduction from an instance of the edge-weighted version to the standard version, that is polynomial in size and time if the edge weights are polynomial in n. Refer to the example in Figure 1. We replace each edge $(x, y \mid z)$ of weight w, with w (unit-weight) edges $(x, y \mid z_1), (x, y \mid z_2), \ldots, (x, y \mid z_w)$. We then add w vertices $z_1^{xy}, z_2^{xy}, \ldots, z_w^{xy}$, and corresponding edges $(z, z_1^{xy} \mid x), (z, z_2^{xy} \mid x), \ldots, (z, z_w^{xy} \mid x)$.

If the original graph dissolves without $(x, y \mid z)$ being deleted (in advance), then somehow x, y become disconnected from z: the same happens in the new graph. The $(x, y \mid z^{xy})$ and $(z, z^{xy} \mid x)$ edges can thus be removed. A similar argument follows assuming that none of the $(x, y \mid z^{xy})$ and $(z, z^{xy} \mid x)$ edges are deleted in advance.

On the other hand, if the original graph requires $(x, y \mid z)$ to be deleted, then in the new graph we can delete the equivalent parallel $(x, y \mid z^{xy}$ edges at the same cost. In the new graph, a solution might delete some of the $(x, y \mid z^{xy})$ edges and some of the $(z, z^{xy} \mid x)$ edges. However, for each i, at least one of $(x, y \mid z_i^{xy})$ and $(z, z_i^{xy} \mid x)$ would need to be deleted to effect a removal of all the edges between x and y, so the cost is still w. Note that if the remainder of the graph has dissolved then the $(z, z^{xy} \mid x)$ edges will too.

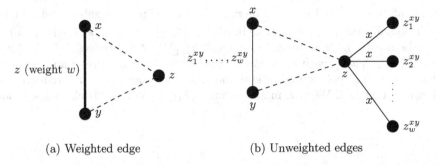

(a) Weighted edge   (b) Unweighted edges

Fig. 1. The reduction from a weighted-edge graph to a unit-edge-cost graph. The dashed lines between x and z, and between y and z, indicate that x and z might be connected: similarly between y and z.

4   MinRTI Is Label Cover-hard

To show that MINDG, and hence MINRTI, is hard, we refer to another problem about a 'process' in a graph. The TARGET SET SELECTION problem asks questions about the spread of influence in graphs. Formally, we are given a connected undirected graph H, with a *threshold* function $t : V \to \{0, 1, \dots, n-1\}$ on the vertices. Each vertex is initially *inactive*. We choose a *target set* to be active initially and then the following step is repeated until no more changes occur:

If at least $t(v)$ neighbours of some inactive vertex v are active, then v becomes active.

In an n-vertex graph, there can be at most $n - 1$ executions of this step. The related optimization problem, TARGET SET SELECTION, is: What is the smallest set of vertices that need to be active initially to ensure that all vertices are active at the end of the procedure?

Chen studied the problem in significant detail [22]. In particular, he showed that for a graph in which $t(v) = 2$ for every vertex, TARGET SET SELECTION cannot be approximated within a ratio $O(2^{\log^{1-\varepsilon} n})$, for every fixed $\varepsilon > 0$, unless NP is in DTIME($n^{\text{polylog}(n)}$). That is, this threshold-2 version of TARGET SET SELECTION (TSS-2) is LABEL COVER-hard [23].

We show that TSS-2 reduces to a special case of MINRTI, proving LABEL COVER-hardness for the general problem. This almost closes the gap between the hardness result and the best approximation known for general MINRTI [13].

Reduction. Let H stand for the TSS-2 instance; we first show how to construct an instance G of MINDG, and then that this is an L-reduction.

Graph G has a central vertex r. For each vertex $v \in H$, there is a an edge $(x_v, y_v \mid r)$ in G of cost 1. Between y_v and r we construct a mechanism to mimic the activation threshold of TSS-2. Let $d(v)$ be v's degree in graph H and let $D(v) = \binom{d(v)}{2}$. Build $D(v) - 1$ intermediate nodes $a_1^v, a_2^v, \dots, a_{D(v)-1}^x$, and define a_0^v to be y_v and $a_{D(v)}^v$ to be r. Match the pairs of neighbours of v in H, with the adjacent pairs of vertices in $a_0^v, \ldots, a_D^v(v)$. That is, each pair (u, u') of neighbours of v in H is assigned some (a_i^v, a_{i+1}^v), so G has edges $(a_i^v, a_{i+1}^v \mid x_u)$ and $(a_i^v, a_{i+1}^v \mid x_{u'})$, each of weight n. Since the maximum cost of an edge is n, this reduction is polynomial in size and time. An example of this reduction is shown in Figure 2. We now map a solution Σ_G for MINDG back to a solution

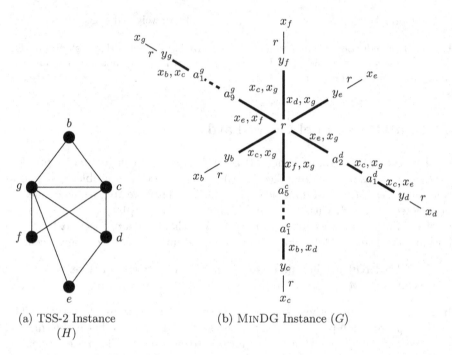

(a) TSS-2 Instance
(H)

(b) MINDG Instance (G)

Fig. 2. An example of the reduction from TSS-2 to MINDG. In the MINDG instance, thick edges have weight n, while thin edges have weight 1.

for TSS-2. If Σ_G contains an edge of cost n, which we regard as a *nonsense* solution, then set all vertices in H to be active initially. Otherwise, for each edge $(x_v, y_v \mid r)$ that is deleted (is in Σ_G), set v to be active initially.

Claim. Excluding nonsense solutions to MINDG, solutions to TSS-2 and MINDG have identical costs.

Proof. Consider a solution Σ_H to TSS-2. For each v that is initially active, we delete $(x_v, y_v \mid r)$. This of course disconnects x_v from the rest of the graph, and so all edges marked x_v can be removed. A vertex u in H is active whenever two of its neighbours are active. Similarly, a vertex x_u in G becomes disconnected from the rest of the graph whenever two of its 'H-neighbours' are disconnected from r, since the edges on the path from r to x_u are labeled with the possible pairs of 'H-neighbours' of u. Therefore a solution Σ_H to TSS-2 leads to a solution of MINDG of the same cost.

Likewise, if the set of cost-1 edge deletions in Σ_G leads to the graph dissolving, the sequence of disconnections of the x-vertices in G is matched by the activation of corresponding vertices in H.

Lemma 2. *The (previous) reduction from TSS-2 to MINDG is approximation-preserving.*

Proof. The requirements for an L-reduction are that $\mathrm{OPT}(G) \leq \alpha\mathrm{OPT}(H)$ for some fixed constant α, and that for some fixed β, for every solution Σ_G to MINDG, the corresponding Σ_H solution to TSS-2 satisfies $\mathrm{cost}(\Sigma_G) - \mathrm{OPT}(G) \geq \beta[\mathrm{cost}(\Sigma_H) - \mathrm{OPT}(H)]$.

The optimum solution to G has the same cost as the optimum solution to H, so the α constant is 1. Except for nonsense solutions, all corresponding solution costs, Σ_H and Σ_G, are equal. Since nonsense solutions cost at least n, and the cost of a solution to TSS-2 is at most n, the β constant is also 1.

The graph generated by the reduction from TSS-2 is a weighted multi-edge tree. Since our conversion from weighted multi-edges to unit-cost multi-edges in Section 3.1 preserves acyclicity, we reach the following conclusion.

Corollary 2. *On multi-edge trees, MINDG is LABEL COVER-hard.*

4.1 Simple Graph Case

The remainder of this paper examines the tractability of MINDG on special classes of graphs. So far, we have shown that for a general multigraph, MINDG is LABEL COVER-hard.

Via a straightforward reduction, an example of which is in Figure 3, we now extend this hardness result to simple graphs. Each multi-edge between two nodes can be converted into a gadget of single edges connecting those two nodes, while maintaining essentially the same dissolving behaviour. The reduction is polynomial in the size of the input: if the original graph has m edges, the resultant graph will have a total of $n + 2m$ nodes and $3m$ edges. To replace the multi-edge $\{(x, y \mid z_1), \ldots, (x, y \mid z_k)\}$, we add vertices $u_1, \ldots, u_k, v_1, \ldots, v_k$ and, for all $i \leq k$, edges $(x, u_i \mid y)$, $(u_i, v_i \mid z_i)$ and $(v_i, y \mid x)$.

Whenever at least one of the z_i-labelled edges is still present, x and y are still 'adjacent'. Once the original graph has dissolved, x and y are completely disconnected, and so the auxiliary edges, $(x, u_i \mid y)$ and $(v_i, y \mid x)$, can be removed.

Lemma 3. *On simple graphs, MINDG is in general LABEL COVER-hard.*

5   On Simple Trees, MINDG Is in P

We now turn to some tractable special cases of MINDG. Assuming that $G(\mathcal{T})$ is a simple tree, we show that MINDG is solved in polynomial time. The key

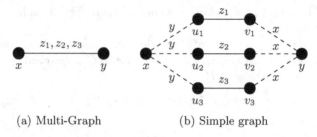

(a) Multi-Graph   (b) Simple graph

Fig. 3. An example of the reduction from a multi-graph to a simple graph. Each multi-edge on the left is turned into an equivalently-connected structure of simple edges on the right.

idea is that in a simple tree, there is a unique path between every edge and the node it is labelled with. Given an instance of MinDG, $G = (V, E)$, we construct a dependency graph $H = (V', E')$, a directed instance of TSS, in the following manner. For every (labelled) edge $e = (x, y \mid z) \in E$, create a node $v_e \in V'$. Now, for every $e \in E$, for every $e' \in E$ which lies on the path between node z and edge $e = (x, y \mid z)$, add to E' an arc from $v_{e'}$ to v_e. Since each edge e is connected to its label z via a unique path, the removal of some edge e' along that path disconnects e from z and thus enables the removal of e.

The graph H is a directed instance of TSS, in which the threshold for each node is one. A node v_e in H is active whenever the edge e has been removed (or deleted initially) from G. The optimal starting active set in H, representing the optimal MinDG solution in G, is trivial: activate a single node in each source strongly connected component (SCC) of H.

Since we search the graph for each edge in G, construction of H requires $O(n^2)$ time; the size of H can in fact be quadratic in n. Kosaraju's algorithm [24], for instance, finds the source SCCs in time proportional to the size of H, so overall the running time is in $O(n^2)$.

6   On Multi-edge Paths, MinDG Is in P

Our solution for simple trees fails in the multi-edge path case, because we are no longer guaranteed to partition the graph by removing a single edge. Instead, we present here a dynamic programming solution. We find optimal solutions for subpaths of the original path, assuming they are isolated from the rest of the graph. For subpaths of length one, the cost is zero. For every larger subpath P, we consider the cost of each possible initial cut, and combine that with costs of the remaining subpaths of P, now isolated. By working in increasing subpath length order, the dynamic program will succeed. We need to record the position of the optimal cut for each subpath in a second matrix, so we can reconstruct the set of edges that constitutes the solution. This process is formalized in Algorithm 1, and Algorithm 2 details the process of reconstructing the set of deleted edges. Assuming the nodes have been enumerated $1, 2, \ldots, n$, each $M_{i,j}$ contains the minimum cost for the section from i to j, and each $K_{i,j}$ likewise holds the position of the optimal cut.

The function $\text{cost}(k, i, \ell)$ is defined as the number of edges between k and $k + 1$ whose labels are in the range $i, \ldots, i + \ell$. This definition ensures that the algorithm is correct. Every solution to MINDG must cause some first cut in the graph. Until this happens, the graph remains connected, therefore an optimal solution need only delete edges between a single pair of adjacent nodes. Once this first cut has occurred, the two subpaths behave independently, as captured by the cost() function. Since an optimal solution is formed from a first cut and optimal solutions on subpaths, the dynamic program is correct. The number of

---

Algorithm 1. MINDG-MULTIPATH$(G(\mathcal{T}))$

1: label the nodes along the path $1, \ldots, n$
2: construct $n \times n$ matrices, M and K
3: initialize the diagonal above the main diagonal of M to 0
4: **for** $\ell = 2$ to $n - 1$ **do**
5:     **for** $i = 1$ to $n - \ell$ **do**
6:         $M_{i,i+\ell} \leftarrow \min_{i \leq k < i+\ell}(\text{cost}(k, i, \ell) + M_{i,k} + M_{k+1,i+\ell})$
7:         $K_{i,i+\ell} \leftarrow \arg\min_{i \leq k < i+\ell}(\text{cost}(k, i, \ell) + M_{i,k} + M_{k+1,i+\ell})$
8:     **end for**
9: **end for**
10: return DeletedEdges$(G, K, 1, n)$

Algorithm 2. DeletedEdges(G, K, i, j)

1: $k \leftarrow K_{i,j}$
2: $\Sigma \leftarrow$ edges $(k, k + 1 \mid z)$, where $i \leq z \leq j$.
3: unite Σ with DeletedEdges(G, K, i, k) and DeletedEdges$(G, K, k + 1, j)$
4: return Σ

---

entries in M is quadratic in n, and for each entry, to evaluate the cost function, we inspect $O(m)$ simple edges. To reconstruct the solution, we only revisit $n - 1$ entries in K, each with $O(m)$ worst-case cost, so we can safely ignore the time taken by Algorithm 2 in our asymptotic analysis. Hence the overall running time is in $O(mn^2) \subseteq O(n^4)$. The space requirement is quadratic, since we only need to store an (integer) cost and position for each subpath. Of course, a similar dynamic program could work on arbitrary multigraphs. Unfortunately, even on a multi-edge tree, the number of subproblems is exponential in n.

7 Wrapping Up

The principal result in this paper is the resolution of the approximability of MINRTI: it is essentially linear in n. In obtaining the hardness reduction, we extended Aho et al's and Bryant's graph interpretations of the rooted tree consistency problem [8, 10]. In the full version of this paper, by describing a reduction from MINDG to a directed version of TSS, we will tighten the nexus between these problems.

As summarized in Table 2, our positive results focus on specific families of (multi-edge) graphs. An interpretation of special cases in terms of the triplet

Table 2. Summary of results for MINDG as a function of graph type

	Path	Tree	Graph
Simple	$O(n^2)$-time	$O(n^2)$-time	LABEL COVER-hard
Multi-Edge	$O(n^4)$-time	LABEL COVER-hard	LABEL COVER-hard

formulation might be more natural, but the DISSOLVING GRAPH formulation is highly effective for proofs.

A related question is determining the tractability of MINDG on multi-edge trees with constant degree. Graphs of constant multi-edge degree represent MIN-RTI inputs in which each item (species) is deemed to be *similar* to at most a constant number of other items.

Finally, the other open problem we would like to explore was introduced by Byrka et al. [13]. Is there a constant-factor approximation algorithm for the dense case of MINRTI?

Acknowledgement. Anthony Wirth thanks the University of Milano-Bicocca for hosting him in July 2011.

References

1. Maddison, D., Schulz, K., Maddison, W.: The Tree of Life Web Project. Zootaxa 1668, 19–40 (2007)
2. Erdős, P., Steel, M., Székely, L., Warnow, T.: A few logs suffice to build (almost) all trees (i). Random Structures and Algorithms 14(2), 153–184 (1999)
3. Jiang, T., Kearney, P., Li, M.: A polynomial time approximation scheme for inferring evolutionary trees from quartet topologies and its application. SIAM Journal on Computing 30(6), 1942–1961 (2001)
4. Snir, S., Yuster, R.: A linear time approximation scheme for maximum quartet consistency on sparse sampled inputs. In: Goldberg, L.A., Jansen, K., Ravi, R., Rolim, J.D.P. (eds.) APPROX/RANDOM 2011. LNCS, vol. 6845, pp. 339–350. Springer, Heidelberg (2011)
5. Snir, S., Yuster, R.: Reconstructing approximate phylogenetic trees from quartet samples. In: SODA 2010: Proceedings of the Twenty-First ACM-SIAM Symposium on Discrete Algorithms, pp. 1035–1044 (2010)
6. Snir, S., Rao, S.: Using Max Cut to enhance rooted trees consistency. IEEE/ACM Transactions on Computational Biology and Bioinformatics 3(4), 323–333 (2006)

7. Snir, S., Rao, S.: Quartets MaxCut: A divide and conquer quartets algorithm. IEEE/ACM Transactions on Computational Biology and Bioinformatics 7(4), 704–718 (2010)
8. Aho, A., Sagiv, Y., Szymanski, T., Ullman, J.: Inferring a tree from lowest common ancestors with an application to the optimization of relational expressions. SIAM Journal on Computing 10(3), 405–421 (1981)
9. Steel, M.: The complexity of reconstructing trees from qualitative characters and subtrees. Journal of Classification 9, 91–116 (1992)
10. Bryant, D.: Building Trees, Hunting for Trees, and Comparing Trees. PhD thesis, Department of Mathematics, University of Canterbury, New Zealand (1997)
11. Jansson, J.: On the complexity of inferring rooted evolutionary trees. Electronic Notes in Discrete Mathematics 7, 50–53 (2001)
12. Wu, B.: Constructing the maximum consensus tree from rooted triples. Journal of Combinatorial Optimization 8(1), 29–39 (2004)
13. Byrka, J., Guillemot, S., Jansson, J.: New results on optimizing rooted triplets consistency. Discrete Applied Mathematics 158(11), 1136–1147 (2010)
14. Guillemot, S., Mnich, M.: Kernel and fast algorithm for dense triplet inconsistency. In: Kratochvíl, J., Li, A., Fiala, J., Kolman, P. (eds.) TAMC 2010. LNCS, vol. 6108, pp. 247–257. Springer, Heidelberg (2010)
15. Byrka, J., Gawrychowski, P., Huber, K., Kelk, S.: Worst-case optimal approximation algorithms for maximizing triplet consistency within phylogenetic networks. Journal of Discrete Algorithms 8(1), 65–75 (2010)
16. Iersel, L., Kelk, S., Mnich, M.: Uniqueness, intractability and exact algorithms: Reflections on level-k phylogenetic networks. Journal of Bioinformatics and Computational Biology 7(4), 597–623 (2009)
17. Jansson, J., Lingas, A., Lundell, E.: A triplet approach to approximations of evolutionary trees. Poster H15 presented at RECOMB 2004: Manuscript obtained from first author's homepage (2004)
18. Gasieniec, L., Jansson, J., Lingas, A., Östlin, A.: On the complexity of constructing evolutionary trees. Journal of Combinatorial Optimization 3(2), 183–197 (1999)
19. Basu, S., Davidson, I., Wagstaff, K.: Constrainted Clustering: Advances in Algorithms, Theory, and Applications. CRC Press (2009)
20. Bansal, N., Blum, A., Chawla, S.: Correlation clustering. Machine Learning 56(1), 89–113 (2004)
21. Kenyon-Mathieu, C., Schudy, W.: How to rank with few errors. In: STOC 2007: Proceedings of the Thirty-Ninth Annual ACM Symposium on Theory of Computing, pp. 95–103 (2007)
22. Chen, N.: On the approximability of influence in social networks. SIAM Journal on Discrete Mathematics 23(3), 1400–1415 (2009)
23. Arora, S., Lund, C.: Hardness of approximations. In: Hochbaum, D. (ed.) Approximation Algorithms for NP-hard Problems, PWS Publishing, Boston (1996)
24. Dasgupta, S., Papadimitriou, C., Vazirani, U.: Algorithms. McGraw-Hill (2008)

Obnoxious Facility Game with a Bounded Service Range*

Yukun Cheng[1,2], Qiaoming Han[2], Wei Yu[1], and Guochuan Zhang[1]

[1] College of Computer Science and Technology, Zhejiang University, Hangzhou, 310027, China
[2] School of Mathematics and Statistics, Zhejiang University of Finance and Economics, Hangzhou 310018, China

Abstract. We study the *obnoxious facility game* with *service range* on a path where each facility is undesirable and has service radius r. In this game there are a number of agents on a path. Each agent tries to be far away from all facilities, but still to be served by a facility. Namely, the distance between an agent and her nearest facility is at most r. The utility of an agent is thus defined as this distance. In a deterministic or randomized *mechanism*, based on the addresses reported by the selfish agents, the locations or the location distributions of facilities are determined. The aim of the mechanisms is to maximize the *obnoxious social welfare*, the total utilities of all agents. The objective of each agent is to maximize her own utility and she may lie if, by doing so, more benefit can be obtained. We are interested in mechanisms without money to decide the facility locations so that the obnoxious social welfare is maximized and all agents are enforced to report their true locations (*strategy-proofness* or *group strategy-proofness*).

In this paper, we give the first attempt for this game on a path to design a group strategy-proof deterministic and randomized mechanism when the service radius $\frac{1}{2} \leq r \leq 1$ by assuming that the path length is one. Depending on the value r, we provide different mechanisms with provable approximation ratios. Lower bounds on any deterministic strategy-proof mechanism are also presented.

Keywords: Algorithmic mechanism design, obnoxious facility location, social choice, service range.

1 Introduction

We study the *obnoxious facility game* with a *service range* that models the following problem in economics. The local government plans to build one or more garbage dumps to serve the local community in a city, represented by a metric

* Research was partially supported by the National Nature Science Foundation of China (No. 11271009, 11271325) and the Nature Science Foundation of Zhejiang Province (No. LQ12A01011).

T.-H.H. Chan, L.C. Lau, and L. Trevisan (Eds.): TAMC 2013, LNCS 7876, pp. 272–281, 2013.

space. Due to the limited service ability, each garbage dump can only serve the residents within its service scope. The government should decide the most appropriate locations to install these garbage dumps based on the home addresses reported by the residents. The utility of each resident is her distance from the nearest garbage dump if she stays in one facility's service scope. Otherwise this resident cannot be served at all and her utility is minus infinity. Since the garbage dump is not enjoyable, every resident wants to be as far from such a facility as possible. But on the other hand the garbage dumps are necessary for the local community, each resident must be stay in at least one garbage dump's service scope in order to get the corresponding service. In our setting the locations of the residents are private information. Residents may report wrong locations to improve their utilities. The core of this game for the government is to design a mechanism (algorithm), that maps the reported home addresses of residents to a set of locations where the facilities will be open, to fulfill the purposes. The goals of the government are twofold: enforcing all residents to report their true home addresses and maximizing the total utilities of all residents, which is called the *obnoxious social welfare*. In this paper we are interested in the mechanism design without money, a topic extensively investigated in economics theory, game theory and public choice theory. From the algorithmic perspective, we would like our mechanisms to be approximately optimal with respect to the obnoxious social welfare, where approximation is defined in the usual sense by looking at the worst case ratio between the social welfare of the optimal solution and the social welfare of the mechanism's solution. Meanwhile, we also want our mechanisms to provide a stronger guarantee by showing *group strategy-proofness*, that is whenever a coalition of agents lies, at least one of the members of the coalition does not gain from the lies.

1.1 Previous Results

The previous work was mainly about the classical facility game without service range, in which each agent (resident) wants to stay as close to a facility as possible. There are two optimization targets being concerned: the social cost and the maximum cost. For the social cost Procaccia and Tennenholtz [11] studied the facility game when all agents are on a line. If only one facility should be located, it is trivial that there exists an optimal group strategy-proof mechanism. For the 2-facility game on a line, they gave an upper bound of $n - 2$ and a lower bound of 1.5 for deterministic strategy-proof mechanisms. Later, Lu et al. [9] obtained an upper bound of $\frac{n}{2}$ and a lower bound of 1.045 for randomized strategy-proof mechanisms. Recently, Lu et al. [8] improved the lower bound for deterministic strategy-proof mechanisms to $\frac{n-1}{2}$ and designed a 4-approximation randomized mechanism in general metric spaces. For the 1-facility game on a circle, it follows directly from the results of Schummer and Vohra [12] that no deterministic strategy-proof mechanism can obtain an approximation ratio better that $\Omega(n)$. But Alon et al.[1] showed a simple randomized mechanism which is group strategy-proof and gave a $(2 - \frac{2}{n})$-approximation ratio. For the 1-facility game with maximum cost, Procaccia and Tennenholtz [11] considered the case that the network is a line and gave the group strategy-proof deterministic and randomized mechanisms that yielded approximation ratios of 2 and $\frac{3}{2}$, respectively. And they also provided the matching strategy-proof lower bounds for the deterministic and randomized mechanisms. When the network is a circle, Alon et al. [2] designed a novel "hybrid" strategy-proof randomized mechanism with a tight approximation ratio of $\frac{3}{2}$. They also showed that no randomized strategy-proof mechanism can provide an approximation ratio better than $2 - o(1)$ even when the network is a tree. For the game to locate more than two facilities, Fotakis and Tzamos [5] considered a variant of the game where an authority can impose on some agents the facilities where they will be served. With this restriction, they proposed a strategy-proof randomized Mechanism whose approximation ratio is linear on the number of facilities. Recently, Escoffier et al. [4] studied a special facility game in which there are $n - 1$ facilities should be located for n agents in a general metric space and in a tree. They provided lower and upper bounds on the approximation ratio of deterministic and randomized mechanisms for the social cost and the maximum cost respectively.

Recently, we first proposed the *obnoxious facility game* in which each facility is not desirable any more and gave several mechanisms on path, tree, circle and general networks [3]. Specially, for a path, a 3-approximation group strategy-proof deterministic mechanism and a group strategy-proof randomized mechanism with tight approximation ratio of $\frac{3}{2}$ were given. For a circle or a tree, two group strategy-proof deterministic mechanisms that provide the approximation ratio of 3, respectively, were shown. Finally, for a general network, a 4-approximation group strategy-proof deterministic mechanism and a 2-approximation group strategy-proof randomized mechanism were derived. Similar to the work of Moulin [10] who characterized all strategy-proof mechanisms for the classical facility game on a path, Han et al. [7] and Ibara et al. [6] characterized all strategy-proof mechanisms for the obnoxious facility game on path independently. They pointed out that there is no strategy-proof mechanisms such that the number of candidates is more than two. In particular, by using the complete characterization, Han [7] provided the matching lower bound of 3 for deterministic group strategy-proof mechanisms when all agents are on a path, tree or circle which showed that the deterministic mechanisms in [3] are the best possible.

1.2   Our Contribution

In this paper we extend the work in [3] by considering the obnoxious facility game with a service radius r on a path. We normalize the path as an interval $[0, 1]$ and focus on the case that $\frac{1}{2} \leq r \leq 1$. It is obvious that this model is the same as the previous one in [3] if $r = 1$ in which one facility suffices. Due to the bounded service range, this game becomes more complicated. First, we need to build more than one facilities in some situations since each facility's service scope may not cover the whole interval. Then, the agent set must be partitioned reasonably corresponding to the facility locations. Second, we know that if there is no restriction of service radius, one of two endpoints of the interval must be an optimal facility location. But now such a nice property does not hold at all once the service range exists. Thus it is difficult for us to evaluate the approximation ratios of mechanisms by applying appropriate upper bound of the social welfare of an optimal solution.

In the next section we introduce some useful notations and some results pertinent to the non-selfish version of this problem. By dealing with different service radius, we give both group strategy-proof deterministic and randomized mechanisms in Sections 3 and 4. For $\frac{1}{2} \leq r < \frac{3}{4}$ and $\frac{3}{4} \leq r \leq 1$, the approximation ratio of the deterministic mechanism is $8r - 1$ and $\frac{r+\frac{1}{2}}{r-\frac{1}{2}}$, respectively, while the approximation ratio of the randomized mechanism is $4r$ and $\frac{r}{r-\frac{1}{2}}$, respectively. Meanwhile, we propose lower bounds of the deterministic strategy-proof mechanisms in Section 3, that is $4r - 1$ if $\frac{1}{2} \leq r < \frac{3}{4}$, $\frac{1}{2r-1}$ if $\frac{3}{4} \leq r < \frac{5}{6}$ and $3r - 1$ when $\frac{5}{6} \leq r < 1$, respectively.

2   Preliminaries

In this section we introduce useful notations and related results pertinent to the obnoxious facility problem on a path with service range in which the locations of all agents are public.

Let $N = \{1, 2, \cdots, n\}$ be the set of agents. All of the agents are located on a path P. For the sake of simplicity, assume that the left endpoint of the path is zero and the right endpoint of the path is one. We regard the path as an interval $I = [0, 1]$. The distance between any two points $x, y \in I$ is $d(x, y) = |x - y|$. Thus for all $x \in I$, $d(x, x) = 0$. The location reported by agent i is $x_i \in I$. Denote $\mathbf{x} = (x_1, x_2, \cdots, x_n)$ to be a location profile.

In the obnoxious facility game with service radius r, a deterministic mechanism outputs a facility set based on a given location profile and thus is a function $f : I^n \to I^p$. Assuming the output of f to be $f(\mathbf{x}) = Y = \{y_1, \cdots, y_p\}$ where y_j, $j = 1, 2, \cdots, p$, is the location of the jth facility, the utility of agent i is her distance to the facility set Y, i.e., $d(x_i, Y) = \min_{1 \leq j \leq p} d(x_i, y_j)$ if this distance is no more than r. Otherwise her utility is minus infinity. Thus

$$u(x_i, f(\mathbf{x})) = \begin{cases} d(x_i, Y) & \text{if } d(x_i, Y) \leq r, \\ -\infty & \text{otherwise .} \end{cases}$$

A randomized mechanism is a function $f \colon I^n \to \Delta(I^p)$, where $\Delta(I^p)$ is the set of distributions over I^p. The utility of agent $i \in N$ is now her expected utility over such a distribution.

Given a location profile $\mathbf{x}$, a feasible facility set $Y \subseteq [0, 1]$ is defined that for any agent's location x_i, $d(x_i, Y) \leq r$, and we define the objective function of feasible facility set Y as follows,

$$F_{\mathbf{x}}(Y) = \sum_{i=1}^{n} d(x_i, Y).$$

Denote $OPT(\mathbf{x})$ to be the optimal one, that is $OPT(\mathbf{x}) = \max_Y F_{\mathbf{x}}(Y)$. The *obnoxious social welfare* of a mechanism f on a location profile $\mathbf{x}$ is defined as the total utility of n agents

$$SW(f, \mathbf{x}) = \sum_{i=1}^{n} u(x_i, f(\mathbf{x})).$$

In the randomized case, this obnoxious social welfare is an expected value. For the obnoxious facility game with service radius r, we are interested in the strategy-proof mechanisms that also do well respect to maximize the obnoxious social welfare. We say a mechanism f has an approximation ratio γ, if for all profile $\mathbf{x} \in I^n$,

$$OPT(\mathbf{x}) \leq \gamma SW(f, \mathbf{x}).$$

Let $\mathbf{x}_{-i} = (x_1, \cdots, x_{i-1}, x_{i+1}, \cdots, x_n)$ be the location profile without agent i. For an agent set $S \subseteq N$, we denote $\mathbf{x}_S$ and $\mathbf{x}_{-S}$ to be the location profiles of agents in and outside S, respectively. Thus we have three equivalent notations: $\mathbf{x} = \langle x_i, \mathbf{x}_{-i} \rangle = \langle \mathbf{x}_S, \mathbf{x}_{-S} \rangle$. For simplicity, we write $f(x_i, \mathbf{x}_{-i}) = f(\langle x_i, \mathbf{x}_{-i} \rangle)$ and $f(\mathbf{x}_S, \mathbf{x}_{-S}) = f(\langle \mathbf{x}_S, \mathbf{x}_{-S} \rangle)$. In the following we formulate the definitions of strategy-proofness and the group strategy-proofness.

Definition 1. *A mechanism for the obnoxious facility game with service range is strategy-proof if no agent can benefit from misreporting her location. Formally, given agent i, profile $\mathbf{x} = \langle x_i, \mathbf{x}_{-i} \rangle \in I^n$, and a misreported location $x_i' \in I$, it holds that*

$$u(x_i, f(x_i, \mathbf{x}_{-i})) \geq u(x_i, f(x_i', \mathbf{x}_{-i})).$$

Definition 2. *A mechanism for the obnoxious facility game with service range is group strategy-proof if for any group of agents, at least one of them cannot benefit if they misreport simultaneously. Formally, given a non-empty set $S \subseteq N$, profile $\mathbf{x} = \langle \mathbf{x}_S, \mathbf{x}_{-S} \rangle \in I^n$, and the misreported location $\mathbf{x}_S' \in I^{|S|}$, there exists $i \in S$, satisfying*

$$u(x_i, f(\mathbf{x}_S, \mathbf{x}_{-S})) \geq u(x_i, f(\mathbf{x}_S', \mathbf{x}_{-S})).$$

In this paper we focus on the obnoxious facility game on interval $[0, 1]$ with service radius $\frac{1}{2} \leq r \leq 1$. Beforehand we should show the characterizations of the optimal solutions for the obnoxious facility problem on paths with service range in which the locations of all agents are public.

Proposition 1. *For the obnoxious facility problem on interval $[0, 1]$ with service radius $\frac{1}{2} \leq r \leq 1$, there is an optimal solution containing at most two facilities.*

Thus in the rest of the paper we only consider an optimal solution containing at most two facilities.

Proposition 2. *Suppose $\{y_1^*, y_2^*\}$ is an optimal solution for a location profile* **x**. *Then we have*

- *if $0 \le y_1^* \le y_2^* < 1 - r$, then $\{y_2^*\}$ is optimal;*
- *if $r < y_1^* \le y_2^* \le 1$, then $\{y_1^*\}$ is optimal;*
- *if y_1^* or $y_2^* \in [1 - r, r]$, then $\{y_1^*\}$ or $\{y_2^*\}$ is optimal.*

Proof. Here we only discuss the first case. The proofs for other cases are similar. Since $\{y_1^*, y_2^*\}$ is feasible for the location profile, any agent with $x_i \in (\frac{y_1^* + y_2^*}{2}, 1]$ should be served by the facility at y_2^* and $d(x_i, y_2^*) \le r$. On the other hand, because $0 \le y_1^* \le y_2^* < 1 - r$, it is obvious that $d(x_i, y_2^*) \le r$ if $x_i \in [0, \frac{y_1^* + y_2^*}{2}] \subseteq [0, 1 - r]$. Thus, we know that $d(x_i, y_2^*) \le r$, $i = 1, 2, \cdots, n$. So the facility at y_2^* can serve all of the agents on $[0, 1]$ and $\{y_2^*\}$ is a feasible solution. Thus

$$F_{\mathbf{x}}(y_1^*, y_2^*) = \sum_{x_i \in [0, \frac{y_1^* + y_2^*}{2}]} d(x, y_1^*) + \sum_{x_i \in (\frac{y_1^* + y_2^*}{2}, 1]} d(x_i, y_2^*) \le \sum_{x_i \in [0, 1]} d(x_i, y_2^*) = F_{\mathbf{x}}(y_2^*)$$

The inequality comes from the fact that for any agent in $[0, \frac{y_1^* + y_2^*}{2}]$, $d(x_i, y_1^*) \le d(x_i, y_2^*)$. Therefore solution $\{y_2^*\}$ is optimal too. □

Furthermore we arrive at the following theorem.

Theorem 1. *For the obnoxious facility problem on interval $[0, 1]$ with service radius $\frac{1}{2} \le r \le 1$, if there are two distinct facilities in an optimal solution at y_1^* and y_2^* with $y_1^* < y_2^*$, then $y_1^* \in [0, 1 - r)$ and $y_2^* \in (r, 1]$.*

3 Deterministic Mechanisms

In this section we propose a group strategy-proof deterministic mechanism for the obnoxious facility game on interval $[0, 1]$ with service radius $\frac{1}{2} \le r \le 1$ and explore the upper and lower bounds of the approximation ratio.

Mechanism 1. Given a location profile **x** on interval $[0, 1]$. Let n_1, n_2, n_3 and n_4 be the number of the agents on $[0, \frac{1}{4}]$, $(\frac{1}{4}, \frac{1}{2}]$, $(\frac{1}{2}, \frac{3}{4})$ and $[\frac{3}{4}, 1]$, respectively.

- For $\frac{1}{2} \le r < \frac{3}{4}$,
  if $n_1 + n_4 \ge n_2 + n_3$ then return $f(\mathbf{x}) = \{\frac{1}{2}\}$; otherwise return $f(\mathbf{x}) = \{0, 1\}$, i.e., pick two facilities at $y_1 = 0$ and $y_2 = 1$, respectively.
- For $\frac{3}{4} \le r \le 1$,
  if $n_1 + n_2 \ge n_3 + n_4$ then return $f(\mathbf{x}) = \{r\}$; otherwise return $f(\mathbf{x}) = \{1 - r\}$.

Note that when $\frac{1}{2} \le r < \frac{3}{4}$, it is possible to build two facilities. It is an obvious difference between the obnoxious facility game with and without service range. The following theorem shows that Mechanism 1 is group strategy-proof.

Theorem 2. *Mechanism 1 is group strategy-proof.*

Proof. It is easy to check that the outputs $\{\frac{1}{2}\}$ and $\{0,1\}$ of Mechanism 1 are both feasible for any agent location profile $\mathbf{x}$ on $[0,1]$. In order to prove the group strategy-proofness of Mechanism 1, let $S \subseteq N$ be a coalition. We must demonstrate that the agents in S cannot all gain strictly by lying. Suppose that any agent in S misreports her location from x_i to x_i'. Let n_1', n_2', n_3' and n_4' be the numbers of agents on $[0,\frac{1}{4}]$, $(\frac{1}{4},\frac{1}{2}]$, $(\frac{1}{2},\frac{3}{4})$ and $[\frac{3}{4},1]$ after deviating, respectively. Denote the new profile to be $\mathbf{x}'$. Following we discuss two cases depending on service radius r.

For $\frac{1}{2} \leq r < \frac{3}{4}$, w.l.o.g., we assume that $n_1 + n_4 \geq n_2 + n_3$. Hence Mechanism 1 outputs the facility location $f(\mathbf{x}) = \{\frac{1}{2}\}$. By Mechanism 1, if $n_1' + n_4' \geq n_2' + n_3'$, then $f(\mathbf{x}') = f(\mathbf{x}) = \{\frac{1}{2}\}$ and $u(x_i, f(\mathbf{x}')) = u(x_i, f(\mathbf{x}))$ for any agent i. But if $n_1' + n_4' < n_2' + n_3'$, then $f(\mathbf{x}) = \{0,1\}$. Under this case, we find that at least one agent with $x_i \in [0,\frac{1}{4}] \cup [\frac{3}{4},1]$ lies to $x_i' \in (\frac{1}{4},\frac{3}{4})$. It is obvious that $d(x_i,\{0,1\}) \leq \frac{1}{4} \leq d(x_i,\frac{1}{2})$, i.e. $u(x_i, f(\mathbf{x}')) \leq u(x_i, f(\mathbf{x}))$. For $\frac{3}{4} \leq r \leq 1$, suppose that $n_1 + n_2 \geq n_3 + n_4$ without loss of generality. So the facility location $f(\mathbf{x}) = \{r\}$ is returned by Mechanism 1. Obviously if $n_1' + n_2' \geq n_3' + n_4'$, then $f(\mathbf{x}') = f(\mathbf{x}) = \{r\}$ and $u(x_i, f(\mathbf{x}')) = u(x_i, f(\mathbf{x}))$ for any agent i. But if $n_1' + n_2' < n_3' + n_4'$ then we have $f(\mathbf{x}') = \{1-r\}$ by Mechanism 1. In this instance, there must be at least one agent in $[0,\frac{1}{2}]$ lying her location to $x_i' \in (\frac{1}{2},1]$. And it is easy to know that

$$u(x_i, f(\mathbf{x}')) = d(1-r, x_i) \leq d(1-r, \frac{1}{2}) + d(\frac{1}{2}, x_i)$$

$$= d(r, \frac{1}{2}) + d(\frac{1}{2}, x_i) = d(r, x_i) = u(x_i, f(\mathbf{x})).$$

The inequality comes from the triangle inequality directly.   □

The following theorem provides the approximation ratio of Mechanism 1.

Theorem 3. *The approximation ratio of Mechanism 1 is*

$$\gamma(r) = \begin{cases} 8r - 1 & \text{if } \frac{1}{2} \leq r < \frac{3}{4}, \\ \frac{r+\frac{1}{2}}{r-\frac{1}{2}} & \text{if } \frac{3}{4} \leq r \leq 1. \end{cases} \tag{1}$$

The proof for the approximation ratio of Mechanism 1 is a little complicated. The main idea is following. Let the optimal solution be $\{y_1^*, y_2^*\}$ with $y_1^* \leq y_2^*$. We should discuss different cases which are obtained by distinguishing all possible locations of y_1^* and y_2^* and by the nice properties shown in Proposition 1, 2 and Theorem 1. For each case, we compute the ratio between the social welfare of the optimal solution and the social welfare of the mechanism's solution under the worst instance and obtain the final result shown in Theorem 3. Specifically, when the service radius $\frac{1}{2} \leq r < \frac{3}{4}$, there are five distinct cases should be discussed: i) $y_1^* = y_2^* = y^* \in [0,\frac{1}{4}]$; ii) $y_1^* = y_2^* = y^* \in [\frac{1}{4},\frac{1}{2}]$; iii) $y_1^* \in [0,\frac{1}{4}]$ and $y_2^* \in [\frac{1}{2},\frac{3}{4}]$; iv) $y_1^* \in [0,\frac{1}{4}]$ and $y_2^* \in [\frac{3}{4},1]$; v) $y_1^* \in [\frac{1}{4},\frac{1}{2}]$ and $y_2^* \in [\frac{1}{2},\frac{3}{4}]$. When the service radius $\frac{3}{4} \leq r \leq 1$, there are three different cases: i) $y_1^* = y_2^* = y^* \in [0,\frac{1}{2}]$; ii) $y_1^* = y_2^* = y^* \in [\frac{1}{2},1]$ and iii) $y_1^* \in [0,1-r)$, $y_2^* \in (r,1]$.

It is clear to check that the approximation ratio shown in (1) is tight for Mechanism 1. When $\frac{1}{2} \le r < \frac{3}{4}$, we consider an instance that there are $\frac{n}{2}$ agents at $\frac{1}{2}$ and $\frac{n}{2}$ agents at $\frac{3}{4}$. Under this case, $\text{OPT}(\mathbf{x}) = (2r - \frac{1}{4})\frac{n}{2}$ where an optimal solution is $y^* = \frac{3}{4} - r \in [0, \frac{1}{4}]$. And by Mechanism 1, the output is $\frac{1}{2}$. Thus the obnoxious social welfare $SW(f, \mathbf{x}) = \frac{1}{8}n = \frac{1}{(8r-1)}\text{OPT}(\mathbf{x})$. When $\frac{3}{4} \le r \le 1$, another instance is studied that there are $\frac{n}{2}$ agents at $\frac{1}{2}$ and $\frac{n}{2}$ agents at r. Obviously the optimal solution is $y^* = 0$ and $\text{OPT}(\mathbf{x}) = (r + \frac{1}{2})\frac{n}{2}$. But the output of Mechanism 1 is $f(\mathbf{x}) = r$ with obnoxious social welfare $SW(f, \mathbf{x}) = (r - \frac{1}{2})\frac{n}{2} = \frac{r - \frac{1}{2}}{r + \frac{1}{2}}\text{OPT}(\mathbf{x})$. We also note that the approximation ratio γ of Mechanism 1 monotonously increases on service radius r when $\frac{1}{2} \le r < \frac{3}{4}$ and decreases when $\frac{3}{4} \le r \le 1$. And γ reaches its maximum value 5 when $r = \frac{3}{4}$ and its minimum value 3 when $r = \frac{1}{2}$ or 1. This result just matches the approximation ratio 3 of Mechanism 1 in [3] in which there is no restriction of service range.

In order to explore the lower bound of any deterministic strategy-proof mechanism, the following proposition is necessary to be stated.

Proposition 3. *Given a profile* $\mathbf{x}$*, facility locations* y_1 *and* y_2 *($y_1 \le y_2$) are returned by a deterministic mechanism* f *for the obnoxious facility game with bounded service radius* r *on a path. If* y_i*,* $i = 1$ *or* 2*, satisfies* $d(y_i, x_j) \le r$ *for any* $j \in \{1, 2, \cdots, n\}$*, then the obnoxious social welfare* $SW(f, \mathbf{x}) \le \sum_{j=1}^{n} d(y_i, x_j)$*.*

Recall that the approximation ratio of deterministic strategy-proof mechanisms is at least 3 for the setting without bounded service range [7]. It is obvious that if $r = 1$, then this case is equivalent to that without the service range which implies the lower bound of the approximation ratio is 3 if $r = 1$. So in the following, we first try our best to compute the lower bounds of approximation ratio for any strategy-proof deterministic mechanism if $\frac{1}{2} \le r < 1$. Similar to the upper bound of approximation ratio shown in (1), the lower bounds depend on the value of r are proposed in Theorem 4.

Theorem 4. *In the obnoxious facility game on* $[0, 1]$ *with bounded service radius* $\frac{1}{2} \le r < 1$*, any strategy-proof deterministic mechanism* f *has approximation ratio of at least*

$$\gamma(r) \ge \begin{cases} 4r - 1 & \text{if } \frac{1}{2} \le r < \frac{3}{4}, \\ \frac{1}{2r-1} & \text{if } \frac{3}{4} \le r < \frac{5}{6}, \\ 3r - 1 & \text{if } \frac{5}{6} \le r < 1. \end{cases} \tag{2}$$

for the obnoxious social welfare.

In order to prove Theorem 4, we should construct different instances based on the value of the service radius r. For each instance if only one agent misreports her location, one possible facility location can be determined by the strategy-proofness. Furthermore, we can compute the upper bound of the obnoxious social welfare by Proposition 3 and obtain the lower bound of the approximation ratios shown in (2).

4 Randomized Mechanisms

In this section we propose a randomized mechanism for the obnoxious facility game with service radius $\frac{1}{2} \le r \le 1$ on a path and explore its group strategy-proofness and approximation ratio.

Mechanism 2. Given a location profile $\mathbf{x}$ on interval $[0, 1]$. Let n_1, n_2, n_3 and n_4 be the number of the agents on $[0, \frac{1}{4}]$, $(\frac{1}{4}, \frac{1}{2}]$, $(\frac{1}{2}, \frac{3}{4})$ and $[\frac{3}{4}, 1]$ respectively.
     When $\frac{3}{4} \le r \le 1$,

- if $n_1 + n_2 > n_3 + n_4$, then return $f(\mathbf{x}) = \{1 - r\}$ and $f(\mathbf{x}) = \{r\}$ with probability $\frac{2}{5}$ and $\frac{3}{5}$, respectively.
- if $n_1 + n_2 = n_3 + n_4$, then return $f(\mathbf{x}) = \{1 - r\}$ and $f(\mathbf{x}) = \{r\}$ with probability $\frac{1}{2}$, respectively.
- if $n_1 + n_2 < n_3 + n_4$, then return $f(\mathbf{x}) = \{1 - r\}$ and $f(\mathbf{x}) = \{r\}$ with probability $\frac{3}{5}$ and $\frac{2}{5}$, respectively.

When $\frac{1}{2} \le r < \frac{3}{4}$, then return $f(\mathbf{x}) = \{\frac{1}{2}\}$ and $f(\mathbf{x}) = \{0, 1\}$ with probability $\frac{1}{2}$ respectively.

The following theorem shows that Mechanism 2 is group strategy-proof and provides the approximation ratio.

Theorem 5. *Mechanism 2 is a group strategy-proof mechanism for the obnoxious facility game with service radius $\frac{1}{2} \le r \le 1$ and its approximation ratio is*

$$\gamma(r) = \begin{cases} 4r & \text{if } \frac{1}{2} \le r < \frac{3}{4}, \\ \frac{r}{r - \frac{1}{2}} & \text{if } \frac{3}{4} \le r \le 1. \end{cases} \tag{3}$$

The proof for the group strategy-proofness of Mechanism 2 is similar to Theorem 2. For the approximation ratio, the key of the proof is how to upper bound the social welfare of $\text{OPT}(\mathbf{x})$ appropriately. Specifically, when $\frac{3}{4} \le r \le 1$, $\text{OPT}(\mathbf{x}) \le r(n_1 + n_2) + \frac{1}{2}(n_3 + n_4)$ if $n_1 + n_2 \ge n_3 + n_4$; Otherwise $\text{OPT}(\mathbf{x}) \le \frac{1}{2}(n_1 + n_2) + r(n_3 + n_4)$. When $\frac{1}{2} \le r < \frac{3}{4}$, we use the upper bound $\text{OPT}(\mathbf{x}) \le rn$.

5 Concluding Remarks

This paper is the first one to study the obnoxious facility game with service range on an interval which has much more real significance. Our goal is to design deterministic and randomized group strategy-proof mechanisms with small approximation ratios and explore the lower bound of approximation ratio for any strategy-proof mechanism. There are a lot of interesting open problems. One of them is a truly intriguing gap between our deterministic upper bound shown in (1) and the lower bound shown in (2). Moreover it is not clear about the result of the lower bound on the approximation ratio for any strategy-proof randomized mechanism.

The other one is to study the obnoxious facility game problem when the service radius $0 < r < \frac{1}{2}$. Note that we need as more facilities to serve all the agents as r is smaller. Hence much more cases should be discussed and such a problem becomes more complicated. We strive to find a general rule corresponding to the value of r.

References

1. Alon, N., Feldman, M., Procaccia, A., Tennenholtz, M.: Strategyproof Approximation Mechanisms for Location on Networks, Computing Research Repository-CORR, abs/0907.2049 (2009)
2. Alon, N., Feldman, M., Procaccia, A., Tennenholtz, M.: Strategyproof Approximation of the Minimax on Networks. Mathematics of Operations Research 35, 513–526 (2010)
3. Cheng, Y., Yu, W., Zhang, G.: Strategy-proof Approximation Mechanisms for an Obnoxious Facility Game on Networks. Theoretical Computer Science (2011), doi: 10.1016/j.tcs.2011.11.041
4. Escoffier, B., Gourvès, L., Thang, N.K., Pascual, F., Spanjaard, O.: Strategy-Proof Mechanisms for Facility Location Games with Many Facilities. In: Brafman, R. (ed.) ADT 2011. LNCS, vol. 6992, pp. 67–81. Springer, Heidelberg (2011)
5. Fotakis, D., Tzamos, C.: Winner-imposing Strategyproof Mechanisms for Multiple Facility Location Games. In: Saberi, A. (ed.) WINE 2010. LNCS, vol. 6484, pp. 234–245. Springer, Heidelberg (2010)
6. Ibara, K., Nagamochi, H.: Characterizing Mechanisms in Obnoxious Facility Game. In: Lin, G. (ed.) COCOA 2012. LNCS, vol. 7402, pp. 301–311. Springer, Heidelberg (2012)
7. Han, Q., Du, D.: Moneyless Strategy-proof Mechanism on Single-sinked Policy Domain: Characerization and Applications (2012) (working paper)
8. Lu, P., Sun, X., Wang, Y., Zhu, Z.: Asymptotically Optimal Strategy-proof Mechanisms for Two-facility Games. In: 11th ACM Conference on Electronic Commerce, pp. 315–324. ACM, New York (2010)
9. Lu, P., Wang, Y., Zhou, Y.: Tighter bounds for facility games. In: Leonardi, S. (ed.) WINE 2009. LNCS, vol. 5929, pp. 137–148. Springer, Heidelberg (2009)
10. Moulin, H.: On strategy-proofness and single peakedness. Public Choice 35(4), 437–455 (1980)
11. Procaccia, A., Tennenholtz, M.: Approximate mechanism design without money. In: 10th ACM Conference on Electronic Commerce (ACM-EC), pp. 177–186. ACM, New York (2009)
12. Schummer, J., Vohra, R.V.: Mechanism design without money. In: Nisan, N., Roughgarden, T., Tardos, E., Vazirani, V. (eds.) Algorithmic Game Theory, ch. 10. Cambridge University Press (2007)

Efficient Self-pairing on Ordinary Elliptic Curves

Hongfeng Wu[1] and Rongquan Feng[2]

[1] College of Sciences, North China University of Technology, Beijing 100144, China
[2] LMAM, School of Mathematical Sciences, Peking University, Beijing 100871, China
whfmath@gmail.com, fengrq@math.pku.edu.cn

Abstract. A self-pairing $e(P, P)$ is a special bilinear pairing where both points are equal. Self-pairings are used in some cryptographic schemes and protocols, such as ZSS shorter signatures and so on. In this paper, We first generalize a result in [28] to any elliptic curve with more simpler final exponentiation. Then we present a new self-pairing on ordinary elliptic curves with short loop length in Miller's algorithm. We also provide examples of self-pairing friendly elliptic curves which are of interest for efficient pairing implementations. Finally, we present explicit formulae for Miller's algorithm to compute self-pairing on ordinary elliptic curves with embedding degree one.

Keywords: Elliptic curve, Self-pairing, Tate pairing, Weil pairing, Pairing based cryptography.

1 Introduction

Pairing-based cryptographic applications have received much attention and were developed rapidly. In order to make these applications practical, pairing computations need to be efficiently carried out. It leads to fast developments of algorithmic foundations of pairings. For this purpose, several efficient pairings such as Weil pairing [3,4,18], Tate pairing [3,4,8,15], Ate pairing [11], twisted Ate pairing [16], optimal pairing [24], self-pairing [29], and pairing lattices [12] etc., have been proposed. Many efficient techniques which speed up pairing computations have been presented, such as shortening the loop length in Miller's algorithm ([7,11,12,15,16,28]), or speeding up the basic doubling and addition steps in Miller's algorithm ([1,2,6]), etc.. Some surveys of pairing computations can be found in [3,4,9].

Some pairings with specific properties are often required in cryptographic applications. A self-pairing $e(P, P)$ is a special bilinear pairing where both points are equal. Self-pairings are used in some cryptographic schemes and protocols, such as short signatures [25,26], ID-based Chameleon hashing schemes [27], and on-line/off-line signature schemes [25], etc.. There are only a few studies for computing self-pairings [19,29]. It is well known that for the Weil pairing $e(P, P) = 1$ for any P. For cryptographic applications, we need to let the latter P map to another independent point for keeping non-degeneracy. Note that the distortion map exists only on supersingular curves and ordinary curves with embedding

T-H.H. Chan, L.C. Lau, and L. Trevisan (Eds.): TAMC 2013, LNCS 7876, pp. 282–293, 2013.

degree one ([14,23]). To implement self-pairing based protocols in practice, it is necessary to match curves which admit self-pairing with an efficient pairing computation algorithm. In this paper, we only consider the self-pairing computation on ordinary curves with embedding degree one.

It was proposed in [28] a super-optimal pairing based automorphism with great efficiency on two families of elliptic curves $y^2 = x^3 + a$ and $y^2 = x^3 + ax$. In this paper, we first generalize the result in [28] to any elliptic curves with more simpler final exponentiation. Then we present a new self-pairing on ordinary curves with embedding degree one. It has more shorter loop length in Miller's algorithm than in [29]. Finally, We apply the new self-pairing to two families of elliptic curves and give the explicit self-paring computation algorithms.

The paper is organized as follows. In Section 2, we provide some background and notations used through this paper. In Section 3, we present a new self-pairing on ordinary elliptic curves with embedding degree one. In Section 4, we apply the new self-pairing to two examples. We draw our conclusion in Section 5. In this paper, the cost of a field inversion, a multiplication, and a squaring are denoted by I, M, and S respectively.

2   Preliminaries

A brief background on pairings is given in this section.

2.1   Tate and Weil Pairing

Let $\mathbb{F}_q$ be a finite field with $q = p^m$ elements, where $p > 3$ is a prime, and E an elliptic curve defined over $\mathbb{F}_q$ with neutral element denoted by O. Let r be a large prime divisor of the group order $\sharp E(\mathbb{F}_q)$ with $\gcd(r, q) = 1$. Let $k > 0$ denote the embedding degree with respect to r, that is, k is the smallest integer such that $r \mid q^k - 1$. For technical reasons we assume that r^2 does not divide $q^k - 1$. We denote by $E[r]$ the r-torsion group of E.

Let $t \in \mathbb{F}_q(E)$ be a fixed local parameter at infinity O. We say that $f \in \mathbb{F}_{q^k}(E)$ is monic [12] if $ft^{-v}(O) = 1$, where v is the order of f at O. In other words this says that the Laurent series expansion of f in terms of t is of the form $f = t^v + O(t^{v+1})$. We will consider monic functions f throughout the paper without further mentioning.

For any point $P \in E(\mathbb{F}_q)[n]$, let $f_{i,P}$ denote a rational function on E with divisor $div(f_{i,P}) = i(P) - (iP) - (i-1)(O)$, and let D_P be a degree zero divisor which is linearly equivalent to $(P) - (O)$, then $div(f_{i,P}) = rD_P$. Assume that μ_r is the set of r-th roots of unit in $\mathbb{F}_{q^k}$. The reduced Tate pairing [8] is defined as follows:

$$T_r : E(\mathbb{F}_q)[r] \times E(\mathbb{F}_{q^k}) \to \mu_r, \quad T_r(P, Q) = f_{r,P}(Q)^{(q^k-1)/r}.$$

Note that $f_{r,P}(Q)^{a(q^k-1)/r} = f_{ar,P}(Q)^{(q^k-1)/r}$ for any integer a. The rational function $f_{r,P}$ can be computed in polynomial time by using Miller's algorithm

([17,18]). Let $r = (r_{l-1}, \cdots, r_1, r_0)_2$ be the binary representation of r, where $r_{l-1} = 1$. Let $l_{R,P}$ and v_R be the rational functions with divisors $div(l_{R,P}) = (R)+(P)+(R+P)-3(O)$ and $div(v_R) = (R)+(-R)-2(O)$, respectively. The Miller's algorithm ([17,18]) for computation $f_{r,P}(Q)$ is as follows:

Algorithm 1. Miller's algorithm

Input: $r = \sum_{i=0}^{l-1} r_i 2^i$, where $r_i \in \{0,1\}$. $P, Q \in E$.
Output: $f_r(Q)$

1: $f \leftarrow 1, R \leftarrow P$
2: **for** $i = l - 2$ down to 0 **do**
3:      $f \leftarrow f^2 \cdot \frac{l_{R,R}(Q)}{v_{2R}(Q)}, \ R \leftarrow [2]R$
4:      **if** $r_i = 1$ **then**
5:         $f \leftarrow f \cdot \frac{l_{R,P}(Q)}{v_{R+P}(Q)}, \ R \leftarrow R + P$
6:      **end if**
7: **end for**
8: **return** f

Suppose that $P, Q \in E[r]$ and $P \neq Q$. Then the Weil pairing [18] is

$$e_r : \ E[r] \times E[r] \to \mu_r, \quad e_r(P,Q) = (-1)^r f_{r,P}(Q)/f_{r,Q}(P).$$

Since r is an odd prime we always have $(-1)^r = -1$. For $k \mid \sharp Aut(E)$ and $s \equiv q$ mod r, the Weil pairing with ate reduction ([12]) with respect to s is given by

$$e_s : \ E[r] \times E[r] \to \mu_r, \quad e_s(P,Q) = w f_{s,P}(Q)/f_{s,Q}(P)$$

for some suitable k-th root of unity $w \in \mathbb{F}_q$.

2.2 Weil Pairing with Automorphism

In [12], an extended Weil pairing with an automorphism is presented. Let s be a primitive n-th root of unity modulo r with $n \mid \text{lcm}(k, \sharp Aut(E))$. Let $u = sq^{-d}$ mod r be some primitive e-th root of unity modulo r with $e \mid \gcd(n, \sharp Aut(E))$ and $d \geq 0$. Define $v = s^{-1} q^d = u^{-1}$ mod r. Let π_q be the Frobenius endomorphism, i.e. $\pi_q : E \to E : (x,y) \mapsto (x^q, y^q)$. If $k > 1$, then $E(\mathbb{F}_{q^k})[r] = \mathbb{Z}/r\mathbb{Z} \times \mathbb{Z}/r\mathbb{Z}$ and there exists a basis P, Q of $E(\mathbb{F}_{q^k})[r]$ satisfying $\pi_q(P) = P$ and $\pi_q(Q) = qQ$. Since the natural map $Aut(E) \to Aut(E[r])$ is injective, there must exist $\alpha \in Aut(E)$ of order e with $\alpha(Q) = uQ$ and $\alpha(P) = vP$. We define $G_1 = \langle P \rangle$ and $G_2 = \langle Q \rangle$. Suppose $n \mid \sharp Aut(E)$, then there is an n-th root of unity $w \in \mathbb{F}_q$ such that

$$e_s : G_1 \times G_2 \to \mu_r,$$

$$(P,Q) \mapsto \prod_{j=0}^{e-1} (-w f_{s,P}(\alpha^j(Q))/f_{s,\alpha^j(Q)}(P))^{v^j}$$

defines a bilinear pairing. Since w is an n-th root of unity of $\mathbb{F}_q$, one can define the powered pairing with final exponentiation $\theta(n)$, where $\theta(n) = n$ if n is even and $\theta(n) = 2n$ if n is odd. That is

$$e_s^{\theta(n)} = (\prod_{j=0}^{e-1} (f_{s,P}(\alpha^j(Q))/f_{s,\alpha^j(Q)}(P))^{v^j})^{\theta(n)}$$

also defines a bilinear pairing. The pairing e_s and $e_s^{\theta(n)}$ are non-degenerate if and only if $s^n \not\equiv 1 \mod r^2$ holds. Note that when r is an odd composite number such that $\gcd(r, q) = 1$, the above conclusion is also correct by the proof in [12].

Let p be a large prime. Consider elliptic curves $E_1 : y^2 = x^3 + B$, where $p \equiv 1 \mod 3$, and $E_2 : y^2 = x^3 + Ax$, where $p \equiv 1 \mod 4$ over $\mathbb{F}_p$. In [28], Zhao etc. proposed a super-optimal omega pairing for above families of elliptic curves with nontrivial automorphisms. Let ϕ be a nontrivial automorphism over E_1 or E_2. For the points $P, Q \in E[r]$, suppose that $\phi(P) = \lambda P$, then the function $\omega(P, Q) = (f_{\lambda,P}(Q)/f_{\lambda,Q}(P))^{p-1}$ defines a non-degenerate bilinear pairing. Actually, we can extend the omega pairing to any elliptic curve with nontrivial automorphisms.

2.3 Self Pairing

In practical implementations, the self-pairing $e(P, P)$ can be designed by *Type 1* pairings ([9]), i.e., it can be constructed on supersingular elliptic curves with even embedding degrees. Let E be the supersingular curves with distortion map ϕ over the ground field $\mathbb{F}_q$ as given in Table 1 of [29], and let r be a large prime dividing the order of $E(\mathbb{F}_q)$. The embedding degree with respect to r is equal to k. Take $P \in Ker(\pi_q - [1]) \cap E[r]$, the self-pairing based on the Weil pairing can be given by

$$e_s(P, P) = f_{r,P}(\phi(P))^{4(q^{k/2}-1)}.$$

Since the distortion maps also exist for ordinary elliptic curves with embedding degree one, a new self-pairing was proposed in [29] on ordinary curves. Koblitz and Menezes first gave the concrete construction of ordinary curves with embedding degree one and analyzed the efficiency of pairing computations on these curves ([14]). Assume that the prime $p = A^2 + 1$. The equation of the elliptic curve E_5 over $\mathbb{F}_p$ is defined by $E_5 : y^2 = x^3 + ax$, where $a = -1$ or $a = -4$ response to $A \equiv 0 \mod 4$ or $A \equiv 2 \mod 4$. The order of $E_5(\mathbb{F}_p)$ is $\sharp E_5(\mathbb{F}_p) = p-1$. the map $\phi : (x, y) \rightarrow (-x, Ay)$ is a distortion map on E_5. Let $P \in E_5(\mathbb{F}_p)$ have prime order r, the self-pairing in [29] is defined as $e_s(P, P) = f_{r,P}(\phi(P))^4$.

3 Self Pairing Functions of Lower Degrees

The next theorem extends Theorems 1 and 2 in [28].

Theorem 1. *Let E be an ordinary elliptic curve over a finite field $\mathbb{F}_q$, r a prime factor of $E(\mathbb{F}_q)$ with $r \mid q - 1$, and let n be the order of λ modulo r*

with $n \mid \sharp Aut(E)$. Let $\varphi \in Aut(E)$ and G_1 and G_2 be two eigenspaces of φ on $E[r]$ such that $\varphi(Q) = \lambda Q$ for any $Q \in G_2$ and $\varphi(P) = \lambda^{-1}P$ for any $P \in G_1$. Set $\theta(n) = n$ if n is even, and $\theta(n) = 2n$ if n is odd. Then

$$\hat{e}_\lambda : G_1 \times G_2 \to \mu_r, \quad (P, Q) \mapsto (f_{\lambda,P}(Q)/f_{\lambda,Q}(P))^{\theta(n)}$$

defines a bilinear pairing. The pairing $\hat{e}_\lambda$ is non-degenerate if and only if $s^n \not\equiv 1$ mod r^2 holds.

Proof. Let $\kappa \equiv \lambda^{-1} \mod r$ and $e_\lambda = \prod_{j=0}^{n-1} \left(\frac{f_{\lambda,P}(\varphi^j(Q))}{f_{\lambda,\varphi^j(Q)}(P)} \right)^{\kappa^j \theta(n)}$. Then e_λ is a bilinear pairing from Section 2.2. Let ψ be the dual isogeny of φ, then $\psi(P) = \lambda P$ for any $P \in G_1$ and $\psi(Q) = \lambda^{-1}Q$ for any $Q \in G_2$. Since

$$div(f_{\lambda,\lambda Q} \circ \varphi) = \varphi^*(div(f_{\lambda,\lambda Q})) = \varphi^*(\lambda(\lambda Q) - (\lambda^2 Q) - (\lambda - 1)(O))$$
$$= \lambda(Q) - (\lambda Q) - (\lambda - 1)(O) = div(f_{\lambda,Q}),$$

we have

$$f_{\lambda,\lambda Q}(\lambda Q)^{\theta(n)} = f_{\lambda,Q}(Q)^{\theta(n)} \quad \text{and} \quad f_{\lambda,\lambda Q}(\lambda^{-1}P)^{\theta(n)} = f_{\lambda,Q}(P)^{\theta(n)}.$$

Similarly,

$$f_{\lambda,\lambda P}(\lambda P)^{\theta(n)} = f_{\lambda,P}(P)^{\theta(n)} \quad \text{and} \quad f_{\lambda,\lambda P}(\lambda^{-1}Q)^{\theta(n)} = f_{\lambda,P}(Q)^{\theta(n)}.$$

Moreover, since all rational functions in this paper are assumed to be chosen as monic functions, $f_{\lambda,P}((\lambda - 1)O) = 1$. Thus

$$\left(f_{\lambda,P}(\lambda Q)f_{\lambda,P}(Q)^{-\lambda}\right)^{\theta(n)} = \left(f_{\lambda,P}(\lambda Q)f_{\lambda,P}(Q)^{-\lambda}f_{\lambda,P}((\lambda - 1)O)\right)^{\theta(n)}$$
$$= \left(f_{\lambda,P}(-div(f_{\lambda,Q}))\right)^{\theta(n)} = \left(f_{\lambda,Q}(-div(f_{\lambda,P}))\right)^{\theta(n)} = \left(f_{\lambda,Q}(\lambda P)f_{\lambda,Q}(P)^{-\lambda}\right)^{\theta(n)}.$$

That is, $\left(\frac{f_{\lambda,P}(\lambda Q)}{f_{\lambda,Q}(\lambda P)} \right)^{\theta(n)} = \left(\frac{f_{\lambda,P}(Q)}{f_{\lambda,Q}(P)} \right)^{\lambda \theta(n)}$. Therefore,

$$\left(\frac{f_{\lambda,P}(\lambda^j Q)}{f_{\lambda,Q}(\lambda^j P)} \right)^{\theta(n)} = \left(\frac{f_{\lambda,\lambda P}(\lambda^{j-1}Q)}{f_{\lambda,Q}(\lambda^j P)} \right)^{\theta(n)} = \left(\frac{f_{\lambda,\lambda P}(Q)}{f_{\lambda,Q}(\lambda P)} \right)^{\lambda^{j-1}\theta(n)}$$

$$= \left(\frac{f_{\lambda,P}(\lambda Q)}{f_{\lambda,Q}(\lambda P)} \right)^{\lambda^{j-1}\theta(n)} = \left(\frac{f_{\lambda,P}(Q)}{f_{\lambda,Q}(P)} \right)^{\lambda^j \theta(n)}.$$

Hence, one gets

$$e_\lambda = \prod_{j=0}^{n-1} \left(\frac{f_{\lambda,P}(\lambda^j Q)}{f_{\lambda,\lambda^j Q}(P)} \right)^{\kappa^j \theta(n)} = \prod_{j=0}^{n-1} \left(\frac{f_{\lambda,P}(Q)}{f_{\lambda,Q}(P)} \right)^{(\lambda\kappa)^j \theta(n)} = \left(\frac{f_{\lambda,P}(Q)}{f_{\lambda,Q}(P)} \right)^{\theta(n)\sum_{j=0}^{n-1}(\lambda\kappa)^j}.$$

Note that $\lambda\kappa \equiv 1 \mod r$, we have $\sum_{j=0}^{n-1}(\lambda\kappa)^j \equiv n \mod r$. From Weil's reciprocity law, we know that

$$\left(\frac{f_{\lambda,P}(Q)}{f_{\lambda,Q}(P)}\right)^{r\theta(n)} = \left(\frac{f_{\lambda,P}(div(f_{r,Q}))}{f_{\lambda,Q}(div(f_{r,P}))}\right)^{\theta(n)} = \left(\frac{f_{r,Q}(div(f_{\lambda,P}))}{f_{r,P}(div(f_{\lambda,Q}))}\right)^{\theta(n)}$$

$$= \left(\frac{f_{r,Q}(P)^\lambda f_{r,P}(\lambda Q)}{f_{r,P}(Q)^\lambda f_{r,Q}(\lambda P)}\right)^{\theta(n)} = 1.$$

Therefore,

$$e_\lambda(P,Q)^{(n^{-1} \mod r)} = \left(\frac{f_{\lambda,P}(Q)}{f_{\lambda,Q}(P)}\right)^{\theta(n)} = \hat{e}_\lambda.$$

Thus $\hat{e}_\lambda$ defines a bilinear pairing. From Section 2.2, the pairing $\hat{e}_\lambda$ is non-degenerate if and only if $s^n \not\equiv 1 \mod r^2$ holds.

From the above theorem, taking $Q = \phi(P)$ with ϕ a distortion map, we get the following theorem.

Theorem 2. *Using the same notations as in Theorem 1, and let ϕ be a distortion map on an ordinary elliptic curve E. Then the self-pairing based on the automorphism φ can be given by*

$$e_{self}(P,P) \triangleq \hat{e}_s(P,\phi(P)) = (f_{\lambda,P}(\phi(P))/f_{\lambda,\phi(P)}(P))^{\theta(n)}.$$

4 Applications

4.1 Self-pairing on $y^2 = x^3 + ax$

Let $p = A^2 + 1$ be a prime. Let $E_5 : y^2 = x^3 + ax$ be an elliptic curve defined over $\mathbb{F}_p$, where $a = -1$ if $A \equiv 0 \mod 4$, or $a = -4$ if $A \equiv 2 \mod 4$. Note that E_5 is an ordinary elliptic curve and $\phi : (x,y) \to (-x, Ay)$ is a distortion map on E_5. Furthermore, $\sharp Aut(E) = 4$ ([21]). Assume that $\sigma \in \mathbb{F}_p$ is an element of order 4. Let the automorphism φ be given by $(x,y) \to (-x, \sigma y)$. Let $P \in E_5(\mathbb{F}_p)$ has prime order r, and λ be the root of equation $x^2 + 1 = 0 \mod r$ such that $\varphi(P) = \lambda^{-1}P$. Then the order of λ modulo r is 4. The self-pairing based on the automorphism φ is

$$e_{self}(P,P) = (f_{\lambda,P}(\phi(P))/f_{\lambda,\phi(P)}(P))^4$$

by Theorem 2. Note that the self-pairing in [29] is $(f_{r,P}(\phi(P))/f_{r,\phi(P)}(P))^4$. Hence, the proposed self-pairing has more shorter loop length in Miller's algorithm.

In Miller's algorithm, $f_{\lambda,P}(\phi(P)$ and $f_{\lambda,\phi(P)}(P)$ are computed simultaneously in each iteration step, and do only one final quotient $f_{\lambda,P}(\phi(P))/f_{\lambda,\phi(P)}(P)$.

Noting that $Q \in \{P, \phi(P)\} \subset E(\mathbb{F}_p)$, the self-pairing can be combined with windowing methods by replacing the computation in step (5) in Miller's algorithm by

$$f \leftarrow f \cdot f_{c,P}(Q) \cdot \frac{l_{R,cP}(Q)}{v_{R+cP}(Q)}, \quad R \leftarrow R + [c]P,$$

where the current window in the binary representation of n corresponds to the value c. The Miller function $f_{c,P}$ is defined via $div(f_{c,P}) = c(P) - ([c]P) - (c-1)(O)$.

In [5], a variant of Miller's algorithm which gives rise to a generically faster algorithm for any pairing friendly curve was given. We can apply this variant of Miller's algorithm to self-pairing computation. In implementations, the variant algorithm saves between 10% in running time in comparison with the usual version of Miller's algorithm.

4.2   Self-pairing on $y^2 = x^3 + b$

Let both $r \equiv 2 \mod 3$ and $p = r^2 + r + 1$ be prime, thus $p \equiv 1 \mod 3$. Choose $\beta \in \mathbb{F}_p$ such that the polynomial $X^6 - \beta$ is irreducible in $\mathbb{F}_p[X]$. Set $b = \beta$ or $b = \beta^5$, then $E_b : y^2 = x^3 + b$ is an ordinary elliptic curve over $\mathbb{F}_p$, $E_b(\mathbb{F}_p) = \mathbb{Z}/r\mathbb{Z} \times \mathbb{Z}/r\mathbb{Z}$ and $\sharp E_b(\mathbb{F}_p) = r^2$. Moreover, $\sharp Aut(E) = 6$. The map $\phi : (x, y) \rightarrow (rx, y)$ is a distortion map on $E_b(\mathbb{F}_p)$. Here is a curve with 256-bits security level:

$$r = 2^{512} + 436711, \quad E_b/\mathbb{F}_p : y^2 = x^3 + 29.$$

The more details can be found in [13].

Let $P \in E_b(\mathbb{F}_p)$ has prime order r, and ρ be an element of order 3 in $\mathbb{F}_p$. Set automorphism φ be given by $(x, y) \rightarrow (\rho x, y)$. Assume that λ is a root of $x^2 + x + 1 = 0 \mod r$ such that $\varphi(P) = \lambda^{-1}P$. Then the order of λ modulo r is 3. The self-pairing based on the automorphism φ is

$$e_{self}(P, P) = (f_{\lambda,P}(\phi(P))/f_{\lambda,\phi(P)}(P))^6$$

by Theorem 2.

Doubling Step. For $P, T \in E_b(\mathbb{F}_p)$, let $l_{P,T}$ denote the line through P and T, and let v_{P+T} denote the line through $P + T$ and $-(R + T)$. In the case of the doubling step of the self-pairings, after initially setting $T = P, f_1 = f_2 = 1$, for each bit of λ we do

$$f_1 \leftarrow f_1^2 \frac{l_{T,T}(\phi(P))}{v_{2T}(\phi(P))},$$
$$f_2 \leftarrow f_2^2 \frac{l_{\phi(T),\phi(T)}(P)}{v_{\phi(2T)}(P)},$$
$$T \leftarrow 2T.$$

Let λ_T be the slope of the tangent line through the point T, then the slope of the tangent line through the point $\phi(T)$ is $r^2 \lambda_T$. Note that $r^3 = 1$ and $r^2 = -r - 1$, it follows that

$$l_{T,T}(\phi(P)) = (y_P - y_T) - \lambda_T(rx_P - x_T),$$
$$l_{\phi(T),\phi(T)}(P) = (y_P - y_T) - r^2 \lambda_T(x_P - rx_T)$$
$$= (y_P - y_T) - \lambda_T(-rx_P - x_P - x_T),$$

and

$$v_{2T}(\phi(P)) = rx_P - x_{2T},$$
$$r^2 v_{\phi(T),\phi(T)}(P) = r^2(x_P - rx_{2T}) = -rx_P - x_P - x_{2T}.$$

For each bit of λ, we do $f_1 \leftarrow f_1^2 \cdot l_{T,T}(\phi(P)) \cdot v_{\phi(T),\phi(T)}(P)$ and $f_2 \leftarrow f_2^2 \cdot l_{\phi(T),\phi(T)}(P) \cdot v_{2T}(\phi(P))$. Since $r^3 = 1 \mod p$ and the final power equals 6, we can replace $v_{\phi(T),\phi(T)}(P)$ by $r^2 v_{\phi(T),\phi(T)}(P)$ in the whole computation. Moreover, we can cache $R_1 = rx_P$. The formulas for the doubling steps in affine coordinates will be given by

$$\lambda = \tfrac{3x_T^2}{2y_T};\ x_{2T} = \lambda^2 - 2x_T;\ y_{2T} = \lambda(x_T - x_{2T}) - y_T;$$
$$t_1 = (y_P - y_T) - \lambda_T(R_1 - x_T), t_2 = (y_P - y_T) - \lambda_T(-R_1 - x_P - x_T),$$
$$v_1 = R_1 - x_{2T}, v_2 = -R_1 - x_P - x_{2T}, f_1 \leftarrow f_1^2 \cdot t_1 \cdot v_2; f_2 \leftarrow f_2^2 \cdot t_2 \cdot v_1;$$

The total cost of the operation for the doubling step in affine coordinates is $1I + 4S + 8M$.

Now we consider the operations for the doubling steps in Jacobian coordinates. In Jacobian projective coordinates, the equation of E_b is $Y^2 = X^3 + bZ^6$. A point is represented as (X_1, Y_1, Z_1) which $Z_1 \neq 0$ corresponds to the affine point (x_1, y_1) with $x_1 = X_1/Z_1^2$ and $y_1 = Y_1/Z_1^3$. To obtain the full speed of pairings on Weierstrass curves it is useful to represent a point by $(X_1, Y_1, Z_1, W_1, \gamma_1)$ with $W_1 = Z_1^2$ and $\gamma_1 = R_1 W_1$. Throughout the loop of Miller's algorithm, the line function is always evaluated at the point P or $\phi(P)$. It is therefore customary to represent this point $P = (x_P, y_P)$ in affine coordinates. Let $T = (X_T, Y_T, Z_T, W_T)$ and $N = 2T = (X_N, Y_N, Z_N, W_N)$. In each bit of λ, the function evaluated in Jacobian coordinates is updated by

$$f_1 \leftarrow f_1^2 \cdot (y_P W_T Z_N - 2Y_T^2 - 3X_T^2(R_1 W_T - X_T)) \cdot (R_1 W_N + W_N + X_N),$$
$$f_2 \leftarrow f_2^2 \cdot (y_P W_T Z_N - 2Y_T^2 - 3X_T^2(-R_1 W_T - W_T - X_T)) \cdot (R_1 W_N - X_N).$$

Note that we can cache $R_1 W_T$ in the last step. The following formulae compute a doubling step in $10M + 9S$.

$$A = X_T^2;\ B = Y_T^2;\ C = B^2;\ D = 2((X_T + B)^2 - A - C);\ E = 3A;\ G = E^2;$$
$$X_N = G - 2D;\ Y_N = E \cdot (D - X_N) - 8C;\ Z_N = (Y_T + Z_T)^2 - B - W_T;$$
$$W_N = Z_N^2;\ t_1 = y_P \cdot W_T \cdot Z_N - 2B;\ t_2 = E \cdot (R_1 W_T - X_T);$$
$$t_3 = E \cdot (R_1 W_T + W_T + X_T);\ H = R_1 \cdot W_N;\ v_1 = H + W_N + X_N;$$
$$v_2 = H - X_N;\ f_1 \leftarrow f_1^2 \cdot (t_1 - t_2) \cdot v_1;\ f_2 \leftarrow f_2^2 \cdot (t_1 + t_3) \cdot v_2.$$

Addition Step. Assume that $T + P = (x_{T+P}, y_{T+P})$. We obtain $\phi(T + P) = (rx_{T+P}, y_{T+P})$. In the case of the addition step of the self-pairings, after initially setting $T = P, f_1 = f_2 = 1$, for each bit of r we do

$$f_1 \leftarrow f_1 \frac{l_{T,P}(\phi(P))}{v_{T+P}(\phi(P))},$$
$$f_2 \leftarrow f_2 \frac{l_{\phi(T),\phi(P)}(P)}{v_{\phi(T+P)}(P)},$$
$$T \leftarrow T + P.$$

Let $\lambda_{T,P}$ be the slope of the line through the points T and P, then the slope of the line through the points $\phi(T)$ and $\phi(P)$ is $\lambda_{T,P}/r$. It follows that

$$l_{T,P}(\phi(P)) = (y_P - y_P) - \lambda_{T,P}(rx_P - x_P) = (1-r)x_P\lambda_{T,P},$$
$$l_{\phi(T),\phi(P)}(P) = (y_P - y_P) - (\lambda_{T,P}/r)(x_P - rx_P) = -(1-r)x_P\lambda_{T,P}/r.$$

Note that we can ignore $-1/r$ due to the final power 6. Therefore, in the addition step, for each bit of r, we only need do update

$$f_1 \leftarrow f_1 \cdot v_{\phi(T+P)}(P),$$
$$f_2 \leftarrow f_2 \cdot v_{T+P}(\phi(P)).$$

The formulas for the addition step in affine coordinates will be given by

$$\lambda = \frac{y_T - y_P}{x_T - x_P};\ x_{T+P} = \lambda^2 - x_P - x_T;\ y_{T+P} = \lambda(x_P - x_{T+P}) - y_P;$$
$$v_1 = r^2(x_P - rx_{T+P}) = -R_1 - x_P - x_{T+P};\ v_2 = R_1 - x_{T+P};$$
$$f_1 \leftarrow f_1 \cdot v_1; f_2 \leftarrow f_2 \cdot v_2;$$

The total cost of the operations for the addition step in affine coordinates is $1I + 1S + 4M$.

Now we consider the operations for the addition step in Jacobian coordinates. Let $T = (X_T, Y_T, Z_T, W_T)$ and $N = T + P = (X_N, Y_N, Z_N, W_N)$. In each bit of λ, the function evaluated in Jacobian coordinates is updated by

$$f_1 \leftarrow f_1^2 \cdot (y_P W_T Z_N - 2Y_T^2 - 3X_T^2(R_1 W_T - X_T)) \cdot (R_1 W_N + W_N + X_N),$$
$$f_2 \leftarrow f_2^2 \cdot (y_P W_T Z_N - 2Y_T^2 - 3X_T^2(-R_1 W_T - W_T - X_T)) \cdot (R_1 W_N - X_N).$$

Note that we can catch $A = y_P^2$ and $R_1 = rx_P$. The following formulae compute an addition step in $9M + 5S$.

$$A = y_P^2;\ B = x_P \cdot W_T;\ D = ((y_P + Z_T)^2 - A - W_T) \cdot W_T;\ H = B - X_T;$$
$$I = H^2; E := 4I; J = H \cdot E; L = (D - 2Y_T); V = X_T \cdot E; X_N = L^2 - J - 2V;$$
$$Y_N = L \cdot (V - X_N) - 2Y_T \cdot J;\ Z_N = (Z_T + H)^2 - W_T - I;\ W_N = Z_N^2;$$
$$U = R_1 \cdot W_N;\ v_1 = U + W_N - X_N;\ v_2 = U - X_N;\ f_1 \leftarrow f_1 \cdot v_1;\ f_2 \leftarrow f_2 \cdot v_2.$$

Pairing Algorithm Using Addition Chain. Throughout the loop of Miller's algorithm, the line function is always evaluated at the point P or $\phi(P)$. Since $Q \in \{P, \phi(P)\} \subset E(\mathbb{F}_p)$, the self-pairing can be combined with windowing methods. Furthermore, the addition step is faster than the doubling step in self-pairing computation on $y^2 = x^3 + b$, thus we can develop the following self-pairing computation algorithm based on addition chain.

Herbaut et al. in [10] proposed a fast and secure point multiplication algorithm based on a particular kind of addition chains (Euclidean addition chains) involving only additions no doubling.

A star addition chain is an addition chain which satisfies: $\forall i, w_i = (i-1, j)$ for some j with $0 \le j \le i-1$. That is to say that for all i we have $v_i = v_{i-1} + v_j$.

In this case we can omit $i - 1$ and just write $w_i = j$. A special addition chain is a star addition chain with $w = (w_3, \cdots, w_s) \in \{0, 1\}^{s-2}$ satisfying :

$$v_0 = 1, \ v_1 = 2, \ v_2 = 3,$$
$$v_i = v_{i-1} \Rightarrow v_{i+1} = v_i + \begin{cases} v_{i-1} & \text{if } w_{i+1} = 0, \\ v_j & \text{if } w_{i+1} = 1. \end{cases}$$

In order to lighten the notations, we will abusively denote $n = (w_3, w_4, \cdots, w_s)$. For example, we can represent 31 to $(1, 0, 0, 1, 1, 0)$, or $(1, 2 \rightarrow 1+2 = 3 \rightarrow 2+3 = 5 \rightarrow 3+5 = 8 \rightarrow 5+8 = 13 \rightarrow 5+13 = 18 \rightarrow 13+18 = 31$. That is, the special addition chain of 31 is $(3, 5, 8, 13, 18, 31)$. Given two points P, Q on an elliptic curve E, an integer n and let $n = (w_3, \cdots, w_s)$ be the special addition chain computing n, it is easy to deduce the following algorithm to compute $f_{n,P}(Q)$.

Algorithm 2. Addition chain algorithm

Input: $P, Q \in E$ and $n = (w_3, \cdots, w_s)$
Output: $[n]P \in E$ and $f_{n,P}(Q)$
1: $(U_1, U_2, U_3) \leftarrow (P, [2]P, [3]P)$
2: $(F_1, F_2, F_3) \leftarrow (f_{1,P}(Q), f_{2,P}(Q), f_{3,P}(Q))$
3: **for** $i = 3$ up to s **do**
4:    **if** $w_i = 0$ **then**
5:       $U_1 \leftarrow U_2, F_1 \leftarrow F_2$
6:    **end if**
7:    $U_2 \leftarrow U_3, F_2 \leftarrow F_3$
8:    $F_3 \leftarrow F_1 \cdot F_2 \cdot \frac{l_{U_1,U_2}}{v_{U_1+U_2}}(Q), U_3 \leftarrow U_1 + U_2$
9: **end for**
10: **return** U_3

Using the algorithm in [10] to compute $U_1 + U_2$ cost $5M + 2S$, updating the functions f_1 and f_2 needs $1S + 3M$. Hence, computing the self-pairing by using the addition chain algorithm costs $8M + 3S$ in each step. But it is an open problem to find minimal special addition chains for any integer, but it showed in [10] a way to find small chains by looking for them in a clever range.

5 Conclusion

In this paper, we first generalize the result in [28] to any elliptic curve with more simpler final exponentiation. Then we present a new self-pairing on ordinary elliptic curves with short loop length in Miller's algorithm. We also provide examples of self-pairing friendly elliptic curves which are of interest for efficient pairing implementations. Finally, we present explicit formulae for Miller's algorithm to compute self-pairing on ordinary elliptic curves with embedding degree one. For the elliptic curve $y^2 = x^3 + b$, we can apply the special addition chain to compute the self-pairing efficiently.

Acknowledgment. Thanks to the referee for his/her suggestions on this paper.

Hongfeng Wu's research was supported by National Natural Science Foundation of China (No. 11101002 and No. 11271129) and Beijing Natural Science Foundation (No. 1132009). Rongquan Feng's research was supported by National Natural Science Foundation of China (No. 10990011 and No. 61170264) and the research fund for the Doctoral Program of Higher Education of China (No. 20100001110007).

References

1. Aranha, D.F., Karabina, K., Longa, P., Gebotys, C.H., López, J.: Faster Explicit Formulas for Computing Pairings over Ordinary Curves. In: Paterson, K.G. (ed.) EUROCRYPT 2011. LNCS, vol. 6632, pp. 48–68. Springer, Heidelberg (2011)
2. Arene, C., Lange, T., Naehrig, M., Ritzenthaler, C.: Faster Computation of the Tate Pairing. Journal of Number Theory 131, 842–857 (2011)
3. Avanzi, R., Cohen, H., Doche, C., Frey, G., Lange, T., Nguyen, K., Vercauteren, F.: Handbook of Elliptic and Hyperelliptic Curve Cryptography. CRC Press (2005)
4. Blake, I.F., Seroussi, G., Smart, N.P.: Advances in Elliptic Curve Cryptography. Cambridge University Press (2005)
5. Boxall, J., El Mrabet, N., Laguillaumie, F., Le, D.-P.: A Variant of Miller's Formula and Algorithm. In: Joye, M., Miyaji, A., Otsuka, A. (eds.) Pairing 2010. LNCS, vol. 6487, pp. 417–434. Springer, Heidelberg (2010)
6. Costello, C., Lange, T., Naehrig, M.: Faster Pairing Computations on Curves with High-Degree Twists. In: Nguyen, P.Q., Pointcheval, D. (eds.) PKC 2010. LNCS, vol. 6056, pp. 224–242. Springer, Heidelberg (2010)
7. Duursma, I., Lee, H.-S.: Tate pairing implementation for hyperelliptic curves $y^2 = x^p - x + d$. In: Laih, C.-S. (ed.) ASIACRYPT 2003. LNCS, vol. 2894, pp. 111–123. Springer, Heidelberg (2003)
8. Frey, G., Rück, H.-G.: A remark concerning m-divisibility and the discrete logarithm in the divisor class group of curves. Math. Comp. 62(206), 865–874 (1994)
9. Galbraith, S.D., Paterson, K., Smart, N.: Pairings for cryptographers. Discr. Appl. Math. 156, 3113–3121 (2008)
10. Herbaut, F., Liardet, P.-Y., Méloni, N., Téglia, Y., Véron, P.: Random Euclidean Addition Chain Generation and Its Application to Point Multiplication. In: Gong, G., Gupta, K.C. (eds.) INDOCRYPT 2010. LNCS, vol. 6498, pp. 238–261. Springer, Heidelberg (2010)
11. Hess, F., Smart, N.P., Vercauteren, F.: The Eta pairing revisited. IEEE Trans. Infor. Theory 52, 4595–4602 (2006)
12. Hess, F.: Pairing Lattices. In: Galbraith, S.D., Paterson, K.G. (eds.) Pairing 2008. LNCS, vol. 5209, pp. 18–38. Springer, Heidelberg (2008)
13. Hu, Z., Xu, M., Zhou, Z.: A Generalization of Verheul's Theorem for Some Ordinary Curves. In: Lai, X., Yung, M., Lin, D. (eds.) Inscrypt 2010. LNCS, vol. 6584, pp. 105–114. Springer, Heidelberg (2011)
14. Koblitz, N., Menezes, A.: Pairing-based cryptography at high security levels. In: Smart, N.P. (ed.) Cryptography and Coding 2005. LNCS, vol. 3796, pp. 13–36. Springer, Heidelberg (2005)
15. Lee, E., Lee, H.-S., Park, C.-M.: Efficient and generalized pairing computation on Abelian varieties. IEEE Trans. Inform. Theory 55(4), 1793–1803 (2009)

16. Matsuda, S., Kanayama, N., Hess, F., Okamoto, E.: Optimised Versions of the Ate and twisted Ate pairings. In: Galbraith, S.D. (ed.) Cryptography and Coding 2007. LNCS, vol. 4887, pp. 302–312. Springer, Heidelberg (2007)
17. Miller, V.S.: Short programs for functions on curves, http://crypto.stanford.edu/miller/miller.pdf
18. Miller, V.S.: The Weil pairing and its efficient calculation. J. Cryptol. 17(44), 235–261 (2004)
19. Park, C.M., Kim, M.H., Yung, M.: A Remark on Implementing the Weil Pairing. In: Feng, D., Lin, D., Yung, M. (eds.) CISC 2005. LNCS, vol. 3822, pp. 313–323. Springer, Heidelberg (2005)
20. Paterson, K.G.: Cryptography from Pairing - Advances in Elliptic Curve Cryptography. Cambridge University Press (2005)
21. Silverman, J.H.: The arithmetic of elliptic curves. Springer, New York (1992)
22. Tso, R., Yi, X., Huang, X.: Efficient and short certificateless signatures secure against realistic adversaries. J. Supercomput. 55(2), 173–191 (2011)
23. Verheul, E.R.: Evidence that XTR is more secure than supersingular elliptic curve cryptosystems. In: Pfitzmann, B. (ed.) EUROCRYPT 2001. LNCS, vol. 2045, pp. 195–210. Springer, Heidelberg (2001)
24. Vercauteren, F.: Optimal pairings. IEEE Trans. Inf. Theory 56, 455–461 (2010)
25. Zhang, F., Chen, X., Susilo, W., Mu, Y.: A New Signature Scheme Without Random Oracles from Bilinear Pairings. In: Nguyên, P.Q. (ed.) VIETCRYPT 2006. LNCS, vol. 4341, pp. 67–80. Springer, Heidelberg (2006)
26. Zhang, F., Safavi-Naini, R., Susilo, W.: An efficient signature scheme from bilinear pairings and its applications. In: Bao, F., Deng, R., Zhou, J. (eds.) PKC 2004. LNCS, vol. 2947, pp. 277–290. Springer, Heidelberg (2004)
27. Zhang, F., Safavi-Naini, R., Susilo, W.: ID-Based Chameleon Hashes from Bilinear Pairings, Cryptology ePrint Archive, Report 2003/208
28. Zhao, C.-A., Xie, D.Q., Zhang, F., Zhang, J., Chen, B.-L.: Computing Bilinear Pairings on Elliptic Curves with Automorphisms. Designs, Codes and Cryptography 58(1), 35–44 (2011)
29. Zhao, C.-A., Zhang, F., Xie, D.Q.: Faster Computation of Self-Pairings. IEEE Transactions on Information Theory 58(5), 3266–3272 (2012)

Grey-Box Public-Key Steganography

Hirotoshi Takebe and Keisuke Tanaka

Department of Mathematical and Computing Sciences, Tokyo Institute of Technology
2-12-1 Ookayama, Meguro-ku, Tokyo 152-8552, Japan

Abstract. Steganography is one of the information-hiding techniques. By encoding secret messages to other documents which are meaningful, parties can send the secret messages without any suspicion. Recently, Liśkiewicz, Reischuk, and Wölfel [TAMC 2011] proposed a grey-box model for the channel setting. This model formalizes a more realistic situation that everyone knows partial information of communication. They constructed some schemes of grey-box steganography in the symmetric-key setting. In this paper, we apply their idea of the grey-box model to the public-key setting, and then construct a scheme of grey-box public-key steganography via a standard public-key encryption scheme. We show that our proposed scheme is steganographically secure if the underlying public-key encryption scheme satisfies indistinguishability from random bits.

Keywords: provable security, public-key cryptography, steganography.

1 Introduction

1.1 Background

Steganography is one of the information-hiding techniques which can be representative solutions of the Prisoners' Problem formalized by Simmons [14]. The Prisoners' Problem is as follows. Two prisoners want to take into consultation secretly to escape from the jail. However, they must communicate through a public channel, and then their conversation is always watched by a warden. If the prisoners send a letter which looks meaningless such a standard ciphertext, the warden may feel suspicious. He may isolate them so that they cannot communicate anymore if the worst. Thus, the prisoners should communicate secretly through a public channel without being suspected by the warden.

A lot of solutions have been proposed for this hard problem, and steganography can be one of them as mentioned before. Intuitively, the sender transforms some real messages like standard encryption, and generates something meaningful documents different from original messages. We call these by *stegotexts*. On the other hand, the standard documents which are not associated with steganography are called by *covertexts*. Since stegotexts seem meaningful as well as covertexts, the warden monitoring the communication channel does not feel suspicious. Only the valid receiver can get hidden messages from the stegotexts.

T.-H.H. Chan, L.C. Lau, and L. Trevisan (Eds.): TAMC 2013, LNCS 7876, pp. 294–305, 2013.

Then the sender and the receiver can complete secret communication without any suspicion.

While the standard cryptography hides the contents of messages by encoding to ciphertexts, the purpose of steganography is to hide the presence of messages hidden in stegotexts. By the characteristic property of steganography, there are many applications such as copyright protection of digital contents, confidential communication, and multimedia database systems (see e.g. [1]).

On steganography, there are several settings for communication channels. Recently, in TAMC 2011, an interesting setting called by *grey-box steganography* was proposed by Liśkiewicz, Reischuk, and Wölfel [11]. They modeled a more realistic situation that all the parties are allowed to know partial information of communication.

1.2 Related Work

Public-key Steganography. Some formal models of steganography were introduced as surveyed in e.g. [1]. For example, there are several models of symmetric-key steganography such as [13,3,8]. In the symmetric-key setting, only the parties priorly sharing some secrets can use the protocols. Namely, any pair of parties needs to share some secrets so that anyone except them cannot detect the secrets. In contrast, public-key steganography allows all parties to communicate steganographically without priorly sharing secrets.

Public-key steganography was first formalized by von Ahn and Hopper [15]. They defined a security notion for public-key steganography, which is steganographic security against the chosen-hiddentext attack (SS-CHA-security). This corresponds to the security of public-key encryption called by indistinguishability against the chosen-plaintext attack (IND-CPA-security). They constructed a scheme of public-key steganography. Their general construction makes use of a public-key encryption scheme. They also defined a security notion for public-key encryption in order to prove SS-CHA-security on their scheme. It is indistinguishability from random bits under the chosen-plaintext attack (IND$-CPA-security). Furthermore, they proposed some public-key encryption schemes satisfying IND$-CPA-security under the RSA assumption and the decisional Diffie-Hellman assumption.

Backes and Cachin [2] defined a new security notion for public-key steganography which is stronger than that of von Ahn and Hopper [15]. It is steganographic security against the adaptive chosen-covertext attack (SS-CCA-security) which seems to be the most general type of security on public-key steganography since this corresponds to indistinguishability against the chosen-ciphertext attack (IND-CCA-security) in the field of standard public-key cryptography. They also defined another security notion called by steganographic security against the replayable adaptive chosen-covertext attack (SS-RCCA-security), which is a relaxed notion of SS-CCA-security. They showed that SS-RCCA-secure schemes can be constructed from RCCA-secure [4] public-key encryption schemes with pseudorandom ciphertexts. Hopper [7] proposed a construction of SS-CCA-secure schemes. This construction relies on the existence of public-key encryption schemes satisfying indistinguishability from random bits under the chosen-ciphertext attack (IND\$-CCA-security). He also proposed the encryption scheme by modifying that of Kurosawa and Desmedt [10] which is the modification of the original Cramer-Shoup scheme [5], and showed that his scheme satisfies IND\$-CCA-security under the decisional Diffie-Hellman assumption.

Communication Channels and the Grey-Box Model. In most of the previous works of steganography such as [15,2,7,12,9,6,8], it is assumed that all the parties have nothing about covertext distributions, and then access to the corresponding oracles to get documents or its information. Namely, an adversary can obtain information of covertext distributions only by sampling from the oracles. This setting can be called by the *black-box* steganography since the oracles behave as black-box. With regard to the black-box steganography, there are some negative results. For example, Lysyanskaya and Meyerovich [12] showed the difficulty of sampling based on the full history and the insecurity of sampling with restricted-length histories. Hundt, Liśkiewicz, and Wölfel [9] gave the construction of sampling oracles with an intractable problem.

In contrast, Liśkiewicz, Reischuk, and Wölfel [11] proposed other formalization called by the *grey-box* steganography. Intuitively, the grey-box channel allows all the parties to know partial information of communication. Partial information means topics, habits, and so on. Since often the topic of communication has been naturally determined when the parties are decided, the grey-box model can be considered as a more realistic setting. They proposed some concrete channel models which belong to the grey-box setting, one of which is a *monomial channel*. They gave the efficient construction of monomial channels so that they can overcome the exponential sampling complexity caused in the black-box steganography. Based on these, they constructed some schemes of grey-box steganography in the symmetric-key setting.

1.3   Our Contribution

As mentioned before, the schemes of [11] belong to the symmetric-key setting which needs some priorly sharing secrets. In this paper, we extend their idea of the grey-box model to the public-key setting, and then construct a scheme of grey-box public-key steganography with respect to monomial channels. The idea of our construction basically follows that of [11]. Concretely, we modify some transformation procedures used to construct the scheme of [11] in order to apply to the public-key setting. Then we construct a scheme of grey-box public-key steganography by composing the modified procedures and a public-key encryption scheme. After that, we show that our proposed scheme of grey-box public-key steganography satisfies SS-CHA-security if the underlying public-key encryption scheme satisfies IND\$-CPA-security.

Here, we would like to emphasize the technical advantage of our scheme. With regard to the scheme of symmetric-key grey-box steganography in [11], the secret key plays a very useful role to analyze the security and the reliability which is similar to the standard correctness property. Then if we just simply apply their idea to the public-key setting in which there is no priorly sharing secrets, it can cause an unavoidable problem that we cannot make use of such analyses directly. However, we propose new procedures for our construction by modifying their ones, and then we can achieve more simply analyses.

2   Preliminaries

Notation. We say that a function $\mu : \mathbb{N} \to \mathbb{R}^+$ is *negligible* in λ if for any $c > 0$, there exists λ_0 such that $\mu(\lambda) < \frac{1}{\lambda^c}$ for any $\lambda > \lambda_0$. We denote the uniform distribution on a set $\{0,1\}^\delta$ by $\mathcal{U}_\delta$. For a probability distribution $\mathcal{D}$ over some domain D, we denote by $x \leftarrow \mathcal{D}$ the action of drawing a sample x according to the distribution $\mathcal{D}$. For an algorithm A, we denote by $y \leftarrow \mathsf{A}(x)$ the event that the algorithm A with an input x returns y as its output.

2.1   Public-Key Encryption

We first define a public-key encryption scheme.

Definition 1 (Public-Key Encryption). *A public-key encryption scheme* $\mathcal{E} = (\mathsf{Gen}, \mathsf{Enc}, \mathsf{Dec})$ *is a tuple of three algorithms.*

- Gen *is a key generation algorithm. On input a security parameter* 1^λ, Gen *returns a pair of* (pk, sk). *pk and sk are public and secret keys, respectively. We write this as* $(pk, sk) \leftarrow \mathsf{Gen}(1^\lambda)$.
- Enc *is an encryption algorithm. On input a public key pk and a message m,* Enc *returns a ciphertext c. We write this as* $c \leftarrow \mathsf{Enc}(pk, m)$.
- Dec *is a decryption algorithm. On input a secret key sk and a ciphertext c,* Dec *returns either a message m or a symbol* $\perp$ *which indicates that the ciphertext c is invalid. We write this as* $m/\perp \leftarrow \mathsf{Dec}(sk, c)$.

We require the correctness property as follows. For any pair of keys $(pk, sk) \leftarrow \mathsf{Gen}(1^\lambda)$, any message m, and any ciphertext $c \leftarrow \mathsf{Enc}(pk, m)$, it holds that $\Pr[m \leftarrow \mathsf{Dec}(sk, c)] = 1$.

Second, we review the property of public-key encryption proposed by von Ahn and Hopper [15]. It is indistinguishability from random bits under the chosen-plaintext attack. Let $\mathcal{E} = (\mathsf{Gen}, \mathsf{Enc}, \mathsf{Dec})$ be a public-key encryption scheme and λ a security parameter. We denote by f a function which implies the length of the ciphertexts of $\mathcal{E}$. Hence we define a distinguishing game under the chosen-plaintext attack against $\mathcal{E}$ by an adversary A and a challenger. We consider the experiments $\mathbf{Exp}^i_{\mathbf{CPA}}$ for $i \in \{0,1\}$ as described below.

$\mathbf{Exp}^i_{\mathbf{CPA}}(1^\lambda)$
1. $(pk, sk) \leftarrow \mathsf{Gen}(1^\lambda)$.
2. A is given pk.
3. A can make a challenge query adaptively. Specifically, A passes a message m^* to the challenger. The challenger passes c^*_i to A as its response.

    4. A outputs a bit γ.

    5. Return γ.

We define c_i^* for $i \in \{0, 1\}$ as follows.

- c_0^*: The challenger computes $c_0^* \leftarrow \mathsf{Enc}(pk, m)$.
- c_1^*: The challenger samples $c_1^* \leftarrow \mathcal{U}_f$.

We define A's advantage against $\mathcal{E}$ by

$$\mathbf{Adv}_{\mathcal{E},A}^{\mathsf{ind\$-cpa}}(\lambda) := |\Pr[\mathbf{Exp}_{\mathbf{CPA},A}^0(1^\lambda) = 1] - \Pr[\mathbf{Exp}_{\mathbf{CPA},A}^1(1^\lambda) = 1]| \ .$$

Definition 2 (IND\$-CPA). *We say that $\mathcal{E}$ is indistinguishable from random bits under the chosen-plaintext attack (IND\$-CPA-secure) if for any probabilistic polynomial-time adversary A, $\mathbf{Adv}_{\mathcal{E},A}^{\mathsf{ind\$-cpa}}(\lambda)$ is negligible in λ.*

In [15], von Ahn and Hopper constructed IND\$-CPA-secure public-key encryption schemes, which are based on the RSA assumption and the decisional Diffie-Hellman assumption.

2.2 Channels

In this section, we first formalize the communication channels in order to review some definitions of public-key steganography.

Intuitively, the communication between the parties follows the distribution relied on the previous communications. For defining this notion, we follow previous works [15,2,7,12,9,6,8,11] on steganography.

We formalize the communication between two parties by a *channel*. Let $\Sigma = \{0,1\}^\sigma$ be a set of documents, we denote that $\Sigma^* = \Sigma \times \Sigma \times \cdots$. We define a channel $\mathcal{C} = \{\mathcal{C}_\mathcal{H} | \mathcal{H} \in \Sigma^*\}$, which is a family of probability distributions on a set of documents Σ, indexed by sequences $\mathcal{H} \in \Sigma^*$. We call the index $\mathcal{H}$ by *history*. For an integer j, we define the distribution $\mathcal{C}_\mathcal{H}^j := \mathcal{C}_\mathcal{H} \times \mathcal{C}_{(\mathcal{H}||d_1)} \times \mathcal{C}_{(\mathcal{H}||d_1||d_2)} \times \cdots \times \mathcal{C}_{(\mathcal{H}||d_1||d_2||\ldots||d_{j-1})}$, where $d_1 \leftarrow \mathcal{C}_\mathcal{H}, d_2 \leftarrow \mathcal{C}_{(\mathcal{H}||d_1)}, \ldots, d_{j-1} \leftarrow \mathcal{C}_{(\mathcal{H}||d_1||d_2||\ldots||d_{j-2})}$. A history $\mathcal{H} = (d_1||d_2||\ldots||d_j)$ is *legal* with respect to $\mathcal{C}$ if for all i, it holds that $\Pr[d_i \leftarrow \mathcal{C}_{(d_1||d_2||\ldots||d_{i-1})}|d_1 \leftarrow \mathcal{C}_\varepsilon, d_2 \leftarrow \mathcal{C}_{d_1}, \cdots, d_{i-1} \leftarrow \mathcal{C}_{(d_1||d_2||\ldots||d_{i-2})}] > 0$ where ε is an empty string.

In the grey-box model proposed by [11], we assume that parties have partial knowledge of channels. In order to formalize this situation, we make use of the notion of concept classes, and define a channel family $\mathcal{F}$ as a subset of $\mathcal{C}$ so that all the channels in $\mathcal{F}$ satisfy some common features. We explain this notion by showing an example as follows.

Monomial Covertext Channel [11]. A monomial can be one of the example of concept classes. We represent a monomial over $\{0,1\}^\sigma$ by a vector $\mathbf{H} = (\mathbf{h}_1, \mathbf{h}_2, \ldots, \mathbf{h}_\sigma) \in \{0, 1, \times\}^\sigma$ where $\times$ is a special symbol called by a free variable. Then we define $\mathbf{H}$ as the subset of $\{0,1\}^\sigma$ such that $\mathbf{H}$ includes all the elements satisfying the following conditions:

- if $\mathbf{h}_i$ is 0, then i-th component is also 0.
- if $\mathbf{h}_i$ is 1, then i-th component is also 1.
- if $\mathbf{h}_i$ is ×, then i-th component is either 0 or 1.

If $\sigma = 5$ and $\mathbf{H} =$ "$0 \times 1 \times 1$" then $\mathbf{H}$ means $\{00101, 00111, 01101, 01111\}$ for example. Namely, the monomial representation of $\mathbf{H}$ indicates common features of all elements in the set $\mathbf{H}$. In the case of the monomial channel, $\mathcal{F}$ consists of some distributions on $\mathbf{H}$.

We denote a channel oracle according to a channel $\mathcal{C}$ with a history $\mathcal{H}$ by $EX_\mathcal{C}(\mathcal{H})$. In the grey-box model, there exist two types of queries. We give the details in the case of the monomial channel as follows.

- Sampling a document s according to $\mathcal{C}_\mathcal{H}$ (denoted by $s \leftarrow EX_\mathcal{C}(\mathcal{H})$).
- Learning a monomial $\mathbf{H}$ according to $\mathcal{C}_\mathcal{H}$ (denoted by $\mathbf{H} := EX_\mathcal{C}(\mathcal{H})$).

Note that the learning query corresponds to knowing partial information. Liśkiewicz et al. [11] gave the construction of the monomial channels, and proposed concrete schemes of steganography in these models.

2.3 Public-Key Steganography

Now we review the definition of public-key steganography and its security notion formalized by von Ahn and Hopper [15].

Definition 3 (Public-Key Steganography). *A scheme of public-key steganography* $\mathcal{S} = (\mathsf{SGen}, \mathsf{SEnc}, \mathsf{SDec})$ *is a tuple of three algorithms.*

- SGen *is a key generation algorithm. On input a security parameter* 1^λ, SGen *returns a pair of* (pk, sk). *pk and sk are public and secret keys, respectively. We write this as* $(pk, sk) \leftarrow \mathsf{SGen}(1^\lambda)$.
- SEnc *is a steganographic encoding algorithm. On input a public key pk, a message m, and a history $\mathcal{H}$, SEnc returns a sequence of some documents* $(s_1, s_2, \ldots, s_l)$ *from the support of* $\mathcal{C}_\mathcal{H}^l$. *We write this as* $(s_1, s_2, \ldots, s_l) \leftarrow \mathsf{SEnc}(pk, m, \mathcal{H})$. *We call* $(s_1, s_2, \ldots, s_l)$ *by a stegotext, and often simply write s.*
- SDec *is a steganographic decoding algorithm. On input a secret key sk, a stegotext $s = (s_1, s_2, \ldots, s_l)$, and a history $\mathcal{H}$, SDec returns either a message m or a symbol $\perp$ which indicates that the stegotext is invalid. We write this as* $m/\perp \leftarrow \mathsf{SDec}(sk, s, \mathcal{H})$.

We require the correctness property as follows. For any pair of keys $(pk, sk) \leftarrow \mathsf{SGen}(1^\lambda)$, any message-history pair $(m, \mathcal{H})$, and any stegotext $s \leftarrow \mathsf{SEnc}(pk, m, \mathcal{H})$, there is a negligible function $\mu(\lambda)$ such that $\Pr[m \leftarrow \mathsf{SDec}(sk, s, \mathcal{H})] \geq 1 - \mu(\lambda)$. We note that this notion in the symmetric-key setting was formalized as *reliability* in [11].

Next, we review the security property of public-key steganography called by steganographic security against the chosen hiddentext attack (SS-CHA-security), which is formalized by von Ahn and Hopper [15]. Let $\mathcal{S} = (\mathsf{SGen}, \mathsf{SEnc}, \mathsf{SDec})$ be a scheme of public-key steganography, λ a security parameter, and $\mathcal{C}$ a channel. We denote by f^* a function which implies the length of the stegotexts of $\mathcal{S}$. Hence we define a distinguishing game under the chosen hiddentext attack against S by an adversary W and a challenger. We consider the experiments $\mathbf{Exp}^i_{\mathbf{CHA}}$ for $i \in \{0, 1\}$ as described below.

$\mathbf{Exp}^i_{\mathbf{CHA}}(1^\lambda)$
1. $(pk, sk) \leftarrow \mathsf{SGen}(1^\lambda)$.
2. W is given pk.
3. W can make a challenge query adaptively. Specifically, W produces a message m^* and a history $\mathcal{H}^*$, and passes them to the challenger. The challenger passes s^*_i to W as its response.
4. W outputs a bit γ.
5. Return γ.

We define s^*_i for $i \in \{0, 1\}$ as follows.

- s^*_0: The challenger computes $s^*_0 \leftarrow \mathsf{SEnc}(pk, m^*, \mathcal{H}^*)$.
- s^*_1: The challenger samples $s^*_1 \leftarrow \mathcal{C}^{f^*}_{\mathcal{H}^*}$.

W can also make a query to the channel oracle adaptively in the above experiments. In the grey-box model, W produces a history $\mathcal{H}$ as a sampling query or a learning query. Then W receives a document s where $s \leftarrow EX_{\mathcal{C}}(\mathcal{H})$ as the response for the sampling query, or a monomial $\mathbf{H}$ where $\mathbf{H} := EX_{\mathcal{C}}(\mathcal{H})$ as that for the learning query in the case of the monomial channel, for example.

We define W's advantage against $\mathcal{S}$ with respect to $\mathcal{C}$ by

$$\mathbf{Adv}^{\mathsf{ss\text{-}cha}}_{\mathcal{S},\mathcal{C},W}(\lambda) := |\Pr[\mathbf{Exp}^0_{\mathbf{CHA},\mathcal{C},W}(1^\lambda) = 1] - \Pr[\mathbf{Exp}^1_{\mathbf{CHA},\mathcal{C},W}(1^\lambda) = 1]|.$$

Definition 4 (SS-CHA). *We say that $\mathcal{S}$ is steganographically secure under the chosen-hiddentext attack with respect to $\mathcal{C}$ (SS-CHA-secure) if for any probabilistic polynomial-time adversary W, $\mathbf{Adv}^{\mathsf{ss\text{-}cha}}_{\mathcal{S},\mathcal{C},W}(\lambda)$ is negligible in λ.*

3   Our Scheme

In this section, we propose a construction for a scheme of grey-box public-key steganography. As mentioned before, our construction basically follows the idea of Liśkiewicz et al. [11].

3.1   Related Algorithms

We review the algorithms proposed by Liśkiewicz et al. [11] for constructing a scheme of grey-box steganography in the symmetric-key setting with respect to monomial channels. Let b be the length of target messages and σ the length of monomials such that $\sigma = bt$ where t is some constant. For a monomial $\mathbf{H} = (\mathbf{h}_1, \mathbf{h}_2, \ldots, \mathbf{h}_\sigma) \in \{0, 1, \times\}^\sigma$, a permutation π on a set $\Pi = \{1, 2, \ldots, \sigma\}$, and $1 \leq j \leq b$, we define $I_\pi(j) := \{\pi(t(j-1)+1), \pi(t(j-1)+2), \ldots, \pi(tj)\}$ as the subsets of Π, and $FV_\pi(j)$ the indices in $I_\pi(j)$ which belong to free variables with respect to $\mathbf{H}$ (i.e. $\mathbf{h}_i = \times$ for all $i \in FV_\pi(j)$). Thus, the algorithms **Monomial-modify** and **Document-decode** are as follows.

Algorithm Monomial-modify$(M, s, \mathbf{H}, K)$
denote a target message $M = (m_1, m_2, \ldots, m_b) \in \{0,1\}^b$
denote a candidate document $s = (s_1, s_2, \ldots, s_{bt}) \in \{0,1\}^{bt}$
denote $\mathbf{H} = (\mathbf{h}_1, \mathbf{h}_2, \ldots, \mathbf{h}_{bt}) \in \{0, 1, \times\}^{bt}$
let π be a permutation specified by a private key K
for $1 \leq j \leq b$
    if $[m_j \neq \bigoplus_{k \in I_\pi(j)} s_k$ and $FV_\pi(j) \neq \emptyset]$
        then $s_{a_j} := 1 - s_{a_j}$ (where $a_j := \min FV_\pi(j)$)
end
return $s = (s_1, s_2, \ldots, s_{bt})$.

Algorithm Document-decode(s, K)
denote $s = (s_1, s_2, \ldots, s_{bt}) \in \{0,1\}^{bt}$
let π be a permutation specified by a private key K
for $1 \leq j \leq b$
    $m_j := \bigoplus_{k \in I_\pi(j)} s_k$
end
return $M = (m_1, m_2, \ldots, m_b)$.

Intuitively, **Monomial-modify** embeds a target message M to a candidate document s in a ratio of 1-bit to t-bits. For a bit m_i from M, it computes a parity of some t-bits of s. If the parity does not match with m_i, then it flips one bit somewhere in the t-bits. **Document-decode** recovers the target message M from s via XOR-operation.

Liśkiewicz et al. [11] constructed a scheme of grey-box steganography in the symmetric-key setting by composing the above algorithms, a symmetric-key encryption scheme, and a pseudorandom permutation.

3.2   Our Construction

Now, we construct a scheme of grey-box public-key steganography with respect to monomial channels. Let $\mathcal{E} = (\mathsf{Gen}, \mathsf{Enc}, \mathsf{Dec})$ be a public-key encryption scheme and λ a security parameter. Thus, our proposed scheme $\mathcal{S} = (\mathsf{SGen}, \mathsf{SEnc}, \mathsf{SDec})$ is as follows.

Algorithm $\mathsf{SGen}(1^\lambda)$
$(pk, sk) \leftarrow \mathsf{Gen}(1^\lambda)$
return (pk, sk).

Algorithm $\mathsf{SEnc}(pk, m, \mathcal{H})$
$c \leftarrow \mathsf{Enc}(pk, m)$
denote $c = (\mathbf{c}_1, \mathbf{c}_2, \ldots, \mathbf{c}_\ell)$ where each $\mathbf{c}_i \in \{0,1\}^b$

for $1 \leq i \leq \ell$
    $\tilde{s} \leftarrow EX_C(\mathcal{H})$ where $\tilde{s} = (\tilde{s}_1, \tilde{s}_2, \ldots, \tilde{s}_{bt}) \in \{0,1\}^{bt}$
    $\mathbf{H} := EX_C(\mathcal{H})$ where $\mathbf{H} = (\mathbf{h}_1, \mathbf{h}_2, \ldots, \mathbf{h}_{bt}) \in \{0,1,\times\}^{bt}$
    $s_i \leftarrow$**Monomial-modify'$(\mathbf{c}_i, \tilde{s}, \mathbf{H})$**
    $\mathcal{H} := \mathcal{H}\|s_i$
end
return $s = (s_1, s_2, \ldots, s_\ell)$.

Algorithm SDec$(sk, s, \mathcal{H})$
denote $s = (s_1, s_2, \ldots, s_\ell)$ where each $s_i \in \{0,1\}^{bt}$
for $1 \leq i \leq \ell$
    $\mathbf{H} := EX_C(\mathcal{H})$ where $\mathbf{H} = (\mathbf{h}_1, \mathbf{h}_2, \ldots, \mathbf{h}_{bt}) \in \{0,1,\times\}^{bt}$
    $(S[1], S[2], \ldots, S[b]) \leftarrow$ **Index-divide$(b, t, \mathbf{H})$**
    denote $s_i = (\tilde{s}_1, \tilde{s}_2, \ldots, \tilde{s}_{bt})$ where each $\tilde{s}_j \in \{0,1\}$
    for $1 \leq j \leq b$
        $c_j := \bigoplus_{k \in S[j]} \tilde{s}_k$
    end
    let $\mathbf{c}_i := (c_1, c_2, \ldots, c_b)$
    $\mathcal{H} := \mathcal{H}\|s_i$
end
let $c := (\mathbf{c}_1, \mathbf{c}_2, \ldots, \mathbf{c}_\ell)$
$m \leftarrow \text{Dec}(sk, c)$
return m.

We give an intuitive explanation for the above algorithms. SGen works in the same way as Gen of $\mathcal{E}$. Given a message m, SEnc produces a vector of stegotexts s which follows the past history $\mathcal{H}$ with accessing to the channel oracle $EX_C(\mathcal{H})$. SEnc samples $\tilde{s}$ as a candidate stegotext at first, then adjusts each bit according to the subroutine algorithm **Monomial-modify'** so that the result s is a correct stegotext of the message m. SDec recovers a message m from a vector of stegotexts s. Namely, SDec works as an invert algorithm of SEnc.

The subroutine algorithms are as follows.

Algorithm Monomial-modify'$(c, s, \mathbf{H})$
denote $c = (c_1, c_2, \ldots, c_b)$ where each $c_i \in \{0,1\}$
denote $s = (s_1, s_2, \ldots, s_{bt})$ where each $s_i \in \{0,1\}$
denote $\mathbf{H} = (\mathbf{h}_1, \mathbf{h}_2, \ldots, \mathbf{h}_{bt})$ where each $\mathbf{h}_i \in \{0,1,\times\}$
$(S[1], S[2], \ldots, S[b]) \leftarrow$ **Index-divide$(b, t, \mathbf{H})$**
for $1 \leq j \leq b$
    let a_j be the minimum index in $S[j]$ such that $\mathbf{h}_{a_j} = \times$
    if $m_j \neq \bigoplus_{k \in S[j]} s_k$ then $s_{a_j} := 1 - s_{a_j}$
end
return $(s_1, s_2, \ldots, s_{bt})$.

Algorithm Index-divide$(b, t, \mathbf{H})$
let $S := \{1, 2, \ldots, bt\}$
let $FV_{\mathbf{H}} := \{a_i \in S \mid \mathbf{h}_{a_i} = \times\}$

denote $FV_{\mathbf{H}} = \{a_1, a_2, \ldots, a_{|FV_{\mathbf{H}}|}\}$ such that $a_i < a_j$ if $i < j$
assume that $|FV_{\mathbf{H}}| \geq b$, then
let $FV_{\mathbf{H}}^b := \{a_1, a_2, \ldots, a_b \mid a_i \in FV_{\mathbf{H}}\}$
    by picking from $FV_{\mathbf{H}}$ in the ascending order
let $\overline{FV_{\mathbf{H}}^b} := \{\bar{a}_i \mid \bar{a}_i \in S \backslash FV_{\mathbf{H}}^b\}$
denote $\overline{FV_{\mathbf{H}}^b} = \{\bar{a}_1, \bar{a}_2, \ldots, \bar{a}_{|\overline{FV_{\mathbf{H}}^b}|}\}$ such that $\bar{a}_i < \bar{a}_j$ if $i < j$
for $1 \leq j \leq b$
    let $S[j] := \phi$ as an empty set
    put $S[j] \leftarrow a_j$ from $FV_{\mathbf{H}}^b$
    put $S[j] \leftarrow \bar{a}_{(j-1)(t-1)+1}, \bar{a}_{(j-1)(t-1)+2}, \ldots, \bar{a}_{j(t-1)}$ from $\overline{FV_{\mathbf{H}}^b}$
end
return $(S[1], S[2], \ldots, S[b])$.

Monomial-modify' is a modified algorithm of the original **Monomial-modify**. The basic idea is similar to that of [11]. Their construction decides where t-bits are computed the parity by a pseudorandom permutation. In contrast, we make use of the subroutine algorithm **Index-divide** for this purpose. **Index-divide**$(b, t, \mathbf{H})$ generates b sets, each of which contains at least one value which belongs to free variables with respect to $\mathbf{H}$. From this construction, it is clear that our scheme satisfies the correctness even if $|FV_{\mathbf{H}}| \geq b$ for all $\mathbf{H}$.

4   Security Proofs

In this section, we give the security proof of our scheme.

Theorem 1. *Suppose that for every $\mathcal{H}$, $\mathcal{C}_{\mathcal{H}}$ is the uniform distribution on each domain. Then our proposed scheme is* SS-CHA-*secure if the underlying public-key encryption scheme is* IND\$-CPA-*secure.*

Proof. Let S be our proposed scheme described in Section 3.2, and W an adversary attacking SS-CHA-security against S. We consider the experiments $\mathbf{Exp}^i$ for $i \in \{0, 1, 2\}$ as described below.

    $\mathbf{Exp}^i(1^\lambda)$
    1. $(pk, sk) \leftarrow \mathsf{SGen}(1^\lambda)$.
    2. W is given pk.
    3. W can make a challenge query adaptively. Specifically, W produces
       a message m^* and a history $\mathcal{H}^*$, and passes them to the challenger.
       The challenger passes s_i^* to W as its response.
    4. W outputs a bit γ.
    5. Return γ.

We define s_i^* for $i \in \{0, 1, 2\}$ as follows.

- s_0^*: The challenger computes $s_0^* \leftarrow \mathsf{SEnc}(pk, m^*, \mathcal{H}^*)$.
- s_1^*: The challenger computes $s_1^* \leftarrow \mathsf{REnc}(pk, m^*, \mathcal{H}^*)$ where REnc is defined as follows.

Algorithm $\mathsf{REnc}(pk, m, \mathcal{H})$
$c \leftarrow \mathcal{U}_{b\ell}$
denote $c = (\mathbf{c}_1, \mathbf{c}_2, \ldots, \mathbf{c}_\ell)$ where each $\mathbf{c}_i \in \{0,1\}^b$
for $1 \leq i \leq \ell$
    $\tilde{s} \leftarrow EX_{\mathcal{C}}(\mathcal{H})$ where $\tilde{s} = (\tilde{s}_1, \tilde{s}_2, \ldots, \tilde{s}_{bt}) \in \{0,1\}^{bt}$
    $\mathbf{H} := EX_{\mathcal{C}}(\mathcal{H})$ where $\mathbf{H} = (\mathbf{h}_1, \mathbf{h}_2, \ldots, \mathbf{h}_{bt}) \in \{0, 1, \times\}^{bt}$
    $s_i \leftarrow$ **Monomial-modify'**$(\mathbf{c}_i, \tilde{s}, \mathbf{H})$
    $\mathcal{H} := \mathcal{H} \| s_i$
end
return $s = (s_1, s_2, \ldots, s_\ell)$.

- s_2^*: The challenger samples $s_2^* \leftarrow \mathcal{C}_{\mathcal{H}^*}^{f^*}$.

Then the SS-CHA-advantage of W is denoted by

$$\mathbf{Adv}_{\mathcal{S},\mathcal{C},W}^{\mathsf{ss\text{-}cha}}(\lambda) = |\Pr[\mathbf{Exp}_{\mathcal{C},W}^0(1^\lambda) = 1] - \Pr[\mathbf{Exp}_{\mathcal{C},W}^2(1^\lambda) = 1]|.$$

Now, we give two claims. From these claims and the triangle inequality, we can obtain the claimed result in Theorem 1.

Claim 1. *It holds that for any W, there exists an adversary A such that*

$$|\Pr[\mathbf{Exp}_{\mathcal{C},W}^0(1^\lambda) = 1] - \Pr[\mathbf{Exp}_{\mathcal{C},W}^1(1^\lambda) = 1]| \leq \mathbf{Adv}_{\mathcal{E},A}^{\mathsf{ind\$\text{-}cpa}}(\lambda).$$

Claim 2. *It holds that for any W,*

$$|\Pr[\mathbf{Exp}_{\mathcal{C},W}^1(1^\lambda) = 1] - \Pr[\mathbf{Exp}_{\mathcal{C},W}^2(1^\lambda) = 1]| = 0.$$

We give the proofs of these claims, and the details are given in the full version.

Proof. (Claim 1, Sketch) The difference of $\mathbf{Exp}^0(1^\lambda)$ and $\mathbf{Exp}^1(1^\lambda)$ is only how to compute c. In $\mathbf{Exp}^0(1^\lambda)$, c is the ciphertext of m according to pk. On the other hand, c is chosen randomly in $\mathbf{Exp}^1(1^\lambda)$. Hence, if the underlying public-key encryption scheme satisfies IND\$-CPA-security, then the value

$$\left|\Pr[\mathbf{Exp}_{\mathcal{C},W}^0(1^\lambda) = 1] - \Pr[\mathbf{Exp}_{\mathcal{C},W}^1(1^\lambda) = 1]\right|$$

is negligibly small. $\square$

Proof. (Claim 2, Sketch) In order to prove this claim, it is sufficient to show that for all s', $\mathcal{H}$, and $\mathbf{H}$,

$$\Pr[\textbf{Monomial-modify'}(c, s, \mathbf{H}) = s'] = \Pr[s' \leftarrow EX_{\mathcal{C}}(\mathcal{H})]$$

where $c \leftarrow \mathcal{U}_{b\ell}$ and $s \leftarrow EX_{\mathcal{C}}(\mathcal{H})$. If the above equation holds, the distribution of s_1^* and that of s_2^* are identical and then Claim 2 immediately holds.

The strategy is as follows. We denote $s' = (s_1', \ldots, s_{bt}')$ where each $s_i' \in \{0,1\}$. Then we estimate the probability $\Pr[s_i' = 0]$ of the left side and that of the right side, and show that they are identical for all i. $\square$

We complete the proof of Theorem 1. $\square$

References

1. Anderson, R.J., Petitcolas, F.A.: On the Limits of Steganography. IEEE Journal of Selected Areas in Communications 16(4), 474–481 (1998)
2. Backes, M., Cachin, C.: Public-Key Steganography with Active Attacks. In: Kilian, J. (ed.) TCC 2005. LNCS, vol. 3378, pp. 210–226. Springer, Heidelberg (2005)
3. Cachin, C.: An Information-Theoretic Model for Steganography. Information of Computation 192(1), 41–56 (2004)
4. Canetti, R., Krawczyk, H., Nielsen, J.B.: Relaxing Chosen-Ciphertext Security. In: Boneh, D. (ed.) CRYPTO 2003. LNCS, vol. 2729, pp. 565–582. Springer, Heidelberg (2003)
5. Cramer, R., Shoup, V.: A Practical Public Key Cryptosystem Provably Secure Against Adaptive Chosen Ciphertext Attack. In: Krawczyk, H. (ed.) CRYPTO 1998. LNCS, vol. 1462, pp. 13–25. Springer, Heidelberg (1998)
6. Dedic, N., Itkis, G., Reyzin, L., Russell, S.: Upper and Lower Bounds on Black-Box Steganography. Journal of Cryptology 22(3), 365–394 (2009)
7. Hopper, N.J.: On Steganographic Chosen Covertext Security. In: Caires, L., Italiano, G.F., Monteiro, L., Palamidessi, C., Yung, M. (eds.) ICALP 2005. LNCS, vol. 3580, pp. 311–323. Springer, Heidelberg (2005)
8. Hopper, N.J., von Ahn, L., Langford, J.: Provably Secure Steganography. IEEE Transactions of Computers 58(5), 662–676 (2009)
9. Hundt, C., Liśkiewicz, M., Wölfel, U.: Provably Secure Steganography and the Complexity of Sampling. In: Asano, T. (ed.) ISAAC 2006. LNCS, vol. 4288, pp. 754–763. Springer, Heidelberg (2006)
10. Kurosawa, K., Desmedt, Y.: A New Paradigm of Hybrid Encryption Scheme. In: Franklin, M. (ed.) CRYPTO 2004. LNCS, vol. 3152, pp. 426–442. Springer, Heidelberg (2004)
11. Liśkiewicz, M., Reischuk, R., Wölfel, U.: Grey-Box Steganography. In: Ogihara, M., Tarui, J. (eds.) TAMC 2011. LNCS, vol. 6648, pp. 390–402. Springer, Heidelberg (2011)
12. Lysyanskaya, A., Meyerovich, M.: Provably Secure Steganography with Imperfect Sampling. In: Yung, M., Dodis, Y., Kiayias, A., Malkin, T. (eds.) PKC 2006. LNCS, vol. 3958, pp. 123–139. Springer, Heidelberg (2006)
13. Mittelholzer, T.: An Information-Theoretic Approach to Steganography and Watermarking. In: Pfitzmann, A. (ed.) IH 1999. LNCS, vol. 1768, pp. 1–16. Springer, Heidelberg (2000)
14. Simmons, G.J.: The Prisoners' Problem and the Subliminal Channel. In: Chaum, D. (ed.) CRYPTO 1983, pp. 51–67. Plenum Press, New York (1984)
15. von Ahn, L., Hopper, N.J.: Public-Key Steganography. In: Cachin, C., Camenisch, J.L. (eds.) EUROCRYPT 2004. LNCS, vol. 3027, pp. 323–341. Springer, Heidelberg (2004)

Linear Vertex-kernels for Several Dense
RANKING r-CONSTRAINT SATISFACTION Problems

Anthony Perez

LIFO, Université d'Orléans
anthony.perez@univ-orleans.fr

Abstract. A RANKING r-CONSTRAINT SATISFACTION problem (ranking r-CSP for short) consists of a ground set of vertices V, an arity $r \geqslant 2$, a parameter $k \in \mathbb{N}$ and a *constraint system* c, where c is a function which maps rankings (*i.e.* orderings) of r-sized sets $S \subseteq V$ to $\{0, 1\}$ [16]. The objective is to decide if there exists a ranking σ of the vertices satisfying all but at most k constraints (*i.e.* $\sum_{S \subseteq V, |S|=r} c(\sigma(S)) \leqslant k$). Famous ranking r-CSPs include FEEDBACK ARC SET IN TOURNAMENTS and DENSE BETWEENNESS [4,15]. In this paper, we prove that so-called l_r-*simply characterized* ranking r-CSPs admit linear vertex-kernels whenever they admit constant-factor approximation algorithms. This implies that r-DENSE BETWEENNESS and r-DENSE TRANSITIVE FEEDBACK ARC SET [15], two natural generalizations of the previously mentioned problems, admit linear vertex-kernels. Both cases were left opened by Karpinksi and Schudy [16]. We also consider another generalization of FEEDBACK ARC SET IN TOURNAMENTS for constraints of arity $r \geqslant 3$, that does not fit the aforementioned framework. Based on techniques from [11], we obtain a 5-approximation and then provide a linear vertex-kernel. As a main consequence of our result, we obtain the first constant-factor approximation algorithm for a particular case of the so-called DENSE ROOTED TRIPLET INCONSISTENCY problem [9].

1 Introduction

Parameterized complexity is a powerful theoretical framework to cope with NP-Hard problems. The aim is to identify some *parameter* k independent from the instance size n, which captures the exponential growth of the complexity to solve the problem at hand. A parameterized problem is said to be *fixed parameter tractable* whenever it can be solved in $f(k) \cdot n^{O(1)}$ time, where f is any computable function [12,18]. In this paper, we focus on *kernelization*. A *kernelization algorithm* (or kernel for short) for a parameterized problem Π is a *polynomial-time* algorithm that given an instance (I, k) of Π outputs an *equivalent* instance (I', k') of Π such that $|I'| \leqslant g(k)$ and $k' \leqslant k$. The function g is said to be the *size* of the kernel, and Π admits a *polynomial kernel* whenever g is a polynomial. A well-known result states that a (decidable) parameterized problem is fixed parameter tractable if and only if it admits a kernel [18]. Observe

T-H.H. Chan, L.C. Lau, and L. Trevisan (Eds.): TAMC 2013, LNCS 7876, pp. 306–318, 2013.
© Springer-Verlag Berlin Heidelberg 2013

that this result provides kernels of super-polynomial size. Recently, several results gave evidence that some parameterized problems *do not* admit *polynomial* kernels (under complexity-theoretic assumptions [7,8]).

We mainly study ranking r-CSPs from the kernelization viewpoint. In a ranking r-CSP, a ground set of vertices V, an arity $r \geqslant 2$ and a set of constraints defined on r-sized subsets $S \subseteq V$ are given. Here, a constraint corresponds to some *allowed* rankings on S. The aim of such problems is to find a linear ranking on V that minimizes the number of constraints ranked in a non allowed manner. We study the decision version of such problems, where the instance comes together with some parameter $k \in \mathbb{N}$ and the aim is to decide if there exists a ranking satisfying *all but at most* k constraints. We consider such problems on *dense instances*, where *every* set of r vertices is a constraint. For instance, FEEDBACK ARC SET IN TOURNAMENTS[1] fits this framework with $r = 2$, any arc uv being satisfied by a ranking σ iff $u <_\sigma v$. Such problems can be equivalenty stated in terms of editing problems: can we *edit* at most k constraints to obtain an instance that admits a ranking satisfying all its constraints?

Related Results. While a lot of kernelization results are known for *graph* editing problems [6,17,20,21], fewer results exist regarding directed graph and hypergraph editing problems. An example of polynomial kernel for a directed graph editing problem is the quadratic vertex-kernel for TRANSITIVITY EDITING [22]. Regarding dense ranking r-CSPs, FEEDBACK ARC SET IN TOURNAMENTS and DENSE BETWEENNESS are NP-Complete [2,3,10] but fixed parameter tractable [4,15], and both admit a linear vertex-kernel [5,19]. Recently, Karpinski and Schudy [16] showed PTASs and subexponential parameterized algorithms for (*weakly*)-fragile ranking r-CSPs. A constraint is (*weakly-*)*fragile* if whenever it is satisfied by one ranking then making one single move (resp. making one of the following moves: swapping the first two vertices, the last two vertices or making a cyclic move) makes it unsatisfied.

Our Results. We introduce so-called l_r-*simply characterized* ranking r-CSPs, and prove that such problems admit linear vertex-kernels whenever they admit constant-factor approximation algorithms (Section 3). Surprisingly, our kernels mainly use a modification of the classical sunflower reduction rule, which usually provides polynomial kernels [4,6,13]. This result implies linear vertex-kernels for r-DENSE BETWEENNESS and r-DENSE TRANSITIVE FEEDBACK ARC SET, two natural generalizations of FEEDBACK ARC SET IN TOURNAMENTS and DENSE BETWEENNESS [16]. Both cases were left opened by Karpinski and Schudy [16]. Finally, we introduce a different generalization of FEEDBACK ARC SET IN TOURNAMENTS for constraints of arity $r \geqslant 3$, which allows more freedom on the satisfiability of a constraint. We mainly focus on the case $r = 3$. We first state that the problem is NP-Complete in this case. Next, based on ideas used for FEEDBACK ARC SET IN TOURNAMENTS [11], we prove that the general case admits a 5-approximation algorithm, and then obtain a linear vertex-kernel (Section 4.3).

---

[1] A tournament is an arbitrary orientation of the complete (undirected) graph.

This result implies a 5-approximation for a particular case of DENSE ROOTED TRIPLET INCONSISTENCY [9]. Notice that finding a constant-factor approximation algorithm for the general case is a well-known open problem [14,19].

2   Preliminaries

Following notations from [16], a ranking r-CSP consists of a ground set of vertices V, an arity $r \geqslant 2$, a parameter $k \in \mathbb{N}$ and a *constraint system* c, where c is a function which maps rankings (*i.e.* orderings) of r-sized sets $S \subseteq V$ to $\{0,1\}$. In a slight abuse of notation, we refer to a set of vertices $S \subseteq V$, $|S| = r$, as a *constraint* (when we are actually referring to c applied to rankings of S). A constraint S is *non-trivial* whenever there exists a ranking σ such that $c(\sigma(S)) = 1$. In the following, we always mean non-trivial constraints when speaking of constraints. A constraint S is *satisfied* by a ranking σ whenever $c(\sigma(S)) = 0$, in which case S is said to be *consistent* w.r.t. σ (we forget the mention *w.r.t.* σ whenever the context is clear). Otherwise, we say that S is *inconsistent*. Similarly, a ranking σ is *consistent* with the constraint system c if it does not contain any inconsistent constraint, and *inconsistent* otherwise. The objective of a ranking r-CSP is to find a ranking of the vertices with *at most* k inconsistent constraints. We consider *dense* instances, where *every* subset of r vertices of V is a constraint. Moreover, we assume that a constraint S can be represented by a subset $sel(S) \subseteq S$ of *selected vertices*, that determine the conditions that a ranking must verify in order to satisfy S.

Let $R = (V, c)$ be an instance of any ranking r-CSP. Given a set of vertices $V' \subseteq V$, we define the instance *induced by* V' (and denote it $R[V']$) as the constraint system c restricted to r-sized subsets of V'. A set of vertices $C \subseteq V$ is a *conflict* if there does not exist any ranking consistent with the instance induced by C. We mainly study the following problems.

---

r-DENSE BETWEENNESS (r-BIT) [16]:

Input: A set of vertices V, an arity $r \geqslant 3$ and a constraint system c, where a constraint $S = \{s_1, \ldots, s_r\}$ contains two *selected vertices* s_i and s_j, $1 \leqslant i < j \leqslant r$, and is satisfied by a ranking σ (*i.e.* $c(\sigma(S)) = 0$) iff $s_i <_\sigma s_l <_\sigma s_j$ or $s_j <_\sigma s_l <_\sigma s_i$ holds for $1 \leqslant l \leqslant r$, $l \neq \{i, j\}$.

Parameter: k.

Output: A ranking σ of V that satisfies all but at most k constraints.

---

r-DENSE FEEDBACK ARC SET (r-DFAS):

Input: A set of vertices V, an arity $r \geqslant 3$ and a constraint system c, where a constraint S contains one selected vertex s and is satisfied by a ranking σ (*i.e.* $c(\sigma(S)) = 0$) iff $u <_\sigma s$ for any $u \in S \setminus \{s\}$.

Parameter: k.

Output: A ranking σ of V that satisfies all but at most k constraints.

An equivalent formulation of these problems is the following: is it possible to *edit* at most k constraints so that there exists a ranking consistent with the new constraint system? By *editing a constraint*, we mean that we modify its set of selected vertices (observe in particular that we do not modify V).

We also consider another generalization of the FEEDBACK ARC SET IN TOURNAMENTS problem, namely r-DENSE TRANSITIVE FEEDBACK ARC SET (r-DTFAS) [16], where a constraint S corresponds to an acyclic tournament and is satisfied by a ranking σ if and only if σ is the transitive ranking of the corresponding tournament (recall that a tournament is acyclic if and only if it admits a transitive ranking σ, *i.e.* a ranking satisfying $u <_\sigma v$ for any arc uv).

Ordered Instances. In the following, we consider instances whose vertices are ordered under some fixed ranking σ (*i.e.* instances of the form $R_\sigma = (V, c, \sigma)$). Given any constraint $S = \{s_1, \ldots, s_r\}$, with $s_i <_\sigma s_{i+1}$ for $1 \leqslant i < r$, $span(S)$ denotes the set of vertices $\{v \in V : s_1 \leqslant_\sigma v \leqslant_\sigma s_r\}$. A constraint S is *unconsecutive* if $|span(S)| > r$, and *consecutive* otherwise. Given $V' \subseteq V$, $R_\sigma[V']$ denotes the instance $R[V']$ ordered under σ. Finally, given a ranking σ over V and an inconsistent constraint S, we say that we *edit S w.r.t.* σ whenever we edit its selected vertices so that it becomes consistent w.r.t. σ.

3   Simple Characterization and Sunflower

We now describe the general framework of our kernelization algorithms, using a modification of the *sunflower rule* together with the notion of *simple characterization*. We first define the notion of sunflower, which has been widely used to obtain polynomial kernels for modification problems [1,4,6,13]. An *editing set* is a set of constraints $\mathcal{F}$ such that one can obtain a consistent instance by editing constraints in $\mathcal{F}$.

Definition 1 (Sunflower). *A* sunflower $\mathcal{S}$ *is a set of conflicts* $\{C_1, \ldots, C_m\}$ *pairwise intersecting in* exactly one *constraint S, called the* center *of $\mathcal{S}$.*

Lemma 1 (Folklore). *Let $R = (V, c)$ be an instance of any ranking r-CSP, and S be the center of a sunflower $\mathcal{S} = \{C_1, \ldots, C_m\}$, $m > k$. Any editing set of size at most k has to edit S.*

Observe that the sunflower rule cannot be applied directly on ranking r-CSPs, $r \geqslant 3$, since it may be the case that there exist several ways to edit the center of a given sunflower. In order to deal with this, we introduce the notion of *simple characterization* for ranking r-CSPs. Roughly speaking, a ranking r-CSP is l_r-*simply characterized* if for any ordered instance, any set of l_r vertices which involve *exactly one* inconsistent constraint is a conflict.

Definition 2 (Simple characterization). *Let Π be a ranking r-CSP, $R_\sigma = (V, c, \sigma)$ be any ordered instance of Π, and $l_r \in \mathbb{N}$. The ranking r-CSP Π is l_r-simply characterized iff any l_r-sized set $C \subseteq V$ such that $R_\sigma[C]$ contains exactly one inconsistent constraint is a conflict.*

Definition 3 (Simple sunflower). *Let $R_\sigma = (V, c, \sigma)$ be an ordered instance of a ranking r-CSP. A sunflower $S = \{C_1, \ldots, C_m\}$ of R_σ is simple if its center is the only inconsistent constraint in $R_\sigma[C_i]$, $1 \leqslant i \leqslant m$.*

Rule 1. *Let Π be a l_r-simply characterized ranking r-CSP. Let $R_\sigma = (V, c, \sigma)$ be an ordered instance of Π and $S = \{C_1, \ldots, C_m\}$, $m > k$, be a simple sunflower of center S. Edit S w.r.t. σ and decrease k by 1.*

Lemma 2. *Rule 1 is sound.*

Proof. Let $\mathcal{F}$ be any editing set of size at most k: by Lemma 1, $\mathcal{F}$ must contain S. Since $|\mathcal{F}| \leqslant k$ and $m > k$, there exists $1 \leqslant i \leqslant m$ such that S is the only constraint edited by $\mathcal{F}$ in $R[C_i]$. Assume that S was not edited w.r.t. σ: since no other constraint has been edited in $R[C_i]$, $R_\sigma[C_i]$ still contains exactly one inconsistent constraint (namely S). Since Π is l_r-simply characterized, it follows that C_i defines a conflict, contradicting the fact that $\mathcal{F}$ is an editing set.   □

The main problem that remains is to compute such a sunflower in polynomial time. The following result will allow us to do so, providing that V contains sufficiently many vertices (w.r.t. parameter k).

Lemma 3. *Let Π be a l_r-simply characterized ranking r-CSP, and $R_\sigma = (V, c, \sigma)$ be an ordered instance of Π with at most $p \geqslant 1$ inconsistent constraints. If $|V| > p(l_r - r) + (l_r - r) \cdot (k + 1) + r$, then there exists a simple sunflower $\{C_1, \ldots, C_m\}$, $m > k$, that can be found in polynomial time.*

Proof. Let S be any inconsistent constraint of R_σ. Since R_σ contains at most p inconsistent constraints, there are at most p disjoint sets P_i, $1 \leqslant i \leqslant p$, such that $|P_i| = l_r - r$ and $R_\sigma[S \cup P_i]$ contains more than one inconsistent constraint. It follows that there exist at least $m \geqslant k + 1$ disjoint sets $\{S_1, \ldots, S_m\}$ of size $l_r - r$ such that: (i) $C_i = S \cup S_i$ contains l_r vertices and (ii) $R_\sigma[C_i]$ contains *exactly one* inconsistent constraint, $1 \leqslant i \leqslant m$. Since Π is l_r-simply characterized, C_i defines a conflict for every $1 \leqslant i \leqslant m$. It follows that $\{C_1, \ldots, C_m\}$ is a simple sunflower of center S.   □

Theorem 1. *Let Π be a l_r-simply characterized ranking r-CSP that admits a q-factor approximation algorithm for some constant $q > 0$. Then Π admits a kernel with at most $k[(q + 1) \cdot (l_r - r)] + l_r$ vertices.*

Proof. Let $R = (V, c)$ be an instance of Π. We start by computing a ranking σ containing p inconsistent constraints using the q-factor approximation algorithm. Observe that we can assume that $p > k$, since otherwise we simply return a small trivial YES-instance. Similarly, we can assume that $p \leqslant qk$, since otherwise we return a small trivial NO-instance. We now consider $R_\sigma = (V, c, \sigma)$ and assume that $|V| > p(l_r - r) + (l_r - r) \cdot (k + 1) + r$: by Lemma 3, it follows that there exists a simple sunflower that can be found in polynomial time, and hence Rule 1 can be applied. Since conditions of Lemma 3 still hold after an application of Rule 1, repeating this process on R_σ implies that every inconsistent constraint must be edited. Hence, since $p > k$, we return a small trivial NO-instance in such a case. This means that $|V| \leqslant qk(l_r - r) + (l_r - r) \cdot (k + 1) + r$, implying the result.   □

4 Simple Characterization of Several Ranking r-CSPs

4.1 3-DENSE BETWEENNESS (BIT)

As a first consequence of Theorem 1, we improve the size of the linear vertex-kernel for BIT from $5k$ [19] to $(2+\epsilon)k+4$ for any $\epsilon > 0$. The result directly follows from the fact that BIT admits a PTAS [16] and is 4-simply characterized [19].

Corollary 1. DENSE BETWEENNESS *admits a kernel with at most* $(2 + \epsilon)k + 4$ *vertices.*

4.2 r-DENSE BETWEENNESS $(r \geqslant 4)$

We now consider the r-BIT problem with constraints of arity $r \geqslant 4$. The main difference with the case $r = 3$ lies in the fact that there is no longer a *unique* way to rank the vertices in order to satisfy all constraints. In particular, this means that the problem is not $(r + 1)$-simply characterized.

Compatible Constraints. However, one can prove that r-BIT is $2r$-simply characterized. To see this, we need the following definition.

Definition 4 (Compatible constraint). *Given an ordered instance* $R_\sigma = (V, c, \sigma)$ *of* r-DENSE BETWEENNESS, *an inconsistent constraint* $S = \{s_1, \ldots, s_r\}$, $s_i <_\sigma s_{i+1}$, $1 \leqslant i < r$, *is* right- (resp. left-)compatible *whenever* $sel(S) = \{s_1, s_l\}$, $2 < l < r$ (resp. $sel(S) = \{s_l, s_r\}$, $1 < l < r - 1$).

Any constraint that does not satisfy Definition 4 is called right- (resp. left-)*incompatible*. The intuition behind Definition 4 is the following: for any vertex u lying after (resp. before) S in σ such that S is the only inconsistent constraint in $R_\sigma[S \cup \{u\}]$, the set $S \cup \{u\}$ *does not* define a conflict (see Figure 1).

Fig. 1. Illustration of the notion of left-compatible constraints (only S is inconsistent). By definition, s_1 is not selected in any constraint, and u and s_r are selected in every constraint but S. Hence swapping s_1 and s_l yields a consistent ranking for $S \cup \{u\}$.

A particular consequence of Definition 4 is that the problem is not $(r + 1)$-simply characterized. Indeed, for instance, any ordered instance on $r+1$ vertices whose only inconsistent constraint $S = \{s_1, \ldots, s_r\}$ is right-compatible does not define a conflict.

The following result comes from definition of compatible constraints.

Observation 2. *Any right- (resp. left-)compatible constraint is left- (resp. right-)incompatible.*

Lemma 4. *The r-BIT problem is $2r$-simply characterized.*

Proof. We use the following result.

Claim 3. *Let $R_\sigma = (V, c, \sigma)$ be an ordered instance of r-BIT, and $C = \{s_1, \ldots, s_{r+1}\}$ be a set of $r + 1$ vertices s.t. $s_i <_\sigma s_{i+1}$, $1 \leqslant i \leqslant r$. Assume that $R_\sigma[C]$ contains exactly one inconsistent constraint S. If one of the following holds:*

(i) S is unconsecutive or,
(ii) S is neither right- nor left-compatible, i.e. $S = \{s_1, \ldots, s_r\}$ and $sel(S) \neq \{s_1, s_l\}$, $2 < l < r$ or $S = \{s_2, \ldots, s_{r+1}\}$ and $sel(S) \neq \{s_l, s_{r+1}\}$, $2 < l < r$.

then C is a conflict.

Let $R_\sigma = (V, c, \sigma)$ be an ordered instance of r-BIT and $C = \{s_1, \ldots, s_{2r}\}$ be a set of $2r$ vertices such that $s_i <_\sigma s_{i+1}$ for $1 \leqslant i < 2r$. Assume that $R_\sigma[C]$ contains exactly one inconsistent constraint S. We need to prove that C is a conflict. By Claim 3, the result holds if S is neither right- nor left-compatible. So we assume that S is right-compatible (the case left-compatible is similar). By Claim 3 we can also assume that the vertices of S are consecutive and are the first of the ranking, since otherwise C is a conflict and we are done (recall that S is left-incompatible by Observation 2). In other words we may assume that $S = \{s_1, \ldots, s_r\}$ and $sel(S) = \{s_1, s_l\}$ for $2 < l < r$. Since S is the only inconsistent constraint in $R_\sigma[C]$, the constraints $S_2 = \{s_l, \ldots, s_r, \ldots, s_{l+r}\}$ and $S_3 = \{s_1, \ldots, s_l, s_r, \ldots, s_{l+r}\}$ (with $|S_3| = r$) have as selected vertices $sel(S_2) = \{s_l, s_{l+r}\}$ and $sel(S_3) = \{s_1, s_{l+r}\}$, respectively. In order to be consistent with S and S_2, any ranking ρ must rank s_r between $\{s_1, s_{l+r}\}$ and s_l, which is inconsistent with the last constraint (which forces s_r to be between s_1 and s_{l+r}). $\square$

Corollary 2. *r-BIT admits a kernel with at most $(2 + \epsilon)rk + 2r$ vertices.*

4.3 r-Dense Transitive Feedback Arc Set (r-DTFAS)

Karpinski and Schudy [16] considered a particular generalization of the Feedback Arc Set in Tournaments problem, where every constraint S corresponds to an acyclic tournament. We show that the r-DTFAS problem admits a linear vertex-kernel as a particular case of *fragile* ranking r-CSP, a notion introduced in [16]. We say that a ranking r-CSP is *strongly-fragile* whenever a constraint is satisfied by *one particular ranking* and no other.

Lemma 5. *Let Π be any strongly-fragile ranking r-CSP, $r \geqslant 3$. Then Π is $(r + 1)$-simply characterized.*

Proof. Let $R = (V, c)$ be any instance of Π, σ be any ranking of V and C be a set of $r + 1$ vertices such that $R_\sigma[C]$ contains exactly one inconsistent constraint S. We need to prove that C is a conflict. Assume for a contradiction that this is not the case, *i.e.* that there exists a ranking ρ consistent with $R[C]$. In particular, there exist two vertices $u, v \in S$ that are such that $u <_\sigma v$ and $v <_\rho u$. Let $S' \neq S$ be any constraint of $R[C]$ such that $\{u, v\} \subset S'$ (observe that S' is well-defined since $r \geqslant 3$). Since S' was consistent in σ and since Π is strongly-fragile, S' is inconsistent in ρ: a contradiction. □

Corollary 3. *Any strongly-fragile ranking r-CSP admits a kernel with at most $(2 + \epsilon)k + (r + 1)$ vertices.*

4.4 r-DENSE FEEDBACK ARC SET (r-DFAS)

As mentioned previously, the r-DTFAS problem deals with constraints that are given by a transitive tournament and are thus satisfied by *one particular ranking* and no other. To allow more freedom on the satisfiability of a constraint, we consider a different generalization of this problem, namely r-DFAS. Recall that in this problem, any constraint S contains a selected vertex s and is satisfied by a ranking σ iff $u <_\sigma s$ for any $u \in S \setminus \{s\}$.

We mainly consider the 3-DFAS problem, which turns out to be equivalent to a particular case of DENSE ROOTED TRIPLET INCONSISTENCY [9], where one is given a set of vertices V, a dense collection $\mathcal{R}$ of rooted binary trees on three vertices and an integer $k \in \mathbb{N}$, and seeks a rooted *binary caterpillar tree*[2] defined over V containing all but at most k trees from $\mathcal{R}$. We thus have the following.

Observation 4. *The 3-DFAS problem is NP-Complete.*

Notice however that the results presented stand for the general case. Now, observe that r-DFAS is not (weakly-)fragile: swapping the first two vertices of any consistent ranking yields a consistent ranking. Hence, we cannot directly apply the PTASs from [16].

Approximation Algorithm. We show that the results needed to obtain a 5-approximation for FEEDBACK ARC SET IN TOURNAMENTS [11] can be generalized to the r-DENSE FEEDBACK ARC SET problem.

Definition 5 (In-degree). *Let $R = (V, c)$ be an instance of r-DFAS, and $v \in V$. The in-degree $In(v)$ of v is the number of constraints where v is selected.*

Algorithm [INC-DEGREE] Order the vertices of R according to their increasing in-degrees.

Theorem 5. INC-DEGREE *is a 5-approximation for r-DFAS.*

We prove Theorem 5 by proving a series of Lemmata. For the sake of simplicity, we let $V = \{1, \ldots, n\}$ in the remaining of this Section.

---

[2] A *binary caterpillar tree* is a rooted binary tree in which every internal node has at least one child that is a leaf.

Definition 6 (Left constraint). *Let $R_\sigma = (V, c, \sigma)$ be an ordered instance of r-DFAS. For any vertex $v \in V$, a* left constraint *$S = \{s_1, \ldots, v, \ldots s_r\}$ is a constraint containing v and vertices before v only (i.e. $s_i \leqslant_\sigma v$, $1 \leqslant i \leqslant r$).*

Given an ordered instance $R_\sigma = (V, c, \sigma)$ of r-DFAS, we let $L_\sigma(v)$ be the set of left constraints containing v, and $l_\sigma(v) = |L_\sigma(v)|$. Moreover, we define $\mathcal{B}_\sigma$ as the set of inconsistent constraints of R_σ, and let $b_\sigma = |\mathcal{B}_\sigma|$. To obtain the algorithm, we need to define a distance function $\mathcal{K}(\rho, \gamma)$ between two rankings ρ and γ. Roughly speaking, $\mathcal{K}(\rho, \gamma)$ gives the number of constraints which are consistent in exactly one out of the two rankings, and thus generalizes the Kendall-Tau distance between two rankings [11]. Formally, we obtain:

$$\mathcal{K}(\rho, \gamma) = \sum_{S \subseteq V, |S| = r} \mathbf{1}_{(c(\rho(S)) = 1 \wedge c(\gamma(S)) = 0) \vee (c(\rho(S)) = 0 \wedge c(\gamma(S)) = 1)}$$

where $\mathbf{1}$ denotes the indicative function. Observe that a constraint S consistent in one out of the two rankings satisfies $c(\gamma(S)) = 0$ *and* $c(\rho(S)) = 1$ or vice versa. The first result gives a bound on the differences between the number of left constraints containing vertices and their in-degree in terms of b_ρ.

Lemma 6. *Let $\rho : V \to V$ be any ranking. The following holds:*

$$2 \cdot b_\rho \geqslant \sum_{v \in V} |l_\rho(v) - In(v)|$$

In the following, we denote by σ_A the ranking returned by INC-DEGREE, and by σ_O the ranking returned by any optimal solution. The following Lemma states that the ranking minimizing the differences between the number of left constraints containing vertices and their in-degree is σ_A.

Lemma 7. *Let $\rho : V \to V$ be any ranking. The following holds:*

$$\sum_{v \in V} |l_\rho(v) - In(v)| \geqslant \sum_{v \in V} |l_{\sigma_A}(v) - In(v)|$$

Lemma 8. *Let $\rho, \gamma : V \to V$ be two rankings. The following holds:*

$$\sum_{v \in V} |l_\rho(v) - l_\gamma(v)| \geqslant |b_\rho - b_\gamma|$$

We are now ready to prove the main result of this section :

Proof (of Theorem 5). By the previous Lemmata, we have the following :

$$
\begin{aligned}
4b_{\sigma_O} &\geqslant \sum_{v \in V} |l_{\sigma_O}(v) - In(v)| + \sum_{v \in V} |l_{\sigma_O}(v) - In(v)| &&\text{(Lemma 6)}\\
&\geqslant \sum_{v \in V} |l_{\sigma_O}(v) - In(v)| + \sum_{v \in V} |l_{\sigma_A}(v) - In(v)| &&\text{(Lemma 7)}\\
&= \sum_{v \in V} (|l_{\sigma_O}(v) - In(v)| + |l_{\sigma_A}(v) - In(v)|)\\
&\geqslant \sum_{v \in V} |l_{\sigma_O}(v) - l_{\sigma_A}(v)|\\
&\geqslant b_{\sigma_A} - b_{\sigma_O} &&\text{(Lemma 8)}
\end{aligned}
$$

Hence we have $b_{\sigma_A} \leqslant 5b_{\sigma_O}$, which implies the result. $\qquad\square$

Corollary 4. INC-DEGREE *is a* 5*-approximation for* DENSE ROOTED TRIPLET INCONSISTENCY *restricted to binary caterpillar trees.*

To the best of our knowledge, this constitutes the first constant-factor approximation algorithm regarding the DENSE ROOTED TRIPLET INCONSISTENCY problem. We would like to mention that finding a constant-factor approximation for the general case constitutes an important open problem [9,19].

Kernelization Algorithm. In order to obtain our kernelization algorithm, we need to study the topology of conflicts that contain exactly one inconsistent constraint. As we shall see, the configuration for r-DENSE FEEDBACK ARC SET is slightly different than the ones previously observed. In particular, the problem is not l_r-simply characterized. However, the addition of a new reduction rule will allow us to conclude as in the other cases.

Lemma 9. *Let* $R_\sigma = (V, c, \sigma)$ *be an ordered instance of* r-DFAS, *and* $C = \{s_1, \ldots, s_{r+1}\}$ *be a set of* $r+1$ *vertices such that* $s_i <_\sigma s_{i+1}$ *for every* $1 \leqslant i \leqslant r$. *Assume that* $R_\sigma[C]$ *contains exactly one inconsistent constraint* S. *Then* C *is a conflict if and only if* S *is unconsecutive or* $S = \{s_2, \ldots, s_{r+1}\}$.

Corollary 5. *There does not exist* $l_r \in \mathbb{N}$ *such that* r-DFAS *is* l_r*-simply characterized.*

Proof. Let $R = (V, c)$ be an instance of r-DFAS and $q \in \mathbb{N}$, $q > r$. Let $C = \{s_1, \ldots, s_q\}$ be any set of vertices ordered under some ranking σ such that $s_i <_\sigma s_{i+1}$ for $1 \leqslant i < q$. Assume that $S = \{s_1, \ldots, s_r\}$ is the only inconsistent constraint of $R_\sigma[C]$. By Lemma 9, we know that C is not a conflict, implying that r-DFAS is not q-simply characterized.   □

We need the following rule, which will imply that the last vertex of any ordered instance of r-DFAS belongs to (at least) one inconsistent constraint.

Rule 2. *Let* v *be any vertex which is selected in every constraint containing it. Remove* v *from* V *and modify the constraint system* c *consequently.*

Lemma 10. *Rule 2 is sound and can be applied in polynomial time.*

Proof. First, observe that any editing set of size at most k for the original instance will yield an editing set for the reduced one. In the other direction, assume that $R_v = (V \setminus \{v\}, c')$ admits an editing set $\mathcal{F}$ of size at most k, and let σ be the consistent ranking obtained after editing the constraints of $\mathcal{F}$. Since adding v to the end of σ does not introduce any inconsistent constraint, $\mathcal{F}$ is also an editing set for the original instance.   □

Notice that a given instance can contain at most one such vertex. We thus iteratively apply Rule 2 until no vertex selected in every constraint containing it remains.

Given any constraint S of an ordered instance $R_\sigma = (V, c, \sigma)$ of r-DFAS, $span^-(S)$ denotes the set containing $span(S)$ and all vertices lying before S in σ. Observe that by Lemma 9, any set $C \subseteq V$ of $r + 1$ vertices such that $R_\sigma[C]$ contains exactly one inconsistent constraint S is a conflict iff $C \subseteq span^-(S)$. In the following, such a conflict will be called *simple*. We use this observation to refine the notion of *simple sunflower* in this case.

Definition 7 (Simple sunflower). *Let $R_\sigma = (V, c, \sigma)$ be an ordered instance of r-DFAS. A sunflower $\mathcal{S} = \{C_1, \ldots, C_m\}$ of center S is simple if (i) S is the only inconsistent constraint in $R_\sigma[C_i]$ and (ii) C_i is a simple conflict, $1 \leqslant i \leqslant r$.*

Observe that in any simple sunflower of center S, $\cup_{i=1}^m C_i \subseteq span^-(S)$ holds.

Rule 3. *Let $R_\sigma = (V, c, \sigma)$ be an ordered instance of r-DFAS and $\mathcal{S} = \{C_1, \ldots, C_m\}$, $m > k$, be a simple sunflower of center S. Edit S w.r.t. σ and decrease k by 1.*

Lemma 11. *Rule 3 is sound.*

Proof. Let $\mathcal{F}$ be any editing set of size at most k: by Lemma 1, $\mathcal{F}$ must contain S. Since $|\mathcal{F}| \leqslant k$, there exists $1 \leqslant i \leqslant m$ such that S is the only constraint edited by $\mathcal{F}$ in $R_\sigma[C_i]$. Assume that S was not edited w.r.t. σ: since no other constraint has been edited in $R_\sigma[C_i]$, $R_\sigma[C_i]$ still contains exactly one inconsistent constraint (namely S). Observe now that, by definition of a simple conflict, $C_i \subseteq span^-(S)$ holds. Hence Lemma 9 implies that C_i is a conflict, contradicting the fact that $\mathcal{F}$ is an editing set. $\square$

Lemma 12. *Let $R_\sigma = (V, c, \sigma)$ be an ordered instance of r-DFAS with at most $p \geqslant 1$ inconsistent constraints, and S be an inconsistent constraint s.t. $|span^-(S)| > p + k + r$. Then S is the center of a simple sunflower $\{C_1, \ldots, C_m\}$, $m > k$.*

Theorem 6. *r-DFAS admits a kernel with at most $6k + r$ vertices.*

Proof. Let $R = (V, c)$ be an instance of r-DFAS reduced under Rule 2. We start by running the constant-factor approximation (Theorem 5) on R, obtaining a ranking $\sigma = v_1 \ldots v_n$ of V with at most p inconsistent constraints. Notice that we may assume $p > k$ and $p \leqslant 5k$, since otherwise we return a small trivial YES- (resp. NO-)instance. Assume now that $|V| > p + k + r$, and let S be any inconsistent constraint containing v_n (thus $span^-(S) = V$). Recall that S is well-defined since R is reduced under Rule 2. By Lemma 11, it follows that S is the center of a simple sunflower $\{C_1, \ldots, C_m\}$, $m > k$. We thus apply Rule 3 and edit S w.r.t. σ. We now apply Rule 2 and repeat this process until we either do not find a large enough simple sunflower or $k < 0$. In the former case, Lemma 11 implies that $|V| \leqslant p + k + r \leqslant 6k + r$, while in the latter case we return a small trivial NO-instance. $\square$

Acknowledgments. Research supported by the AGAPE project (ANR-09-BLAN-0159). The author would like to thank Christophe Paul, Stéphan Thomassé, Sylvain Guillemot and Mathieu Chapelle for helpful discussions and comments.

References

1. Abu-Khzam, F.N.: A kernelization algorithm for d-hitting set. J. Comput. Syst. Sci. 76(7), 524–531 (2010)
2. Ailon, N., Alon, N.: Hardness of fully dense problems. Inf. Comput. 205(8), 1117–1129 (2007)
3. Alon, N.: Ranking tournaments. SIAM J. Discrete Math. 20(1), 137–142 (2006)
4. Alon, N., Lokshtanov, D., Saurabh, S.: Fast fast. In: Albers, S., Marchetti-Spaccamela, A., Matias, Y., Nikoletseas, S., Thomas, W. (eds.) ICALP 2009, Part I. LNCS, vol. 5555, pp. 49–58. Springer, Heidelberg (2009)
5. Bessy, S., Fomin, F.V., Gaspers, S., Paul, C., Perez, A., Saurabh, S., Thomassé, S.: Kernels for feedback arc set in tournaments. JCSS 77(6), 1071–1078 (2011)
6. Bessy, S., Perez, A.: Polynomial kernels for proper interval completion and related problems. In: Owe, O., Steffen, M., Telle, J.A. (eds.) FCT 2011. LNCS, vol. 6914, pp. 229–239. Springer, Heidelberg (2011)
7. Bodlaender, H.L., Downey, R.G., Fellows, M.R., Hermelin, D.: On problems without polynomial kernels. JCSS 75(8), 423–434 (2009)
8. Bodlaender, H.L., Jansen, B.M.P., Kratsch, S.: Cross-composition: A new technique for kernelization lower bounds. In: STACS. LIPIcs, vol. 9, pp. 165–176 (2011)
9. Byrka, J., Guillemot, S., Jansson, J.: New results on optimizing rooted triplets consistency. Discrete Applied Mathematics 158(11) (2010)
10. Charbit, P., Thomassé, S., Yeo, A.: The minimum feedback arc set problem is NP-hard for tournaments. Combinatorics, Probability & Computing 16(1), 1–4 (2007)
11. Coppersmith, D., Fleischer, L., Rudra, A.: Ordering by weighted number of wins gives a good ranking for weighted tournaments. In: SODA, pp. 776–782 (2006)
12. Downey, R.G., Fellows, M.R.: Parameterized Complexity. Springer (1999)
13. Guillemot, S., Havet, F., Paul, C., Perez, A.: On the (Non-)Existence of Polynomial Kernels for P_l-Free Edge Modification Problems. Algorithmica 65(4), 900–926 (2013)
14. Guillemot, S., Mnich, M.: Kernel and fast algorithm for dense triplet inconsistency. In: Kratochvíl, J., Li, A., Fiala, J., Kolman, P. (eds.) TAMC 2010. LNCS, vol. 6108, pp. 247–257. Springer, Heidelberg (2010)
15. Karpinski, M., Schudy, W.: Faster algorithms for feedback arc set tournament, kemeny rank aggregation and betweenness tournament. In: Cheong, O., Chwa, K.-Y., Park, K. (eds.) ISAAC 2010, Part I. LNCS, vol. 6506, pp. 3–14. Springer, Heidelberg (2010)
16. Karpinski, M., Schudy, W.: Approximation schemes for the betweenness problem in tournaments and related ranking problems. In: Goldberg, L.A., Jansen, K., Ravi, R., Rolim, J.D.P. (eds.) RANDOM 2011 and APPROX 2011. LNCS, vol. 6845, pp. 277–288. Springer, Heidelberg (2011)
17. Kratsch, S., Wahlström, M.: Two edge modification problems without polynomial kernels. In: Chen, J., Fomin, F.V. (eds.) IWPEC 2009. LNCS, vol. 5917, pp. 264–275. Springer, Heidelberg (2009)
18. Niedermeier, R.: Invitation to Fixed Parameter Algorithms. Oxford Lecture Series in Mathematics and Its Applications. Oxford University Press, USA (2006)
19. Paul, C., Perez, A., Thomassé, S.: Conflict packing yields linear vertex-kernels for k-FAST, k-DENSE RTI and a related problem. In: Murlak, F., Sankowski, P. (eds.) MFCS 2011. LNCS, vol. 6907, pp. 497–507. Springer, Heidelberg (2011)

20. Thomassé, S.: A $4k^2$ kernel for feedback vertex set. ACM Transactions on Algorithms 6(2) (2010)
21. van Bevern, R., Moser, H., Niedermeier, R.: Kernelization through tidying. In: López-Ortiz, A. (ed.) LATIN 2010. LNCS, vol. 6034, pp. 527–538. Springer, Heidelberg (2010)
22. Weller, M., Komusiewicz, C., Niedermeier, R., Uhlmann, J.: On making directed graphs transitive. In: Dehne, F., Gavrilova, M., Sack, J.-R., Tóth, C.D. (eds.) WADS 2009. LNCS, vol. 5664, pp. 542–553. Springer, Heidelberg (2009)

On Parameterized and Kernelization Algorithms for the Hierarchical Clustering Problem[*]

Yixin Cao[1] and Jianer Chen[2,3]

[1] Computer & Automation Research Inst., Hungarian Academy of Sciences, Hungary
[2] School of Information Science & Engineering, Central South University, P.R. China
[3] Department of Computer Science and Engineering, Texas A&M University, USA

Abstract. HIERARCHICAL CLUSTERING is an important problem with wide applications. In this paper, we approach the problem with a formulation based on weighted graphs and introduce new algorithmic techniques. Our new formulation and techniques lead to new kernelization algorithms and parameterized algorithms for the problem, which significantly improve previous algorithms for the problem.

1 Introduction

Many human activities can be described as a model of data collection and data analysis. The second stage is *discovering knowledge in data*, in which one of the most common tasks is to classify a large set of objects based on the collected information. This is called the *clustering problem*, and has incarnation in many disciplines, including biology, archaeology, geology, geography, business management, and social sciences [8,12,14,15].

In this paper, we are focused on the *hierarchical clustering* problem, which is to recursively classify a given data set into a tree structure in which leaves represent the objects and inner nodes represent clusters of various granularity degrees. We start with some definitions. For an integer $n \geq 1$, let $[n] = \{1, 2, \ldots, n\}$. An $n \times n$ symmetric matrix D is a *distance M-matrix* if $D_{ii} = 0$ for all i and $1 \leq D_{ij} \leq M + 1$ for $i \neq j$. A distance M-matrix D is an *ultrametric M-matrix* if it satisfies the *ultrametric property*: for any three i, j, k in $[n]$, $D_{ij} \leq \max\{D_{ik}, D_{jk}\}$. An *$M$-hierarchical clustering* $\mathcal{C}$ of a set X of n objects, which can be simply given as $X = [n]$, is a rooted tree with the objects of X as leaves at level 0, the root at level $M + 1$, and a path of length exactly $M + 1$ from the root to any leaf. If we define a distance function $d_{\mathcal{C}}$ for the objects in X based on $\mathcal{C}$ such that for any two x and y in X, $d_{\mathcal{C}}(x, y)$ is the height of the subtree rooted at the lowest common ancestor of x and y, then the distance on the objects of X forms an ultrametric M-matrix $D^{\mathcal{C}}$ [2]. It is also known that every ultrametric M-matrix induces an M-hierarchical clustering [2].

If the distance function $d_{\mathcal{C}}$ is precise, then it is easy to construct the hierarchical clustering $\mathcal{C}$ [11]. Unfortunately, there are seldom, if any, data collection methods that can exclude possibilities of errors. As a consequence, the distance

---

[*] Supported in part by the US NSF under the Grants CCF-0830455 and CCF-0917288.

T.-H.H. Chan, L.C. Lau, and L. Trevisan (Eds.): TAMC 2013, LNCS 7876, pp. 319–330, 2013.

matrix D formed by the distance function in general is not ultrametric, i.e., it contains inconsistent information. An important task in hierarchical clustering is to "correct" the errors and achieve data consistency. Formally, the hierarchical clustering problem we are concerned with is defined as follows:

M-HIERARCHICAL CLUSTERING

Given (D, k), where D is a distance M-matrix and k is an integer (i.e., the parameter), is there an ultrametric M-matrix D' such that the difference $d(D, D')$ is bounded by k?

Here $d(D, D')$ is defined as $d(D, D') = \sum_{1 \leq i < j \leq n} |D_{ij} - D'_{ij}|$.

The problem is NP-complete [13]. Polynomial-time approximation algorithms for the problem have been studied [2,3,17]. On the negative side, the problem is known to be APX-hard [1].

The special case $M = 1$ of the problem, in the name of CLUSTER EDITING which remains NP-hard [16], also has independent interest and applications. In fact, the algorithmic results mentioned above for general M-HIERARCHICAL CLUSTERING [2,3,17] are generalizations of algorithms on CLUSTER EDITING.

In practice, the ratio of errors is often low, i.e., the parameter k in the problem instance can be small, so we may achieve data consistency with a relatively small amount of correction. This observation has motivated the study of parameterized algorithms for the problem, which are algorithms running in time $f(k)n^{\mathcal{O}(1)}$ for a function f. Thus, for small parameter values k, such algorithms may solve the problem effectively. A closely related approach is to study *kernelization algorithms* for the problem, which, on an instance (D, k), produces in polynomial time an instance (D', k') such that $k' \leq k$, that the *kernel size* $|D'|$ is small, and that (D, k) is a yes-instance if and only if (D', k') is a yes-instance. Here the kernel size $|D'|$ is defined to be the cardinality of the object set on which the distance matrix D' is given. Many parameterized algorithms (e.g., [5]) and kernelization algorithms (e.g., [6,9]) for the CLUSTER EDITING problem have been developed. Guo *et al.* [10] studied the M-HIERARCHICAL CLUSTERING problem, and proposed an $\mathcal{O}^*(3^k)$-time parameterized algorithm for the problem[1], and an $\mathcal{O}(Mn^5)$-time kernelization algorithm with a kernel size bounded by $(2M + 4)k$. These are currently the best results for M-HIERARCHICAL CLUSTERING.

Based on an *edge-cut* technique, which is much simpler than the *critical cliques* technique used in previous work [7,9], we recently proposed a kernelization algorithm for weighted CLUSTER EDITING [6]. In this paper, we show that this technique can also be applied to HIERARCHICAL CLUSTERING to achieve significant improvements. Most importantly and a bit surprisingly, our kernelization algorithm based on the technique yields a kernel of size $2k$ for the M-HIERARCHICAL CLUSTERING problem, which is independent of the value M. The main results of this paper are summarized as follows:

Theorem 1. *For the M-HIERARCHICAL CLUSTERING problem, there exist an $\mathcal{O}^*(1.82^k)$-time parameterized algorithm, and an $\mathcal{O}(Mn^2)$-time kernelization algorithm that produces a kernel of size bounded by $2k$.*

---

[1] Following the convention, we are using $\mathcal{O}^*(f(k))$ to denote a bound $f(k)n^{\mathcal{O}(1)}$.

2   Cutting Lemmas and HIERARCHICAL CLUSTERING

An ultrametric matrix D^u satisfies $D_{ij}^u \leq \max\{D_{ik}^u, D_{jk}^u\}$ for all i, j, k. Two observations on an ultrametic matrix D^u are: (1) if $D_{ij}^u \neq D_{jk}^u$, then $D_{ik}^u = \max\{D_{ij}^u, D_{jk}^u\}$, and (2) if $D_{ij}^u = D_{jk}^u$ then $D_{ik}^u \leq D_{ij}^u$.

For two matrices D and S of the same size, denote by $D + S$ the pairwise element addition of the matrices.[2] A matrix S is a *solution* to a distance matrix D if $D + S$ is ultrametric. The cost of the solution S is defined as $c(S) = \sum_{1 \leq i < j \leq n} |S_{ij}|$. For a distance matrix D, denote by $c^*(D)$ the minimum solution cost over all solutions to D.

When $M = 1$, M-HIERARCHICAL CLUSTERING degenerates to CLUSTER EDITING, which can be formulated as a graph-theoretical problem that asks for the minimum number of edge additions/deletions to transform a given graph into a disjoint union of cliques [7,9]. The advantage of this formulation is that many powerful graph theory techniques, such as modular decompositions and edge cuts, become useful and applicable in solving the problem. In the following, we introduce a graph-theoretical formulation for general M-HIERARCHICAL CLUSTERING, and show how powerful graph theory techniques can be applied.

Let D be an $n \times n$ distance M-matrix on the object set $X = [n]$. For each positive integer t, $1 \leq t \leq M$, the graph $G_D^t = (X, E_D^t)$ for level t is defined such that $E_D^t = \{(u, v) : u, v \in X, D_{uv} \leq t\}$.

Proposition 1. *Let D be a distance M-matrix. Then each of the M graphs G_D^t defined as above is a disjoint union of cliques if and only if D is ultrametric.*

Thus, a solution to M-HIERARCHICAL CLUSTERING corresponds to M solutions, each to a graph G_D^t considered as an instance of CLUSTER EDITING. Based on this observation, Guo *et al.* [10] applied the $2k$ kernel for unweighted CLUSTER EDITING [7] to obtain a kernel of size $\mathcal{O}(Mk)$ for M-HIERARCHICAL CLUSTERING. The drawback of this formulation is that in the graph G_D^t, the object distances $\{1, 2, \ldots t\}$ are indistinguishable, which can directly introduce a multiplicative error upto $t - 1$. In order to improve this formulation, we consider a weighted version of the graph G_D^t so that the values of different distances are respected. Formally, for each $t \in [M]$, we define a weight function π_D^t on the pairs of vertices in the graph G_D^t, as follows: for each pair u, v in X:

$$\pi_D^t(uv) = \begin{cases} t + 1 - D_{uv} & \text{if } D_{uv} \leq t, \\ D_{uv} - t & \text{if } D_{uv} > t. \end{cases}$$

The two cases here correspond to edges and anti-edges in the graph G_D^t, respectively. The weight function π_D^t always gives a positive integer in $[M]$. As an example, in the graph G_D^M, for each object pair of distance $d \leq M$, an edge with

---

[2] In this paper we directly use matrices as the base of our operations, instead of the $n(n - 1)/2$ vectors used in previous research. This will make our discussion easier. Observe that when counting the cost, we count only the upper-triangle of the matrix.

weight $M + 1 - d$ is created; while for an object pair of distance $M + 1$, no edge is created and the weight of the pair is $(M + 1) - M = 1$.

The graph G_D^t will be called the *t-perspective graph*. A *t-clique* is a clique in the t-perspective graph G_D^t. To *t-split* two disjoint subsets X_1 and X_2 of objects in X in G_D^t is to increase the distance to at least $t + 1$ for objects between X_1 and X_2 – for each pair $u \in X_1$ and $v \in X_2$ such that $D_{uv} \leq t$, set $D_{uv} = t + 1$. Similarly, to *t-merge* a subset X' of objects in X in G_D^t is to decrease the distance between each pair of objects in X' to at most t – if $D_{uv} > t$ for $u, v \in X'$, then set $D_{uv} = t$. In the following, we will be focused on the M-perspective graph.

Now we are ready to generalize the cutting lemmas in [6] to HIERARCHICAL CLUSTERING. For an $n \times n$ matrix F, and a pair of index subsets $I, J \subseteq [n]$, denote by $F|_{I,J}$ the submatrix of F determined by the row index I and the column index J. We write $F|_I$ as a shorthand for $F|_{I,I}$. By definition, for an ultrametric matrix D', the submatrix $D'|_I$ for any index subset I is also ultrametric. For a distance matrix D, the submatrix $D|_I$ for any index subset I can be regarded as an instance of HIERARCHICAL CLUSTERING (where the cost $c^*(D|_I)$ is defined naturally). Moreover, a solution S to the distance matrix D restricted to the index subset I is a solution to $D|_I$, though the optimality may not transfer.

Lemma 1. *Let D be a distance M-matrix for the object set $X = [n]$, let $\mathcal{P} = \{X_1, X_2, \ldots, X_p\}$ be a partition of X, and let $E_\mathcal{P}$ be the set of edges in G_D^M whose two ends belong to two different parts in $\mathcal{P}$. Then $\sum_{i=1}^p c^*(D|_{X_i}) \leq c^*(D) \leq \pi_D^M(E_\mathcal{P}) + \sum_{i=1}^p c^*(D|_{X_i})$.*

Proof. Let S be an optimal solution to D. As noted above, for $1 \leq i \leq p$, $S|_{X_i}$ is a solution to the submatrix $D|_{X_i}$, which implies that $c^*(D|_{X_i}) \leq c(S|_{X_i})$. Thus, $\sum_{i=1}^p c^*(D|_{X_i}) \leq \sum_{i=1}^p c(S|_{X_i}) \leq c(S) = c^*(D)$.

For the second inequality, suppose that we increase all inter-part distance to $M + 1$, that is, to M-split all parts in $\mathcal{P}$ by removing all edges in $E_\mathcal{P}$ in the graph G_D^M, then apply an optimal solution S_i' to each submatrix $D|_{X_i}$. Then we will obviously end up with a solution S' to the matrix D, whose cost is

$$\pi_D^M(E_\mathcal{P}) + \sum_{i=1}^p c(S_i') = \pi_D^M(E_\mathcal{P}) + \sum_{i=1}^p c^*(D|_{V_i}),$$

which is no less than $c^*(D)$. This concludes the lemma. $\qquad\square$

If there is a partition such that all inter-part pairs have distance $M + 1$, then $\pi_D^M(E_\mathcal{P}) = 0$ and Lemma 1 gives

Corollary 1. *Let D be a distance M-matrix for the object set $X = [n]$, and let $\mathcal{P} = \{X_1, X_2, \ldots, X_p\}$ be a partition of X. If $D_{uv} = M + 1$ for each pair u and v that belong to different parts of $\mathcal{P}$, then $c^*(D) = \sum_{i=1}^p c^*(D|_{X_i})$.*

When $p = 2$, i.e. the partition is $\mathcal{P} = \{Y, \overline{Y}\}$, where Y is a subset of X and $\overline{Y} = X \setminus Y$, the edge set $E_\mathcal{P}$ becomes the cut $\langle Y, \overline{Y} \rangle$ (i.e., the set of edges with exactly one end in Y), whose weight will be denoted by $\gamma_D^M(Y)$. Lemma 1 gives

Corollary 2. *For any subset Y of X, we have $c^*(D|_Y) + c^*(D|_{\overline{Y}}) \leq c^*(D) \leq c^*(D|_Y) + c^*(D|_{\overline{Y}}) + \gamma_D^M(Y)$.*

This suggests the following lower bound for $\gamma_D^M(Y)$.

Lemma 2. . *Let S be an optimal solution to a distance M-matrix D for the object set $X = [n]$. For any subset Y of X, $c(S|_{Y,\overline{Y}}) \leq \gamma_D^M(Y)$.*

Proof. The solution S can be divided into three disjoint parts: $S|_Y$, $S|_{\overline{Y}}$, and $S|_{Y,\overline{Y}}$. By Corollary 2 (note $c^*(D) = c(S)$),

$$c(S) = c(S|_Y) + c(S|_{\overline{Y}}) + c(S|_{Y,\overline{Y}}) \leq c^*(D|_Y) + c^*(D|_{\overline{Y}}) + \gamma_D^M(Y). \tag{1}$$

Since $S|_Y$ is a solution to the submatrix $D|_Y$ and $S|_{\overline{Y}}$ is a solution to the submatrix $D|_{\overline{Y}}$, we have $c(S|_Y) \geq c^*(D|_Y)$ and $c(S|_{\overline{Y}}) \geq c^*(D|_{\overline{Y}})$, which combined with (1) gives immediately $c(S|_{Y,\overline{Y}}) \leq \gamma_D^M(Y)$. □

For a distance M-matrix D, it is intuitive that the objective ultrametric M-matrix should have its largest element bounded by $M + 1$. This intuition can be formally proved in the following lemma, which also verifies the validity of the definition of the M-HIERARCHICAL CLUSTERING problem.

Lemma 3. *Let D be a distance M-matrix, and let S' be an optimal solution to D. Then the matrix $D' = D + S'$ is an ultrametric M-matrix (i.e., the largest element in the matrix D' has a value bounded by $M + 1$).*

Proof. We prove the lemma by contradiction that $d' = \max_{1 \leq i < j \leq n}\{D'_{ij}\} > M + 1 \geq \max_{1 \leq i < j \leq n}\{D_{ij}\}$. Consider the following matrix S'':

$$S''_{ij} = \begin{cases} S'_{ij} & \text{if } D'_{ij} < d', \\ S'_{ij} - 1 & \text{if } D'_{ij} = d'. \end{cases}$$

Since $d' > \max_{1 \leq i < j \leq n}\{D_{ij}\}$, if $D'_{ij} = d'$ then $S'_{ij} > 0$. Thus, $c(S'') < c(S')$.

Applying solutions S' and S'' to D, we get two different matrices D' and $D'' = D + S''$. By the above construction, for all $t < d' - 1$, the t-perspective graphs for D' and D'' are the same, which are unions of disjoint cliques. For $t = d' - 1$, the t-perspective graph $G_{D''}^t$ for D'' is a single clique. Thus, by Proposition 1, D'' is ultrametric, so S'' is a solution to D. However, this contradicts the facts that $c(S'') < c(S')$ and that S' is an optimal solution to D. □

Without loss of generality, we will always assume in the rest of this paper that a distance M-matrix has at least one element of value $M + 1$.

3   A Kernel of Size $2k$

To better understand our kernelization algorithm, we start with one that produces a kernel of size $4k$, and discuss the difficulty for improving it. The second part is devoted to overcoming the difficulty and achieving the kernel of size $2k$.

Warming up: a kernel of size $4k$

Fix a distance M-matrix D for $X = [n]$. For an object v in X, devote by $N_v = \{u : D_{uv} < M + 1\}$ the closed neighborhood of v in the graph G_D^M.

A simple but important fact about a solution S of cost bounded by k to the distance M-matrix D is that at most $2k$ different objects in X have some of their distances to other objects changed. As a consequence, if we are also able to bound the number of objects that are not affected by S, we get a kernel. For such an unaffected object v, the v-th row of S consists of only 0's. Thus, in the ultrametric matrix $D' = D + S$, for any two objects u and w in X, where $u \in N_v$, the distance D'_{uw} must satisfy (note $D_{vu} = D'_{vu}$ and $D_{vw} = D'_{vw}$):

$$D'_{uw} \begin{cases} \leq \max(D_{vu}, D_{vw}) \leq M & \text{if } u, w \in N_v; \\ = \max(D_{vu}, D_{vw}) = M + 1 & \text{if } u \in N_v, w \notin N_v. \end{cases} \tag{2}$$

This is a necessary (but not sufficient) condition for a solution S to avoid v. If (2) is not satisfied by D, then $D|_{N_v}$ must be modified by S. To measure the cost of the modification, we introduce a number of functions, as follows:

$$\delta(v) = |\{(u, w) : u, w \in N_v, u < w \text{ and } D_{uw} = M + 1\}|,$$
$$\gamma(v) = \sum_{u \in N_v, w \notin N_v} (M + 1 - D_{uw}) \qquad \text{(i.e., } \gamma(v) = \gamma_D^M(N_v)),$$
$$\rho(v) = 2\delta(v) + \gamma(v).$$

We say that the neighborhood N_v is *reducible* if $\rho(v) < |N_v|$.

We describe two reduction rules on a reducible neighborhood N_v. The first given in the following lemma claims that N_v can be put into a single M-clique.

Lemma 4. *For an object v with N_v reducible, there is an optimal solution S^* to D such that the maximum distance in $(D + S^*)|_{N_v}$ is bounded by M.*

Lemma 4 gives the rule for our first reduction rule immediately:

Rule 1. *For an object v in X such that N_v is reducible, replace every element $M + 1$ in the submatrix $D|_{N_v}$ by M, and decrease the parameter k by $\delta(v)$.*

After Rule 1, we have $\delta(v) = 0$ and $\rho(v) = \gamma(v)$. Now consider $D|_{N_v, \overline{N}_v}$.

Rule 2. *On a reducible N_v on which Rule 1 has been applied, for each object $x \notin N_v$ with $\sum_{u \in N_v} (M + 1 - D_{xu}) \leq |N_v|/2$, M-split x from N_v.*

Lemma 5. *Rule 2 is safe.*

After Rules 1-2, the neighborhood N_v has a very simple structure: there is at most one "pendent" object in $\overline{N}_v$ that is still attached to N_v, as shown by the following lemma.

Lemma 6. *For a reducible N_v on which Rules 1-2 have been applied, there is at most one object $x \notin N_v$ such that $D|_{N_v, x}$ have values not equal to $M + 1$.*

Proof. By the condition of Rule 2, any object x in $\overline{N}_v$ that still has distance smaller than $M + 1$ to some objects in N_v after the application of Rule 2 must satisfy $\sum_{u \in N_v}(M + 1 - D_{xu}) > |N_v|/2$. To prove the lemma, suppose on the contrary that there are two such objects x and y. Then we have

$$\gamma(v) = \gamma_D^M(N_v) \geq \sum_{u \in N_v}(M + 1 - D_{xu}) + \sum_{u \in N_v}(M + 1 - D_{yu}) > |N_v|.$$

This contradicts that N_v is reducible and $\rho(v) = 2\delta(v) + \gamma(v) < |N_v|$.   □

Now we are ready to describe our kernelization algorithm.

The Kernelization Algorithm. For each object v for which the set N_v is reducible
1. decrease value $M + 1$ in $D|_{N_v}$ to M and decrease k accordingly;
2. for each element $x \notin N_v$ such that $\sum_{u \in N_v}(M + 1 - D_{xu}) \leq |N_v|/2$, set all values in $D|_{N_v,x}$ to $M + 1$ and decrease k accordingly.

Note that there is only one condition tested by the algorithm, which is checked only once and is independent of the parameter k.

This kernelization algorithm is applied iteratively, starting from the highest level M. In each run, we take each object set obtained in the splitting in the previous run and apply the kernelization algorithm, until there is no object set on which the kernelization algorithm is applicable. Therefore, the kernel consists of a set of object sets, each forms an independent instance of HIERARCHICAL CLUSTERING. To analyze the size of the final kernel, we count the relation between the object sets and the minimum number of modifications required to make the distance matrix ultrametric. Because our counting does not depend on the value of M, this ratio holds for all subsets that form independent instances of HIERARCHICAL CLUSTERING, and therefore for the entire object set X.

Lemma 7. *Let (D, k) be an instance of M-HIERARCHICAL CLUSTERING on which the kernelization algorithm is not applicable. If the size of D is larger than $4k$, then there is no solution to D of cost bounded by k.*

Proof. Let matrix S be an optimal solution to the distance M-matrix D. For each pair $v, w \in X$, we divide the cost $|S_{vw}|$ into two halves and distribute them evenly to v and w. By this procedure, each object v gets a "cost" $cost(v) = \frac{1}{2}\sum_{u \in X \setminus \{v\}}|S_{uv}|$. The total cost of S is equal to $\sum_{v \in X} cost(v)$. We count the cost on each object, and pay special attention to objects with cost 0.

For two objects u, v with $D_{uv} = M + 1$, if there exists another object x such that $D_{ux} \leq M$ and $D_{vx} \leq M$, then at most one of the objects u and v can has cost 0: to make u, v, x satisfy the ultrametric property, at least one of the distances D_{uv}, D_{ux}, D_{vx} must be changed. Let $Z_S = \{v_1, v_2, \ldots, v_r\}$ be the set of objects with cost 0. For two objects $v_i, v_j \in Z_S$, either $D_{v_i v_j} = M + 1$ and every other object has distance $M + 1$ to at least one of v_i, v_j; or $D_{v_i v_j} \leq M$ and any other object has distance $M + 1$ to v_i if and only if it has distance $M + 1$ to v_j. As a result, the two neighborhoods N_{v_i} and N_{v_j} in G_D^M are either the same (when $D_{v_i v_j} \leq M$) or disjoint (when $D_{v_i v_j} = M+1$). Thus, without loss of generality, we can assume that all neighborhoods in $\{N_{v_1}, N_{v_2}, \ldots, N_{v_r}\}$ are pairwise disjoint. Let $N_S = N_{v_1} \cup N_{v_2} \cup \cdots \cup N_{v_r}$.

Since column $D|_{X, v_i}$ for $v_i \in Z_S$ is unchanged by the solution S, S must decrease the distance $M+1$ between any pair of objects in N_{v_i} to M, and increase the distance between objects in N_{v_i} and $\overline{N}_{v_i}$ to $M+1$. These operations have cost $\delta(v_i) + \gamma(v_i)$. Accordingly, the total cost on the objects in N_{v_i} is $\delta(v_i) + \gamma(v_i)/2 = \rho(v_i)/2$. If N_{v_i} is not reducible, then $\rho(v_i)/2 \geq |N_{v_i}|/2$. On the other hand, if N_{v_i} is reducible, then by Lemma 6, there can be at most one object $x \in \overline{N}_{v_i}$ that has distance bounded by M to some objects in N_{v_i}. According to Rule 2, in this case $\rho(v_i) \geq \gamma(v_i) > |N_{v_i}|/2$. Thus, the cost $\rho(v_i)/2$ is always strictly larger than $|N_{v_i}|/4$. From this analysis, we get

$$\sum_{v \in N_S} cost(v) = \sum_{i=1}^{r} \sum_{v \in N_{v_i}} cost(v) \geq \sum_{i=1}^{r} |N_{v_i}|/4 = |N_S|/4. \tag{3}$$

On the other hand, each object $w \notin N_S$ bears a cost at least $1/2$. Thus

$$\sum_{w \in X \setminus N_S} cost(w) \geq |X \setminus N_S|/2. \tag{4}$$

Combining (3) and (4) shows that the cost of the optimal solution S to D is

$$\sum_{v \in X} cost(v) = \sum_{v \in N_S} cost(v) + \sum_{v \in X \setminus N_S} cost(v) \geq |N_S|/4 + |X \setminus N_S|/2 \geq |X|/4.$$

Thus, if $|X| > 4k$, then the distance M-matrix D has no solution of cost $\leq k$.   □

Destination: a kernel of size $2k$

The main trouble to further improve the kernel size $4k$ given above is that for a reducible N_v, conflicts in N_v are only settled at level M, while conflicts may still occur at lower levels. Our idea for tackling this trouble is: if the cost to fix N_v at lower levels is large enough, we then use it in the counting to complement the deficiency; otherwise we will find another rule to reduce N_v.

We first consider the case where no pendent object $x \notin N_v$ (as described in Lemma 6) exists for N_v. In this case, N_v has been completely resolved in the perspective graph for level M, and we can treat $D|_{N_v}$ as an independent instance for the $(M-1)$-HIERARCHICAL CLUSTERING problem, and continue to apply the Kernelization Algorithm (note that the Kernelization Algorithm does not depend on the value of the parameter k).

The case where the pendent object $x \notin N_v$ exists for N_v is more involved. After previous steps, the M-clique in the final solution is contained in $N_v \cup \{x\}$. Thus, the $(M-1)$-clique containing v is a subset of $N_v \cup \{x\}$. Since $D_{vx} = M+1$, $N_v^{M-1} \subseteq N_v$, where N_v^{M-1} is the neighborhood of the object v in the $(M-1)$-perspective graph G_D^{M-1}. If N_v^{M-1} is reducible in the $M-1$ level, we can apply again the Kernelization Algorithm. We can continue this procedure until

- we meet the first t such that N_v^t is not reducible;
- we meet the first t such that N_v^t gets isolated; or
- we hit the ground when $t = 1$.

In the first situation, we stop. In the second situation, we apply the Kernelization Algorithm to N_v^t as an independent instance. Thus, we only need to deal with the last situation, for which there is a pendent object x^t for N_v^t at each level t. At level $t = 1$, let $N_1 \subset N_v^1$ be the objects with distance 1 to x^1, and let $N_2 \subset N_v^1$ be the objects with distance 2 to x^1. Obviously, $N_v^1 = N_1 \cup N_2$.

Rule 3. *Let v be an object such that N_v^t is reducible and Rules 1-2 have been applied for all levels t. Pick any subset $N_{12} \subseteq N_1$ with $|N_{12}| = |N_2|$, and remove $N' = N_{12} \cup N_2$. For levels $t \geq 2$, increase total distance from x^t to $N_v^t - N'$ by $2|N_2| - 2|\{u \in N_{12} : D_{ux^t} \leq t\}|$, by arbitrarily choosing objects from $N_v^t - N'$ and increasing their distances to x^t to no more than $t + 1$.*

We first verify the validity for Rule 3. Since x^1 survives Rule 2, more objects in N_v^1 have distance 1 to x^1 than those with distance 2, i.e., $|N_1| \geq |N_2|$, which shows the existence of the subset N_{12}. For an upper level $2 \leq t \leq M$, the required increments in distance between x^t and $N_v^t - N'$ is

$$(|N_{12}| + |N_2| - |\{u \in N_{12} : D_{ux^t} \leq t\}|) - |\{u \in N_{12} : D_{ux^t} \leq t\}|,$$

where the first parenthesis constitutes a set of objects with distance $\geq t + 1$ to x^t. This condition is always satisfied, since x^t survived Rule 2.

Lemma 8. *Rule 3 is safe.*

Now we are ready to show that the Kernelization Algorithm has a kernel of size $2k$. In the second situation, we treat N_v^t as an independent instance and apply the Kernelization Algorithm. Since M is finite, we will eventually reach an instance at a lower level on which the second situation no longer holds, where the instance either is already internally ultrametric, or contains no reducible objects anymore. For the latter case, the internal cost is at least twice the number of objects in it, and we can use it to make up the deficiency in upper-level counting. For the former case, we use the following reduction rule that is almost the same as Rule 3, whose safeness follows from a similar argument as that for Rule 3.

Rule 4. *Let v be an object such that N_v^t is reducible and Rules 1-2 have been applied for all levels $t \geq T$. If N_v^T gets separated and $D|_{N_v^T}$ is ultrametric, then remove N_v^T, and from level $t = T$ to level M, increase total distance from x^t to $N_v^t - N_v^T$ by $2|N_v^T|$, by arbitrarily choosing an object from $N_v^t - N_v^T$ and increasing its distance to x^t to no more than $t + 1$.*

Summarizing the above discussions, we conclude with a kernel bound for the Kernelization Algorithm, which was claimed in the second part of Theorem 1.

Theorem 2. *Let (D, k) be an instance of the M-HIERARCHICAL CLUSTERING problem on which the Kernelization Algorithm has been applied. If the size of the distance M-matrix D is larger than $2k$, then no solution to the distance M-matrix D has its cost bounded by k.*

4   An Improved Parameterized Algorithm

Inspired by the formulation and usage of perspective graphs in the last section, one might want to solve the M-HIERARCHICAL CLUSTERING problem in a level-by-level way, by picking an algorithm for the CLUSTER EDITING problem, applying the algorithm to the M-perspective graph G_D^M, then applying the algorithm to the resulting instances at level $M - 1$, and so on. However, this greedy approach does not always work: it can be shown that the set of the operations in an optimal solution to the perspective graph at a higher level for CLUSTER EDITING may not be a subset of the set of the operations in *any* optimal solution to the original instance for M-HIERARCHICAL CLUSTERING.

On the other hand, this negative result does offer some useful information: it indicates that if we want to use an algorithm for CLUSTER EDITING, we cannot use it as a black-box – we must know its internal mechanism.

The HIERARCHICAL CLUSTERING problem, for which the CLUSTER EDITING problem is a special case, can be resolved by the following *breaking conflict-triangle* process: to convert a distance matrix D into an ultrametric matrix, for any three indices i, j, k, if D_{ij}, D_{ik}, and D_{jk} do not satisfy the ultrametric property, then at least one of them must change its value. It naturally suggests a 3-way branching search process for an optimal solution to D, which leads to an $\mathcal{O}^*(3^k)$-time algorithm for the problem. For CLUSTER EDITING, there have been several improved results, following the basic outline of breaking conflict-triangles. With the help of more careful branching steps and more complicated analysis techniques, the current best algorithm for CLUSTER EDITING takes time $\mathcal{O}^*(1.62^k)$ [5]. On the other hand, there has been no non-trivial parameterized algorithm for the general M-HIERARCHICAL CLUSTERING problem.

Instead of adapting a single particular algorithm for the CLUSTER EDITING problem to solve the M-HIERARCHICAL CLUSTERING problem, we go one step further. We show that *any* parameterized algorithm for the CLUSTER EDITING problem, provided it is based on branching on breaking conflict-triangles, can be adapted to solve the M-HIERARCHICAL CLUSTERING problem, with the same time complexity as far as the exponential part is concerned. Indeed, what we show is a *meta algorithm*, which takes as an input, in addition to an instance $I_H = (D, k)$ of the M-HIERARCHICAL CLUSTERING problem, an algorithm for the CLUSTER EDITING problem, and returns an optimal solution to I_H.

The algorithm **Meta-HC** given in Figure 1 shows how a meta algorithm is implemented that adapts an algorithm for CLUSTER EDITING to solve HIERARCHICAL CLUSTERING. We give some explanations on the algorithm.

In step 6 of the algorithm, we decrease each element value $M + 1$ to M and mark the value *forbidden*. This step simplifies the presentation of the algorithm. In particular, after each iterative run, we do not break the instance into smaller subinstances and solve them separately. Instead, we still treat it as a single instance. Note that if there is no conflict-triangle at level M, then the edges with distance $M + 1$ partition the objects with no conflict-triangles. Thus, decreasing them uniformly by 1 at level $M - 1$ will not create new conflict-triangles. On the other hand, in step 3 of the algorithm when we call the algorithm $\mathcal{A}$ for CLUSTER

Algorithm Meta-HC($D, k, \mathcal{A}$)
INPUT: a distance matrix D, an integer k, an algorithm $\mathcal{A}$ for CLUSTER EDITING
OUTPUT: an ultrametic matrix D' such that $d(D, D') \leq k$ if such D' exists

1    $M = \max_{1 \leq i < j \leq n} \{D_{ij}\} - 1$;
2   construct the M-perspective graph G_D^M and the weight function π^M;
3   **for** each solution S returned by $\mathcal{A}(G_D^M, \pi^M, k)$ **do**
4   **if** ($M == 1$) **then return** $D + S$;
5        $D = D + S$, $k = k - c(S)$;
6   **for** each $D_{ij} = M + 1$ **do** $D_{ij} = M$ and mark it as "forbidden";
7   **call Meta-HC($D, k, \mathcal{A}$).**

Fig. 1. The Meta-Algorithm

EDITING, the tags of "forbidden" will be discarded. Thus, the distance of an edge that gets decreased in a turn can be further decreased later. On the other hand, by the procedure, no repeated increment on a single edge can happen.

Now we are ready to present the main result of this section:

Theorem 3. *Let $\mathcal{A}$ be an algorithm for the weighted* CLUSTER EDITING *problem, such that it breaks conflict-triangles by all possible ways and uses branching to count the time complexity. Then there is an algorithm $\mathcal{A}'$ for the* HIERARCHICAL CLUSTERING *problem whose time complexity is $\mathcal{O}^*(MT(n))$, where $T(n)$ is the time complexity of the algorithm $\mathcal{A}$.*

Proof. We show that for a given instance (D, k) of the HIERARCHICAL CLUSTERING problem, if there are solutions to D with cost bounded by k, then the algorithm **Meta-HC** will always find one.

By Proposition 1, a solution to D must break all conflict-triangles in all perspective graphs. Since the algorithm $\mathcal{A}$ tries all possible ways to achieve this, one of the branches must apply a correct operation. Also note that no conflict-triangles can be created at a higher level when the algorithm is working on a lower level. In the iterative way, conflict-triangles at all levels are broken.

We need to ensure that we never make *counteracting* operations on any edge e, by decreasing its distance in a run then increasing it in a later run, or the inverse. In the t-th run, if the distance of e is increased, then it gets a new value $t + 1$, and marked "forbidden" forever. On the other hand, if the distance of e is decreased in the run, then its value becomes t, marked as "forbidden", and keeps stable in this run. Therefore, in later runs ($t - 1$-th or lower), it can only be further decreased, but never increased.

Now consider the time complexity of the algorithm. The extra operations introduce a factor of a low degree polynomial function to the complexity, which can be ignored under the notation $\mathcal{O}^*$. The dominating part is the M calls to the algorithm $\mathcal{A}$. Thus, the algorithm **Meta-HC** runs in time $\mathcal{O}^*(MT(n))$. $\square$

To use Theorem 3, we notice that the algorithm proposed by Böcker, Briesemeister, and Bui [5] runs in time $\mathcal{O}^*(1.82^k)$ and satisfies the conditions of Theorem 3. This gives immediately the following theorem, which was claimed in the first part of Theorem 1.

Theorem 4. *There is an $\mathcal{O}^*(1.82^k)$-time parameterized algorithm for the* M-HIERARCHICAL CLUSTERING *problem.*

References

1. Agarwala, R., Bafna, V., Farach, M., Paterson, M., Thorup, M.: On the approximability of numerical taxonomy (fitting distances by tree metrics). SIAM J. Comput. 28(3), 1073–1085 (1999)
2. Ailon, N., Charikar, M.: Fitting tree metrics: hierarchical clustering and phylogeny. SIAM J. Comput. 40(5), 1275–1291 (2011)
3. Ailon, N., Charikar, M., Newman, A.: Aggregating inconsistent information: ranking and clustering. J. ACM 55(5), Article 23, 1–27 (2008)
4. Böcker, S.: A golden ratio parameterized algorithm for cluster editing. J. Discret Algorithms 16, 79–89 (2012)
5. Böcker, S., Briesemeister, S., Bui, Q., Truss, A.: Going weighted: parameterized algorithms for cluster editing. Theor. Comput. Sci. 410, 5467–5480 (2009)
6. Cao, Y., Chen, J.: Cluster editing: kernelization based on edge cuts. Algorithmica 64, 152–169 (2012)
7. Chen, J., Meng, J.: A $2k$ kernel for the cluster editing problem. Journal of Computer and System Sciences 78, 211–220 (2012)
8. Gan, G., Ma, C., Wu, J.: Data Clustering: Theory, Algorithms, and Applications. ASASIAM Series on Statistics and Applied Probability. SIAM (2007)
9. Guo, J.: A more effective linear kernelization for cluster editing. Theor. Comput. Sci. 410, 718–726 (2009)
10. Guo, J., Hartung, S., Komusiewicz, C., Niedermeier, R., Uhlmann, J.: Exact algorithms and experiments for hierarchical tree clustering. In: Proc. 24th AAAI Conference on Artificial Intelligence, AAAI 2010, pp. 457–462 (2010)
11. Gusfield, D.: Algorithms on Strings, Trees, and Sequences: Computer Science and Computational Biology. Cambridge University Press, New York (1997)
12. Jain, A., Murty, M., Flynn, P.: Data clustering: a review. ACM Comput. Surv. 31, 264–323 (1999)
13. Krivanek, M., Moravek, J.: NP-hard problems in hierarchical-tree clustering. Acta Informatica 23(3), 311–323 (1986)
14. Larose, D.: Discovering Knowledge in Data: An Introduction to Data Mining. John Wiley & Sons (2005)
15. Manning, C., Raghavan, P., Schütze, H.: Introduction to Information Retrieval. Cambridge University Press (2008)
16. Shamir, R., Sharan, R., Tsur, D.: Cluster graph modification problems. Discrete Appl. Math. 144, 173–182 (2004)
17. van Zuylen, A., Williamson, D.: Deterministic pivoting algorithms for constrained ranking and clustering problems. Mathematics of Operations Research 34(3), 594–620 (2009)

Vector Connectivity in Graphs*

Endre Boros[1], Pinar Heggernes[2], Pim van 't Hof[2], and Martin Milanič[3]

[1] RUTCOR, Rutgers University, New Jersey, USA
endre.boros@rutcor.rutgers.edu
[2] Department of Informatics, University of Bergen, Norway
{pinar.heggernes,pim.vanthof}@ii.uib.no
[3] UP IAM and UP FAMNIT, University of Primorska, Koper, Slovenia
martin.milanic@upr.si

Abstract. Motivated by challenges related to domination, connectivity, and information propagation in social and other networks, we initiate the study of the VECTOR CONNECTIVITY problem. This problem takes as input a graph G and an integer k_v for every vertex v of G, and the objective is to find a vertex subset S of minimum cardinality such that every vertex v either belongs to S, or is connected to at least k_v vertices of S by disjoint paths. If we require each path to be of length exactly 1, we get the well-known VECTOR DOMINATION problem, which is a generalization of the famous DOMINATING SET problem and several of its variants. Consequently, our problem becomes NP-hard if an upper bound on the length of the disjoint paths is also supplied as input. Due to the hardness of these domination variants even on restricted graph classes, like split graphs, VECTOR CONNECTIVITY seems to be a natural problem to study for drawing the boundaries of tractability for this type of problems. We show that VECTOR CONNECTIVITY can actually be solved in polynomial time on split graphs, in addition to cographs and trees. We also show that the problem can be approximated in polynomial time within a factor of $\ln n + 2$ on all graphs.

1 Introduction and Motivation

Connectivity between parts of a graph via disjoint paths is one of the best studied subjects in graph theory and graph algorithms, where NETWORK FLOW and DISJOINT PATHS and many of their variants are among the most well-known problems. In this paper, we introduce, motivate, and study a natural network problem, which we call VECTOR CONNECTIVITY. Given a graph $G = (V, E)$ and a vector $\boldsymbol{k}$ indexed by the vertices of G, such that $\boldsymbol{k} = (k_v : v \in V)$ and k_v is between 0 and the degree of v for each vertex $v \in V$, the task of VECTOR CONNECTIVITY is to find a set $S \subseteq V$ of minimum cardinality that satisfies the

---

* This work is supported by the Research Council of Norway (197548/F20) and by the Slovenian Research Agency (research program P1–0285 and research projects J1–4010, J1–4021, BI-US/12–13–029 and N1–0011: GReGAS, supported in part by the European Science Foundation).

T.-H.H. Chan, L.C. Lau, and L. Trevisan (Eds.): TAMC 2013, LNCS 7876, pp. 331–342, 2013.
© Springer-Verlag Berlin Heidelberg 2013

following: every vertex v of G is either in S or is connected to at least k_v vertices of S via paths that pairwise intersect in no other vertex than v.

In VECTOR CONNECTIVITY there is no restriction on the lengths of the involved disjoint paths. If each path is restricted to be of length exactly 1, we get the well-known VECTOR DOMINATION problem; this problem was introduced by Harant et al. [10] as a generalization of the classical problems DOMINATING SET and VERTEX COVER. The DOMINATING SET problem and its variants have been studied extensively, as they naturally appear in a wide variety of theoretical and practical applications. This has led to a vast amount of papers and several books on domination, e.g., [11,12]. DOMINATING SET and hence VECTOR DOMINATION are also among the toughest NP-hard problems as they remain NP-hard on various classes of graphs, such as planar graphs of maximum degree 3, bipartite graphs, and most interesting for our study: split graphs [6,12]. The popularity and the difficulty of these domination problems, the connection between VECTOR DOMINATION and VECTOR CONNECTIVITY, and the question whether allowing paths of unbounded length rather than direct edges or bounded-length paths can result in tractability, are among the motivations for studying the VECTOR CONNECTIVITY problem.

Chlebík and Chlebíková [1] showed that DOMINATING SET, and consequently VECTOR DOMINATION, cannot be approximated in polynomial time within a factor of $(1 - \epsilon) \ln n$ for any constant $\epsilon > 0$ unless NP $\subseteq$ DTIME$(n^{O(\log \log n)})$, even when restricted to the class of bipartite graphs or split graphs. On the positive side, Cicalese et al. [3] presented a greedy algorithm for VECTOR DOMINATION with approximation factor $\ln(2\Delta) + 1$, where Δ denotes the maximum degree of the input graph. Moreover, they showed that the problem can be solved in polynomial time on trees and cographs. If one asks for disjoint paths of bounded length rather than direct edges, it is not known in general whether the problem can be approximated within a factor of $O(\log n)$. This gives another motivation to study the unbounded-length paths case, which is exactly the VECTOR CONNECTIVITY problem.

In this paper, we show that VECTOR CONNECTIVITY can be approximated within a factor of $\ln n + 2$ in polynomial time on general graphs, which we find interesting due to the known and unknown approximation results mentioned above. Furthermore, we show that VECTOR CONNECTIVITY can be solved in polynomial time on split graphs, cographs, and trees. We find in particular the tractability result on split graphs surprising, as it is in contrast with the aforementioned NP-hardness and inapproximability results for the DOMINATING SET problem on split graphs. Furthermore, these intractability results imply that if paths are required to be of length at most an input bound then the problem remains NP-hard on split graphs. However, split graphs do not have any induced paths of length 4 or more. Hence our positive result on split graphs implies that the bounded-length path version of VECTOR CONNECTIVITY, which is a generalization of VECTOR DOMINATION, is solvable in polynomial time on split graphs if the bound is at least 3.

Note that the classes of split graphs, cographs, and trees are all subclasses of perfect graphs, but they are not contained in each other. They form some of the most studied graph classes on which many algorithms have been given, and they play the main role in several books, e.g., in the monograph on perfect graphs by Golumbic [7], and in the monograph by Mahadev and Peled [14] on threshold graphs, which form a subclass of both split graphs and cographs.

Before we proceed to the technical part presenting and proving our results, we end this section by mentioning another motivation, which comes from information propagation in social networks. One famous problem of this type is TARGET SET SELECTION (see, e.g., [2,13,15]), where every vertex v has a threshold t_v such that v gets activated if at least t_v of its neighbors are activated, and the task is to select a minimum cardinality vertex subset that results in the activation of all vertices eventually. The practical application behind this problem is the desire by manufacturers to give away their products to a selected small group of people, based on the scenario that every potential customer will decide to buy the product if he or she has enough friends who possess the product. Another possible scenario can be that every potential customer will decide to buy the product only if he or she has enough independent ways to learn about the product. VECTOR CONNECTIVITY fits into this scenario if we assume that information spreads freely along the paths of the network.

2   Definitions and Notation

Unless otherwise stated, we work with undirected simple graphs $G = (V, E)$, where V is the set of vertices, E is the set of edges, and $|V|$ is denoted by n. We use standard graph terminology. In particular, the degree of a vertex v in G is denoted by $d_G(v)$, the maximum degree of a vertex in G is denoted by $\Delta(G)$, and $V(G)$ refers to the vertex set of G. For a given rooted tree T, we write T_v to denote the subtree rooted at vertex v, including vertex v.

Given a graph $G = (V, E)$, a set $S \subseteq V$ and a vertex $v \in V \setminus S$, a v–S fan of order k is a collection of k paths $P_1, \ldots, P_k$ such that (1) every P_i is a path connecting v to a vertex of S, and (2) the paths are pairwise vertex-disjoint except at v, i.e., for all $1 \leq i < j \leq k$, it holds that $V(P_i) \cap V(P_j) = \{v\}$. Given an integer-valued vector $\mathbf{k} = (k_v : v \in V)$ with $k_v \in \{0, 1, \ldots, d_G(v)\}$ for every $v \in V$, a *vector connectivity set* for $(G, \mathbf{k})$ is a set $S \subseteq V$ such that there exists a v–S fan of order k_v for every $v \in V \setminus S$. We say that k_v is the *requirement* of vertex v. The minimum size of a vector connectivity set for $(G, \mathbf{k})$ is denoted by $\kappa(G, \mathbf{k})$.

The VECTOR CONNECTIVITY problem is the problem of finding a vector connectivity set of minimum size, and can be formally stated as follows:

VECTOR CONNECTIVITY
Input: A graph $G = (V, E)$ and a vector $\mathbf{k} = (k_v : v \in V) \in \mathbb{Z}_+^V$
    with $k_v \in \{0, 1, \ldots, d_G(v)\}$ for all $v \in V$.
Task: Find a vector connectivity set for $(G, \mathbf{k})$ of size $\kappa(G, \mathbf{k})$.

For every $v \in V$ and every set $S \subseteq V \setminus \{v\}$, we say that v is k-*connected* to S if there is a v–S fan of order k in G. Hence, given an instance $(G, \boldsymbol{k})$ of Vector Connectivity, a set $S \subseteq V$ is a vector connectivity set for $(G, \boldsymbol{k})$ if and only if every $v \in V \setminus S$ is k_v-connected to S. For a subset $A \subseteq V$, we write $\boldsymbol{k}|_A$ to denote the sub-vector of $\boldsymbol{k}$ indexed by elements of A, and $\mathbf{1}_A$ denotes the all-one vector indexed by elements of A. We let $\sigma(v, A)$ to denote the maximum order of a v–A fan in G. In other words, $\sigma(v, A) = \max\{s \mid v \text{ is } s\text{-connected to } A\}$. Let $B \subseteq A$, let $\mathcal{A}$ be a v–A fan and let $\mathcal{B}$ be a v–B fan. We say that $\mathcal{A}$ *contains* $\mathcal{B}$ if the collection of paths in $\mathcal{B}$ is a subcollection of the paths in $\mathcal{A}$.

A set of vertices in a graph is a *clique* if they are all pairwise adjacent, and it is an *independent set* if no two of them are adjacent. A graph is a *split graph* if its vertex set can be partitioned into a clique C and an independent set I, where (C, I) is called a *split partition* of G. Split graphs can be recognized and a split partition can be computed in linear time [9].

For two vertex-disjoint graphs G_1 and G_2, $G_1 \oplus G_2$ denotes the *disjoint union* of G_1 and G_2, i.e., $G_1 \oplus G_2 = (V(G_1) \cup V(G_2), E(G_1) \cup E(G_2))$, and $G_1 \otimes G_2$ denotes the *join* of G_1 and G_1, i.e., the graph obtained by adding to $G_1 \oplus G_2$ all edges of the form $\{uv \mid u \in V(G_1), v \in V(G_2)\}$. The class of *cographs* is defined recursively through the following operations: a single vertex is a cograph; if G_1 and G_2 are vertex-disjoint cographs, then $G_1 \oplus G_2$ is a cograph; if G_1 and G_2 are vertex-disjoint cographs, then $G_1 \otimes G_2$ is a cograph. Cographs, split graphs, and trees are not related to each other inclusive-wise.

A well-known characterization of cographs is via cotrees. A *cotree* T of a cograph G is a rooted tree with two types of interior nodes, $\oplus$-nodes and $\otimes$-nodes, that has the following property: there is a bijection between the vertices of G and the leaves of T such that two vertices u and v are adjacent in G if and only if the lowest common ancestor of the leaves u and v in T is a $\otimes$-node. In particular, every node t of T corresponds to an induced subgraph of G, which is the disjoint union or the join of the subgraphs of G corresponding to the children of t. A graph is a cograph if and only if it has a cotree [4]. Cographs can be recognized and a cotree can be generated in linear time [5,8]. For our purposes, it is convenient to use the binary version of a cotree, which is commonly used for algorithms on cographs: the recursive definition of cographs implies that we can assume the cotree to be binary. We will call this a *nice* cotree. Clearly, given a cotree of a cograph, a nice cotree can be obtained in linear time.

3   A Polynomial-Time Approximation Algorithm

In this section, we show that Vector Connectivity can be approximated in polynomial time by a factor of $\ln n + 2$ on all graphs. We will achieve this by showing that Vector Connectivity can be recast as a particular case of the well-known Minimum Submodular Cover problem, which will allow us to apply a classical approximation result due to Wolsey [17].

First, we recall some definitions and results about submodular functions, hypergraphs and matroids that we will use in our proofs (see, e.g., [16]). Given a finite set U, a function $g : 2^U \to \mathbb{Z}_+$ is *submodular* if for every $X, Y \subseteq U$ with $X \subseteq Y$ and every $x \in U \setminus Y$, we have that $g(Y \cup \{x\}) - g(Y) \leq g(X \cup \{x\}) - g(X)$. An instance of the (unweighted) MINIMUM SUBMODULAR COVER problem consists of a set U and an integer-valued, non-decreasing, submodular function $g : 2^U \to \mathbb{Z}_+$. The objective is to pick a set $S \subseteq U$ of minimum cardinality such that $g(S) = g(U)$.

A *hypergraph* is a pair $H = (U, \mathcal{E})$ where U is a finite set of *vertices* and $\mathcal{E}$ is a set of subsets of U, called *hyperedges*. A *matroid* is a hypergraph $M = (U, \mathcal{F})$ such that $\mathcal{F}$ is nonempty and closed under taking subsets, and its elements, called *independent sets*, satisfy the following "exchange property": for every two independent sets A and B such that $|A| < |B|$, there exists an element of B whose addition to A results in a larger independent set. Given a matroid $M = (U, \mathcal{F})$, the *rank function* of M is the function that assigns to every subset S of U the maximum size of an independent set contained in S. The rank function of every matroid is submodular (see, e.g., [16]). A *gammoid* is a hypergraph $\Gamma = (U, \mathcal{E})$ derived from a triple (D, S, T) where $D = (V, A)$ is a digraph and $S, T \subseteq V$, such that $U = S$ and a subset S' of S forms a hyperedge if and only if there exist $|S'|$ vertex-disjoint directed paths in D connecting S' to a subset of T. Every gammoid is a matroid (see, e.g., [16]).

For any instance $(G = (V, E), \boldsymbol{k})$ of VECTOR CONNECTIVITY, we define a function $f : 2^V \longrightarrow \mathbb{Z}_+$ as follows:

$$f(X) = \sum_{v \in V} f_v(X), \text{ where } X \subseteq V, \text{ and}$$

$$f_v(X) = \begin{cases} \min\{\sigma(v, X), k_v\} & \text{if } v \notin X; \\ k_v & \text{if } v \in X. \end{cases} \tag{1}$$

Observe that a set $S \subseteq V$ satisfies $f(S) = f(V)$ if and only if S is a vector connectivity set for $(G, \boldsymbol{k})$. Consequently, Lemma 1 below immediately implies that VECTOR CONNECTIVITY is a special case of MINIMUM SUBMODULAR COVER.

Lemma 1. *Let $(G = (V, E), \boldsymbol{k})$ be an instance of* VECTOR CONNECTIVITY. *Then the function $f : 2^V \longrightarrow \mathbb{Z}_+$, given by (1), satisfies the following properties:*

(i) $f(\emptyset) = 0$;
(ii) f is integer-valued, i.e., $f(X) \in \mathbb{Z}_+$ for every $X \subseteq V$;
(iii) f is non-decreasing, i.e., $f(X) \leq f(Y)$ whenever $X \subseteq Y \subseteq V$;
(iv) f is submodular.

Proof. It is easy to verify that properties (i)–(iii) hold. In order to show that f is submodular, it suffices to show that all the functions $f_v(\cdot)$ are submodular, that is, that for all $X \subseteq Y \subseteq V$ and for all $w \in V \setminus Y$,

$$f_v(Y \cup \{w\}) - f_v(Y) \leq f_v(X \cup \{w\}) - f_v(X). \tag{2}$$

Suppose first that $f_v(Y) = k_v$. Then $f_v(Y \cup \{w\}) = k_v$ and the left-hand side of inequality (2) is equal to 0. Hence inequality (2) holds since f_v is non-decreasing.

Now suppose that $f_v(Y) < k_v$, which implies that $f_v(Y) = \sigma(v, Y)$. If $f_v(X \cup \{w\}) = k_v$, then $f_v(Y \cup \{w\}) = k_v$ and inequality (2) holds due to the fact that f_v is non-decreasing. In what follows, we assume that $f_v(X \cup \{w\}) < k_v$, which implies that $f_v(X \cup \{w\}) = \sigma(v, X \cup \{w\})$, and also that $v \neq w$. Since $\sigma(v, Y \cup \{w\}) \leq \sigma(v, Y) + 1 \leq k_v$, we have, by the definition of $f_v(\cdot)$ and using the fact that $v \notin Y \cup \{w\}$, the equality $f_v(Y \cup \{w\}) = \sigma(v, Y \cup \{w\})$. Since f_v is non-decreasing, $f_v(Y) < k_v$ implies that $f_v(X) < k_v$, and hence $f_v(X) = \sigma(v, X)$. Inequality (2) then simplifies to

$$\sigma(v, Y \cup \{w\}) - \sigma(v, Y) \leq \sigma(v, X \cup \{w\}) - \sigma(v, X). \tag{3}$$

Hence, in order to prove Lemma 1, it suffices to show that inequality (3) holds for any fixed vertex $v \in V$, i.e., that the function $g_v : 2^{V \setminus \{v\}} \longrightarrow \mathbb{Z}_+$, defined by $g_v(W) = \sigma(v, W)$ for all $W \subseteq V \setminus \{v\}$, is submodular. Consider the gammoid Γ derived from the triple $(D, V \setminus \{v\}, N_G(v))$ where D is the digraph obtained from G by replacing each edge with a pair of oppositely directed arcs. Since Γ is a gammoid, it is a matroid. It follows directly from the definition that function g_v is equal to the rank function of Γ. Therefore, the function g_v is submodular, which completes the proof of Lemma 1. □

Theorem 1. VECTOR CONNECTIVITY *can be approximated within a factor of* $\ln n + 2$ *in polynomial time.*

Proof. Let $(G = (V, E), \boldsymbol{k})$ be an instance of VECTOR CONNECTIVITY with $|V| = n$. From the definition of the function f, given by (1), it follows that a set $S \subseteq V$ satisfies $f(S) = f(V)$ if and only if S is a vector connectivity set for $(G, \boldsymbol{k})$. Hence, an optimal solution to the VECTOR CONNECTIVITY problem is provided by a minimum size subset $S \subseteq V$ such that $f(S) = f(V)$, i.e., by an optimal solution for MINIMUM SUBMODULAR COVER. An approximation to such a set S can be found in the following way.

Let $\mathbb{A}$ denote the natural greedy strategy which starts with $S = \emptyset$ and iteratively adds to S the element $v \in V \setminus S$ such that $f(S \cup \{v\}) - f(S)$ is maximum, until $f(S) = f(V)$ is achieved. The maximum order of a v–S fan can be computed in polynomial time using an easy reduction to the well-known MAXIMUM FLOW problem, and thus the function f is polynomially computable. Therefore, the greedy strategy can be implemented in polynomial time. Moreover, Wolsey [17] proved that if f satisfies the four properties listed in Lemma 1, then algorithm $\mathbb{A}$ is an $H(\tau)$-approximation algorithm for MINIMUM SUBMODULAR COVER, and consequently for VECTOR CONNECTIVITY, where $H(j) = \sum_{i=1}^{j} \frac{1}{i}$ denotes the j-th harmonic number, and $\tau = \max_{y \in V} f(\{y\}) - f(\emptyset)$. For every $y \in V$, we have

$$f(\{y\}) = \sum_{v \in V \setminus \{y\}} f_v(\{y\}) + f_y(\{y\}) \leq n - 1 + k_y \leq n + \Delta(G).$$

Since $f(\emptyset) = 0$, this implies $\tau \leq n + \Delta(G)$. Hence, algorithm $\mathbb{A}$ is an $H(n + \Delta(G))$-approximation algorithm for VECTOR CONNECTIVITY. Since $H(n) \leq \ln n + 1$ for $n \geq 1$, we can further bound the approximation ratio ρ of $\mathbb{A}$ from above as follows:

$$\rho \le H(n + \Delta(G)) \le \ln(n + \Delta(G)) + 1 \le \ln(2n) + 1 = \ln n + \ln 2 + 1 \le \ln n + 2,$$

yielding the desired result. □

4   A Polynomial-Time Algorithm for Split Graphs

Recall that the VECTOR DOMINATION problem on split graphs is both NP-hard and hard to approximate within a factor of $(1 - \epsilon) \ln n$ for any constant $\epsilon > 0$. In this section, we give a polynomial-time algorithm to solve the VECTOR CONNECTIVITY problem on split graphs. Our algorithm is based on the following lemma.

Lemma 2. *Let (G, k) be an instance of VECTOR CONNECTIVITY, where G is a split graph. Let S be any set of vertices in G such that $k_u \ge k_v$ for every pair of vertices $u \in S$ and $v \in V(G) \setminus S$. Then there exists a v–S fan of order $\min\{k_v, |S|\}$ for every $v \in V(G) \setminus S$.*

Proof. Let (C, I) be a split partition of $G = (V, E)$, and for convenience let $S_I = S \cap I$ and $S_C = S \cap C$. We will call the vertices of $V \setminus S$ *free vertices*. Let v be a free vertex of G. We first show that every vertex $u \in S_I$ has at least $k_v - |S_C|$ free neighbors. To see this, let $u \in S_I$. It is obvious that u has at least $d_G(u) - |S_C|$ free neighbors. Since $u \in S$ and $v \in V \setminus S$, we have that $k_v \le k_u$. This, together with the assumption that $k_v \le d_G(v)$ for every $v \in V$, implies that u has at least $d_G(u) - |S_C| \ge k_u - |S_C| \ge k_v - |S_C|$ free neighbors.

Suppose that v is a vertex of C. Every vertex of S_C is a neighbor of v, and thus v is $\min\{k_v, |S_C|\}$-connected to S_C. If $k_v \le |S_C|$ then the lemma follows, so assume that $k_v > |S_C|$. Recall that every vertex $u \in S_I$ has at least $k_v - |S_C|$ free vertices in its neighborhood. Let $S' \subseteq S_I$ be any subset of S_I such that $|S'| = \min\{k_v - |S_C|, |S_I|\}$. Let G' be the bipartite subgraph of G obtained from the subgraph of G induced by $S' \cup (N_G(S') \setminus S_C)$ by deleting all edges of the form $\{xy \mid x, y \in N_G(S')\}$. Since $|S''| \le |N_{G'}(S'')|$ for every subset $S'' \subseteq S'$, Hall's Theorem implies that there is a matching M in G' that saturates S'. Let Y be the set of endpoints of M that are not in S'. Then $Y \subseteq C$, and it is possible that $v \in Y$. Since both v and all the vertices of Y belong to the clique C, v can reach at least $|S'| = \min\{k_v - |S_C|, |S_I|\}$ vertices of S_I via disjoint paths that do not contain vertices of S_C, using the edges of M. Consequently, v is $\min\{k_v, |S|\}$-connected to S, and the lemma follows.

Suppose now that v is a vertex of I. Since $k_v \le d_G(v)$, v is $\min\{k_v, |S_C|\}$-connected to S_C. Let $\mathcal{P}_C$ be a v–S_C fan of order $\min\{k_v, |S_C|\}$ that is of smallest total path length. In particular, every path in $\mathcal{P}_C$ is of length 1 or 2. If $k_v \le |S_C|$ then the lemma follows, so assume that $k_v > |S_C|$. In this case, exactly $|S_C|$ neighbors of v are used by the paths in $\mathcal{P}_C$. However, v has at least $d_G(v) - |S_C| \ge k_v - |S_C|$ additional neighbors that are free vertices in C. Furthermore, we already proved that every $u \in S_I$ has at least $k_v - |S_C|$ free vertices in its neighborhood.

Each such vertex is either a neighbor of v or a neighbor of a neighbor of v. Thus v can reach at least $\min\{k_v - |S_C|, |S_I|\}$ vertices of S_I via disjoint paths that intersect each other and the paths of $\mathcal{P}_C$ only in vertex v. This shows that v is $\min\{k_v, |S|\}$-connected to S, and the lemma follows.   □

Lemma 2 implies that we can sort the vertices of G by their k-values in non-increasing order, and greedily pick vertices from the start of the sorted list to be in S until we have a vector connectivity set. This is formalized in the proof of the following theorem.

Theorem 2. VECTOR CONNECTIVITY *can be solved in polynomial time on split graphs.*   □

5   A Polynomial-Time Algorithm for Cographs

In this section we show that VECTOR CONNECTIVITY can be solved in polynomial time on cographs. We will in fact solve the following more general variant of VECTOR CONNECTIVITY. For a graph $G = (V, E)$, an integer-valued vector $k = (k_v : v \in V)$, and an integer ℓ, we say that a set $S \subseteq V$ is a *vector connectivity set for* (G, k, ℓ) if S is a vector connectivity set for (G, k) such that $v \in S$ whenever $k_v \geq \ell$. Let us denote by $\kappa(G, k, \ell)$ the minimum size of a vector connectivity set for (G, k, ℓ). Since $S = V$ is a vector connectivity set for (G, k, ℓ), the above parameter is well defined and satisfies $\kappa(G, k, \ell) \leq |V|$. Clearly, the following relation holds, and hence solving the described variant indeed also solves VECTOR CONNECTIVITY.

Lemma 3. $\kappa(G, k) = \kappa(G, k, \max_{v \in V} k_v + 1)$.

In order to simplify the presentation of our algorithm, we assume in this section that in the input to the VECTOR CONNECTIVITY problem and its variant mentioned above, requirements k_v are allowed to be negative. If $k_v < 0$, no condition is imposed on vertex v, and it can be treated the same as if $k_v = 0$.

The first lemma below is an easy observation.

Lemma 4. *Let* $G_1 = (V_1, E_1)$ *and* $G_2 = (V_2, E_2)$ *be two graphs such that* $V_1 \cap V_2 = \emptyset$, *and let* $G = G_1 \oplus G_2$. *Then it holds that*

$$\kappa(G, k, \ell) = \kappa(G_1, k|_{V(G_1)}, \ell) + \kappa(G_2, k|_{V(G_2)}, \ell).$$

Lemma 5. *Let* $G_1 = (V_1, E_1)$ *and* $G_2 = (V_2, E_2)$ *be two graphs such that* $V_1 \cap V_2 = \emptyset$, *and let* $G = G_1 \otimes G_2$. *Let* $n_1 = |V_1|$ *and* $n_2 = |V_2|$, *and let* $\mathcal{F} = \{0, 1, \ldots, n_1\} \times \{0, 1, \ldots, n_2\}$. *For every integer* ℓ, *it holds that*

$$\kappa(G, k, \ell) = \min_{(i,j) \in \mathcal{F}} f(i, j)$$

with

$$f(i, j) = \max\left\{\kappa\left(G_1, k^{1ij}, \ell_1^{ij}\right), i\right\} + \max\left\{\kappa\left(G_2, k^{2ij}, \ell_2^{ij}\right), j\right\},$$

where

- $k^{1ij} = k|_{V_1} - \min\{i + j, n_2\} \cdot 1_{V_1}$,
- $k^{2ij} = k|_{V_2} - \min\{i + j, n_1\} \cdot 1_{V_2}$,

$$- \ell_1^{ij} = \min\{\ell, i+j+1\} - \min\{i+j, n_2\},$$
$$- \ell_2^{ij} = \min\{\ell, i+j+1\} - \min\{i+j, n_1\}.$$

$\square$

Theorem 3. VECTOR CONNECTIVITY *can be solved in polynomial time on cographs.*

Proof. Consider the input $(G = (V, E), \boldsymbol{k})$ to the VECTOR CONNECTIVITY problem, where G is a cograph and $n = |V|$. By Lemma 3, computing the value of $\kappa(G, \boldsymbol{k})$ is equivalent to computing the value of $\kappa(G, \boldsymbol{k}, K)$ with $K = \max_{v \in V} k_v + 1$. We compute this value as follows. First, we compute a nice cotree T of G. We traverse T bottom up, processing a node only after all its children have been processed. When processing a node t of T, we compute all $O(n^2)$ values of

$$\kappa(H, \boldsymbol{k}|_{V(H)} - i \cdot \boldsymbol{1}_{V(H)}, \ell), \quad i \in \{0, 1, \ldots, n\}, \quad \ell \in \{0, 1, \ldots, K\},$$

where H is the induced subgraph of G corresponding to the subtree T_t. For every leaf of the cotree, corresponding to a single vertex v of G, each of the $O(n^2)$ values can be computed in $O(1)$ time as follows:

$$\kappa((\{v\}, \emptyset), k_v - i, \ell) = \begin{cases} 0 & \text{if } k_v - i \leq \min\{\ell - 1, 0\}, \\ 1 & \text{otherwise.} \end{cases}$$

Depending on whether an internal node t is a $\oplus$-node or a $\otimes$-node, we can use Lemma 4 or Lemma 5 to compute each of the $O(n^2)$ values of $\kappa(H, \boldsymbol{k}|_{V(H)} - i \cdot \boldsymbol{1}_{V(H)}, \ell)$ in time $O(n^2)$. Hence, each internal node of the modified cotree can be processed in time $O(n^4)$, yielding an overall time complexity of $O(n^5)$, since a cotree has $O(n)$ nodes.

A minimum vector connectivity set can also be computed in the stated time. In addition to the values of $\kappa(H, \boldsymbol{k}|_{V(H)} - i \cdot \boldsymbol{1}_{V(H)}, \ell)$ at each node of the cotree, we need to store also a minimum vector connectivity set achieving each of these values. These sets can be computed recursively as follows. For an internal node t with corresponding subgraph H, let H_1 and H_2 denote the subgraphs of G corresponding to the two children of t in T.

- If H corresponds to a leaf of T, then $V(H) = \{v\}$ for some $v \in V$, and a minimum vector connectivity set for $(H, k_v - i, \ell)$ is either empty or $\{v\}$, depending on whether $k_v - i \leq \min\{\ell - 1, 0\}$ or not.
- If $H = H_1 \oplus H_2$, then a minimum vector connectivity set for $(H, \boldsymbol{k}|_{V(H)} - i \cdot \boldsymbol{1}_{V(H)}, \ell)$ is given by the union of minimum vector connectivity sets for $(H_1, \boldsymbol{k}|_{V(H_1)} - i \cdot \boldsymbol{1}_{V(H_1)}, \ell)$ and $(H_2, \boldsymbol{k}|_{V(H_2)} - i \cdot \boldsymbol{1}_{V(H)}, \ell)$.
- If $H = H_1 \otimes H_2$, then a minimum vector connectivity set S for $(H, \boldsymbol{k}|_{V(H)} - i \cdot \boldsymbol{1}_{V(H)}, \ell)$ can be computed in $O(n^2)$ time: first compute a pair (I, J) minimizing the function f defined in Lemma 5 (with H, H_1, H_2 in place of G, G_1, G_2, respectively), and then take the union of minimum vector connectivity sets S_1' and S_2' for $(H_1, \boldsymbol{k}^{1IJ}, \ell_1^{IJ})$ and $(H_2, \boldsymbol{k}^{2IJ}, \ell_2^{IJ})$, together with some extra vertices if necessary so that $|S \cap V(H_1)| \geq I$ and $|S \cap V(H_2)| \geq J$.

Finally, let us remark that all the instances $(H, \mathbf{k}', \ell)$ for which $\kappa(H, \mathbf{k}', \ell)$ must be evaluated in order to compute the value of $\kappa(G, \mathbf{k}) = \kappa(G, \mathbf{k}, \max_{v \in V} k_v + 1)$ satisfy the property that for all $v \in V(H)$, either $k'_v \leq d_H(v)$ or $k'_v \geq \ell$. This can be proved by induction on the distance of a node t representing H from the root of the cotree T, and assures that the values of $\kappa(H, \mathbf{k}', \ell)$ are well defined for such instances. This completes the proof of Theorem 3.   □

6 A Polynomial-Time Algorithm for Trees

We have seen that VECTOR CONNECTIVITY is solvable in polynomial time on cographs and split graphs. These two graph classes do not contain graphs with long induced paths. In particular, cographs are equivalent to graphs that do not have induced paths of length 3 or more [5], and it is easy to observe that split graphs do not contain induced paths of length 4 or more. In this section, we give a polynomial-time algorithm to solve the VECTOR CONNECTIVITY problem on trees, a graph class that allows the existence of arbitrarily long induced paths.

Theorem 4. VECTOR CONNECTIVITY *can be solved in polynomial time on trees.*

Proof. Let $(T, \mathbf{k})$ be an instance of VECTOR CONNECTIVITY, where $T = (V, E)$ is a tree. We assume that T has at least two vertices and is rooted at an arbitrary vertex r. Since the requirements of the vertices do not change during the execution of the algorithm, we will simply speak of a vector connectivity set for T_v instead of a vector connectivity set for $(T_v, \mathbf{k}|_{V(T_v)})$, for every $v \in V$.

The idea of the algorithm is to construct a vector connectivity set for T of minimum size, starting from the leaves of T and processing a vertex only after all its children have been processed. At any step of the algorithm, let $S \subseteq V$ be the set of vertices that have thus far been chosen to belong to the solution. For any vertex v of T, we define $S_v = S \cap V(T_v)$. When processing a vertex v, the algorithm computes the values $f(v)$, $n(v)$ and $r(v)$, which are defined as follows. For every vertex $v \in V$, $f(v) = 1$ if the subtree T_v contains at least one vertex of S; otherwise $f(v) = 0$. The value $r(v)$ denotes the number of children w of v for which $f(w) = 1$. Note that if $f(w) = 1$ for a child w of vertex v, then v is 1-connected but not 2-connected to S_w, regardless of how many vertices S_w contains. Furthermore, v is 1-connected to $S \setminus V(T_v)$ if S contains a vertex outside T_v. We let $n(v)$ denote whether or not a vertex in T_v "needs" an additional path to a vertex outside of T_v, indicated by 1 or 0, for every $v \in V$. More precisely, $n(v) = 0$ if for every vertex $w \in V(T_v)$, there is a w–S_v fan of order k_w in T_v, i.e., every vertex w of T_v, including v itself, is k_w-connected to S_v and hence also to S. On the other hand, $n(v) = 1$ if there is a vertex $w \in V(T_v)$ such that there is a w–S_v fan of order $k_w - 1$ but no w–S_v fan of order k_w in T_v.

We now describe the algorithm in detail. Initially, we set $S = \emptyset$. Let $v \in V$ be a leaf of T. We set $r(v) = 0$. If $k_v = 0$, then we set $f(v) = 0$ and $n(v) = 0$. If $k_v = 1$, then we set $f(v) = 0$ and $n(v) = 1$.

Next, let v be a vertex that is not a leaf and not the root, and assume that the children of v have all been processed. For every child w of v, if $n(w) = 1$ and v has a child $w' \neq w$ such that $f(w') = 1$, then we set $n(w) = 0$. We then compute $r(v)$ by adding up the f-values of all the children of v. If $k_v \leq r(v)$, then we set $n(v) = 1$ if v has a child w with $n(w) = 1$, and we set $n(v) = 0$ otherwise. If $k_v = r(v) + 1$, then we set $n(v) = 1$. If $k_v \geq r(v) + 2$, then we add v to S and set $n(v) = 0$. In each of the above cases, we set $f(v) = 1$ if $r(v) \geq 1$ or if v is added to S, and we set $f(v) = 0$ otherwise.

Finally, let v be the root of T. We set $n(v) = 0$. If $k_v \leq r(v)$, then we perform the following check: if v has a child w such that $n(w) = 1$ and $f(w') = 0$ for every other child $w' \neq w$ of v, then we add v to S. If $k_v \geq r(v) + 1$, then we add v to S. The algorithm outputs the set S and terminates as soon as the root has been processed.

The correctness of the algorithm can be shown by observing that for every $v \in V$, the following three statements are true immediately after v is processed, where p denotes the parent of v in T.

(i) If $n(v) = 0$, then S_v is a vector connectivity set for T_v.
(ii) If $n(v) = 1$, then $S_v \cup \{p\}$ is a vector connectivity set for the subtree of T induced by $V(T_v) \cup \{p\}$;
(iii) There is no vector connectivity set S' for T such that $|S' \cap V(T_v)| < |S_v|$.

Since these statements hold for the root of T, the set S constructed by the algorithm is a vector connectivity set for T of minimum size. The observation that all steps of the algorithm can be performed in polynomial time completes the proof of Theorem 4.   □

7   Concluding Remarks

In this paper, we initiated the study of the VECTOR CONNECTIVITY problem, which opens a research path with many interesting questions. The most prominent of these questions is of course the computational complexity of VECTOR CONNECTIVITY on general input graphs. Could it be that the problem is polynomial-time solvable on all graphs, or is its tractability heavily dependent on either the absence of long induced paths or on a tree-like structure of the input graph? On which other graph classes is VECTOR CONNECTIVITY solvable in polynomial time? Does VECTOR CONNECTIVITY admit a polynomial-time constant-factor approximation algorithm on general graphs?

Another interesting variant of the problem can be obtained by allowing the requirement k_v of each vertex v to be arbitrarily large, in which case a vertex v with $k_v > d_G(v)$ is forced to be in every vector connectivity set. Is it perhaps easier to prove this variant to be NP-hard in general? Note that the algorithms given in this paper, except the algorithm for split graphs, work in polynomial time also for this variant. Is this variant polynomial-time solvable on split graphs?

Acknowledgements. We are indebted to Ferdinando Cicalese and Ugo Vaccaro for initial discussions during which the VECTOR CONNECTIVITY problem was discovered and research on it started. We are also grateful to Nicola Apollonio for helpful discussions, and to two anonymous referees for their careful reading of the paper and for comments that helped simplify the proof of Lemma 1.

References

1. Chlebík, M., Chlebíkova, J.: Approximation hardness of dominating set problems in bounded degree graphs. Information and Computation 206, 1264–1275 (2008)
2. Chen, N.: On the approximability of inuence in social networks. SIAM Journal on Discrete Mathematics 23, 1400–1415 (2009)
3. Cicalese, F., Milanič, M., Vaccaro, U.: On the approximability and exact algorithms for vector domination and related problems in graphs. Discrete Applied Mathematics (2012), doi:10.1016/j.dam.2012.10.007
4. Corneil, D.G., Lerchs, H., Stewart Burlingham, L.: Complement reducible graphs. Discrete Applied Mathematics 3, 163–174 (1981)
5. Corneil, D.G., Perl, Y., Stewart, L.K.: A linear recognition algorithm for cographs. SIAM J. Comput. 14, 926–934 (1985)
6. Garey, M.R., Johnson, D.S.: Computers and Intractability. W.H. Freeman and Co., New York (1979)
7. Golumbic, M.C.: Algorithmic Graph Theory and Perfect Graphs, 2nd edn. Annals of Discrete Mathematics, vol. 57. Elsevier (2004)
8. Habib, M., Paul, C.: A simple linear time algorithm for cograph recognition. Discrete Applied Mathematics 145, 183–197 (2005)
9. Hammer, P.L., Simeone, B.: The splittance of a graph. Combinatorica 1, 275–284 (1981)
10. Harant, J., Prochnewski, A., Voigt, M.: On dominating sets and independent sets of graphs. Combinatorics, Probability and Computing 8, 547–553 (1999)
11. Haynes, T.W., Hedetniemi, S., Slater, P.: Fundamentals of Domination in Graphs. Marcel Dekker, New York (1998)
12. Haynes, T.W., Hedetniemi, S., Slater, P.: Domination in Graphs: Advanced Topics. Marcel Dekker, New York (1998)
13. Kempe, D., Kleinberg, J., Tardos, E.: Maximizing the spread of inuence through a social network. In: Proc. 9th ACM KDD, pp. 137–146. ACM Press (2003)
14. Mahadev, N., Peled, U.: Threshold graphs and related topics. Annals of Discrete Mathematics 56 (1995)
15. Nichterlein, A., Niedermeier, R., Uhlmann, J., Weller, M.: On tractable cases of target set selection. Social Network Analysis and Mining (2012)
16. Schrijver, A.: Combinatorial Optimization. Polyhedra and Efficiency. Volumes A–C. Algorithms and Combinatorics, vol. 24. Springer, Berlin (2003)
17. Wolsey, L.A.: An analysis of the greedy algorithm for the submodular set covering problem. Combinatorica 2, 385–393 (1982)

Trees in Graphs with Conflict Edges or Forbidden Transitions

Mamadou Moustapha Kanté*, Christian Laforest**, and Benjamin Momège***

Clermont-Université, Université Blaise Pascal, LIMOS, CNRS, France
{mamadou.kante,laforest,momege}@isima.fr

Abstract. In a recent paper [*Paths, trees and matchings under disjunctive constraints*, Darmann et. al., Discr. Appl. Math., 2011] the authors add to a graph G a set of *conflicts*, i.e. pairs of edges of G that cannot be both in a subgraph of G. They proved hardness results on the problem of constructing minimum spanning trees and maximum matchings containing no conflicts. A *forbidden transition* is a particular conflict in which the two edges of the conflict must be incident. We consider in this paper graphs with forbidden transitions. We prove that the construction of a minimum spanning tree without forbidden transitions is still $\mathcal{NP}$-Hard, even if the graph is a complete graph. We also consider the problem of constructing a maximum tree without forbidden transitions and prove that it cannot be approximated better than $n^{1/2-\varepsilon}$ for all $\varepsilon > 0$ even if the graph is a star. We strengthen in this way the results of Darmann et al. concerning the minimum spanning tree problem. We also describe sufficient conditions on forbidden transitions (conflicts) to ensure the existence of a spanning tree in complete graphs. One of these conditions uses *graphic sequences*.

1 Introduction

In some practical situations, classical graphs are not complex enough to model all the constraints. For example, a city map can be modelled by a graph where streets are edges. However a car cannot always follow any route on this map. In some points it can be forbidden to turn left or right for example. This means that some paths in the graph are not valid. In the graph of a city with such restrictions, finding a spanning tree containing no restriction would be useful to ensure the connectivity between any pair of locations. The cars could travel on this tree submitted to no forbidden transitions. In this paper we investigate this kind of problem from a pure theoretical point of view.

In the following paragraphs we give the main definitions, notations and concepts that will be used throughout the paper. We also give some bibliographical references on related results.

* M.M. Kanté is supported by the French Agency for Research under the DORSO project.

** Ch. Laforest is supported by the French Agency for Research under the DEFIS program TODO, ANR-09- EMER-010.

*** B. Momège has a PhD grant from CNRS and région Auvergne.

T-H.H. Chan, L.C. Lau, and L. Trevisan (Eds.): TAMC 2013, LNCS 7876, pp. 343–354, 2013.
© Springer-Verlag Berlin Heidelberg 2013

Specific Notions and Notations. In this paper, we only consider undirected, unweighted and simple graphs. We refer to [2] for definitions and undefined notations. The vertex set of a graph G is denoted by V_G and its edge set by E_G. An edge between u and v in a graph G is denoted by uv. A *tree* is an acyclic connected graph and a *star* is a tree with a distinguished vertex adjacent to the other vertices. A complete graph (resp. star) with n vertices is denoted by K_n (resp. S_n). A path (or a cycle) of G is *Hamiltonian* if it contains all the vertices of G exactly once (all paths and cycles are elementary here).

If G is a graph, a *conflict* is a pair $\{e_1, e_2\}$ of edges of G. A conflict $\{e_1, e_2\}$ is called a *forbidden transition* if e_1 and e_2 are incident. In a forbidden transition $\{uv, vw\}$, the vertex v is called its *centre* and the vertices u and w its *extremities*. We denote by $(G, \mathcal{C})$ (resp. $(G, \mathcal{F})$) a graph G with a set of conflicts $\mathcal{C}$ (resp. with a set of forbidden transitions $\mathcal{F}$). (We use the notation $\mathcal{C}$ to denote conflicts and $\mathcal{F}$ for forbidden transitions.) A *spanning tree* T in $(G, \mathcal{C})$ is a spanning tree in G without conflicts, i.e., for any e, e' of T, $\{e, e'\} \notin \mathcal{C}$ (similarly for the other subgraph notions).

The *spanning tree problem without conflicts* (STWC) is, given $(G, \mathcal{C})$, constructs a spanning tree T in $(G, \mathcal{C})$, if one exists, otherwise say NO. We define similarly the *spanning tree problem without forbidden transitions* (STWFT). Similarly, the *Hamiltonian path (or cycle) problem without forbidden transitions* is denoted by HPWFT (or HCWFT). The problem of constructing a tree without forbidden transitions of maximum size will be denoted by MTWFT.

Works Involving Forbidden Transitions. Graphs with forbidden transitions have already been investigated and several problems known to be polynomial in graphs have been shown to be intractable in graphs with forbidden transitions. For instance it is proved in [10] that knowing whether there exists a path between two nodes avoiding forbidden transitions is $\mathscr{N}\mathscr{P}$-complete and a line between tractable and intractable cases have been identified. The problem of finding two-factors[1] is considered in [4] and a dichotomy between tractable and intractable instances is also given. In a very recent paper [7] we propose an exact exponential time algorithm that checks the existence of paths without forbidden transitions between two vertices; we also generalise the notion of cut in such graphs.

It is worth noticing that the $\mathscr{N}\mathscr{P}$-hardness of the connectedness of two vertices in graphs with forbidden transitions does not imply the $\mathscr{N}\mathscr{P}$-hardness of STWFT. Indeed, a classical result in graph theory (see [2]) states that a graph is connected if and only if it contains a spanning tree. Unfortunately, this is not the case anymore if we take into account $\mathcal{F}$ and STWFT. The simplest proof of this fact is the following. Consider a complete graph K_n with $n \geq 3$ where each possible transition is forbidden. Each pair of vertices is connected by a path with one edge, i.e. without forbidden transitions, but any spanning tree contains at least two edges (since $n \geq 3$) thus a forbidden transition.

Works Involving Conflicts. Since forbidden transitions are special cases of conflicts, the $\mathscr{N}\mathscr{P}$-hard problems considered in [4,10] remains $\mathscr{N}\mathscr{P}$-hard in

---

[1] A subgraph such that for any vertex its in-degree and its out-degree is exactly one.

graphs with conflicts. But notice that dichotomy theorems in [4,10] are no longer valid when dealing with conflicts. Some tractable cases of the path problem have been investigated in [8]. Another set of problems have been considered in the literature. For example authors of [9] considered the problem of constructing a scheduling such that two conflicting tasks cannot be executed on the same machine or packing problems under the condition that two conflicting items cannot be packed together. In [3] the authors proved that STWC is $\mathcal{NP}$-complete. They proved also similar results for the maximum matching problem.

Summary. We prove in Section 2 that STWFT is $\mathcal{NP}$-complete and characterise (in)tractable cases. Hence our result is stronger since we prove the hardness for a more restrictive type of conflicts. We go further by proving the hardness even in complete graphs with forbidden transitions. We also show that HPWFT and HCWFT are also $\mathcal{NP}$-complete in complete graphs with forbidden transitions. We furthermore prove that MTWFT cannot be approximated, even in stars with forbidden transitions. In Section 3.1 we adapt and use a result on graphical sequences to give a sufficient condition to ensure an always YES instance for STWFT in polynomial time when restricting instances to $(K_n, \mathcal{F})$. In Section 3.2, we prove that STWFT is polynomial in $(K_n, \mathcal{F})$ where each vertex is in a bounded number of forbidden transitions. We also prove that if each edge is involved in at most one conflict, then there always exist an Hamiltonian path in $(K_n, \mathcal{C})$. Finally, in Section 3.3 we describe a polynomial time process to transform any instance $(G, \mathcal{C})$ into an instance $(G_f, \mathcal{C}_f)$ containing less edges and conflicts and ensuring that $(G, \mathcal{C})$ is a YES instance for STWC iff $(G_f, \mathcal{C}_f)$ is.

2 Hardness Results

2.1 $\mathcal{NP}$-Hardness of STWFT, HPWFT and HCWFT

If $(G, \mathcal{C})$ is a graph with conflicts, we can associate with it a *conflict* graph that has as edge set $\mathcal{C}$ and as vertex set edges involved in $\mathcal{C}$. A 2-*ladder* is a disjoint union of edges and a 3-*ladder* is a disjoint union of paths with 3 vertices. We recall the following from [3].

Theorem 1 ([3]). STWC *is $\mathcal{NP}$-complete, even if conflict graphs are 3-ladders. However, STWC is polynomial in $(G, \mathcal{C})$ with 2-ladder conflict graphs.*

A slight modification of the proof of Theorem 1 gives the following result for forbidden transitions. Its proof is given for completeness.

Theorem 2. STWFT *is $\mathcal{NP}$-complete, even in bipartite graphs with 3-ladders as conflict graphs. STWFT is polynomial in $(G, \mathcal{F})$ with 2-ladder conflict graphs.*

Proof. The second statement follows directly from Theorem 1.

As in the proof of the $\mathcal{NP}$-completeness of STWC, we will reduce the (3,B2)-SAT to STWFT with 3-ladders as forbidden transitions. We recall that a (3,B2)-SAT instance is a 3-SAT instance such that each variable occurs exactly four times, twice positive and twice negated.

Let I be an instance of the (3,B2)-SAT with m clauses $C_1, \ldots, C_m$ and n variables $X_1, \ldots, X_n$. We let $(G, \mathcal{F})$ with

$$V_G := \{r\} \cup \{c_j \mid C_j \text{ is a clause}\} \cup \{x_i, \overline{x_i}, r_i, s_i \mid X_i \text{ is a variable}\},$$
$$E_G := \{rr_i, r_i x_i, r_i \overline{x_i}, x_i s_i, \overline{x_i} s_i \mid i \in \{1, \ldots, n\}\} \cup$$
$$\{x_i c_j \mid X_i \text{ occurs positively in } C_j\} \cup \{\overline{x_i} c_j \mid X_i \text{ occurs negatively in } C_j\}.$$

G is bipartite (colour "black" the vertices: $c_1, \ldots, c_m, r_1, \ldots, r_n, s_1, \ldots, s_n$ which are independent and "white" the other also independent remaining vertices). The structure of the graph G is the same as in [3] but the conflicts we define are now forbidden transitions.

$$\mathcal{F} := \bigcup_{i \in \{1, \ldots, n\}} \{\{c_j x_i, x_i s_i\}, \{c_k x_i, x_i s_i\} \mid c_j x_i \in E_G \text{ and } c_k x_i \in E_G\} \cup$$
$$\bigcup_{i \in \{1, \ldots, n\}} \{\{c_j \overline{x_i}, \overline{x_i} s_i\}, \{c_k \overline{x_i}, \overline{x_i} s_i\} \mid c_j \overline{x_i} \in E_G \text{ and } c_k \overline{x_i} \in E_G\}$$

One easily checks that the conflict graph of $\mathcal{F}$ is a 3-ladder. We now prove that I is satisfiable iff there exists a spanning tree of $(G, \mathcal{F})$.

Assume I is satisfiable. Then there is a mapping $\delta : \{X_1, \ldots, X_n\} \to \{0, 1\}$ such that each clause is satisfied and for each clause there is a variable X_i such that $\delta(X_i)$ allows to satisfy it. Let T be the graph formed with the following edges:

$$\bigcup_{i \in \{1, \ldots, n\}} \{rr_i, r_i x_i, r_i \overline{x_i}\} \cup \bigcup_{i \in \{1, \ldots, n\}} \{s_i x_i \mid \delta(X_i) = 1\} \cup$$
$$\bigcup_{i \in \{1, \ldots, n\}} \{s_i \overline{x_i} \mid \delta(X_i) = 0\} \cup$$
$$\bigcup_{i \in \{1, \ldots, n\}} \{x_i c_j, x_i c_k \mid \delta(X_i) = 1 \text{ and } X_i \text{ occurs positively in } C_j \text{ and in } C_k\} \cup$$
$$\bigcup_{i \in \{1, \ldots, n\}} \{\overline{x_i} c_j, \overline{x_i} c_k \mid \delta(X_i) = 0 \text{ and } X_i \text{ occurs negatively in } C_j \text{ and in } C_k\}$$

One checks that T spans G, is acyclic, is connected and does not contain a forbidden transition. Therefore, T is a spanning tree of $(G, \mathcal{F})$.

Assume now that $(G, \mathcal{F})$ has a spanning tree T without forbidden transitions. For each $i \in \{1, \ldots, n\}$ for which exactly one the edges $x_i s_i$ or $\overline{x_i} s_i$ is in E_T, we do the following assignment $\delta : \{X_1, \ldots, X_n\} \to \{0, 1\}$

$$\delta(X_i) := \begin{cases} 1 & \text{if } \overline{x_i} s_i \in E_T \\ 0 & \text{if } x_i s_i \in E_T. \end{cases}$$

The other variables X_i receive arbitrary assignments. We claim that the assignment δ satisfies the instance I. Let us consider any clause C_j. There exists some

$i \in \{1, \ldots, n\}$ such that $x_i c_j \in E_T$ or $\overline{x_i} c_j \in E_T$. If $x_i c_j \in E_T$, then $x_i s_i \notin E_T$ and therefore $\overline{x_i} s_i \in E_T$. By the definition of δ, we have $\delta(X_i) = 1$ and then C_j is satisfied by $\delta(X_i)$. Similarly, if $\overline{x_i} c_j \in E_T$, we have $\overline{x_i} s_i \notin E_T$ and then $x_i s_i \in E_T$. Again, by the definition of δ, we have $\delta(X_i) = 0$, so C_j is satisfied by $\overline{\delta(X_i)}$. We conclude that I is satisfied by δ.   □

If we take as parameter the number of conflicts a vertex (an edge) is involved in graphs with forbidden transitions, Theorem 2 gives a sharp line between tractable and intractable cases. We leave open the question for a dichotomy between tractable and intractable cases with respect to conflict graphs as done in [4,10]. In the following, we show that when restricting to complete graphs, the $\mathcal{NP}$-completeness of STWFT remains true. For any $(G, \mathcal{F})$ such that G has $n \geq 3$ vertices, construct the complete graph K_n with the same set of vertices than G (and with all possible edges) and $\overline{\mathcal{F}(G)} := \mathcal{F} \cup \{\{e, f\} \mid e \in E_{K_n} \setminus E_G, f \in E_{K_n}, e \neq f$ and e and f incident in $K_n\}$.

Lemma 1. *T is a spanning tree of* $(G, \mathcal{F})$ *iff T is a spanning tree of* $(K_n, \overline{\mathcal{F}(G)})$.

Proof. It is clear that any spanning tree of $(G, \mathcal{F})$ is also a spanning tree of $(K_n, \overline{\mathcal{F}(G)})$. Conversely, assume now that T is a spanning tree without forbidden transitions of $(K_n, \overline{\mathcal{F}(G)})$. Since $n \geq 3$, T does not contain any "non-edge" of G, otherwise T would contain a forbidden transition. Therefore, T is a spanning tree without forbidden transitions of $(G, \mathcal{F})$.   □

From Theorem 2 and Lemma 1, we can prove the following.

Theorem 3. STWFT *is* $\mathcal{NP}$-*complete in complete graphs with forbidden transitions.*

It is well-known that K_n contains many Hamiltonian paths or cycles that can be computed in polynomial time (if $n \geq 3$). This is not the case in complete graphs with forbidden transitions.

Theorem 4. HPWFT *and* HCWFT *are* $\mathcal{NP}$-*complete in complete graphs with forbidden transitions.*

Proof. We reduce the Hamiltonian path (or cycle) problem to HPWFT (or HCWFT). Let G be an n-vertex graph without forbidden transitions with $n \geq 3$ and let $(K_n, \overline{\mathcal{F}(G)})$ be the complete graph with forbidden transition associated with it (see definition of $\overline{\mathcal{F}(G)}$ before Lemma 1). One can easily show that G contains an Hamiltonian path (or cycle) if and only if $(K_n, \overline{\mathcal{F}(G)})$ contains an Hamiltonian path (or cycle) containing no forbidden transitions. But the problem of determining whether a graph G contains an Hamiltonian path or cycle is $\mathcal{NP}$-complete (see [5]).   □

2.2   Inapproximability of MTWFT

The previous results show that constructing a spanning tree without forbidden transitions is a hard problem. We investigate here the optimisation version. Given $(G, \mathcal{F})$, we denote by $\alpha(G, \mathcal{F})$ the maximum size of a tree in $(G, \mathcal{F})$. Notice that if $(G, \mathcal{F})$ is a YES instance for STWFT, then $\alpha(G, \mathcal{F}) = |V|$.

Theorem 5. *Let $(G, \mathcal{F})$ and let n be the number of vertices of G. Then $\alpha(G, \mathcal{F})$ cannot be approximated with a ratio better than $n^{1/2-\varepsilon}$ for all $\varepsilon > 0$ even if G is a star.*

Proof. We will reduce the maximum clique problem to MTWFT in stars. Let G be a graph with n vertices. Construct the star G' with vertex set $V_G \cup \{r\}$ and edge set $\{ru \mid u \in V_G\}$ (r is a new vertex), and let $\mathcal{F} := \{\{ru, rv\} \mid uv \notin E_G\}$. We claim that T is a tree of size k in $(G', \mathcal{F})$ if and only if $T \setminus r$ induces a clique of size $k - 1$ in G.

Let T be a tree of size k in $(G', \mathcal{F})$. Hence, T is a star S_k with distinguished vertex r and $k-1$ other vertices from V_G. As T contains no forbidden transitions, for all u and all v in $T \setminus r$, we have $uv \in E_G$. Therefore, $T \setminus r$ induces a clique of size $k - 1$ in G.

Conversely, let $C := \{u_1, \ldots, u_k\}$ be a clique of size k in G. Then in G' none of the edges $ru_1, \ldots, ru_k$ is involved in a pair in $\mathcal{F}$. Therefore, $C \cup \{r\}$ induces a tree of size $k + 1$ in $(G', \mathcal{F})$.

Now, using the fact that one cannot construct a clique of maximum size in an n-vertex graph with a better approximation ratio than $n^{1/2-\varepsilon}$ for all $\varepsilon > 0$ (see [1]) we get the desired result.   □

3   Constructive Results in Complete Graphs

From proof of Theorem 2, we deduce the $\mathcal{NP}$-completeness of STWFT even if the number of forbidden transitions an edge or vertex is involved is bounded. However, the reduction in the proof of Theorem 3 does not preserve this property. We will see in this Section 3.2 that bounding the number of conflicts an edge or vertex is involved implies a polynomial time algorithm for STWFT in complete graphs. We will also provide some other sufficient conditions.

3.1   A Sufficient Condition to Contain a STWFT

For $(K_n, \mathcal{F})$, we will construct a graph G with the same set of n vertices and containing only the edges of G that are not in any forbidden transitions of $\mathcal{F}$. If G is connected, then we are done since we can just take any spanning tree of G, and it will be of course a spanning tree of $(K_n, \mathcal{F})$. So, we will assume that G is not connected and let us denote by $C_1, \ldots, C_k$ its $k > 1$ connected components and n_i the number of vertices of C_i. In the following we will also use C_i to denote the set of the n_i vertices of the component C_i. Some components are not necessarily complete graphs and some of them may be composed of only one vertex.

We call *Edge Between Components*, noted EBC, an edge having its two extremities in two different components. The general idea is the following. If it is possible to connect the k components of G using EBC, in a "meta-tree" of components, in such a way that each vertex of each component is incident to at most one such EBC, then $(K_n, \mathcal{F})$ contains a spanning tree. Indeed, this "meta-tree" is a tree of the k components, it is connected and cycle-free. Inside each component, it is sufficient to take any spanning tree. These k trees connected by these EBC form a spanning tree T of $(K_n, \mathcal{F})$ (T is connected and is cycle-free). Indeed, T contains no forbidden transitions because, by construction, edges of components are part of no forbidden transitions and EBC are pairwise non incident by construction. In Theorem 7 we give a sufficient condition under which it is possible to do this construction. Before going further, we need some notions, definitions and preliminary results.

A sequence $n_1, \ldots, n_k$ of positive integers $(n_i \geq 1)$ is a SDT (*Sequence of Degrees Tree*) if there exists a tree T of k vertices denoted by $u_1, \ldots, u_k$ such that $d_T(u_i) \leq n_i$. We will use the following theorem.

Theorem 6 ([6]). *Let $n_1, \ldots, n_k$ be a sequence of positive integers, $k \geq 2$. There exists a tree with k vertices having degrees $n_1, \ldots, n_k$ if and only if $\sum_{i=1}^{k} n_i = 2k - 2$.*

We underline the fact that the proof of Theorem 6 in [6] describes a polynomial time algorithm to construct the tree from the sequence.

Lemma 2. *The sequence of positive integers $n_1, \ldots, n_k$, $k \geq 2$, is a SDT if and only if $\sum_{i=1}^{k} n_i \geq 2k - 2$.*

Proof. We suppose first that $\sum_{i=1}^{k} n_i \geq 2k - 2$ and we show that $n_1, \ldots, n_k$ is a SDT. We decrease the value of some n_i (keeping them strictly positive) to obtain a sum equal to $2k - 2$. This operation can easily be done in polynomial time. Then we can apply Theorem 6 on this new sequence. In the corresponding tree T the degrees are less than $n_1, \ldots, n_k$ and hence this sequence is a SDT.

Let us show now that if $n_1, \ldots, n_k$ is a SDT, then $\sum_{i=1}^{k} n_i \geq 2k - 2$. As $n_1, \ldots, n_k$ is a SDT, there exists a tree T whose k vertices $u_1, \ldots, u_k$ are such that $d_T(u_i) \leq n_i$ (for $i = 1, \ldots, k$). But, it is well-known that in any graph, the sum of degrees of vertices is equal to two times the number of edges and in a tree the number of edges is equal to the number of vertices minus 1. This gives here $\sum_{i=1}^{k} d_T(u_i) = 2(k - 1)$ and we get the expected inequality. $\square$

Theorem 7. *Let $(K_n, \mathcal{F})$ and let $n_1, \ldots, n_k$ be the number of vertices of the k connected components induced by the n vertices of K_n and all the edges that are not in any forbidden transitions. If $\sum_{i=1}^{k} n_i \geq 2k - 2$, then $(K_n, \mathcal{F})$ contains a spanning tree that can be constructed in polynomial time.*

Proof. If $\sum_{i=1}^{k} n_i \geq 2k - 2$, then by Lemma 2 there exists a tree T_C with k vertices $u_1, \ldots, u_k$ such that $d_{T_C}(u_i) \leq n_i$. Now, replace each vertex u_i by the connected component C_i having n_i vertices. For each edge of T_C, between C_i

and C_j choose a vertex u in C_i and a vertex v in C_j and connect them by an EBC (this EBC exists since G is a complete graph). These two vertices u and v will not be used in any other connections between components. The number of vertices in each component is sufficient to ensure that property. Now construct in each component C_i any tree spanning its n_i vertices. The whole graph composed of these k trees plus the selected EBC forms a spanning tree of $(K_n, \mathcal{F})$. Note that, as the proofs of Theorem 6 and Lemma 2 are constructive and polynomial, there is a polynomial time algorithm to construct it.   □

3.2 Other Sufficient Conditions for a Polynomial Testing

In this section we study the case where each vertex is the extremity of a limited number of forbidden transitions. We have shown in Theorem 4 that deciding whether $(K_n, \mathcal{F})$ contains an Hamiltonian Path is $\mathcal{NP}$-complete. We first notice that when each edge is in at most one conflict, it is possible to construct one (recall that in [3] the authors proved a polynomial testing for STWC in such graphs).

Theorem 8. *Let $(K_n, \mathcal{C})$ be such that the conflict graph associated with $\mathcal{C}$ is a 2-ladder. Then $(K_n, \mathcal{C})$ contains an Hamiltonian path and it can be constructed in polynomial time.*

Proof. We construct the Hamiltonian path in $(K_n, \mathcal{C})$ step by step, by adding one by one the vertices and keeping the property that the chosen vertices form a path in $(K_n, \mathcal{C})$. We begin with any two vertices of K_n. Suppose now that we have constructed a path in $(K_n, \mathcal{C})$ with $p \geq 2$ vertices, denoted by H. We denote by a and b the two extremities of H and a' (resp. b') the unique neighbour of a (resp. b) in H. Consider a vertex c outside H.

Case 1. If one of $\{a'a, ac\}$ or $\{b'b, bc\}$ is not a conflict, then we can add c as a new extremity (by adding the edge ac or bc) of H that becomes a path with $p+1$ vertices in $(K_n, \mathcal{C})$.

Case 2. In the other cases, this means that $\{a'a, ab\}$ is not a conflict (otherwise the edge $a'a$ would be involved into 2 conflicts) and similarly for $\{ab, bc\}$. One can construct a path H' composed in this order of: $b', \ldots, a', a, b, c$ which is a path with $p+1$ vertices in $(K_n, \mathcal{C})$.

We can therefore conclude that $(K_n, \mathcal{C})$ contains an Hamiltonian path. Since at each step the construction can be done in polynomial time, an Hamiltonian path can be constructed in polynomial time.   □

We now look at the case where each vertex is in a limited number of forbidden transitions.

Fact 1. *Let $(K_n, \mathcal{F})$ be given. We suppose that each vertex is the extremity of at most k forbidden transitions and that $n \geq k+1$. If $(K_n, \mathcal{F})$ contains a tree T with $k+1$ vertices, then one can extend it to a spanning tree of $(K_n, \mathcal{F})$ in polynomial time.*

Proof. If $n = k + 1$ then T is already a spanning tree of $(K_n, \mathcal{F})$. If $n > k + 1$, consider a vertex u of K_n outside T. As u is the extremity of at most k forbidden transitions and as T contains $k + 1$ vertices, T contains a vertex v which is the centre of no forbidden transitions of extremity u. One can complete T by connecting u to v. This induces no forbidden transitions in this new tree that now has $k+2$ vertices. We do the same process with this new tree containing $k+2$ vertices, etc. At each step one can connect any vertex outside the current tree to this tree by adding no forbidden transitions. One can continue until obtaining a spanning tree. If the initial T is given, then completing it into a spanning tree can be done in polynomial time. □

Fact 2. *Let k be a fixed positive integer. There is a polynomial time algorithm that constructs in $(G, \mathcal{F})$ a tree with $k+1$ vertices if and only if there exists one.*

Proof. We apply here a brute force method: Generate all the subsets with $k + 1$ vertices; Each subset induces a graph with $k+1$ vertices; Test in each such graph all the possible spanning trees. When one such tree without forbidden transitions is found, stop and return it. If none is found, this means that there is no such tree (since the process is exhaustive).

Generating all the subsets with $k + 1$ vertices can be done in $O(n^{k+1})$. We generate all the trees with $k + 1$ vertices in such induced subgraphs. There are at most $(k + 1)^{k-1}$ trees (this is a well-known result, see [2]). Each such tree can be generated and tested in polynomial time. But, as k is a constant, $(k + 1)^{k-1}$ is also a constant. The whole process is a polynomial time algorithm able to construct a required tree if and only if there exists one. □

Theorem 9. *Let k be a fixed positive integer. Let $(K_n, \mathcal{F})$ be given. If each vertex of K_n is the extremity of at most k forbidden transitions, then in polynomial time one can decide whether there exists a spanning tree in $(K_n, \mathcal{F})$ and, in this case, construct one in polynomial time.*

Proof. If K_n contains at most $k + 1$ vertices, then the technique used in the proof of Fact 2 shows that one can determine and construct in polynomial time a spanning tree in $(K_n, \mathcal{F})$ if and only if there exists one. Let us consider now the case where K_n contains more than $k + 1$ vertices.

Suppose that the algorithm in the proof of Fact 2 constructs a tree T. Thanks to Fact 1 one can extend it into a spanning tree of $(K_n, \mathcal{F})$. Both operations are done in polynomial time.

Suppose now the opposite, that is the algorithm in Fact 2 constructs no tree. In this case, $(K_n, \mathcal{F})$ contains no spanning tree. Indeed, if it contains one, T, then one can easily extract from T a tree on $k + 1$ vertices, without forbidden transitions. This is in contradiction with Fact 2. □

We notice that it is challenging to reduce the time complexity of the procedure in Fact 2 from $O(n^{k+1})$ to $O(f(k) \cdot n^c)$ for some constant c not depending in k and n. One just notices that a naive local search from a given vertex that maximises at each step the number of neighbours of the current vertex will not work since one can reduce the problem of finding a maximum clique in a graph to a such local search.

Theorem 10. *Let k be a fixed positive integer and let $(K_n, \mathcal{F})$ be given. If each vertex of K_n is the extremity of at most k forbidden transitions and if $n \geq 2k+1$, then $(K_n, \mathcal{F})$ necessarily contains a spanning tree that can be constructed in polynomial time.*

Proof. The global idea is the following: First construct a tree T without forbidden transitions on $k + 1$ vertices in polynomial time and then use Fact 1 with this tree T to extend it and finish the proof.

If u is a vertex of K_n we denote by $Ext(u)$ the set of vertices of K_n which are centres of a forbidden transition having u as an extremity. Hence, by hypothesis, for all u, $|Ext(u)| \leq k$.

Let us construct T by selecting first its vertices. Take any vertex u_1. Take any vertex $u_2 \neq u_1$ which is not in $Ext(u_1)$: $u_2 \notin Ext(u_1) \cup \{u_1\}$, etc., take u_{i+1} a vertex not already taken and not in $Ext(u_i)$; $u_{i+1} \notin Ext(u_i) \cup \{u_1, \ldots, u_i\}$, etc. until obtaining a vertex u_{k+1}. One can always choose at each step a new vertex u_{i+1} because u_{i+1} is any vertex outside the set $Ext(u_i) \cup \{u_1, \ldots, u_i\}$ but $|\{u_1, \ldots, u_i\}| \leq k$ and $|Ext(u_i)| \leq k$; as $n \geq 2k+1$, u_{i+1} can always be selected.

The tree T is then the path $u_1, u_2, \ldots, u_{k+1}$ spanning the selected vertices. The only transitions in T are of the form $\{u_{i-1}u_i, u_i u_{i+1}\}$. But such a transition is not forbidden since u_{i+1} was selected outside $Ext(u_i)$ (if the transition $\{u_{i-1}u_i, u_i u_{i+1}\}$ is forbidden, this means that u_{i+1} would have been the extremity of a forbidden transition with centre u_i, which is not the case by construction).

The path/tree T contains no forbidden transitions and has $k + 1$ vertices. This tree T always exists and can be constructed in polynomial time. We end the construction of the spanning tree by using the polynomial time constructing process of Fact 1 while T is given here.   □

3.3 Simplification of $(G, \mathcal{C})$

We describe a process to simplify an instance $(G, \mathcal{C})$ (if it is possible) by suppressing edges and conflicts to obtain a new (reduced) instance $(G_f, \mathcal{C}_f)$ in which there is a spanning tree if and only if there is a spanning tree in $(G, \mathcal{C})$. $(G_f, \mathcal{C}_f)$ is constructed iteratively, step by step. Let $G_0 = G$ and $\mathcal{C}_0 = \mathcal{C}$. For each i, we let H_i be the subgraph of G_i composed of all the vertices of G and only edges that are not involved in a conflict in $\mathcal{C}_i$ and let S_i be the set of edges of G_i that are in a conflict of $\mathcal{C}_i$ **and** whose two extremities are in the same connected component of H_i. The shape of the algorithm is in Fig. 1.

As at each step some edges are removed, the algorithm terminates and is polynomial. We denote by $(G_f, \mathcal{C}_f)$ its final result. The graph G_f contains all the vertices of the initial graph G and a subset of its edges. Moreover $\mathcal{C}_f \subseteq \mathcal{C}$.

Theorem 11. $(G, \mathcal{C})$ *contains a spanning tree if and only if $(G_f, \mathcal{C}_f)$ contains a spanning tree.*

$i = 1$;

while $(S_{i-1} \neq \emptyset)$ do

   let G_i be obtained from G_{i-1} by deleting edges in S_{i-1};

   let C_i be obtained from C_{i-1} by removing conflicts involving at least an edge in S_{i-1};

   $i = i + 1$;

endwhile

return (G_{i-1}, C_{i-1});

Fig. 1. Simplification of (G, C)

Proof. If (G_f, C_f) contains a spanning tree T, then T is also a spanning tree of (G, C). Indeed, T covers all the vertices of G and it contains no conflicts of C. Otherwise, assume that two edges e and e' T are in conflict in C. As $\{e, e'\} \notin C_f$, this means that the conflict was eliminated during the construction of (G_f, C_f). However the algorithm removes a conflict only if one of its edges is removed. Hence e and e' cannot be both in G_f; Contradiction.

Consider now a spanning tree T of (G, C). This tree T covers all the vertices of G. Let us denote by $C_1, \ldots, C_k$ the k connected components of G_f in which all the edges that are in at least one conflict of C_f are removed. Let uv be any edge of T that has its two extremities into two different such connected components. Let us show that the edge uv is in G_f. If not, this means that it was removed at some step in the construction of (G_f, C_f), say step i and the two vertices u and v are in a same connected component of H_i since the algorithm deletes an edge involved in a conflict only if its two extremities are in a same connected component of H_i. However, the algorithm guarantees that an edge in a connected component of H_i will be always kept in the remaining steps $j > i$. So the edge uv is an edge in G_f.

Let I_C be the set of edges of T that are in G_f and have their two extremities between two connected components $C_1, \ldots, C_k$. Consider now the graph G' composed of all the vertices of G and of all the edges of I_C and all the edges of the connected components $C_1, \ldots, C_k$. This graph is clearly connected. Moreover, it contains no conflicts of C_f. Indeed, let us consider any pair e and e' of edges of G'.

Case 1. There exist C_i and C_j such that $e \in C_i$ and $e' \in C_j$ (we may have $i = j$). By construction this means that they are not involved in a conflict of C_f.

Case 2. e and e' are both in I_C. As I_C is a set of edges of a tree without conflicts in C, edges e and e' do not form a conflict in $C_f \subseteq C$.

Case 3. One edge, say e, is in a connected component and the other one, e', is in I_C. As e is in a connected component, by construction it is not involved in a conflict of C_f.

It is then easy to construct any spanning tree of G' (with a BFS for example) that is a spanning tree in (G_f, C_f). □

If $(G, \mathcal{C})$ is given, let $G(\overline{\mathcal{C}})$ be the graph containing all the vertices of G but only the edges that are not involved in a conflict of $\mathcal{C}$.

Corollary 1. *If $G_f(\overline{\mathcal{C}_f})$ is connected, then $(G, \mathcal{C})$ contains a spanning tree.*

It is easy to give instances $(G, \mathcal{C})$ in which $G(\overline{\mathcal{C}})$ is not connected while $G_f(\overline{\mathcal{C}_f})$ is connected and thus contains a trivial solution (any spanning tree). This simplification process leads to transform some instances that seem to be too complicated to solve but that are in fact trivial.

References

1. Ausiello, G., Crescenzi, P., Gambosi, G., Kann, V., Marchetti-Spaccamela, A., Protasi, M.: Complexity and Approximation. Springer (1999)
2. Bondy, A., Murty, U.S.R.: Graph Theory. Springer London Ltd. (2010)
3. Darmann, A., Pferschy, U., Schauer, J., Woeginger, G.J.: Paths, trees and matchings under disjunctive constraints. Discrete Applied Mathematics 159(16), 1726–1735 (2011)
4. Dvořák, Z.: Two-factors in orientated graphs with forbidden transitions. Discrete Mathematics 309(1), 104–112 (2009)
5. Garey, M.R., Johnson, D.S.: Computers and Intractability: A Guide to the Theory of NP-Completeness. W. H. Freeman & Co., New York (1979)
6. Gupta, G., Joshi, P., Tripathi, A.: Graphics sequences of trees and a problem of frobenius. Czechoslovak Mathematical Journal 57(132), 49–52 (2007)
7. Kanté, M.M., Laforest, C., Momège, B.: An exact algorithm to check the existence of (elementary) paths and a generalisation of the cut problem in graphs with forbidden transitions. In: van Emde Boas, P., Groen, F.C.A., Italiano, G.F., Nawrocki, J., Sack, H. (eds.) SOFSEM 2013. LNCS, vol. 7741, pp. 257–267. Springer, Heidelberg (2013)
8. Kolman, P., Pangrác, O.: On the complexity of paths avoiding forbidden pairs. Discrete Applied Mathematics 157(13), 2871–2876 (2009)
9. Pferschy, U., Schauer, J.: The knapsack problem with conflict graphs. Journal of Graph Algorithms and Applications 13(2), 233–249 (2009)
10. Szeider, S.: Finding paths in graphs avoiding forbidden transitions. Discrete Applied Mathematics 126(2-3), 261–273 (2003)

Author Index

Ackerman, Nathanael L.   133
Agarwal, Manu   84
Ahn, Hee-Kap   52
Angel, Eric   10
Asano, Tetsuo   32

Bae, Sang Won   52
Bampis, Evripidis   10
Birks, Martin   20
Bley, Andreas   72
Boros, Endre   331
Bournez, Olivier   169
Brand, Michael   181

Cao, Yixin   319
Chau, Vincent   10
Chen, Jianer   319
Chen, Lin   1
Cheng, Siu-Wing   121
Cheng, Yukun   272
Chester, Andrew   260

Dondi, Riccardo   260

Elmasry, Amr   32

Feng, Rongquan   282
Freer, Cameron E.   133
Fung, Stanley   20

Graça, Daniel S.   169

Han, Qiaoming   272
Hashemi, S. Mehdi   72
Heggernes, Pinar   331
Higashikawa, Yuya   121
Hutter, Marcus   212

Ibarra, Oscar H.   156

Jaiswal, Ragesh   84

Kanté, Mamadou Moustapha   343
Katajainen, Jyrki   32
Katoh, Naoki   121
Kratochvíl, Jan   108
Kulkarni, Raghav   224

Laforest, Christian   343
Lattimore, Tor   212
Letsios, Dimitrios   10
Li, Mingfei   42
Liu, Nan   236
López-Ortiz, Alejandro   193

Ma, Chu Chung Christopher   42
Merkle, Wolfgang   144
Milanič, Martin   331
Momège, Benjamin   343

Nagamochi, Hiroshi   96
Ni, Guanqun   121
Ning, Li   42

Pal, Arindam   84
Perez, Anthony   306
Popa, Alexandru   62
Pouly, Amaury   169

Qiao, Youming   224

Rezapour, Mohsen   72

Salinger, Alejandro   193
Seki, Shinnosuke   156
Son, Wanbin   52
Stephan, Frank   144
Su, Bing   121
Sun, Xiaoming   224
Suzuki, Akira   248

Takebe, Hirotoshi   294
Tanaka, Keisuke   294
Teutsch, Jason   144

Uchizawa, Kei   248
Ueckerdt, Torsten   108

van 't Hof, Pim   331

Wang, Wei   144
Wirth, Anthony   260
Wu, Hongfeng   282

Xiao, Mingyu   96
Xu, Yinfeng   121

Yang, Yue   144
Ye, Deshi   1
Yu, Wei   272

Zhang, Guochuan   1, 272
Zhou, Xiao   248
Zhu, Daming   236
Zimand, Marius   205